W9-CMS-907

MITOCHONDRIA

The Wiley Bicentennial–Knowledge for Generations

Each generation has its unique needs and aspirations. When Charles Wiley first opened his small printing shop in lower Manhattan in 1807, it was a generation of boundless potential searching for an identity. And we were there, helping to define a new American literary tradition. Over half a century later, in the midst of the Second Industrial Revolution, it was a generation focused on building the future. Once again, we were there, supplying the critical scientific, technical, and engineering knowledge that helped frame the world. Throughout the 20th Century, and into the new millennium, nations began to reach out beyond their own borders and a new international community was born. Wiley was there, expanding its operations around the world to enable a global exchange of ideas, opinions, and know-how.

For 200 years, Wiley has been an integral part of each generation's journey, enabling the flow of information and understanding necessary to meet their needs and fulfill their aspirations. Today, bold new technologies are changing the way we live and learn. Wiley will be there, providing you the must-have knowledge you need to imagine new worlds, new possibilities, and new opportunities.

Generations come and go, but you can always count on Wiley to provide you the knowledge you need, when and where you need it!

William J. Pesce
President and Chief Executive Officer

Peter Booth Wiley
Chairman of the Board

MITOCHONDRIA

Second Edition

IMMO E. SCHEFFLER
Department of Biology
University of California, San Diego

WILEY-LISS

A JOHN WILEY & SONS, INC., PUBLICATION

Cover Image: The author wishes to express great appreciation for the splendid cover image provided by Guy Perkins, PhD, Director of Tomography in the National Center for Microscopy and Imaging Research at the School of Medicine, University of California, San Diego, California.

Published by John Wiley & Sons, Inc., Hoboken, New Jersey
Published simultaneously in Canada

For general information on our other products and services or for technical support, please contact our Customer Care Department within the United States at (800) 762-2974, outside the United States at (317) 572-3993 or fax (317) 572-4002.

Wiley also publishes its books in a variety of electronic formats. Some content that appears in print may not be available in electronic formats. For more information about Wiley products, visit our web site at www.wiley.com.

Anniversary Logo Design: Richard J. Pacifico

Library of Congress Cataloging-in-Publication Data:

Scheffler, Immo E.
Mitochondria / Immo E. Scheffler.—2nd ed.
p. ; cm.
Includes bibliographical references and index.
ISBN 978-0-470-04073-7 (cloth)
1. Mitochondria. 2. Mitochondrial pathology. I. Title.
[DNLM: 1. Mitochondria. 2. DNA, Mitochondrial. 3. Genes, Mitochondrial. QU 350 S317m 2008]
QH603.M5S34 2008
571.6′57—dc22

2007017348

Printed in the United States of America

10 9 8 7 6 5 4 3 2

CONTENTS

PREFACE

While I can say that I was naïve about the task confronting me in starting the first edition of the book on mitochondria, I had no illusions about the challenge of following it with a second edition. Was there in fact a need for a second edition?

The greatest encouragement came from the reception of the first edition and from the many favorable and supportive comments from friends and fellow scientists over the past eight years. As projected, the book filled the need of a diverse group of researchers and students to gain a broad perspective on the subject and also served as an introduction to many intricacies of this broad field. It was intended to explain basic principles and insights that would not become obsolete, although many more details, supporting data, and specific examples could be anticipated to be published as the book was written and during the following years. While the first edition undoubtedly contained mistakes and omissions, there were no major errors and misinterpretations that kept me awake at night wishing to publish an immediate correction or retraction. Thus, the plan for a second edition could mature over several years, with careful considerations about what to modify and what to add.

From the outset it was clear that the focus had to be sharpened to keep the volume within bounds. Thus, most attention was paid to progress in the understanding of mammalian mitochondria. Information from the first edition about mitochondria in other organisms (fungi, protozoa, plants) was retained to illustrate the diversity of mitochondria and some of the unique aspects of mitochondria in these organisms. And, as pointed out before, pioneering studies in microorganisms such as yeasts have continued to provide important clues and insights that have guided the exploration of mitochondria in higher eukaryotes. Consequently, multiple references to such recent studies are included.

What is really new since the appearance of the first edition? First of all, literally hundreds of new, complete sequences of mtDNA from many different organisms have been added to the database, providing raw material for addressing questions on the evolution of this organelle over both long and shorter time scales. The relationship of hydrogenosomes and mitosomes to mitochondria has been clarified.

Many more details have been added to the understanding of the biogenesis of mitochondria, including the replication of mtDNA, the transcription of the mitochondrial genome, and pathways for protein import into various membranes and compartments of the organelle. An important new addition to this subject is the elucidation of the biogenesis of the iron–sulfur centers in mitochondria, one of the most essential functions contributed to the cell exclusively by this organelle.

The dynamic behavior of mitochondria, already apparent many years ago, is an active area in which much progress has been achieved with regard to components involved in fission and fusion, but much more needs to be learned, especially about the control of these processes. A functional connection between some of these proteins governing these normal morphological processes and their role in apoptosis is becoming apparent. Controversy or uncertainty continues in a detailed understanding of the two phenomena relating mitochondria to apoptosis: What is the mechanism of the release of cytochrome c and other proteins from the intermembrane space, and what is the nature of the pore/channel responsible for the permeability transition (breakdown of the (inner) membrane potential)?

High-resolution structures of some of the five complexes of the electron transport chain were included in the last edition of this book. The structure of complex II (succinate–ubiquinone oxidoreductase) from yeast and mammals can now be added to the list, and significant progress has been made with the recent publication of the structure of the peripheral subcomplex of the prokaryotic complex I. However, the role of ~31 "ancillary" subunits in the mammalian/eukaryotic complex continues to be a challenging puzzle. Significant progress has also been made in the elucidation of the complete structure of complex V. The structure of the c-ring has been solved for related enzymes, allowing more informed guesses about the translocation of protons and the mechanism inducing rotations. Some significant question marks remain. The existence of supercomplexes in the electron transport chain has been firmly established, challenging us to seek an understanding of their physiological significance.

An explosion of proteomic studies has emphasized the view that mitochondria are not just the powerhouse of the cell producing more or less ATP, depending on demand, but their composition is highly variable from tissue to tissue, presumably because of the many other functions contributed by mitochondria to the specialized cell types. Characterizing mitochondria from rat liver may have limits when one is interested in mitochondria from neurons. Major progress has been made in the identification and structural characterization of the large family of transporters, although the specificity of many family members remains to be established. The traffic of metabolites and ions firmly integrates mitochondria into many cellular activities. In this context the spotlight has in recent years been on potential regulatory mechanisms involving protein kinases, phosphatases, and their targets inside the mitochondria. Many intriguing observations suggest the presence of such mechanisms, but signaling pathways and specific targets remain to be identified and understood. For example, how (and why) does a defect in the mitochondrial PINK-1 kinase cause Parkinson's disease?

An avalanche of publications (>75,000 total, more than 3000 in 2006–2007 in PubMed) on reactive oxygen species (ROS) seems to lead to the paradox that "the more facts we learn, the less we understand the process we study." The reader can be referred to an illuminating essay by Y. Lazebnik (*Cancer Cell*, 2002) entitled "Can a biologist fix a radio?—Or, what I learned while studying apoptosis." The mitochondrial electron transport chain continues to be thought of as the major source of single electrons for superoxide formation, with complex II now added to complexes I and III as a source of the single electron. Then what happens? A low level of ROS may be a beneficial regulatory signal, but at higher levels it becomes a potentially pathological "oxidative stress" that can affect proteins, lipids, and DNA, whatever one's preference and particular focus. Experiments with animal models in which scavenger enzymes have been elevated or knocked out continue to emphasize the effects of ROS on life spans and well-being.

Finally, nitric oxide has emerged as a potential player in the control of mitochondrial activity. Its postulated functions range from being (1) an inhibitor of cytochrome oxidase, (2) a reaction with ROS to form peroxynitrite and nitrosothiols in proteins, and (3) a regulator of mitochondrial biogenesis via activation of guanylate cyclase.

Instead of "oxidative stress," one can also consider the "redox state" of a cell which could be reflected, for example, in NAD^+/NADH or GSH/GSSG ratios. Mitochondria are likely to be key players in regulating the relative levels of these small molecules; and in addition to modulating metabolic activity, they thus may influence chromatin remodeling and gene expression through NAD^+-dependent deacetylations of histones, for example. Such a mechanism provides a very direct cross-talk between mitochondria and the nucleus.

The number of "mitochondrial diseases" continues to expand, depending on the definition of the term. The original emphasis was on defects in oxidative phosphorylation due to mutations in mitochondrial DNA. Not surprisingly, nuclear mutations affecting subunits of the complexes of the electron transport chain must be included. However, where does one draw the line? In this book, some attention is given to (a) diseases resulting from defective mtDNA maintenance or replication and (b) Friedreich's ataxia, a problem in mitochondrial iron "metabolism." Defects in cardiolipin metabolism can also lead to a mitochondrial disease, because of the unique association of this lipid with mitochondrial membrane complexes. Finally, the involvement of mitochondria in the development of neurodegenerative diseases such as Alzheimer's disease, Parkinson's disease, Huntington's disease, and amyotrophic lateral sclerosis (ALS) and in the general process of senescence (aging) remains an intriguing and attractive hypothesis, although a distinction must be made between a primary effect (due to specific mutations/polymorphisms in mtDNA) and secondary effects of aged mitochondria on other cellular housekeeping mechanisms such as proteasome activity (protein quality control).

There are great expectations for progress from a new (?) approach referred to as systems biology. Even without the hype, it is clear that a higher level of understanding biological phenomena must be derived not from understanding a single enzyme, or even a single pathway, but from an understanding of a larger ensemble of proteins and their activities. Mitochondria represent such an ensemble of proteins, contained within a well-defined compartment. Their individual functions are becoming clear now, but an understanding of the extensive and multifaceted integration of mitochondrial and cellular activities will be a challenge for some time to come. From the perspective of a systems biologist:

A genome, organized into chromosomes that are condensed into nucleosomes, is expressed by the action of enhanceosomes, transcriptosomes, and splicosomes as a transcriptome, and with the help of ribosomes the transcriptome is turned into a proteome. Chromosomes, consisting mostly of autosomes, but also X and Y chromosomes, are duplicated by a replisome. Members of the proteome are organized in higher-order structures such as peroxisomes, lysosomes, endosomes, and so on. Undesirable members of the proteome are attacked by proteasomes (degradomes). Syndromes arise when specific members of the proteome are absent or misbehave. A large number of metabolomes are responsible for providing the required energy and raw materials, while signalosomes or kinomes regulate the creation of order out of chaos. Under certain conditions, constituents of the signalosome activate the apoptosome to organize the return to chaos. Somewhere in all of this ***mitochondria*** *play an absolutely essential role.*

I would like to thank again many individuals too numerous to mention for helpful discussions over the years. Thanks to those who contributed original figures to the first edition, allowing me to use them again. New figures were generously provided by G. Cecchini, J. M. Shaw, L. A. Sazanov, E. Pebay-Peyroula, R. B. Gennis, T. Meier, and F. Foury.

Finally, I would like to thank several individuals (Thomas H. Moore, Ian Collins, and Danielle Lacourciere) associated with the publisher, John Wiley & Sons, for much encouragement and assistance in the publication of this book.

IMMO E. SCHEFFLER

University of California, San Diego

PREFACE TO FIRST EDITION

When talking to friends and colleagues about writing a book about mitochondria, they naturally assumed that I was editing a book about mitochondria, and the usual follow-up question was about the other contributors. My response, that there were none, caused a pause, a look of incredulity, and this reaction also immediately raised doubts in my mind whether I was in over my head. I knew it was going to be a challenge!

There certainly was never any doubt that I would have enough material to include. There are close to 80,000 references with mitochondria as a keyword in the Medline database. That was just one of the challenges: Is it possible to gain command and perspective on such a vast range of studies covering anthropology, biochemistry and biophysics, cell biology, diseases, evolution, forensics, genetics, history, and more. The reader will have to be the judge.

I was "dragged" into the subject more than 25 years ago, and that was several years after I thought I had had my last course on intermediary metabolism in my life. A mutant hunt with Chinese hamster cells yielded auxotrophs for carbon dioxide, which on further analysis turned out to be respiration-deficient mutants, the first such mutants identified for mammalian cells in tissue culture. A search of NSF files by one of Senator Proxmire's underlings in search of a candidate for a "Golden Fleece of the Month" award uncovered one of my successful grant applications, and I subsequently had a telephone call from the *National Enquirer* asking why I was interested in respiration-deficient Chinese. Several witty answers occurred to me only after I had hung up. In those days I unfortunately did not know that just about every neurodegenerative disease and even aging in general could potentially be explained with the help of defective mitochondria. The subject has grown on me over the years, and I certainly find it fascinating, even if mitochondrial defects cannot explain all such problems.

One of the most memorable quotes by one of my most esteemed undergraduate teachers was: “We learn more and more about less and less until we know everything about nothing.” An ever-expanding horizon in the biological sciences makes specialization inescapable and necessary. Accepting the challenge of writing this book was my personal act of rebellion against this trend. Quite obviously, an active participation in research requires the individual researcher to focus. But does that mean that researchers interested in complex V (ATP synthase) no longer talk to researchers investigating complex IV (cytochrome oxidase)? Anecdotal reports from the Gordon Conference on Electron Transport suggest that we are not too far from such a situation.

Mitochondria have been pivotal in the development of some of the most profound ideas in modern Biology. Thermodynamics and life (Bioenergetics) are intricately linked through mitochondria, and the Chemiosmotic Hypothesis is surely one of the great intellectual triumphs of twentieth-century science. They occupy a central position in any discussion of metabolism. It is fair to say that the evolution of Cell Biology from Anatomy and Physiology received a major impulse from the visualization of mitochondria by electron microscopy, as well as from their physical isolation by differential centrifugation. When DNA was discovered in mitochondria, “molecular” biologists took notice and geneticists had to pay attention. The evolution of eukaryotes owes much to the conversion of a prokaryotic symbiont to a subcellular organelle, now known as mitochondria. Today, understanding the evolution of large populations of mitochondrial genomes in somatic cells of a complex organism remains a major challenge: How does it happen and what are the consequences?

A historical perspective and an emphasis on the evolution of our understanding of the properties and role of mitochondria have guided the writing of several chapters. I have tried to give credit where credit is due. Several reviewers pointed out to me missing references or seminal papers. Nevertheless, no claim is made to have pursued every concept to its roots. Time has obscured many details from the past, and the explosion of knowledge in the past two or three decades has involved many laboratories almost simultaneously. It is relatively easy to identify many giants in the field, but it is not possible for an outsider to identify them in all the areas covered.

This book is intended first of all to remind the reader of the fundamental importance of mitochondria in the understanding of life and the eukaryotic cell, and the aim is to ignite in the serious, beginning student an interest in the many fascinating aspects of this organelle. There are also many mature investigators, especially in the medical sciences but also in other fields, who will have rediscovered mitochondria later in their careers when addressing problems related to diseases such as cardiomyopathy and neurodegenerative disease, developmental processes such as apoptosis, cancer, and senescence, or male sterility in plants. This volume is an attempt to provide such individuals with a single source of information on the major aspects of the biochemistry, cell biology, and molecular genetics of mitochondria. It is hoped that this book will help in establishing a broad perspective on mitochondrial biology and its many facets and that at the same time it will provide sufficient detail and depth to make it a reference of some lasting value. The reader should be able to find a more than superficial treatment of the various topics with reference to diverse organisms and should use this treatise and its key references as a convenient springboard into the voluminous past literature, as well as future publications. While more potentially relevant studies are being published as this is being written, they should not make the basic principles and insights obsolete.

There was some thought of limiting the discussion to mitochondria of one organism—for example, mammals. In the future this may be the only way of coping with the wealth of information being accumulated. Clearly, information such as sequence polymorphisms in mitochondrial genomes in vast numbers of representatives of different human ethnic groups will be most efficiently stored and disseminated by the World Wide Web. This treatise will be more concerned with the nature of information and data to be collected, using illustrative examples, and it will attempt to demonstrate how such data can serve in addressing fundamental biological questions about the role of mitochondria in health and disease, in understanding the evolution of humans and other organisms, and in defining their role as the powerhouse of the fundamental unit of life, the cell.

Model systems with lower eukaryotes and comparative studies have provided invaluable guidance and insights on many aspects of mitochondrial biology. For example, the budding yeast, *Saccharomyces cerevisiae*, has been the organism of choice for numerous pioneering studies. At the same time one should not develop the notion that all problems related to mitochondria or oxidative phosphorylation can be addressed in yeast. It is true that there is a significant amount of conservation of structure and function, especially with regard to electron transport and oxidative phosphorylation. On the other hand, some of the most striking differences between different organisms can be seen in the size of the mitochondrial genomes and in the organization of the genes. Maintenance, repair, recombination, and expression of the different genomes are not nearly as conserved and similar as one might have initially suspected. Undoubtedly, there are lessons for the evolutionary biologist, and one may wonder whether important functional characteristics and regulatory mechanisms are also affected. Certainly, the bizarre organization and expression of genetic information in kinetoplasts of trypanosomes, for example, has raised many interesting questions.

This book could serve as a textbook for an advanced graduate course, but it is not written primarily as a "textbook," with a glossary and review questions and exercises. The extensive, but by no means exhaustive, bibliography has been assembled to give the interested reader a start in exploring past investigations. In each bibliography a selected few references to recent reviews, books, or book chapters are highlighted which should be consulted for comprehensive and authoritative treatments of specialized subjects. Of necessity, such sources also had to substitute as references for numerous additional original papers. The author wishes to apologize in advance for any serious omissions or misattribution of credit.

The author owes a debt of gratitude to numerous individuals who were encouraging and helpful in many ways. Several individuals reviewed entire chapters of the book and provided valuable constructive criticisms for many sections. Special thanks and credit should go to Drs. W. Allison, S. Brody, J. Fee, Y. Hatefi, L. Simpson, M. Stoneking, C. Wills, and T. Yagi for their help, encouragement, and many constructive comments. Two anonymous reviewers also provided most valuable and appreciated suggestions for corrections and improvements. Drs. J. Singer, G. Palade, G. Perkins, L. Simpson, L. B. Chen, and J. Nunnari are owed thanks for contributing original photographs, but I am specially indebted to Dr. D. Fawcett, who responded from his retirement in Montana with a generous collection of photographs from his book *The Cell*. The contributors of the original photographs are acknowledged in the figure legends, and I thank the publisher, the W. B. Saunders Company, for permission to reproduce these electron micrographs.

I am particularly grateful to R. Chan, of EmbiTec, San Diego, whose financial support of the laboratory allowed me to take time out from endless grant applications long enough to get the bulk of this material onto paper. Graduate students and postdoctoral fellows in the laboratory over the years have also contributed in numerous ways: by providing data to think about, by asking challenging questions, and by being an essential and lively part of an environment in which learning and research can thrive.

IMMO E. SCHEFFLER

University of California, San Diego

1

HISTORY

Summary

Major landmarks in the history of studies on mitochondria are presented here. Ideas do not evolve in a vacuum, and insights grow on a rich base of observations and interpretations of the past. While some may be of interest only to the historian of science, history often connects the facts as we can present them today, and seeing facts in the context of history enriches them for us.

A review by Cowdry (1918), quoted by Lehninger (1), contains more than a dozen terms referring to structures that we now identify as mitochondria: blepharoblasts, chondriokonts, chondriomites, chondrioplasts, chondriosomes, chondriospheres, fila, fuchsinophilic granules, Korner, Fadenkorper, mitogel, parabasal bodies, plasmasomes, plastochondria, plastosomes, vermicules, sarcosomes, interstitial bodies, bioblasts, and so on. Chondros (Greek), grain (English), and Korn (German) are descriptions of the morphology of distinct structures noted inside of cells by microscopists beginning about 1850. Improvements in staining yielded more accurate morphological descriptions, and the grains were in some tissues seen as threads (Faden in German, mitos in Greek), hence Fadenkorper, or mitochondria (Benda, ca. 1898). In 1888 (R. A. Kollicker), by microdissecting some of these granules out of insect muscle and observing them to swell in water, microscopists reached the conclusion that the mitochondria had a membrane.

With our current knowledge it is astonishing to learn that Altman in around 1890 referred to mitochondria as bioplasts in a book on Elementarorganismen, in which he proposed that these granules were autonomous, elemental living units, forming bacteria-like colonies in the cytoplasm of the host cell. A wild, lucky guess, or an idea ahead of its time?

For a while, in the early part of this century, cytologists were anxious to provide their geneticist colleagues with a concrete entity to which they could assign a role in the transmission of genetic information, and mitochondria were favored for a short time. Another function, challenging their role in genetics, was also proposed as early as 1912 (Kingsbury); and as retold by Lehninger, Warburg in 1913 found granular, insoluble subcellular structures to be associated with respiration. But another 30–40 years of intense and painstaking biochemical analyses were required to lead to the characterization of mitochondria as the "powerhouse of the cell." Some of the most illustrious names in the early history of Biochemistry can be found among contributors of the highlights of the evolution of this concept. The discoveries of cytochromes, iron–porphyrin compounds (heme), flavin and pyridine nucleotides, and various dye-reducing dehydrogenases fall into the period between 1920 and the 1930s, and the formulation of the citric acid cycle by Krebs was one of the crowning achievements of the study of metabolism and respiration in muscle preparations.

Although ATP was discovered in 1931 (Lohmann), it took 10 more years to demonstrate its general role beyond muscle, and during this period Warburg and Meyerhof described what is now referred to as substrate level phosphorylation (ATP synthesis coupled to the enzymatic oxidation of compounds such as glyceraldehyde phosphate), in contrast to oxidative phosphorylation, first shown by Kalckar to be firmly coupled to respiration (1937–1941).

This was also a time when biochemists proceeded to grind up tissue, filter it through cheese cloth, perhaps even centrifuge the mixture at some indeterminate speed, and then discard everything but the supernatant. Insoluble particles constituted a nuisance and an insurmountable obstacle in the purification of an enzyme, and students were urged not to "waste clean thoughts on impure enzymes." The study of mitochondria required the isolation and purification of larger quantities of the organelle, and the first attempts in this direction were doomed by the use of unsuitable buffers and media for cell suspension and breakage.

Cell Biology metamorphosed out of the older science of Cytology when two powerful new methodologies were perfected and applied to the study of biological tissues. Pioneering advances and applications were made in parallel and synergistically at the same institution during the 1940s: At the Rockefeller Institute, Claude and his colleagues began to use the centrifuge as a sophisticated analytical tool for the fractionation of subcellular structures (differential centrifugation), a technique that De Duve (2) would later describe as "exploring cells with a centrifuge." Subcellular particles were fractionated reproducibly and with increasing resolution to achieve pure fractions. At the same time, careful biochemical characterizations were carried out. Important concepts to emerge were the recognition of the polydisperse population of particles (size variation) and the postulate of biochemical homogeneity. Microsomes and mitochondria were represented by overlapping populations of granules of different size, but at the large end were almost pure mitochondria and at the small end were almost pure microsomes (which were later resolved further into lysosomes and peroxisomes (2)). The discovery that ~0.3 M sucrose greatly stabilized mitochondria (Hogeboom, Schneider, Palade, 1948) greatly aided their isolation from liver in a morphologically intact form.

The criterion of morphological intactness could not have been applied if at the same time the groups led by Porter and Palade had not pursued the application of the electron microscope to the exploration of cells. Particles could now be viewed and compared *in situ* and after isolation by differential centrifugation, thereby increasing confidence that the particle was intact and therefore most likely completely functional.

Nevertheless, isolating and looking at a particle does not immediately give many clues about its biological function, although remarkably prescient guesses and deductions had been made from staining experiments. An approach from a different direction finally led to the full appreciation of the role of mitochondria in respiration. Lehninger (and independently Leloir and Munoz) in the period 1943–1947 had focused on the oxidation of fatty acids in liver homogenates and found the activity to be dependent on an insoluble component that was sensitive to osmotic conditions. The newly established conditions for mitochondrial isolation were applied by Kennedy and Lehninger to prove that fatty acid oxidase activity of the liver was found almost exclusively in mitochondria. The same investigators then extended the biochemical characterization of these organelles by demonstrating that (a) the reactions of the citric acid cycle can be carried out in mitochondria at a rate that can account for most, if not all, of the activity found in liver cells and (b) such reactions were accompanied by the synthesis of ATP (oxidative phosphorylation).

Just as a cell was recognized to be much more than a "bag of enzymes," it soon became clear that many of the enzymes catalyzing the biochemical reactions observed in mitochondria were not simply contained within this organelle in soluble form by the mitochondrial membrane. In fact, the potential complexity of this organelle became apparent from the early electron microscopic observations that revealed the existence of an inner and an outer membrane, with the inner membrane often highly folded (termed cristae by Palade). Topologically, one can therefore distinguish two spaces inside the mitochondrion: the intermembrane space and the matrix. However, a full appreciation of the significance of this compartmentalization was probably not achieved until much later.

The enzymes for fatty acid oxidation and for the citric acid cycle (with the exception of SDH) were found to be soluble in the mitochondrial matrix. The enzymes responsible for oxidation of NADH and electron transport to oxygen were insoluble and localized to the inner membrane (cristae). A more detailed description of their characterization will be deferred to a later chapter, but even an abbreviated historical introduction should mention two accomplishments of the 1960s. First, the overall arrangement of the components of the electron transport chain and the flow of electrons from dehydrogenases to flavoproteins to various nonheme iron–sulfur centers and cytochromes and finally to oxygen, first glimpsed by Keilin in the 1930s, was established by a combination of spectroscopic studies and the use of specific inhibitors such as rotenone, antimycin, and cyanide in various laboratories, with that of B. Chance deserving special mention. D. E. Green was another influential investigator in the 1950s and 1960s. While his ideas about an "elementary particle" within mitochondria have not stood up to the test of time, his Institute for Enzyme Research was the training ground for a number of prominent researchers in the subsequent decades. A very informative, entertaining, and highly personalized account has been written by one of the pioneers in the field, E. Racker (3). Second, the efforts of Hatefi and colleagues culminated in the fractionation and characterization of five multisubunit complexes from the inner membrane, four of which are involved in the respiratory chain, and the fifth was identified as the site of the phosphorylation of ADP to ATP (4, 5). Among the high points in the past decades in the biochemical and structural analysis of these complexes is undoubtedly the achievement of high-resolution structures for complex II (6–8), complex III (9), complex IV (cytochrome oxidase) (10), and complex V (ATP synthase) (11–14) by X-ray diffraction and other biophysical means. A Nobel Prize in Chemistry has been awarded for the elucidation of the structure and function of the F_1-ATP synthase to J. Walker of Cambridge and P. Boyer at UCLA. After the initial breakthroughs, similar structures from different organisms and at higher resolution have quickly followed.

The distinction between substrate-level phosphorylation and oxidative phosphorylation is made the subject of examination questions for thousands of undergraduate students every year. The former is straightforward to explain in terms of enzyme kinetics and the coupling of exergonic and endergonic enzyme-catalyzed reactions. The challenge to explain oxidative phosphorylation has preoccupied some of the best biochemists for a good part of their career. An ingenious solution, offered by P. Mitchell (15), was slow to be accepted, but it eventually revolutionized our thinking about bioenergetics, membranes, membrane potentials, active transport, ion pumps, and "vectorial metabolism." A detailed understanding of the structure of membranes, along with an understanding of the relationship between lipids and integral membrane proteins, was a prerequisite (16). The chemiosmotic hypothesis has found universal acceptance in explaining not only oxidative phosphorylation in mitochondria, but also aspects of photosynthesis in chloroplasts and light-driven phosphorylations in bacteria.

While failing to live up to early expectations of mitochondria as the carriers of all hereditary information, a most important discovery was made in 1963 when one of the first definite identifications of DNA in mitochondria was made (17). This discovery had in many ways been anticipated by the discovery of non-mendelian, cytoplasmic inheritance in yeast by the Ephrussi laboratory (18). Ramifications of this discovery are wide. It renewed and strengthened interest in the evolutionary origin of mitochondria. The problem of understanding how two genomes, nuclear and mitochondrial genes, interact in the biogenesis of this organelle is still an acute one. Changes in mitochondrial DNA sequences are believed to represent a molecular clock on a time scale that appears particularly suitable for human evolution; and provocative ideas and speculations have centered around deductions from such sequence comparisons, with implications for primate and human evolution ("the Mitochondrial Eve") and for the spread of human populations by migrations. Controversies still center around the question whether this clock is more accurate than a sundial (N. Howell). Because of the high degree of polymorphisms in human mitochondrial DNA, forensic investigations utilize comparisons of mitochondrial DNA sequences from victims and suspects. Where there is DNA, there must be mutations. An explosion of publications in the last two decades, triggered by the pioneering work of Wallace and his colleagues (19) and Holt et al. (20), has described mutations in mitochondrial DNA which are directly responsible for human genetic diseases (myopathies and neuropathies), while speculations go even further in relating accumulating defects in mitochondrial DNA to a variety of ailments accompanying aging and senescence. "Mitochondrial diseases" are now recognized to be caused by both mitochondrial and nuclear mutations.

The final chapter on mitochondria has only started to be written during the past decade when it was recognized that mitochondria are not merely the "powerhouse of the cell," but are much more intimately involved in the function, life, and death of a cell. They supply ATP, but they are also critically involved in maintaining the cellular redox potential and ionic conditions in the cytosol (e.g., Ca^{2+}). They are the target of numerous cellular signaling pathways, and in turn they can be triggered to release factors/proteins that initiate the process of apoptosis. They are the major source of reactive oxygen species (ROS); and while these can be highly injurious at high concentrations, they also have a positive regulatory function at lower concentrations. Proteomic studies have emphasized the diverse composition of mitochondria in different tissues and cell types, and understanding the tissue-specific behavior of mitochondria remains one of the challenges of the future. It is the key to understanding the many still puzzling features of mitochondrial diseases. A book published in 2005 entitled *Power, Sex, Suicide—Mitochondria and the Meaning of Life* (N. Lane, Oxford University Press) reflects the hype associated with mitochondrial studies nowadays. Mitochondria may not help in explaining the "meaning of life," but understanding them is definitely essential for understanding a healthy life and many pathological conditions. Nowadays they also occupy a prominent position in discussions on aging.

Thus, mitochondria occupy a central position in our understanding of the cell, the "basic unit of life," and from the very beginning their study not only has contributed to more details on ATP production, but also has led to fundamental insights covering the entire spectrum of biological sciences from anthropology to biophysics to cell biology, genetics, and physiology. Their demonstrated role in disease, and their implied role in neurodegenerative developments in advanced age makes them particularly fashionable now, but a history of biochemistry is unthinkable without reference to mitochondria and the monumental discoveries made in the course of their study.

REFERENCES

1. Lehninger, A. L. (1965) *The Mitochondrion—Molecular Basis of Structure and Function*, W. A. Benjamin, New York.
2. De Duve, C. (1975) *Science* **189**, 186–194.
3. Racker, E. (1976) *A New Look at Mechanisms in Bioenergetics*, Academic Press, New York.
4. Hatefi, Y., Galante, Y. M., Stiggal, D. L., and Ragan, C. I. (1979) *Methods Enzymol.* **56**, 577–602.
5. Hatefi, Y. (1985) *Annu. Rev. Biochem.* **54**, 1015–1069.
6. Sun, F., Huo, X., Zhai, Y., Wang, A., Xu, J., Su, D., Bartlam, M., and Rao, Z. (2005) *Cell* **121**, 1043–1057.
7. Yankovskaya, V., Horsefield, R., Tornroth, S., Luna-Chavez, C., Miyoshi, H., Leger, C., Byrne, B., Cecchini, G., and Iwata, S. (2003) *Science* **299**, 700–704.
8. Cecchini, G. (2003) *Annu. Rev. Biochem.* **72**, 77–109.
9. Xia, D., Yu, C.-A., Kim, H., Xia, J.-Z., Kachurin, A. M., Zhang, L., Yu, L., Deisenhofer, J., Yu, C. A., and Xian, J. Z. (1997) *Science* **277**, 60–66.
10. Tsukihara, T., Aoyama, H., Yamashita, K., Tomizaki, T., Yamaguchi, H., Shinzawa-Iyoh, K., Nakashima, R., Yaono, R., Yoshikawa, S., Yamashita, E., and Shinzawa-Itoh, K. (1996) *Science* **272**, 1136–1144.
11. Abrahams, J. P., Leslie, A. G. W., Lutter, R., and Walker, J. E. (1994) *Nature* **370**, 621–628.
12. Walker, J. E., Collinson, I. R., Van Raaij, M. J., and Runswick, M. J. (1995) *Methods Enzymol.* **260**, 163–190.
13. Van Raaij, M. J., Abrahams, J. P., Leslie, A. G. W., and Walker, J. E. (1996) *Proc. Natl. Acad. Sci. USA* **93**, 6913–6917.
14. Dickson, V. K., Silvester, J. A., Fearnley, I. M., Leslie, A. G., and Walker, J. E. (2006) *EMBO J.* **25**, 2911–2918.
15. Mitchell, P. (1961) *Nature* **191**, 144–148.
16. Singer, S. J. and Nicolson, G. L. (1972) *Science* **175**, 720–731.
17. Nass, S. and Nass, M. M. K. (1963) *J. Cell. Biol.* **19**, 593–629.
18. Slonimski, P. P. and Ephrussi, B. (1949) *Ann. Inst. Pasteur* **77**, 47.
19. Wallace, D. C., Singh, G., Lott, M. T., Hodge, J. A., Schurr, T. G., Lezza, A. M. S., Elsas, L. J., II, and Nikoskelainen, E. K. (1988) *Science* **242**, 1427–1431.
20. Holt, I. J., Harding, A. E., and Morgan Hughes, J. A. (1988) *Nature* **331**, 717–719.

2

EVOLUTIONARY ORIGIN OF MITOCHONDRIA

We can't rewind the tape of life and replay it to see what happens next time, alas, so the only way to answer questions about such huge and experimentally inaccessible patterns is to leap boldly into the void with the risky tactic of deliberate oversimplification.

—From *Darwin's Dangerous Idea: Evolution and the Meanings of Life*
by Daniel C. Dennett

Summary

Hypotheses proposing a relationship between mitochondria and bacteria have been around almost since their discovery, but speculations were bolstered considerably after the detection of DNA in mitochondria. Today the Endosymbiotic Theory of the origin of mitochondria is widely accepted. As more and more complete mtDNA sequences from diverse organisms have become available, refinements in phylogenetic relationships have been achieved; and in particular, the identification of the closest living prokaryotic species has become possible. There is also an emerging consensus about the monophyletic origin of mitochondria: A singular event in evolution gave rise to this organelle, rather than multiple independent symbiotic partnerships. Nevertheless, a surprising divergence can also be observed between vertebrates, plants, fungi, protozoa, and so on, with regard to the number of mitochondrial genes retained, their organization, and their expression.

Soon after mitochondria were first glimpsed in cells of higher organisms about 120 years ago, the idea that they were somehow related to bacteria was expressed in a more or less explicit form (Altmann, 1890; Mereschovsky, 1910; Portier, 1918; Wallin, 1927; Lederberg, 1952; see reference 1 for a complete bibliography). The idea continued to float about without much further experimental stimulus for several decades, until the discovery of DNA in this organelle gave it a major boost. Then, thoughts about the origin of mitochondria and chloroplasts culminated in an extremist statement of the Serial Endosymbiotic Theory, that "The eukaryotic 'cell' is a multiple of the prokaryotic 'cell' " (Taylor, 1974), and L. Margulis became a particularly eloquent and forceful popularizer of ideas on symbiosis in eukaryotic cellular evolution. In 1925 such ideas had been dismissed as "too fantastic for present mention in polite biological society" (Wilson, 1925). The review of a book by Portier on the subject by an anonymous reviewer in *Nature* contained the following: ". . . Prof. Portier is good humored enough to quote the paradox that a theory is not of value unless it can be demonstrated to be false. We have no hesitation in prophesying that his theory will attain that value . . .". Today, while there may still be snickering in "polite biological society" about the proposed derivation of eukaryotic cilia and flagella from symbiotic spirochetes, the prokaryotic ancestry of mitochondria is discussed in most textbooks, including those written by the "establishment." Formally, it is necessary at least to consider two alternative hypotheses concerning the evolutionary origin of mitochondria. The autogenous origin hypothesis (direct filiation) proposes that mitochondria arose from within a single cell by a process of intracellular compartmentalization and functional specialization. When DNA was discovered in these organelles, the assumption had to be added that genetic information from a single nuclear genome became distributed between these two compartments. The second hypothesis suggests that mitochondria originated from a symbiotic relationship between a proto-eukaryotic cell and a primitive prokaryote capable of oxidative phosphorylation. Only the second hypothesis is being seriously considered today.

A critical review of the arguments for and against the endosymbiont hypothesis published by Gray and Doolittle in 1982 (2) made an attempt to define the kind of data on which proof for the endosymbiont hypothesis must rest, and it concluded that "the case for an endosymbiont origin of mitochondria was more problematic and considerably less certain" than the case for the origin of chloroplasts (plastids) from photosynthetic eubacteria (cyanobacteria). A huge volume of additional data began to accumulate during the 1980s which was periodically reviewed by Gray and others, culminating in an exhaustive and authoritative review and evaluation by Gray in 1992 (3). The author came to the conclusion that reevaluating alternate hypotheses was pointless in view of the massive and compelling evidence in favor of the endosymbiont hypothesis.

The greatest boost to the hypothesis of the endosymbiotic origin of mitochondria came from the discovery of DNA in mitochondria (4), as well as from the generalization that all mitochondria in all eukaryotes contained DNA (however, see below for exceptions). Another statement, that all eukaryotes contain mitochondria, is also no longer completely tenable, since eukaryotes lacking mitochondria have been identified—for example, those placed in the kingdom of Archezoa (5). One idea was that they may represent surviving relatives of the earliest amitochondriate eukaryotes. However, evidence is increasingly favoring the proposal that these eukaryotes had harbored mitochondria at one time, but had lost them in their adaptation to certain environments (see below). It may in fact be impossible to find a relative of the proto-eukaryotic host cell. Instead, one plausible scenario envisions the origin of eukaryotes from the combination of two bacteria, one acting as the host and the other as the symbiont. In other words, the postulated symbiosis defines the origin of the first eukaryotic cell (6, 7). Eukaryotic cells can be defined as having a nucleus, with its outer membrane contiguous with the endoplasmic reticulum, a Golgi complex, various other membrane-enclosed subcellular organelles, and structures referred to as the cytoskeleton including actin filaments, intermediate filaments, and microtubules, in addition to mitochondria. No prokaryotic organisms are known that have some or all of these particular structures (8). However, a search for a "mitotic apparatus" in microorganisms has uncovered proteins that are distant relatives of tubulin and actin. Before the genetic information in mtDNA was completely analyzed, it was shown that mitochondrial DNA could be mutated and that such "cytoplasmic" mutations had serious consequences—not surprisingly affecting the capacity of the cells to carry out oxidative phosphorylation and respiration (9). Thus, the functional importance of this genome was proved. These early genetic investigations were primarily performed with the yeast *Saccharomyces cerevisiae*, and many others have followed. Mitochondrial DNA encoded very specific and unique proteins, although it was obvious almost from the start that the mitochondrial genome was far too small to encode information sufficient for a free-living organism. The loss of genetic information from primordial proto-mitochondria was compensated by functions encoded by the nucleus, but at this point one may have to distinguish between two possibilities. An obvious one is that proto-mitochondrial genes were transferred to the nucleus, and there is persuasive evidence for such events to have occurred. It should also be obvious, however, that symbiosis implied that the bacteria destined to become mitochondria had to invade a host cell, a proto-eukaryote, which prior to this evolutionary event was equally capable of an independent existence. The distinction between the two cells in the simplest view is that the proto-eukaryote already had its DNA compartmentalized in a nucleus with a nuclear membrane, while the proto-mitochondrion resembled a bacterium without a nucleus. One can speculate that the free-living proto-eukaryote had the capacity for anaerobic glycolysis and fermentation, while the prokaryote contributed the electron transport system. Both must have been capable of DNA replication, transcription, protein synthesis, and the biosynthesis of various lipids to form a membrane. Krebs cycle enzymes may also have been present in both for the purpose of interconverting short carbon compounds and amino acids. Therefore, the initial association between these two cells must have led to a considerable amount of redundancy of genetic information. The redundant genes in the mitochondrion could simply be lost, others were transferred to the nucleus, and only a very small number had to remain in the mitochondrion.

As pointed out by Gray (3), it is absolutely necessary to start with the presumed common ancestor of both eukaryotic and prokaryotic cells and then to consider distinct traits (genes) that evolved specifically in one of the groups. Only such genes constitute a unique and novel contribution to the symbiotic relationship and distinguish the symbiont from the host. From the latest proteomic studies it has been estimated that only 14–16% of the modern mitochondrial proteome has an origin that can be traced back to the bacterial endosymbiont. The large majority of the very diverse mitochondrial proteins were recruited to this organelle because it assumed many more functions in addition to ATP production (10).

It is not the intent here to speculate about the origin of life and the minimal number of genes required for an independent existence, regardless of the ultimate energy source. Arguments similar to the above apply to the hypothesized origin of chloroplasts (plastids) from photosynthetic bacteria. Light as an energy source has been available since the beginning of the earth. Presumably the evolution of oxygenic photosynthesis in prokaryotes (cyanobacteria) preceded respiration. Respiration requires oxygen, and oxygen did not become abundant until massive quantities were liberated by photosynthesis. If photosynthesis was performed exclusively by prokaryotes during this early period of life on earth, it is possible to speculate that mitochondria were acquired by a eukaryote before the acquisition of photosynthetic prokaryotes which subsequently evolved into chloroplasts in plants. Thus, the mitochondria of plants, animals, and fungi may all have a common ancestor and hence similar properties. The review by Embley and Martin (8) includes one of the most thorough and up-to-date discussions with a new tree describing the most plausible phylogenetic relationships among the major groups of eukaryotes, including species harboring hydrogenosomes and mitosomes. The view that the evolution of cells with mitochondria is causally linked to the increase in atmospheric oxygen ~2 Gyr ago is now being challenged by new information of anaerobic eukaryotes using protons (to form H_2), or other carbon compounds as the ultimate electron acceptors (8). Fascinating discussions with a much broader perspective on the effect of oxygen on the "earth's metabolic evolution" can be found in the paper by Raymond and Segre (11) and commentary by Falkowski (12). An analysis referred to as metabolic network expansion that makes use of 6836 reactions involving 5057 distinct compounds in 70 different organisms in the KEGG database leads to the conclusion that the availability of oxygen in the earth's atmosphere caused a major proliferation of new reactions and metabolic pathways resulting in a very much expanded pool of metabolites.

The arrival of oxygen on earth was not an unmixed blessing. Apart from the fact that uncontrolled respiration is combustion ("fire") with all of its destructive aspects, even a highly controlled reaction such as respiration can generate "reactive oxygen species" (ROS) such as superoxide, the hydroxyl radical, and hydrogen peroxide, whose potential for destruction of biological macromolecules may provide a basis of an explanation for an age-related deterioration of the mitochondrial genomes and perhaps even for senescence and death (13, 14). According to F. Jacob, "the most important inventions of evolution are sex and death." Did death from free radical bombardments precede programmed cell death, or apoptosis, as it is now called? Is the mechanism of apoptosis absolutely dependent on mitochondrial functions (see Chapter 7)?

It is more fruitful to pursue the molecular fossil record for a proto-mitochondrion. Phylogenetic data support an origin of mitochondria from the alpha-proteobacterial order Rickettsiales (15–18). All mitochondrial genomes encode two ribosomal RNAs, and most encode 20 or more tRNAs (a notable exception may be the protozoans such as trypanosomes). As discussed in more detail elsewhere in this book, mammalian mtDNA also encodes 13 peptides for complexes of the OXPHOS system, and additional mitochondrial genes are found in other organisms. From the point of view of evolution, it is particularly illuminating to consider the four genes for complex II: They are all nuclear in mammals, but diverse organisms have been found recently in which the mtDNA still carries one or more of the same genes. The iron protein, Ip, and the flavoprotein, Fp, of complex II constitute the enzyme succinate dehydrogenase (SDH) of the citric acid cycle. SDH is attached to the inner mitochondrial membrane via two integral membrane proteins, or anchor proteins, C_{II-3} and C_{II-4}. Ip and Fp are remarkably well conserved in evolution at the protein sequence level and even at the nucleotide sequence level from prokaryotes to eukaryotes. This has made it possible to identify and clone the corresponding genes from many organisms, starting with bacteria (19). As a first approximation and from a biochemist's simplest perspective, the complex in *Escherichia coli* looks very much like the complex in mammalian mitochondria. In bacteria such as *E. coli* the four genes form an operon and are coordinately expressed. In mammals the four genes (SDHA, SDHB, SDHC, and SDHD) are dispersed over various chromosomes. The SDHB (Ip) and SDHC (C_{II-3}) genes are found on the mitochondrial genome in the red algae *Chondrus crispus* (20, 21), and three of the four genes (SDHB, SDHC, and SDHD) are found on the mitochondrial genome of another red alga, *Porphyra purpurea*, and on a phylogenetically distant heterotrophic zooflagellate, *Reclinomas americana* (22). Apparently, the transfer of these genes from the mtDNA to the nuclear genome has not (yet) occurred, and these organisms represent examples of "intermediates" in the evolution of a maximally reduced mitochondrial genome. In full agreement with the same train of thought, *Chondrus crispus* mtDNA still retains four genes for ribosomal proteins, while certain plant mtDNAs encode even more ribosomal proteins. In vertebrates all mitochondrial ribosomal protein genes have been translocated to the nucleus. Studies such as these have provided supporting arguments in favor of a monophyletic origin of mitochondria (22). The explosion of information on genomic sequences from many organisms has provided additional examples (6, 23–27). The diversity of genes still found in mitochondrial genomes can be explained by a random gene loss over time from a common ancestral genome.

When the complete sequence of the mitochondrial genome of the freshwater protozoon *Reclinomonas americana* was completed, some rather surprising findings strengthened the hypothesized link between mitochondria and eubacteria-like endosymbionts ("the mitochondrion that time forgot") (28, 29). With 69,034 basepairs the genome is not exceptional in size, but the existence of at least 97 genes on this genome was unexpected, since many of the other mitochondrial genomes now known have far fewer genes, even when their size was an order of magnitude greater (see Chapter 4). The *Reclinomonas* mtDNA encodes 23 components of the electron transport chain and oxidative phosphorylation apparatus (compared to 13 in mammals), 18 mitoribosomal proteins (none in mammals, some in plants), 27 distinct tRNAs, 3 ribosomal RNAs, including a 5S RNA not found in mammals, an RNA component of ribonuclease P (used in RNA processing (see Chapter 4), and four genes of a multisubunit RNA polymerase resembling a eubacterial-type polymerase. The latter finding is particularly intriguing and unique so far. A preliminary interpretation is that a eubacterial multisubunit RNA polymerase may have been lost during evolution from all other mitochondria (analyzed so far) and replaced by a single-subunit enzyme, encoded by the nuclear genome, with similarities to the bacteriophage T3/T7 enzyme (see Chapter 4). Since then a large number of genes encoding mitochondrial enzymes for transcription and DNA replication have been identified and analyzed in fungi, animals, and plants, leading Shutt and Gray (30) to hypothesize that another early event in the evolution of the eukaryotic cell (and almost coincident with the establishment of endosymbiosis) was the acquisistion of several replication genes from an ancestor of T-odd bacteriopages. Furthermore, these authors propose that a phage RNA polymerase initially functioning in primer formation for DNA replication evolved into the present day mitochondrial RNA polymerase.

Ribosomal RNA sequence comparisons have served as a predominant source of data in the construction of phylogenetic trees and evolutionary relationships, and rRNAs have been referred to as the "ultimate molecular chronometer" (31). These ancient, ubiquitous, and functionally identical molecules and their genes have helped to reevaluate taxonomic subdivisions between existing life forms on earth. It is not the intent here to enter the debate about how many kingdoms there should be (the eukaryotes, Animalia, Plantae, Fungi, Protista, and the prokaryotes, Eubacteria and Archaebacteria), but to emphasize a major distinction between archaebacteria, eubacteria, and eukaryotes. Various analyses have shown a closer relationship of archaebacteria to eukaryotes, as well as many similarities in genome organization and gene expression between plastid (chloroplast) and eubacterial genomes, in support of the endosymbiotic origin of plastids. A case for a eubacterial origin of mitochondria was more difficult to make on the basis of the "molecular biology as practiced in mitochondria" (3), but rRNA sequence analyses strongly favor the same conclusion. Various aspects such as information content and organization of the genome, transcription, and translation will be covered in separate chapters when evolutionary comparisons will be instructive, but not central to the discussion.

Mitochondria were found to have their own machinery for protein synthesis. A full treatment of this subject is reserved for another chapter. However, a frequently quoted example in favor of the endosymbiont hypothesis can be mentioned here. Cytoplasmic protein synthesis employs larger ribosomal subunits with more protein subunits compared to mitochondrial ribosomal subunits, which in turn resemble bacterial ribosomes (32, 33). Cycloheximide is a specific inhibitor of cytoplasmic protein synthesis, whereas chloramphenicol inhibits the bacterial and mitochondrial translation machinery (34). The distinction based on the differential sensitivity to these two inhibitors has been the basis for many elegant experiments elucidating the specific mitochondrial translation products. On the other hand, a closer look at these similarities in ribosomal properties in bacteria and mitochondria made it apparent that they were rather superficial, hiding the uniqueness and variability of mitochondrial ribosomes (for a detailed discussion see Gray and Doolittle (2)).

Phylogenetic trees based on rRNA comparisons (for the small as well as for the large subunit) have not only proved convincing in relating plant mitochondria to eubacteria, but have also helped to identify the living eubacteria that are the closest relative of mitochondria: Members of the α-subdivision of purple bacteria had been suspected initially on biochemical grounds, and rRNA sequence comparisons strengthen this conclusion considerably. In contrast, similar comparisons with animal, ciliate, and fungal mitochondrial rRNA sequences are less convincing, because of significant primary sequence divergences. Nevertheless, the eubacterial ancestry of mitochondria is not seriously challenged by such data. Left open was the question of the monophyletic origin of mitochondria—that is, whether the symbiosis between proto-eukaryotes and a eubacterial species was a singular event in evolution, or occurred more than once. In particular, did plant mitochondria arise independently of animal mitochondria, for example? The absence of a common root for current rRNA trees had been interpreted by some authors in support of a biphyletic or multiphyletic origin (see reference 3 for a summary of the evidence and interpretation), but the same authors also raised the possibility of artifactual trees generated by the choice of algorithms, the selection of sequences, and other methodological subtleties. The most recent studies on complex II genes in the red algae, *Porphyra purpurea*, and the zooflagellate, *Reclinomonas americana,* have favored a monophyletic lineage for mitochondria (22).

Has the limit been reached in the reduction of the mitochondrial genome (35)? It is generally agreed that the remaining peptides encoded by mtDNA are extremely hydrophobic peptides that would not be readily imported by the mitochondrial translocation machinery. On the other hand, two hydrophobic anchor peptides of complex II are encoded by nuclear genes in most organisms examined so far. Theoretical arguments have been forwarded predicting that the mitochondrial genome need ultimately retain only two genes encoding peptides in addition to the ribosomal RNAs and tRNAs (36–38). The two genes singled out are the apo-cytochrome b and the COX1 genes encoding peptides with more than three or four hydrophobic transmembrane segments. Claros et al. (37) define a property referred to as mesohydrophobicity (average hydrophobicity over an extended stretch of a polypeptide chain) to make quantitative predictions. The hypothesis was also tested by constructing a series of chimeric genes expressed in the nucleus yielding cytoplasmically synthesized peptides. These peptides had a mitochondrial targeting sequence, a reporter enzyme, and one or more of the eight transmembrane segments of apocytochrome b. A complete interpretation of these results, however, also requires a more detailed understanding of how such integral membrane proteins with multiple transmembrane segments are localized in the inner mitochondrial membrane, an understanding that is still expanding today (see Chapter 4).

Some recent discoveries provide convincing evidence that the complete loss of the mitochondrial genome has in fact occurred in evolution, raising several new questions (35). Such mitochondria would not be expected to be able to support respiration. What would be their function if oxidative phosphorylation was no longer observed? And how would such an organelle still be recognized as a mitochondrion? An emerging consensus is that several phylogenetically distinct lineages of eukaryotes may have undergone a reductive evolution in the course of adapting to anaerobic or micro-aerobic environments (8, 39–44). In other words, the original mitochondrial endosymbiont has evolved independently in diverse species inhabiting such environments, resulting in a collection of morphologically, genetically, and functionally heterogeneous organelle variants that include anaerobic and aerobic mitochondria, hydrogenosomes, and mitosomes (45). The best-studied example is the parasitic flagellate protist *Trichomonas vaginalis*, an air tolerating anaerobe which has an energy producing organelle referred to as a hydrogenosome. Hydrogenosomes have been identified in a wide variety of organisms (rumen-dwelling ciliates, fungi, free living ciliates), and by several criteria such hydrogenosomes are believed to be highly modified mitochondria (16, 41, 42, 46). They generally lack DNA, are devoid of cytochromes and most citric acid cycle enzymes, and produce hydrogen by the use of enzymes usually found in anaerobes (pyruvate : ferredoxin oxidoreductase, hydrogenase). They make ATP by substrate-level phosphorylations involving succinyl CoA synthetase, a Krebs cycle enzyme. However, they have a double membrane, they import proteins post-translationally, and they divide by fission. Further evidence for the relationship of hydrogenosomes to mitochondria comes from the characterization of *Trichomonas* nuclear genes of the heat-shock proteins Hsp10, Hsp60, and

Hsp70 (47). These proteins are found in the hydrogenosome, and by phylogenetic analysis these Hsps have been found to be among the "most reliable tracers of the eubacterial ancestry of both the mitochondrion and the chloroplast" (35). Thus, the evidence supports a common origin for hydrogenosomes and mitochondria, rather than the hypothesis of an independent endosymbiotic event involving an anearobic eubacterium yielding a hydrogenosome (48). A hydrogenosome with a genome has been identified in the anaerobic ciliate *Nyctotherus ovalis*, which thrives in the hindgut of cockroaches (49). More recently it was shown that it is a rudimentary genome encoding components of a mitochondrial electron transport chain (39). The authors interpret the *Nycotherus* hydrogenosome as a "true missing link" between mitochondria and hydrogenosomes. As discussed by Gray (50), the hydrogenosome's past is still muddied by recent findings in the genome-less hydrogenosome of *Trichomonas vaginalis*. Two groups arrived at conflicting evolutionary positions, based on the sequence analysis of two nuclear genes encoding *T. vaginalis* homologues of the mitochondrial complex I (51, 52).

Such evidence may strengthen arguments over which genes were originally present in the proto-mitochondria. However, at this time the more challenging questions to ask are questions about the evolution of more individual differences between organisms: the size of the mitochondrial genome, the organization of the remaining genes, and some rather unusual mechanisms in gene expression. Plants have mitochondrial genomes that are one to two orders of magnitude larger than metazoan mitochondrial genomes, without a proportionate increase in the number of genes. What is this "junk DNA" doing and where did it come from? Mitochondrial plasmids and transposon-like parasitic DNA have invaded even mitochondrial genomes. Understanding the evolution of the mind-boggling RNA editing process in trypanosomes (see Chapter 4) is a major challenge by itself. Mitochondrial DNA sequences continue to to be altered by mutations, providing an ever-richer database for evolutionary studies—not about the evolution of mitochondria per se, but about the much more recent evolution of closely related species such as the primates, including humans (see Chapter 8).

REFERENCES

1. Margulis, L. (1981) *Symbiosis in Cell Evolution*, W. H. Freeman, San Francisco.
2. Gray, M. W. and Doolittle, W. F. (1982) *Microbiol.Revs.* **46**, 1–42.
3. Gray, M. W. (1992) *Int. Rev. Cytol.* **141**, 233–357.
4. Nass, S. and Nass, M. M. K. (1963) *J. Cell. Biol.* **19**, 593–629.
5. Cavalier-Smith, T. (1981) *Biosystems* **14**, 461–481.
6. Gray, M. W., Burger, G., and Lang, B. F. (1999) *Science* **283**, 1476–1481.
7. Vellai, T. and Vida, G. (1999) *Proc. R. Soc. Lond. B Biol. Sci.* **266**, 1571–1577.
8. Embley, T. M. and Martin, W. (2006) *Nature* **440**, 623–630.

9. Slonimski, P. P. and Ephrussi, B. (1949) *Ann. Inst. Pasteur* **77**, 47.
10. Gabaldon, T. and Huynen, M. A. (2004) *Biochim. Biophys. Acta* **1659**, 212–220.
11. Raymond, J. and Segre, D. (2006) *Science* **311**, 1724–1725.
12. Falkowski, P. G. (2006) *Science* **311**, 1724–1725.
13. Loeb, L. A., Wallace, D. C., and Martin, G. M. (2005) *Proc. Natl. Acad. Sci. USA* **102**, 18769–18770.
14. Ames, B. N., Shigenaga, M. K., and Hagen, T. M. (1995) *Biochim. Biophys. Acta Mol. Basis Dis.* **1271**, 165–170.
15. Andersson, S. G. E., Zomorodipour, A., Andersson, J. O., Sicheritz-Pontén, T., Alsmark, U. C. M., Podowski, R. M., Näslund, A. K., Eriksson, A. S., Winkler, H. H., and Kurland, C. G. (1998) *Nature* **396**, 133–140.
16. Andersson, S. G., Karlberg, O., Canback, B., and Kurland, C. G. (2003) *Philos. Trans. R. Soc. Lond. B. Biol. Sci.* **358**, 165–79.
17. Emelyanov, V. V. (2003) *Arch. Biochem. Biophys.* **420**, 130–141.
18. Emelyanov, V. V. (2003) *Eur. J. Biochem.* **270**, 1599–1618.
19. Gould, S. J., Subramani, S., and Scheffler, I. E. (1989) *Proc. Natl. Acad. Sci. USA* **86**, 1934–1938.
20. Viehmann, S., Richard, O., Boyen, C., and Zetsche, K. (1996) *Curr. Genet.* **29**, 199–201.
21. Leblanc, C., Boyen, C., Richard, O., Bonnard, G., Grienenberger, J.-M., and Kloareg, B. (1995) *J. Mol. Biol.* **250**, 484–495.
22. Burger, G., Lang, B. F., Reith, M., and Gray, M. W. (1996) *Proc. Natl. Acad. Sci. USA* **93**, 2328–2332.
23. Gray, M. W., Lang, B. F., and Burger, G. (2004) *Annu. Rev. Genet.* **38**, 477–524.
24. Burger, G., Gray, M. W., and Lang, B. F. (2003) *Trends Genet.* **19**, 709–716.
25. Lang, B. F., Gray, M. W., and Burger, G. (1999) *Annu. Rev. Genet.* **33**, 351–397.
26. Adams, K. L. and Palmer, J. D. (2003) *Mol. Phylogenet. Evol.* **29**, 380–395.
27. Bullerwell, C. E. and Gray, M. W. (2004) *Curr. Opin. Microbiol.* **7**, 528–534.
28. Palmer, J. D. (1997) *Nature* **387**, 454–455.
29. Lang, B. F., Burger, G., O'Kelly, C. J., Cedergren, R., Golding, G. B., Lemieux, C., Sankoff, D., Turmel, M., and Gray, M. W. (1997) *Nature* **387**, 493–497.
30. Shutt, T. E. and Gray, M. W. (2006) *Trends Genet.* **22**, 90–95.
31. Woese, C. R. (1987) *Microbiol. Rev.* **51**, 221–271.
32. Linnane, A. W., Lamb, A. J., Christodoulou, C., and Lukins, H. B. (1968) *Proc. Natl. Acad. Sci. USA* **59**, 1288–1293.
33. Lamb, A. J., Clark-Walker, G. D., and Linnane, A. W. (1968) *Biochim. Biophys. Acta* **161**, 415–427.
34. Towers, N. R., Dixon, H., Kellerman, G. M., and Linnane, A. W. (1972) *Arch. Biochem. Biophys.* **151**, 361–369.
35. Palmer, J. D. (1997) *Science* **275**, 790–791.
36. Popot, J.-L. and deVitry, C. (1990) *Annu. Rev. Biophys. Biophys. Chem.* **19**, 369–403.
37. Claros, M. G., Perea, J., Shu, Y., Samatey, F. A., Popot, J. L., and Jacq, C. (1995) *Eur. J. Biochem.* **228**, 762–771.

38. Claros, M. G. (1995) *Comp. Appl. Biosci.* **11**, 441–447.
39. Boxma, B., de Graaf, R. M., van der Staay, G. W., van Alen, T. A., Ricard, G., Gabaldon, T., van Hoek, A. H., Moon-van Der Staay, SY, Koopman, W. J., Van Hellemond, J. J., Tielens, A. G., Friedrich, T., Veenhuis, M., Huynen, M. A., and Hackstein, J. H. (2005) *Nature* **434**, 74–79.
40. Atteia, A., van Lis, R., Van Hellemond, J. J., Tielens, A. G., Martin, W., and Henze, K. (2004) *Gene* **330**, 143–148.
41. Van Hellemond, J. J., Van Der Klei, A., Van Weelden, S. W., and Tielens, A. G. (2003) *Philos. Trans. R. Soc. Lond. B. Biol. Sci.* **358**, 205–215.
42. Tielens, A. G. M., Rotte, C., Van Hellemond, J. J., and Martin, W. (2002) *Trends Biochem. Sci.* **27**, 564–572.
43. Embley, T. M., van der, G. M., Horner, D. S., Dyal, P. L., Bell, S., and Foster, P. G. (2003) *IUBMB. Life* **55**, 387–395.
44. Embley, T. M., Van Der Giezen, M., Horner, D. S., Dyal, P. L., and Foster, P. (2003) *Philos. Trans. R. Soc. Lond. B. Biol. Sci.* **358**, 191–203.
45. Van Der Giezen, M. and Tovar, J. (2005) *EMBO Rep.* **6**, 525–530.
46. Andersson, S. G. and Kurland, C. G. (1999) *Curr. Opin. Microbiol.* **2**, 535–541.
47. Bui, E. T. N., Bradley, P. J., and Johnson, P. J. (1996) *Proc. Natl. Acad. Sci. USA* **93**, 9651–9656.
48. Van Der Giezen, M., Birdsey, G. M., Horner, D. S., Lucocq, J., Dyal, P. L., Benchimol, M., Danpure, C. J., and Embley, T. M. (2003) *Mol. Biol. Evol.* **20**, 1051–1061.
49. Akhmanova, A., Voncken, F., van Alen, T., Van Hoek, A., Boxma, B., Vogels, G., Veenhuis, M., and Hackstein, J. H. P. (1998) *Nature* **396**, 527–528.
50. Gray, M. W. (2005) *Nature* **434**, 29–31.
51. Dyall, S. D., Yan, W., Delgadillo-Correa, M. G., Lunceford, A., Loo, J. A., Clarke, C. F., and Johnson, P. J. (2004) *Nature* **431**, 1103–1107.
52. Hrdy, I., Hirt, R. P., Dolezal, P., Bardonova, L., Foster, P. G., Tachezy, J., and Embley, T. M. (2004) *Nature* **432**, 618–622.

3

STRUCTURE AND MORPHOLOGY. INTEGRATION INTO THE CELL

Summary

A detailed morphological description of mitochondria became possible only with the perfection of the electron microscope, when their double membranes and the existence of cristae were established. The shape and size of mitochondria is quite variable, and most significantly, there is an amazing variation in the number and structure of the cristae. While it is clear that cristae serve to increase the surface area of the inner membrane, the shape of cristae as revealed by cross sections viewed in the electron microscope can still be a puzzle. More refined three-dimensional images of outer and inner mitochondrial membranes and cristae have recently been achieved by electron tomography, but the full significance of novel findings (cristae junctions) remains to be generalized and even fully explained.

A very challenging and exciting new area of research is concerned with the dynamics of the mitochondrial morphology and the distribution of mitochondria within specialized cells. They have to divide by fission; in many organisms they also constantly fuse with each other, and their interaction with the cytoskeleton assures their distribution in dividing and differentiating cells, as well as their optimal localization within cells. The elucidation of components of the fusion and fission apparatus has been a frontier where major progress has been achieved in the past decade.

3.1 STRUCTURE AND MORPHOLOGY

When the nineteenth-century morphologists with their light microscopes discovered grains (chondria) and filaments (mito), they could not be sure that they were all looking at the same functionally distinct structure in different cells, and undoubtedly they were not. Nevertheless, variability in number and shape must have been apparent. A more systematic approach became possible when selective staining methods were used (fuchsin, Altman, 1890; crystal violet, Benda, 1898; and Janus green, Michaelis, 1898). The capacity of mitochondria to reduce Janus green, an indicator of oxidation–reduction reactions, should have been a major clue, which was evidently missed at the time or, rather, not interpreted fully.

Major progress in the characterization and understanding of the morphology and ultrastructure of mitochondria was not possible until the techniques for electron microscopy and specimen preparation were perfected, but the use of the light microscope in the study of mitochondria was revived in the past few decades by two technical innovations. First, mitochondria could be viewed with the fluorescence microscope in permeabilized cells after staining with specific antibodies and secondary antibodies conjugated to fluorescent dyes. Another very important discovery was the observation that certain rhodamine derivatives and other lipophilic "vital dyes" would stain mitochondria, because the dye became concentrated in these organelles by an uptake mechanism driven by the membrane potential (1). The biophysical basis of this phenomenon will be addressed in a later chapter. Rhodamine staining can be performed with live cells; and in combination with time-lapse photography, and more recently image intensification and videotaping, it has opened up a window on the dynamic behavior of mitochondria (changes in size and shape), their distribution in the cytosol, and their redistribution following certain perturbations of the cytoskeleton, for example.

While the resolution of the light/fluorescence microscope is limited, it does offer a view of the entire cell and the global distribution of mitochondria within it (see Section 3.2 for further discussion). With the increased understanding of protein targeting sequences, it has become possible to target recombinant reporter proteins such as the green fluorescent protein (GFP) to the mitochondrial matrix, or even the intermembrane space (see, for example, references 2 and 3). Such "labeled" mitochondria can be studied in relation to other subcellular structures identified by probes with a second chromophore.

Mitochondria have no fixed size, but typically in hepatocytes and fibroblasts they have a sausage-like shape with dimensions of 3–4 μm (length) and ~1 μm (diameter); as a first approximation they do in fact resemble small bacteria in size and shape. The number of mitochondria per cell is a parameter that appears to vary significantly from cell type to cell type, and this will become an important consideration when mitochondrial biogenesis is considered in the context of cell differentiation and specialization (see also below). Estimates from serial sections of cells yield values in the range of a few hundred to a few thousand per cell—for example, ~800 per hepatocyte. Specialized vertebrate cells show significant variations. The human oocyte is estimated to contain >100,000 mitochondria, while spermatozoa have relatively few (but these may be exceptionally large). A somewhat simplified view of mitochondria in some sperm cells is that of a small number of highly elongated mitochondria wrapped around the axoneme of the sperm tail. In lower eukaryotes the number can also be substantial, for example, 500,000 mitochondria have been found in the giant amoebae *Chaos chaos*. The actual number is most likely not the most relevant parameter to consider. Of far more significance may be the mitochondrial volume as a fraction of the cellular volume, or even the total surface area of the mitochondrial cristae per cell (see below). Clearly, details in the morphology depend on the organism. Another interesting number from a geneticist's point of view is the number of mitochondrial genomes per cell. This subject will be taken up in a later chapter, but it can be mentioned here that there are multiple mtDNAs per mitochondrion and therefore thousands of mtDNAs per cell.

Morphological studies of trypanosomes drew attention to a structure at the base of the flagellum referred to as kinetoplast. Over 70 years ago the kinetoplast was found to be stainable with Janus green (oxidation–reduction) and, shortly thereafter, to give a positive Feulgen reaction (nucleic acid). Later, electron microscopic examination revealed cristae similar to those found in mitochondria, and the kinetoplast was recognized to be a highly specialized mitochondrion. The first definite report of DNA in this organelle appeared in 1960 (4), but the kinetoplast was considered to be so specialized and exceptional that the finding was not immediately generalized to all mitochondria.

A new chapter in the study of the morphology of mitochondria began with the perfection of the electron microscope as a tool for biological studies. The design of the microscope was a physics and engineering problem. The preparation of a sufficiently thin specimen, a slice of a single cell many times thinner than the diameter of the cell, was a challenge that required a more empirical approach. In sum, the tissue had to be fixed, dehydrated, embedded "in plastic," stained, and sectioned to give a specimen that could hold together in a high vacuum, provide contrxast in the final image based on differential diffraction of the electron beam (uptake of electron-dense stains such as uranyl acetate and osmium tetroxide by different constituents of the cell), and still reflect biological reality—that is, meaningful images of subcellular structures. The number of skeptics was rapidly diminished by the spectacular images produced by the pioneers at the Rockefeller Institute. A new view of the morphology of individual mitochondria emerged from the studies of Palade and his colleagues, and of Sjostrand, who not only confirmed that mitochondria are membrane-bound organelles, but recognized that there were two membranes: an outer membrane and an inner membrane which was convoluted and folded into *cristae mitochondriales* (Palade) (Figures 3.1 and 3.2). It should also be reemphasized that the development of buffers for cell fractionation studies

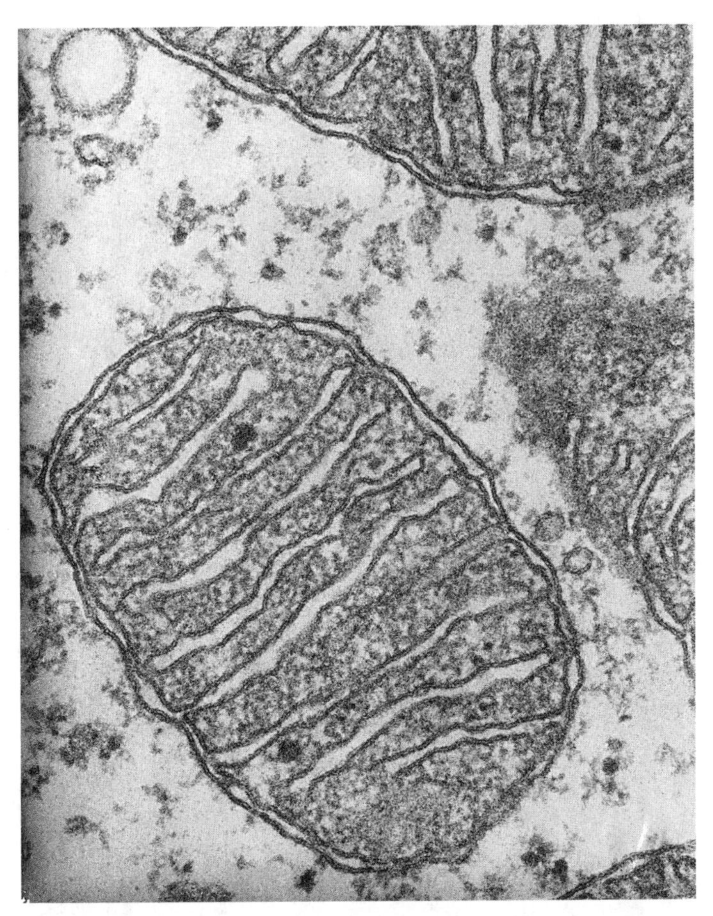

Figure 3.1 Mitochondrion from centroacinar cell of pancreas (×190,000). (Courtesy of Dr. G. Palade.)

yielding highly purified, intact mitochondria was an important contribution. Such isolated organelles could be viewed by electron microscopy and shown to have a morphology similar to those viewed *in situ*.

A plausible explanation made already at the time was that this topology vastly increased the surface area of the inner membrane. The full significance of the compartmentalization into an intermembrane space and a matrix could not be fully appreciated until later, but in this chapter, attention will be largely confined to a discussion of observable morphology of individual mitochondria. It is evident that the size and shape of mitochondria seen in the electron microscope depends on the plane of sectioning relative to their long axis.

As more examples of mitochondria in different tissues were examined, distinguishing features soon became apparent. The matrix was not homogeneous and exhibited a "fine granularity," and frequently there were distinct crystalline inclusions or small particles of high electron density (Figure 3.3). Their number

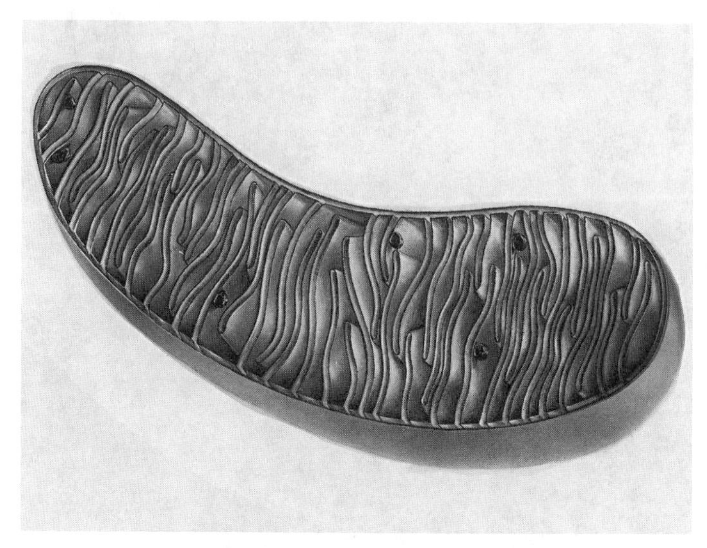

Figure 3.2 Schematic view of mitochondrion showing the outer and inner membranes and the folded cristae. (Drawing provided by Dr. D. Fawcett.)

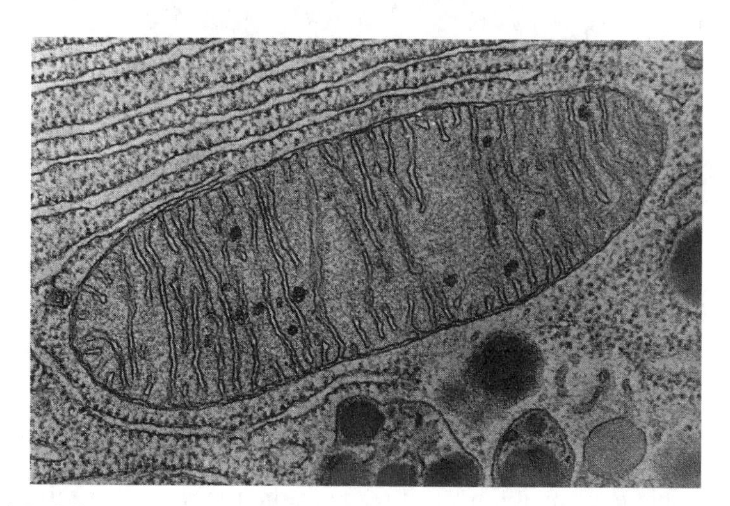

Figure 3.3 Mitochondrion showing electron-dense inclusions. (Original photograph of K. Porter; from *The Cell*, by D. W. Fawcett, 1981, W. B. Saunders Co.)

could in some instances be shown depend on the metabolic state of the tissue observed. The high electron density may be the result of a sequestration and storage of calcium. Other studies have suggested that the electron density is the consequence of the retention of osmium tetroxide by a phospholipoprotein that may bind calcium *in vitro*, but may not have such a role *in vivo*.

Most notable was the variable appearance of the cristae (Figure 3.4). They were often lamellar in appearance, but also tube-like in other cells, and a

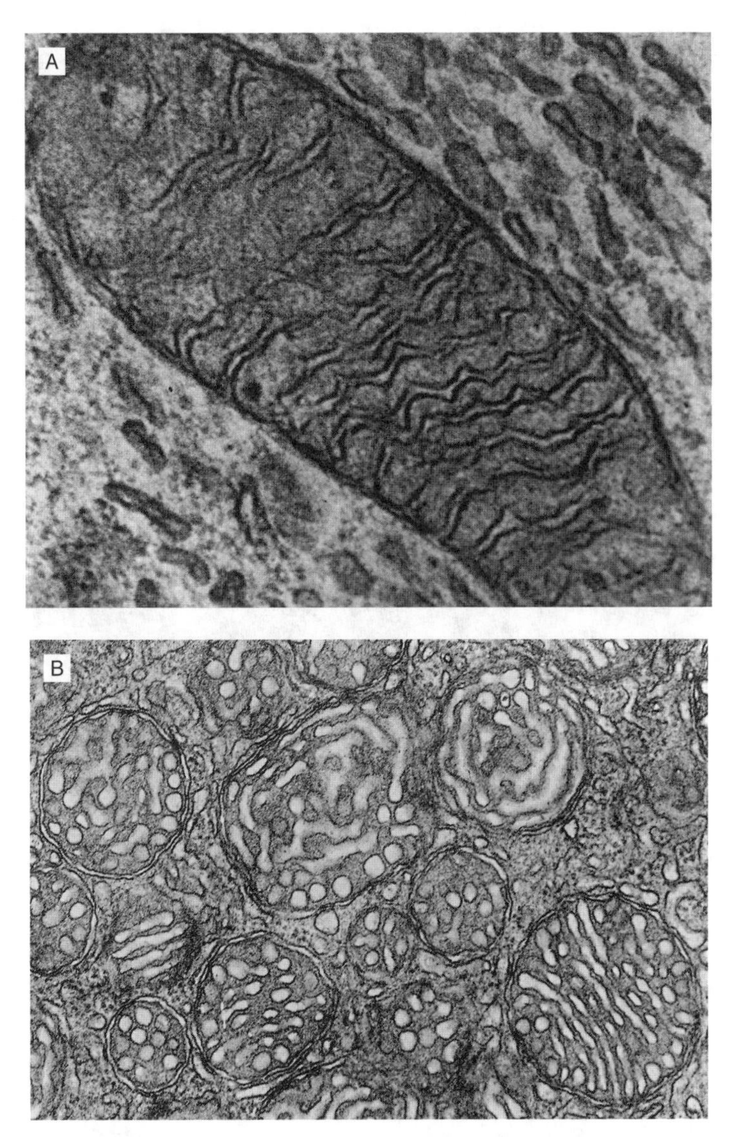

Figure 3.4 (A) Mitochondria with angular cristae from ventricular cardiac muscle. (Original photograph of J. P. Revel; from *The Cell*, by D. W. Fawcett, 1981, W. B. Saunders Co.) (B) Mitochondria with tubular cristae in amoeba. (Original photograph of T. Pollard; from *The Cell*, by D. W. Fawcett, 1981, W. B. Saunders Co.) (C) Mitochondria from adrenal cortex. (Original photograph of D. Friend; from *The Cell*, by D. W. Fawcett, 1981, W. B. Saunders Co.) (D) Mitochondrion from astrocyte. (Original photograph of K. Blinziger; from *The Cell*, by D. W. Fawcett, 1981, W. B. Saunders Co.) (E) Mitochondria from fish pseudobranch. (Original photograph of J. Harb and I. Copeland; from *The Cell*, by D. W. Fawcett, 1981, W. B. Saunders Co.) (F) Mitochondria from adrenal cortex. (Steroid secreting cells). (Original photograph of D. Fawcett; from *The Cell*, by D. W. Fawcett, 1981, W. B. Saunders Co.)

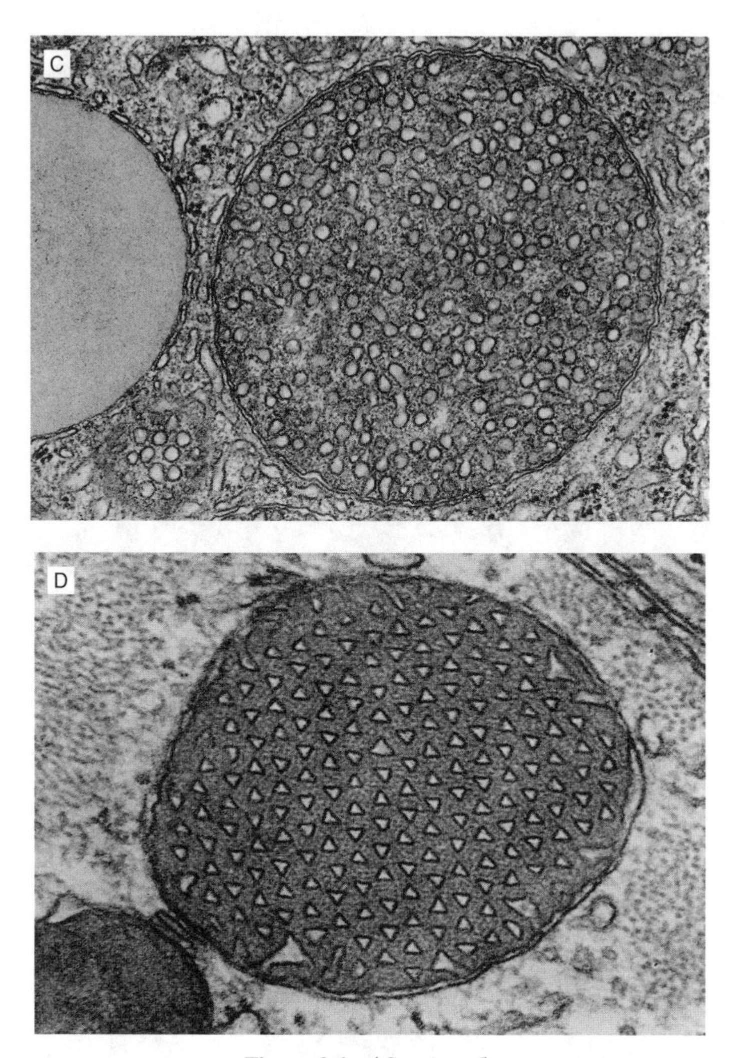

Figure 3.4 (*Continued*)

complete description required many images, and ideally, serial sections. Thin sections in the electron microscope still require an interpretation in three dimensions, and frequently the continuity of a membrane is not seen in a given section. Thus, tubes become "free" circular (vesicular) membranes in cross section, and sheet-like foldings of the inner membrane in some sections may appear as flattened sacs floating in the matrix. In the current view the inner membrane is continuous, and there appears to be no functional reason to have membrane vesicles within the matrix analogous to the thylakoid membranes (grana) of chloroplasts.

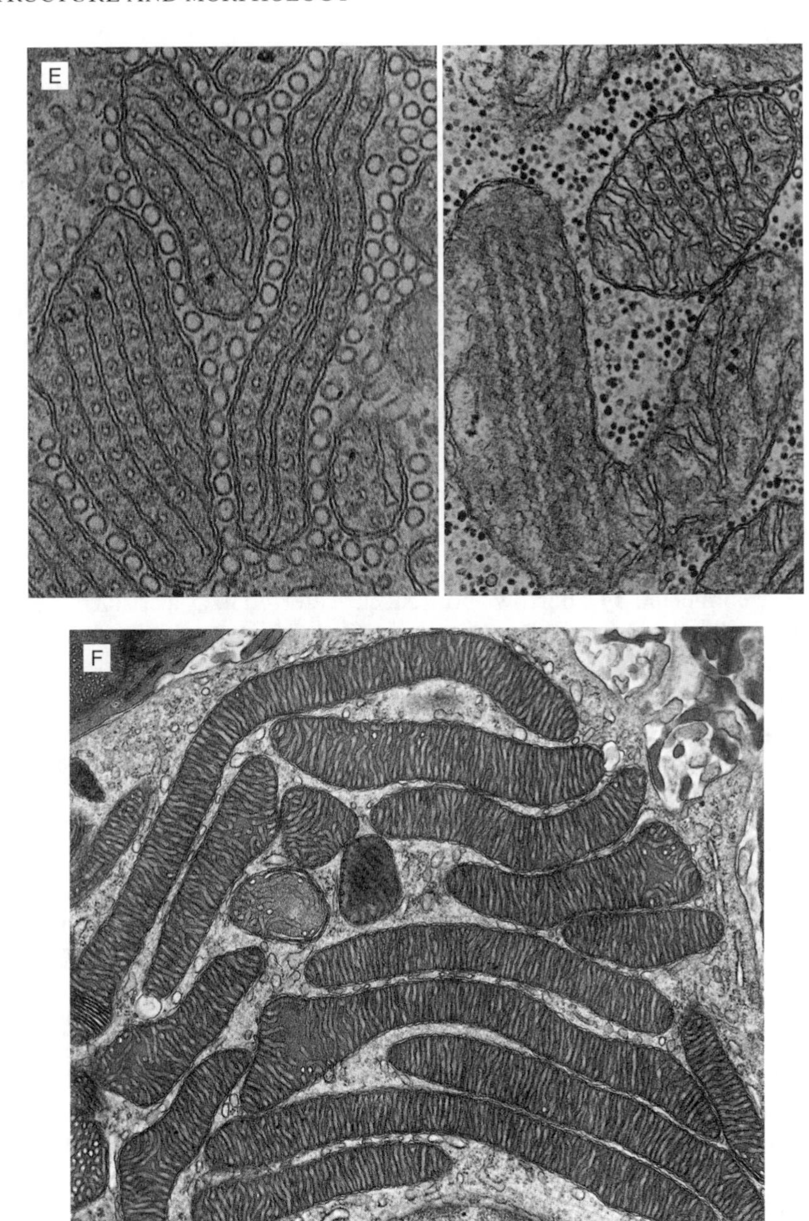

Figure 3.4 (*Continued*)

The interpretation of cristae topology from thin-section electron microscopy by Palade and Sjostrand postulated that the cristae are sheet-like invaginations of the inner membrane into the matrix of the mitochondrion (Figure 3.2). A maximum number of cristae can be accommodated if the sheet-like membranes are arranged in parallel, but it seems that the models fudge the issue of what happens to the image if a different plane of sectioning were presented. The T-shaped junction cannot be continuous around the entire circumference of the mitochondrion. Recent advances in electron tomography have made it possible to view complex 3-D objects like mitochondria in thick sections and reconstruct a high-resolution 3-D image. Data are collected by recording 2-D images over a wide range of tilt angles, and 3-D reconstructions are achieved by computer-assisted imaging. A 3-D resolution in the range of 5–10 nm can be achieved, hence images of the outer and inner membranes of mitochondria can be viewed at a hitherto unobtainable resolution (5, 6). Only a limited number of tissues have been examined by this powerful methodology (Figure 3.5A,B).

Of particular interest and significance is the conclusion by Perkins et al. (5) that cristae consist of a lamellar portion similar to that envisioned before, but the continuity with the inner boundary membrane (which lies close to the outer membrane) is achieved by a relatively small number of narrow tube-like connections. In other words, though continuous, the inner membrane juxtaposed to the outer membrane and the inner membrane of the lamellar structures are connected by a narrow "bottleneck" referred to as a cristae junction. Topologically, the intermembrane space is still continuous with the "inside" of the lamellae, but the authors raise the question whether the bottleneck can constitute a barrier and hence serve to separate two compartments: intracristal and intermembrane. It may also serve to segregate patches of inner membrane with differing protein compositions. Studies exploring the implications of these morphological observations have been challenging (7–12). The tubular structure of the cristae junctions requires a mechanistic explanation. On the one hand, theoretical models have considered the local composition of the lipid bilayer and the packaging of lipids to create the required curvature (13, 14). Other speculations have centered on specialized proteins required for maintaining cristae junctions. Attempts to purify cristae junctions have not yet met with success. More recently, several proteins have been identified and implicated in the control of cristae formation, but their relationship to cristae junctions is unclear. The inner-membrane protein mitofilin is a candidate requiring serious attention (15). Mitofilin is localized in the space between the inner boundary membrane and the outer membranes, where it can assemble into a large multimeric protein complex. A perhaps more surprising development is the discovery in yeast that the ATP synthase (complex V), and specifically the subunits e and g play a role in generating mitochondrial cristae morphology. The dimerization or oligomerization of the ATP synthase appears to be a critical step in this mechanism (16–18). The appearance of the cristae and cristae junctions has been shown to depend on the conditions under which the mitochondria are prepared for tomography (12). They are likely to be highly dynamic *in vivo*, responding to changes in the metabolic state, ADP/ATP ratios, ion fluxes, and matrix swelling. There is evidence that in some but not all cells, inner membranes are being remodeled during apoptosis, possibly as part of a mechanism for mobilizing cytochrome c release (19, 20).

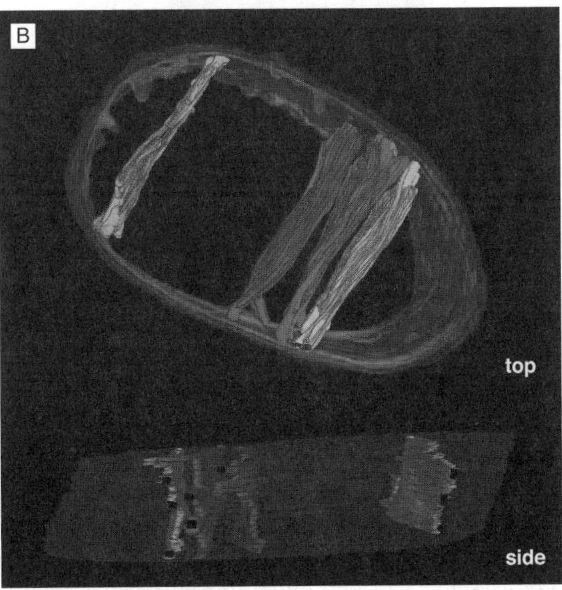

Figure 3.5 (A) Mitochondria seen by electron microscope tomography. Cells were derived from chick CNS. [The original figure was provided by G. Perkins, University of California, San Diego; with permission from Academic Press (Perkins et al. (5)).] (B) The 3-D reconstruction was achieved after the volume was segmented manually contouring into the regions bounded by the outer, inner, and cristal membranes. [For details the reader is referred to the original publication (Perkins et al. (5).] In the bottom of the figure, individual lamellar sections of crista are shown, together with the tubular portions which connect the disks to the inner boundary membrane. (The photographs were generously provided by Dr. G. Perkins, University of California, San Diego; with permission by Academic Press.) See color plates.

The suggestion has been made that cristae junctions represent diffusional barriers between the IMS and the cristal compartment. If the barrier were complete, a great similarity to chloroplasts would emerge, where the thylakoid membranes enclose a compartment distinct from the space between the inner and outer membranes. Such a view might also imply a difference in the biochemical composition of different domains of the inner membrane. At the moment, such speculations have limited experimental support, but a promising beginning has been made by Vogel et al. (21). These authors show that proteins are distributed in an uneven, yet not exclusive, manner between the inner boundary membrane and cristae membranes.

The number and morphology of cristae are likely to reflect the response of the mitochondria to the energy demands of the cell. Highly folded, lamellar cristae with a large surface area are typically found in muscle cells and neurons, where the respiratory rate is the greatest. These observations clearly raise questions about the control of cristae formation and the expression of both nuclear and mitochondrial genes encoding mitochondrial proteins. A telling example, although perhaps an extreme one, comes from the study of mitochondrial morphology in the yeast *Saccharomyces cerevisiae*. When grown in the presence of a nonfermentable substrate such as glycerol or ethanol, the cells are dependent on oxidative phosphorylation, and hence they have mitochondria easily recognizable even by the non-expert. When grown in the presence of glucose, the biogenesis of the inner membrane is severely repressed; and the morphology, but not the number of the mitochondria, changes dramatically. The details of the mechanism of "glucose repression" of respiration will be presented in a later chapter. A similar change in the abundance and morphology of an intracellular organelle can be observed for peroxisomes—for example, in the yeast *Piccia pastoris*, where growth on methanol or fatty acids induces the appearance of large peroxisomes. Free-living microorganisms are exquisitely attuned to their environment and capable of adjusting and optimizing their metabolism to the ever-changing conditions.

The question about the number and shape of mitochondria in yeast, *Saccharomyces cerevisiae*, took a different turn when a series of serial sections were analyzed to reveal that in the extreme there may be only one mitochondrion per cell, existing as a highly branched, continuous structure. A single section alone will show what appear to be discrete mitochondria with spherical or elongated shapes (22, 23). (See Section 3.2 for further details.)

In multicellular organisms, mitochondria are likely to have a more invariable morphology in a given tissue, but different tissues have distinct energy needs. It may well be that an altered mitochondrial morphology is also related to functions of mitochondria not strictly linked to respiration. Steroid secreting cells such as Leydig cells have mitochondria with an unusually wide range of sizes and often a very distinct array of tubular cristae (24). Perhaps the most puzzling display of cristae morphology is one that occurs sporadically (?) in many tissues of diverse species. These cristae appear to be made of parallel, hexagonal arrays of tubules with a triangular cross section. This may be an extreme manifestation of cristae with sharp angulations regularly spaced along their length, which are found in mitochondria of cardiac cells, among others (24).

It is evident that the more subtle aspects of the variations in cristae morphology can generally not be explained in terms of some sweeping generalization about surface area and respiratory activity. The inner mitochondrial membrane is an exceptional membrane with respect to the protein : lipid ratio. While in most membranes this ratio is close to 50:50, conducive to the image of "proteins floating in a sea of lipids," the inner mitochondrial membrane has a ratio of ~75:25, suggestive of relatively densely packed proteins. Researchers focused on the inner mitochondrial membrane were among the most reluctant in their acceptance of the "fluid mosaic model" as a model for all biological membranes (J. Singer, personal communication). However, numerous studies have provided strong experimental support for the view that the inner membrane is in a fluid state rather than a solid state (for a review see references 25 and 26). Thus, the proteins of this membrane cannot be exclusively responsible for its folding. Note, however, the role of ATP synthase oligomerization mentioned above (17). The discovery of supercomplexes formed by complexes of the electron transport chain (see Section 5.2) also requires considerable rethinking about the complexes floating in a sea of lipids.

If the outer membrane has a fixed surface area and hence can confine the contents to a fixed volume, the inner membrane with a larger surface area must necessarily fold to fit into the available volume. Some order to this folding may be provided by the periodic attachment of the inner membrane to the outer membrane. Such rivet-like attachment points have been observed, and there is now agreement that such attachment points may be made up of the components that are responsible for the import of peptides made in the cytosol. They may be formed transiently by the interaction of integral membrane proteins in the outer membrane with others in the inner membrane, thus aligning two channels through both membranes. Further details on such a proposed structure and its function in protein import will be presented in a later chapter.

Formation and maintenance of tubular cristae or cristae with a triangular cross section presents a greater challenge to the imagination. Theoretical studies on the conformation of membranes can lead to stimulating speculations (27), but experimental verification is still a major unachieved goal. The morphology of membranes such as the plasma membrane and the Golgi membranes is demonstrably determined by interactions with cytoskeletal elements, but there is no convincing evidence for the existence of such structures within mitochondria. Nevertheless, our view of the mitochondrial matrix as an amorphous soup of soluble enzymes is most likely too simple-minded. Mitofilin in the IMS may represent the first in a number of structural proteins required for organizing the cristae (15).

In summary, the variation in mitochondrial morphology has been abundantly documented. Three-dimensional reconstructions from serial sections and more recently by confocal microscopy (even in live yeast cells) have removed some of the last ambiguities of interpretation of thin sections used in electron microscopy. A general correlation of the total area of the inner membrane (density of cristae) with the capacity for oxidative phosphorylation is apparent, but some of the more detailed aspects of cristae morphology in relation to function and activity are still relatively obscure. Even less is known about the control of this morphology by nuclear genes in the course of development and differentiation. As will be discussed in the following section, mitochondrial shapes and distributions within a cell may be controlled by extra-mitochondrial, cytoskeletal elements, but the establishment of the particular, cell-specific interior morphology remains mysterious and therefore a challenge for the future.

3.2 INTEGRATION INTO THE CELL

3.2.1 Distribution in the Cytosol

So far in the discussion it has sufficed to consider mitochondria floating around in a cell, with highly diffusible molecules entering and exiting the organelle. This view is certainly too simplistic, and there are now numerous observations indicating that mitochondria have a nonrandom distribution in cells; they can either be restrained in preferred positions, or move or be moved to sites where their output is needed. They are not only not static in their positions within a cell, but they are also quite changeable in their shape. Individual mitochondria can undergo shape changes, or shapes and sizes can be altered by processes of fusion or fission. Thus, some of the dynamics may be due to internal or intrinsic driving forces. Other movements or dislocations are quite clearly due to interactions of mitochondria with the cytoskeleton and the possible involvement of molecular motors.

Studies on the dynamics of shape and position of mitochondria began early in this century before their identity had even been established. Stains were used, but the effect of these stains on the "viability" of these organelles was often not clear. Later (1950s) the phase contrast microscope opened up new ways of exploring live cells and the visible subcellular structures whose identity was, however, not always fully established. When immunofluorescence techniques were perfected and specific antisera against various identified organelles and cytoskeletal elements became available, another significant expansion of opportunities for the study of mitochondria was achieved. However, the need for permeabilization of cells also imposed a limitation: Static images had to be interpreted that could provide only hints of the dynamics in living cells. The first indications that mitochondria interact with the cytoskeleton were made in immunofluorescence studies with distinct labels for mitochondria and microtubules (28, 29). The most powerful technology became available in the 1980s when fluorescence microscopy and the discovery of new specific dyes with minimal harmful effects on mitochondria were combined to study mitochondrial distributions and dynamics in living cells. A landmark paper describing the laser dye rhodamine 123 was published by Chen's group in 1980 (30), illustrating the distribution of mitochondria in a variety of mammalian cell types. It also already gave descriptions of alterations observed in cell transformation, and as the consequence of the disruption of microtubules by colchicine. Many studies followed and have been expertly evaluated and summarized at various times (31–34). A following discussion (Chapter 5) will address in more detail the use of lipophilic cationic dyes to measure the membrane potential and the possibility of heterogeneity within a mitochondrial population of a cell with respect to staining, hence membrane potential, reflecting a difference in metabolic activity. The present discussion will deal with the distribution of mitochondria within cells and the mechanisms by which characteristic or functionally meaningful distributions are achieved or maintained. The review by Bereiter-Hahn (34), in particular, includes many historical references and is comprehensive in its coverage of all aspects of the dynamics of mitochondria in living cells.

Like in so many other areas of cell biology, the green fluorescent protein (GFP) has been used as a versatile reporter to observe the dynamic behavior and fluctuating distribution of mitochondria in various cell types. When produced as a chimeric protein with a mitochondrial targeting sequence, it has been localized into the mitochondrial matrix of yeast cells (35) and HeLa cells (36) and in the intermembrane space (37). Combined with powerful new imaging technology and computers, high-resolution 3-D images can be recorded almost continuously with live cells. The conclusions on the dynamic aspects of mitochondria will be presented below. In studies of HeLa cells with differential labeling of mitochondria and the endoplasmic reticulum (ER), the authors reported close appositions between mitochondria and the ER, and they estimated that about 5–20% of the mitochondrial surface was involved in such contacts (38). A possible significance is that at such sites, microdomains are created where Ca^{2+} is released from stores in the ER mediated by IP_3. However, caution is in order, because both mitochondria and the ER network can become attached to microtubules via kinesin or dynein-like motors, and thus their physical proximity may be coincidental. Furthermore, the density of these subcellular structures also makes apparent contacts quite likely. A broader discussion of the interaction of the smooth endoplasmic reticulum with mitochondria has been presented recently (39).

It is clear that a consideration of a single generic cell does not suffice in discussions of the broad range of observations made on mitochondrial distributions in cells, and the reader is referred to specialized reviews on the subject (33, 34). Some general principles will be emphasized.

Mammalian cells in tissue culture (fibroblasts, cancer cells, etc.) have been popular for study because of ease of access, and important deductions have been made. When certain highly differentiated vertebrate or invertebrate cells are investigated, special considerations have to be included. Unique situations may be encountered in cells of lower eukaryotes, with the kinetoplast of trypanosomes being one specific example. The discussion of events associated with cell divisions and the cell cycle usually ignores mitochondria, but it is obvious that there must be some mechanism assuring that each daughter cell obtains half of the mitochondria of the dividing cell. This creates a special problem for the budding yeast, *Saccharomyces cerevisiae*, to give one example, which will be addressed below.

What dictates the distribution of mitochondria within a cell? In large and highly asymmetrical cells, mitochondria may have to be localized in regions where the ATP supply has to be high. Thus, mitochondria wrapped around or packed along the axonemes of sperm tails can be easily rationalized. (Figure 3.6).

In neurons, mitochondria have to be moved and distributed along the often considerable length of axons and into the growth cone (40, 41). Skeletal muscle cells require a very uniform distribution of mitochondria in the myofibrils along their entire length to provide a uniform supply of ATP to all the sarcomeres (Figure 3.7). A particularly impressive example of a highly regular arrangement of mitochondria in striated muscle cells is found in insect flight muscles.

Rod cells in the retina appear to have their mitochondria concentrated in the cell body closest to the outer segment. In round yeast cells or in a gamete of *Chlamydomonas* the mitochondria are distributed around the cell periphery, perhaps to maximize their access to oxygen. Thus it has been speculated

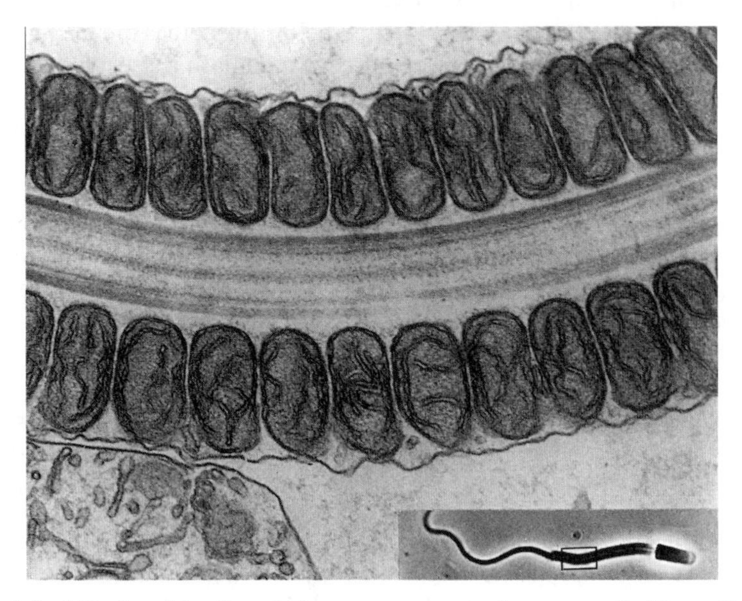

Figure 3.6 Mitochondria aligned along an axoneme of a sperm tail. (From *The Cell*, by D. W. Fawcett, 1981, W. B. Saunders Co.)

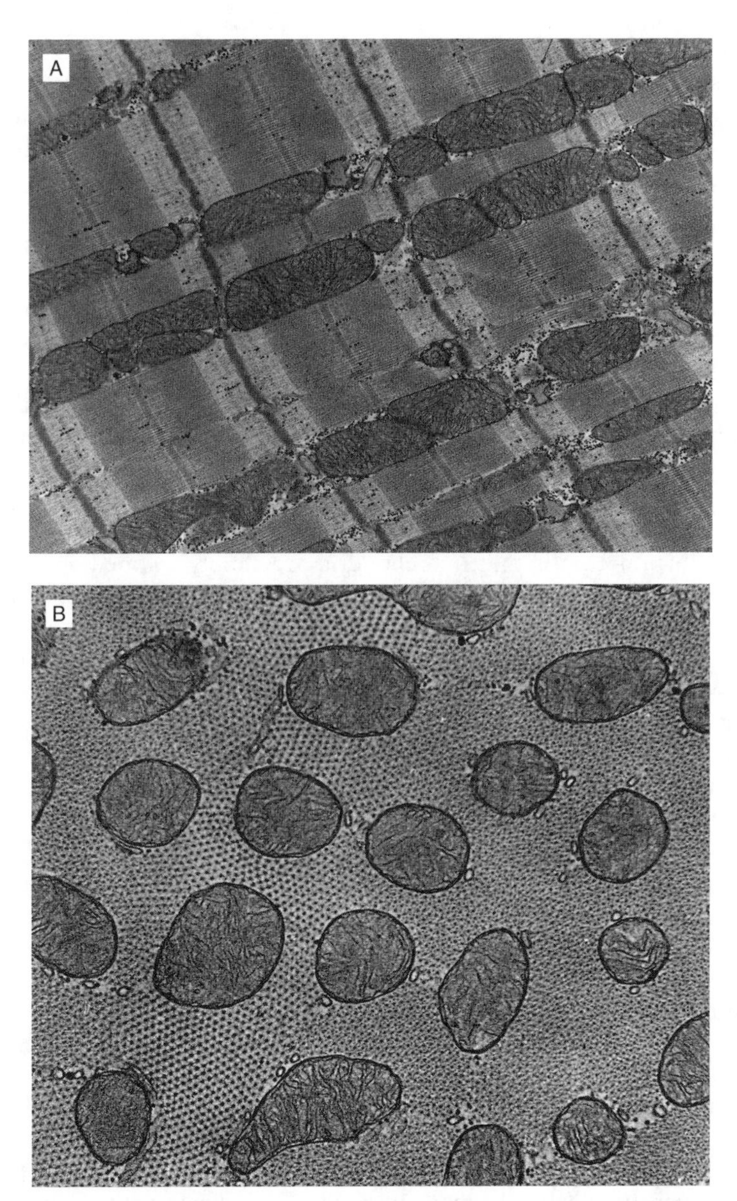

Figure 3.7 (A) Mitochondria distributed in skeletal muscle between sarcomeres. (From *The Cell*, by D. W. Fawcett, 1981, W. B. Saunders Co.) (B) Cross section of cardiac muscle showing uniform distribution of mitochondria. (From *The Cell*, by D. W. Fawcett, 1981, W. B. Saunders Co.)

that ATP (and/or ADP) gradients might be responsible for influencing mitochondrial distributions. Alternatively, the clustering of mitochondria within a cell might create gradients of other metabolites and ions such as calcium and cause transient, highly localized perturbations, including pH gradients.

3.2.2 Interaction with Cytoskeleton

Mitochondria often tend to have an elongated, thread-like appearance, and when viewed in live cells they appear to be roughly aligned along their long axes. When they were colocalized with cytoskeletal structures, their association with microtubules became immediately apparent in several cell types. In fact, mitochondria surrounding the meiotic spindle in insect cells were noted before the composition of the mitotic and meiotic spindles was known. Observations in the laboratory of Singer (28, 29) (Figure 3.8) and of Chen (30) provided definite proof, and many more observations in diverse cell types have followed these early studies.

One interpretation is that mitochondria are simply trapped and partially aligned within this cytoskeletal array of microtubules and are further dispersed by their own "intrinsic motility" (shape changes, fission, and fusion; see below). At the other end of the spectrum of interpretations, it has been argued that molecular motors attached to the outer membrane of mitochondria can become engaged with microtubules. In other words, dynein- and kinesin-like molecules have been looked for, initially based on the differential sensitivity of these motors to specific inhibitors. The evidence clearly points to an ATPase activity bridging the outer mitochondrial membrane and microtubules (34).

Before discussing the locomotion of mitochondria along cytoskeletal structures, it is relevant to consider whether the elongated shape of mitochondria may be related to or even dependent on their attachment to filamentous or long tubular structures, in the same way as the flattening of cells on a substrate is dependent on interactions between integral membrane proteins in the plasma membrane and extracellular matrix elements. In other words, is the sausage shape of mitochondria observed *in vivo* retained when mitochondria are isolated and highly purified *in vitro*? Unfortunately, the isolation of mitochondria tends to break up mitochondria, and the resulting mitochondrial preparations typically consist of smaller, spherical or only slightly elongated vesicles. A more gentle test of the dependence of mitochondrial morphology on intact microtubules can be made by treating intact cells with colchicine, and one of the earliest experiments of this type showed clearly that the distribution or orientation of mitochondria was perturbed, but their threadlike appearance was not altered by such a treatment. Therefore, some internal mechanism must be active that controls the shape. In the absence of an internal mitochondrial skeleton, the answer has to be found in understanding the maintenance of the varying curvature of the membranes. Under some conditions, however, defined by specific mutations in extra-mitochondrial proteins, giant spherical mitochondria can be observed in yeast (see below).

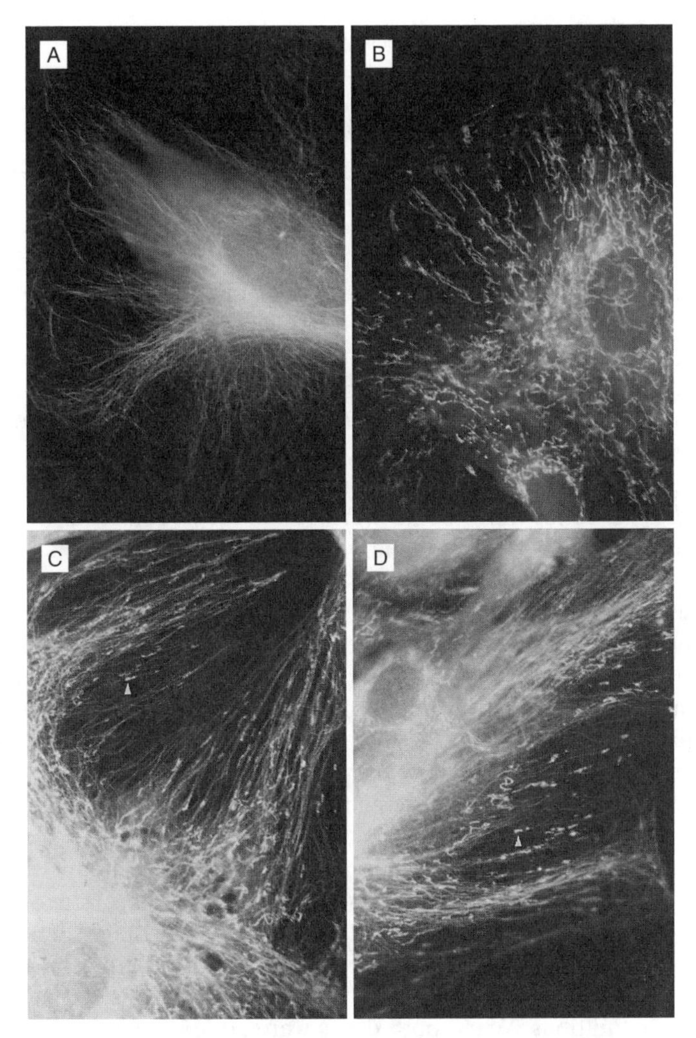

Figure 3.8 Correlation of microtubule alignment and mitochondria (27, 28). Antibodies against beef heart cytochrome oxidase and against chick tubulin were used in indirect immunofluorescence microscopy to localize microtubules and mitochondria in permeabilized cells. (A) Normal rat kidney (NRK) cell stained with antitubulin antibody and rabbit anti-chick tubulin IgG. (B) NRK cell stained with rabbit anti-cytochrome oxidase antibody and goat anti-rabbit IgG. (C, D) NRK cells double-stained with anti-tubulin and anti-cytochrome oxidase antibodies. The axis of elongated mitochondria (*arrow*) is always parallel to the microtubule axis.

Neurons, axons, and axonal transport have received the maximum of attention, since both retrograde and anterograde movement of mitochondria in axons has been observed by video-enhanced microscopy in intact axons as well as in extruded axoplasms (41). Closely related studies with systems of highly aligned microtubules of uniform polarity have been possible in unsheathed nutritive tubes of insect ovaria and in permeabilized reticulopodia of the giant amoeba *Reticulomyxa*. (The primary references to these studies are found in reference 34.) A characteristic aspect of such movements is the saltatory nature of the displacements typical for motor-driven movements on cytoskeletal elements, and not compatible with simple Brownian motion. Unresolved problems are related to the bidirectional displacements, the specific speeds observed, the effectiveness (or lack of it) of known inhibitors such as vanadate (a dynein inhibitor in vertebrates), and the possible contribution of additional mechanisms (e.g., an actin-based motility).

A series of reports from the laboratory of Hollenbeck serves to illustrate the experimental approaches and the conclusions drawn from the observations (42–48). In one study, chick sympathetic neurons were cultured in the continuous presence of either cytochalasin E to eliminate microfilaments (F-actin, MF), or in the presence of vinblastine or nocodazole to eliminate microtubules (MTs). Mitochondrial movements were followed by computer-aided video analysis after staining the organelle with rhodamine 123. In such neurons, mitochondria continued to move with saltatory movements in both directions in neurites with MTs, but no MFs, but failed even to enter the neurites in the absence of MTs but presence of MFs. The dependence of mitochondrial movement on MTs was confirmed, and it appeared that MFs by themselves were insufficient to support the movement of mitochondria into the neurites. Interesting and partially contrasting results were obtained when the neurons were first cultured in the absence of any drug to allow for the normal development of axons. Subsequently, the mature neurons were treated with either cytochalasin E or with vinblastine. The drugs disrupted the cytoskeletal structures as before, but in this case the axons already contained mitochondria. Now the average mitochondrial velocity increased in both directions in axons lacking only MFs, but net directional transport decreased. Surprisingly, in axons with MFs but no MTs, mitochondria also moved in both directions at a reduced average velocity and excursion length; net retrograde transport was favored under these conditions. When both drugs were added to eliminate both MTs and MFs, but leaving neurofilaments, no movement was observed. The studies clearly support an actin-based motility of mitochondria over shorter distances and a faster MT-based motility over long distances. They suggest that mitochondrial movement in neurites is a complex process involving several motors (kinesins, dyneins, and myosins), each with characteristic intrinsic properties and presumably able to fine-tune mitochondrial distributions for optimal functioning of the axon (47). The discussion must also include the nature of various linkers, adaptors, or anchors on the surface of the organelle as reviewed by Hollenbeck (46). Finally, attention has now shifted to cellular signals modulating the activity of the motors and their attachment to mitochondria. An assortment of kinases, a GTPase (Miro in *Drosophila*), G proteins, and growth factors (NGF) have been shown in a variety of model systems to influence axonal transport of mitochondria (see reference 47 for a recent collection of references). Miller and colleagues have made the very interesting observation that mitochondria with high membrane potentials are transported primarily toward the growth cone, while mitochondria with low potential were transported toward the cell body (49). This may reflect an aspect of quality control by turnover (mitophagy; see below).

There have been numerous investigations aimed at establishing an interaction between mitochondria and intermediate filaments. A co-distribution has been observed in a few instances (see reference 34 for references), but since there are few systems with highly organized, regular arrays of such filaments, model systems and especially cell-free systems are lacking. Some observations point to a role of desmin in anchoring mitochondria in skeletal muscle cells to preserve their relative positions during repeated contraction–relaxation cycles, but a more definite proof is lacking. In non-muscle cells deliberate attempts to demonstrate an alignment of mitochondria with intermediate filaments have failed.

Further impetus to relate mitochondria and intermediate filaments has come from pionering studies in Yaffe's laboratory (50) with the yeast *Saccharomyces cerevisiae*. Taking a genetic approach, this laboratory has isolated mutants defective in mitochondrial inheritance; that is, these mutants fail to transfer mitochondria into daughter buds. A gene for an essential protein was thus identified (*MDM1*), and the Mdm1 protein was found to have modest sequence similarity to animal intermediate filament proteins. *In vitro* this protein can form intermediate filament-like (10 nm) structures, while *in vivo* immunochemical staining reveals a punctate distribution of the antigen. A further discussion will be deferred to one of the following sections.

3.3. THE DYNAMICS OF MITOCHONDRIAL MORPHOLOGY

3.3.1 Mitochondrial Shape Changes

The phenomenon referred to as "intrinsic mitochondrial motility" (34) has been observed in animal cells, in fungal hyphae, in algal cells, and in the cytoplasms of plant cells where cytoplasmic streaming is absent to allow longer-term observations. A mitochondrion appears to be able to slither, creep, or slide at rates ranging from 2 to 30 μm/min in the direction of its long axis. The displacement is accompanied by subtle shape changes: slight thickening or contractions along the axis, possibly reflecting localized changes in the internal arrangement of the cristae. Microinjection studies with ATP and/or ADP have shown that these reagents can immobilize mitochondria; and in general, the influence of metabolic conditions on mitochondrial motility have been interpreted in favor of intrinsic activities driving their displacements (34).

Intrinsic dynamic changes in mitochondria are most strongly indicated by the well-documented studies on mitochondrial fusions and fissions. Such events can be followed in the phase-contrast microscope by time-lapse photography

(34); and in some cases, observations on live cells have been elegantly related to follow-up examinations with the electron microscope (see Figure 3.9).

As a result of the combination of the power of genetics with the *in vivo* visualization of mitochondria in the yeast *Saccharomyces cerevisiae*, an explosion of papers has occurred in the past few years, contributing to an understanding of mitochondrial shape changes (fusion, fission, and tubulation). A variety of novel proteins were identified, and their characterization was frequently extended to mammalian cells by "homology cloning." In other words, genetic studies in yeast have generally provided guidance for studies in mammalian cells. Many, but not all, proteins characterized in yeast have been identified in mammalian cells, either because they cannot yet be recognized or because there are mechanistic differences between organisms. The growth of the relevant literature dictates that reference is made primarily to review articles (51–58). The recent reviews by Okamoto and Shaw (59) and by Hoppins et al. (59a) are noteworthy for being up-to-date, authoritative, comprehensive, and illustrated with highly informative drawings. Table 3.1 and Figures 3.11 and 3.12 represent summaries of information from these reviews.

Historically, fusion-defective mitochondria were first discovered in sterile *Drosophila* males (see below). Subsequently, the discovery and observations of yeast mutants with defective mitochondrial morphologies defined three major mechanisms/pathways in genetic and biochemical terms. When fusion is interrupted, ongoing fission produces a large number of mitochondrial fragments. When fission is defective, ongoing fusion leads to extensive interconnected nets. A third pathway that so far is unique to fungi has been called tubulation; defects in this pathway cause tubular mitochondria to form large spheres.

Serial sections and image reconstructions have revealed that in the yeast *Saccharomyces cerevisiae* a cell contains a small number of mitochondria existing in a highly branched, reticulated shape, such that a section viewed in the electron microscope gives the appearance of multiple mitochondria (Figure 3.10). One to ten such reticular structures are expanded below the cortex of the cell (22, 23). A state of the art reexamination of this feature and its dynamic aspects in live cells was pioneered by Nunnari and colleagues (35), who modified a green fluorescent protein (GFP) for import into mitochondria to be able to use wide-field fluorescence microscopy for the acquisition of three-dimensional images over long time intervals. The use of GFP avoids the problems of photobleaching. Beautiful 3-D images were obtained, but more significantly, time-resolved images from single optical sections yielded quantitative data on fusion and fission events. While the authors are cautious to state that the resolution does not unambiguously allow the distinction between fusion and close juxtaposition, many events leading to stable connections were documented. During vegetative growth of wild-type cells, 0.39 ± 0.14 fission events per minute and 0.40 ± 0.08 fusion events per minute were observed, indicative of a highly dynamic situation. The matching figures indicate a steady state in which a continuous network undergoes constant fragmentation and

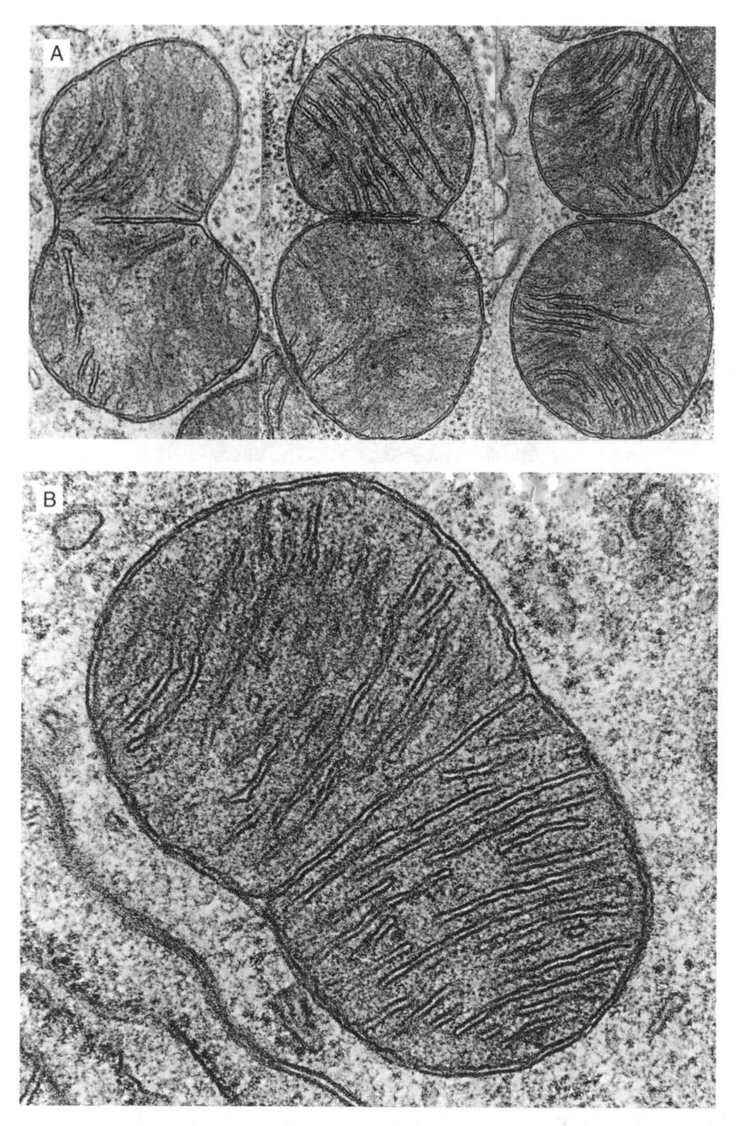

Figure 3.9 (A) Dividing mitochondrion. (Original photograph of T. Kanaseki; from *The Cell*, by D. W. Fawcett, 1981, W. B. Saunders Co.) (B) Successive stages in a dividing mitochondrion. (Original photograph of T. Kanaseki; from *The Cell*, by D. W. Fawcett, 1981, W. B. Saunders Co.)

re-assembly. They confirmed previous observations that at least one tip (end) was involved, where the putative fusion machinery may be located.

A similar study with GFP in HeLa mitochondria (36) also obtained high-resolution 3-D images at 30-s intervals, and it was concluded that these mitochondria also formed a connected, continuous, and highly dynamic network

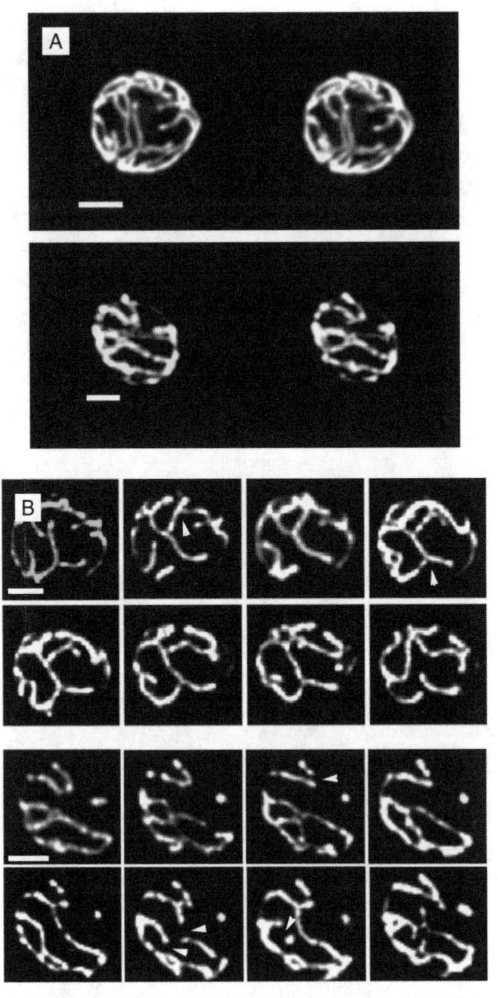

Figure 3.10 Mitochondria in living yeast cells (35). Green fluorescent protein was modified to be imported into mitochondria. Wide-field fluorescence microscopy was used for the acquisition of three-dimensional images over long time intervals. (A) Three-dimensional images of yeast mitochondria in wild-type cells grown on galactose. (B) Fission and fusion of mitochondria. Time-resolved images from a single optical section of 0.2 μm. Time intervals from left to right in 3-min intervals.

exhibiting growth and retraction, as well as fusions to other portions of the network. Photobleaching of a small area within a cell was followed by a rapid recovery (5–20 min) of GFP fluorescence within that area.

3.3.1.1 Fission Fission is the postulated mechanism for mitochondrial proliferation. The insertion of new components into the matrix and all the membranes will cause mitochondria to grow, and eventually there must be a signal for such a mitochondrion to divide, although insertion of lipids and proteins into the inner membrane alone may simply increase the number of cristae. Mitochondrial fission appears a priori to be an essential event in proliferating cells, since mitochondria have to multiply and ultimately be divided between daughter cells. A surprising finding is that mitochondria and peroxisomes share a significant number of components of their division machinery (60). Some spectacular electron micrographs (24) showing bilaminar septa during progressive stages of mitochondrial divisions leave one with little doubt about the overall mechanism, but they raise new questions about the control of this process which remain largely unanswered. Since an inner and an outer membrane are involved, the process is not a simple pinching-in-half of a membrane vesicle, and it may be more analogous to the division of a bacterium, where the plasma membrane and the cell wall represent separate structures surrounding the cytosol. Microscopic observations suggest that fission is preceded by stretching, but constrictions may appear at one site transiently and disappear without the initiation of fission (34).

There is an alternative view of the division of mitochondria, based on observations with a select number of organisms. In the unicellular red alga *Cyanidioschyzon merolae* a structure referred to as an MD ring (mitochondrion dividing apparatus) was observed. This ring-like structure was found on the outside of mitochondria, and it was interpreted to function in a manner analogous to the contractile ring in cytokinesis (61). The authors claimed that cytochalasin B inhibited "mitochondriokinesis," and they deduced the presence of actin filaments in the ring. However, these filaments could not be stained with anti-actin antibodies or with phalloidin. Because such rings have not been observed in any other organism, the generality and significance of these observations must remain in doubt.

The fission mechanism in yeast requires a complex of at least three major proteins that is localized on the outer mitochondrial membrane, with major domains exposed to the cytosol. Fis1p (hFis1p in humans) is an integral membrane protein with a single transmembrane domain. Its C-terminus is anchored in the outer membrane, and its N-terminal domain in the cytosol contains an array of six -helices forming a tetratricopeptide-like fold with a binding pocket for interactions with either of two closely related components, Mdv1p and Caf4p. Structural studies suggest that these can form homo-oligomeric structures. Fis1p is distributed uniformly on mitochondria and required for the association of Mdv1p (or Caf4p) on the mitochondrial surface. Together, Fis1p-Mdv1p (in yeast) assist in the incorporation of the dynamin-related GTPase, Dnm1p, in a fission complex that can be visualized by punctate staining on mitochondria. The assembly pathway and its regulation remain to be elucidated in detail. A significant amount of information has, however, been accumulated on dynamin and dynamin-related proteins (DRP1 or DLP1 in humans) (62–64). Dnp1 behaves like a soluble protein when purified, and it contains a GTPase domain at the N-terminal and a C-terminal GTPase effector domain.

As first recognized in the study of endocytosis, the protein assembles at the location of the constriction where a membrane vesicle buds from its parent membrane. A plausible and favored model suggests that the dynamin or dynamin-like proteins aggregate in a ring-like structure at the site of the membrane constriction and ultimately pinch off the vesicle. A tubular structure like a mitochondrion could be pinched into two pieces at such a site. GTP hydrolysis represents a likely driving force, but a detailed mechanism is still elusive. For example, it is not clear whether constriction is achieved by a contraction of a dynamin ring due to conformational changes, or whether the ring contracts with the concomitant loss of dynamin monomers (62). Further speculations in the case of mitochondria include the formation of an initial, local constriction with the help of the inner-membrane protein Mdm33p, or a stretching of mitochondria by molecular motors interacting with cytoskeletal elements (59). Clearly, the mechanism must include the outer and inner membranes in a concerted fashion, and there must be signals for when and where to assemble the fission apparatus.

3.3.1.2 Fusion Collisions brought about by the movement of the organelles may lead to fusions, especially when at least one tip makes contact with

TABLE 3.1 Genes/Proteins Involved in Mitochondrial Fusion and Fission (59)

Yeast Gene/Protein	Human Homologue	Localization in Mitochondria	Function
Fzo1	Mnf1/2	Outer membrane (IMP)	Fusion
Mgm1	OPA1	IMS or inner membrane (IMP?)	Fusion
Ugo1		Outer membrane (IMP)	Fusion
Mdm30		Cytosol/outer membrane	Fzo1 degradation
Pcp1	hPARL	Inner membrane (IMP)	Mgm1p processing
Dnm1	Drp1/DLP1	Cytosol/outer membrane (peripheral)	Fission
Fis1	hFis1	Outer membrane (IMP)	Fission
Mdv1		Cytosol/outer membrane (peripheral)	Fission
Caf1		Cytosol/outer membrane (peripheral)	Fission
Mmm1		Outer membrane-inner membrane spanning	Tubulation
Mdm10		Outer membrane (IMP)	Tubulation
Mdm12		Outer membrane (IMP)	Tubulation
Mdm31		Inner membrane (IMP)	Tubulation
Mdm32		Inner membrane (IMP)	Tubulation
Mdm33		Inner membrane (IMP)	Inner membrane fission?
Gem1	Miro-1/2	Outer membrane (IMP)	Ca^{2+} signaling?

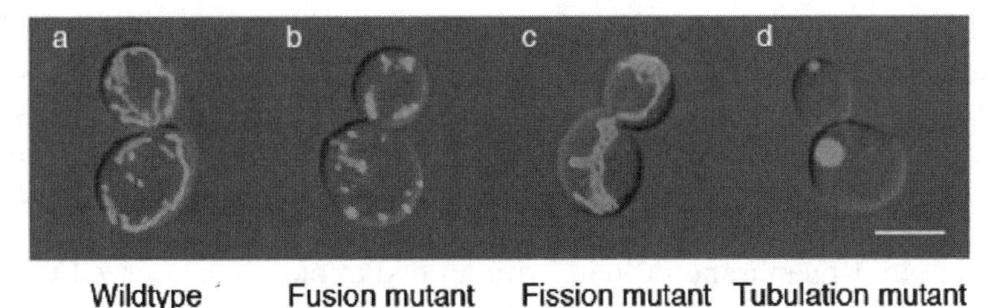

Figure 3.11 GFP targeted to the matrix in wild-type yeast cells and in various mutants: fusion mutant (*fzo1*Δ), fission mutant (*dnm1*Δ), tabulation mutant (*mmm1*Δ); GFP fluorescence images were superimposed on images from a differential interference contrast microscope. (From the review of Okamoto and Shaw (59).) See color plates.

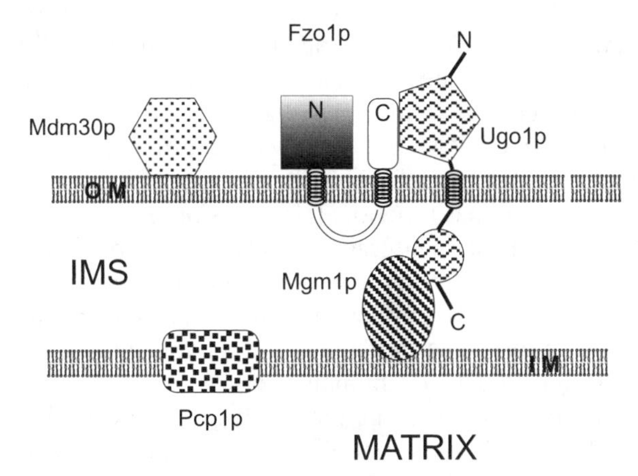

Figure 3.12 Schematic representation of fusion proteins in the outer (OM) and inner (IM) membrane. (From the review of Okomoto and Shaw (59).)

another, or with the side. Contact sites and the creation of contact zones are inferred from observations with the electron microscope, but unfortunately such images represent a static situation that cannot be further resolved in time. In contrast to the secretory pathway, the membrane fusions occur between identical organelles, although some asymmetry between docking vesicle and target membrane may be transiently created when a mitochondrial tip encounters a side. Needless to say, fusion involves the coordinate but presumably sequential fusion of biochemically distinct outer and inner membranes. It is a unique process that excludes all other intracellular membranes. The same mechanisms responsible for the dynamic nature of cristae will operate subsequently to rearrange the cristae in the fused organelle.

Mitochondrial fusion is a developmentally programmed mechanism in spermatogenesis in fruit fly males. The analysis of a mutation causing male sterility led to the discovery of a gene, fuzzy onion (Fzo), named after the appearance of a large structure formed when mitochondria fail to fuse normally during spermatogenesis. Advances in genomics greatly aided in the search for homologous genes in yeast (Fzo) and humans (Mfn1 and Mfn2—mitofusin). When the gene was found in yeast, the characterization of the Fzo protein and its function led to some rapid advances and greatly stimulated the search for additional genes required for the fusion pathway. The Fzo1 protein is an integral outer-membrane protein, with an evolutionarily conserved GTPase domain on the cytoplasmic side. Two transmembrane segments are connected by a short linker exposed to the intermembrane space. The presence of a temperature-sensitive allele in yeast prevents mitochondrial fusions at the nonpermissive temperature, and the mitochondria become increasingly fragmented. A second GTPase essential for fusion was discovered unexpectedly by studies of a gene required for mitochondrial genome maintenance in yeast. The Mgm1p is a GTPase related to the family of dynamins; the corresponding mammalian gene is OPA1, initially named for a mutation in OPA1 that causes autosomal dominant optic atrophy (65). It is an integral inner-membrane protein with its functional domain localized in the intermembrane space. Another dynamin-like GTPase has already been introduced in the context of fission (see above). A third protein, Ugo1p, so far identified only in fungi, spans the outer membrane and has functional domains on the inside to interact with Mgm1p, and on the outside to interact with Fzo1p. Fzo1p, Ugo1p, and Mgm1p thus form a fusion complex connecting the inner and outer membranes (53, 59). A recent development identifies two functional isoforms of Mgm1p produced by alternative topogenesis. The sorting is controlled by a novel protein, Ups1p (in yeast) and PRELI (in humans) (66).

The most dramatic progress in the study of mitochondrial fusion was achieved recently when Meeusen et al. (67) developed an *in vitro* mitochondrial fusion assay. Effectively, two populations of mitochondria were labeled with two different chromophores in the matrix, mixed, concentrated, and brought into contact by centrifugation, and fusion could be observed by the mixing of the chromophores in the fused mitochondria. For the first time, conditions in this *in vitro* system could be chosen to restrict the fusion only to the outer membranes. Such a step required functional Fzo1 proteins on both membranes, suggesting that cytoplasmic domains of the Fzo1p bind to each other in a homotypic interaction. In support, X-ray crystallographic data reveal that the C-terminal domains of Mfn1 (the human homologue) form a dimeric, antiparallel coiled-coil structure (68). Endogenous levels of GTP are sufficient, and the GTPase activity of Fzo1 is required, since fusion did not take place when temperature-sensitive Fzo1 proteins were present. Most intriguingly, outer-membrane fusion requires the proton gradient component of the inner membrane (ΔpH), but not the electrochemical component ($\Delta\Psi$). Thus, when mitochondria are brought close together physically, the Fzo1 proteins act as tethers, similar to the SNARE proteins in the secretory pathway. *In vivo* this initial interaction is likely brought about by an interaction with cytoskeletal elements and ATP-dependent molecular motors. The final steps in fusion and the use of the proton gradient are still mysterious. To achieve inner-membrane fusion, two additional requirements must be met: A sufficiently large electrochemical gradient ($\Delta\Psi$) must be present across the inner membrane, and elevated levels of GTP must be available for hydrolysis. This observation clearly implicates the second GTPase (Mgm1p/OPA1) localized in the IMS. It is also consistent with a role of Mgm1/Opa1 in maintenance of cristae morphology and ATP synthase assembly (69, 70).

A major challenge for the future will be to understand the control of fission and fusion. In wild-type cells these processes must be balanced, except when the number of mitochondria increases in preparation for cell division. At one level, one expects control over the synthesis and turnover of the major proteins introduced above. Undoubtedly, other participating proteins will be added to the picture, some of them universal and some species-specific. Their assembly into fusion and fission complexes is also likely to be under some control mechanism. For example, Mgm1p exists in two forms, a "precursor" and a mature form that are both required for fusion. After import and cleavage of the targeting signal, the precursor is processed by the inner-membrane rhomboid protease Pep1. The exposure of the cleavage site in turn is dependent on matrix ATP levels. As usual, our understanding is more complete in yeast compared to mammalian cells, where it has been suggested that different variants of OPA1 are created by variation in splicing instead of proteolytic processing (53, 71). Jeyaraju et al. (72) have investigated the human orthologue called PARL (presenilin-associated rhomboid-like protein) in animal cells. PARL has an unrelated, vertebrate-specific N-terminal domain. The protein is an integral inner-membrane protein with the N-terminal on the matrix side. After import and processing, a second cleavage by PARL activity (in trans) occurs. The highly conserved sequence of this domain, along with the presence of potential Ser/Thr phosphorylation sites, suggested to these authors to test alanine substitutions at these residues. To summarize a much more detailed analysis, the authors provided evidence for the involvement of the PARL protease in the control of mitochondrial morphology. The non-phosphorylated, cleavable form of the enzyme was able to induce mitochondrial fragmentation when transiently overexpressed in HEK 293 cells. In contrast, aspartate substitutions at these positions to mimic phosphorylation prevented this process.

In another recent study it was shown that the F-box protein Mdm30 binds to Fzo1 and mediates its proteolysis by a proteasome-independent mechanism (73). This represents a novel mechanism for Fzo1 degradation in yeast in addition to the mechanism involving ubiquitinylation and proteasome-dependent turnover.

Fusion events involving mitochondria that are developmentally programmed have been characterized in many systems (74). A later section will detail some salient insights into this phenomenon in the context of cell differentiation.

One should also consider the behavior of the mitochondria and the mitochondrial genomes when two cells fuse, as in mating of haploid yeast cells. This aspect will be considered in more detail below. The observation of constant fusions and fissions of mitochondria in a mammalian somatic cell has implications for the geneticist. In the absence of additional information (see below) the dynamic behavior leads one to conclude that mitochondrial genomes in a heteroplasmic cell are continuously mixed and redistributed, with the result that a given mitochondrion contains both types of genomes at any time. It may serve as a mechanism to "homogenize" mitochondrial genomes among the many mitochondria in a cell. The implications for our understanding of human mitochondrial diseases will be discussed in a later chapter. In some organisms it may also serve to promote recombination between mitochondrial genomes.

3.3.2 Distribution During Cell Division

When a cell contains a large number of mitochondria more or less evenly distributed throughout the cytoplasm, and doubling in number and volume during the cell cycle, no significant conceptual problem exists for distributing these mitochondria evenly into the daughter cells during cytokinesis. Furthermore, if mitochondria become attached evenly to the newly formed microtubules of the mitotic half spindles, their segregation is easy to visualize. The problem during the cell cycle, therefore, is to assure a doubling of mitochondrial mass and number and to maintain a constant ratio of nuclear to mitochondrial genes. Cell differentiation during development, along with mitochondrial proliferation and turnover in terminally differentiated cells like muscle and neurons, poses a different problem.

A special problem arises when cell division occurs not by the classical process of cytokinesis in which a larger cell is partitioned into two cells either by a contractile ring, as in animal cells, or by the assembly of a phragmoplast in plant cells. In the budding yeast *Saccharomyces cerevisiae* a bud is started and enlarged during the cell cycle, and all the constituents for a new viable cell have to be transported through a narrow connection between the bud and the mother cell. How this is achieved for the nuclear chromosomes has been a subject of study for some time, but the problem of equipping the bud with mitochondria has been illuminated by pioneering studies starting in the laboratory of M. Yaffe (see reference 50 for an expert review). Taking a genetic approach, the Yaffe laboratory has identified yeast ts mutants that are defective in the transfer of mitochondria and nuclei into growing daughter buds at the nonpermissive temperature. Others have contributed over the years, and the subject has started to merge with the general topic of mitochondrial morphology and dynamics in yeast and other eukaryotes (52, 59). A pathway that may be unique to fungi has been referred to as the tabulation pathway.

The original *mdm1-1* allele, along with additional *mdm1* alleles isolated later, defines a gene encoding a protein (Mdm1p) with several intriguing properties. The protein is essential for viability. With some mutant proteins the problem of mitochondrial distribution can be separated from nuclear distribution. The protein sequence has some similarity to intermediate filament proteins of vertebrates, and it forms 10-nm filaments *in vitro*. Filament formation fails at the nonpermissive temperature *in vitro* with the protein encoded by the original *mdm1-1* (ts) allele. On the other hand, indirect immunofluorescence microscopy with anti-Mdm1p antibodies reveals punctate staining distributed throughout the cytoplasm of wild-type cells. With mutant alleles affecting mitochondrial inheritance, the mitochondrial morphology in the mother cells is also abnormal. Instead of the usual reticulated, tubular wild-type structure, one finds small, round mitochondria in clusters. These observations invite speculation that a cytoplasmic filamentous network or scaffold exists which is responsible for the maintenance of the mitochondrial morphology of yeast discussed above, and that it also plays a major role in guiding presumably threadlike mitochondrial structures through the connection into the bud. It should be mentioned that fungi do not have the typical intermediate filament proteins of higher eukaryotes. The yeast filaments made from Mdm1p may represent an alternate solution to a problem that reached an evolutionary dead end, and they were replaced by proper intermediate filaments in more advanced organisms. The mdm1 mutation does not appear to affect actin- and tubulin-based cytoskeletal functions.

Mdm1p may not be the only protein involved in this peculiar yeast cytoskeletal structure. Additional mutants isolated in the Yaffe lab have led to the characterization of other *MDM* genes (mitochondrial distribution and morphology), of which *MDM14* encodes a 35-kDa protein that appears to interact with the Mdm1p and is distributed in the cytoplasm in a similar fashion. Not unexpectedly, on second thought, some *mdm* mutants are affected in proteins in the outer mitochondrial membrane (*mdm10, mdm12*) (50, 75). (See Figure 3.13.) Clearly, association with a cytoskeletal element implies the existence of a protein domain on the mitochondrial surface that must at least bind to the scaffold, and perhaps even have motor-like properties. A related gene product in the outer membrane is the Mmm1p (maintenance of mitochondrial morphology) identified by Burgess et al. (76). Depletion of these proteins has pronounced effects on mitochondrial morphology, and null mutations yield a temperature-sensitive phenotype. As pointed out by Berger and Yaffe (50), the fact that these null mutants grow at the permissive temperature on glucose (albeit slowly) suggests that a bypass or backup pathway must exist to assure mitochondrial transmission to daughter buds. Nevertheless, mdm10 and mdm12 mutants lose mitochondrial DNA and generate respiration-deficient cells at higher-than-normal rates. New members have been added to the *MMM* and *MDM* groups of genes (52, 59, 77, 78). The corresponding proteins have been localized in the inner membrane (Mdm31p and Mdm32p), or in (or on) the outer membrane (Mdm10p, Mdm12p, Mmm1p, Mmm1p). The Mmm1p is most unusual in spanning both the inner and outer membrane, with its N-terminus in the matrix and its C-terminus on the cytoplasmic side. It acts like a rivet for these two membranes, and at the same time it forms a complex with the Mdm10p and Mdm12p (the MMM complex) and transiently interacts with

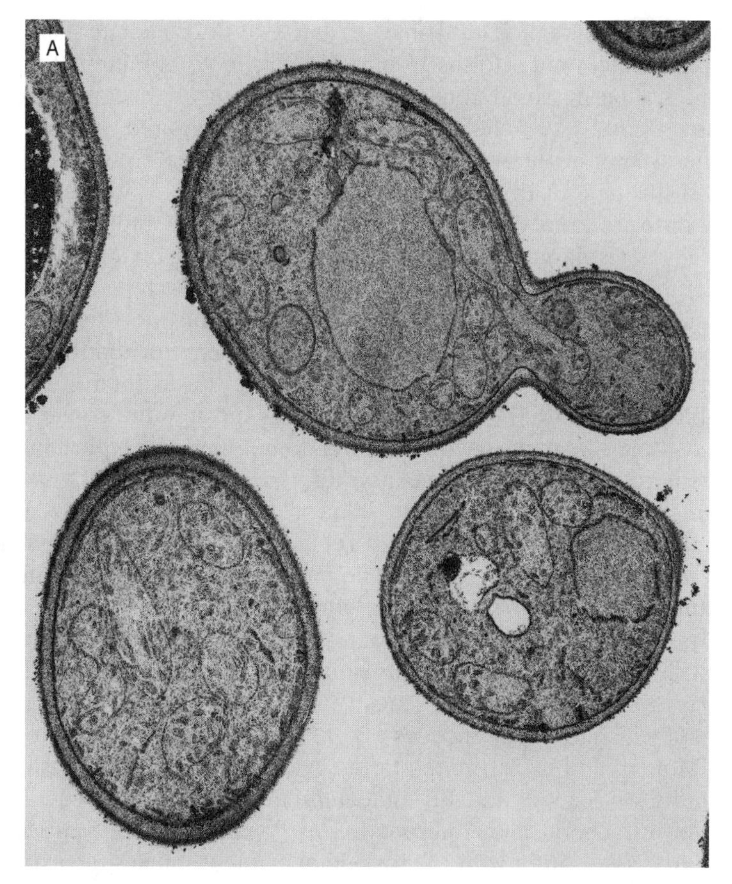

Figure 3.13 (A) Mitochondria in a budding, wild type yeast cell. Note the movement of one elongated mitochondrion into the bud. (B) Abnormal, giant spherical mitochondria in yeast cells with the *mdm10* mutation (50, 75). (Photographs provided by M. Yaffe, University of California, San Diego.)

Mmm2p. Based on phenotypes exhibited by mutants lacking the MMM complex, the complex is implicated in at least three significant processes: attachment of mitochondria to actin, formation of tubular mitochondria, and the the anchoring of the mtDNA nucleoids to the inner membrane. The first is likely to be relevant for moving mitochondria around in the cytosol and into the bud during cell division. The second could also involve attachment to a cytoskeletal element (still undefined; Mdm1p?), or it could include the construction of an external scaffold that constrains mitochondria to a tubular shape. In its absence the mitochondria turn into spheres. Finally, the N-terminal domain of Mmm1p in the matrix has been colocalized, together with other proteins (e.g., Mgm101p) and mtDNA, in a large structure referred to as the mtDNA nucleod. It is the site of both mtDNA replication and transcrip-

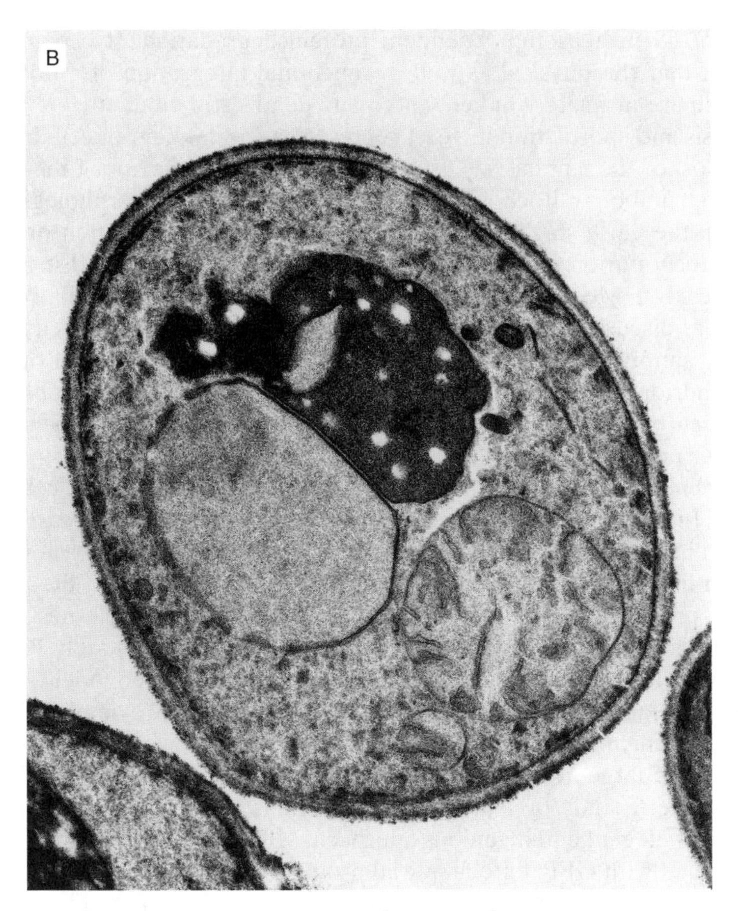

Figure 3.13 (*Continued*)

tion, to be discussed in a later chapter. In the present context it is important to recognize that daughter cells must inherit mitochondria with mtDNA. Therefore, dysfunction of the MMM complex results not only in abnormal mitochondrial shape and inheritance, but also in severe mtDNA instability.

The overall potential complexity of the processes under consideration is illustrated by the failure of mitochondrial inheritance in the *mdm2* mutant. In this mutant the *OLE1* gene is affected, encoding a fatty acid desaturase (79). Mitochondrial membrane fluidity evidently affects mitochondrial transmission, but it may be too simple-minded to think of a mitochondrion squeezing through a tiny pore connecting mother cell and bud. A much more exhaustive screen of 768 yeast mutants (~2/3 of all essential genes) has uncovered 119 genes required for maintenance of mitochondrial morphology (56). They encode proteins involved in ergosterol biosynthesis, mitochondrial protein import, actin-dependent transport processes, vesicular trafficking, and ubiquitin/26S proteasome-dependent protein degradation. It becomes quite apparent that the physical as well as functional integration of mitochondria into a cell are intimately linked, and many details still elude us.

Meiosis and sporulation in yeast represent a special kind of cell division in which most of the emphasis is usually placed on the behavior of the chromosomes. What about mitochondria? One study has defined morphological transitions in the early stages followed by a dramatic fragmentation of a single large mitochondrion into small spherical structures that were distributed to the spores (80). Mechanisms for such events remain largely speculative at this time. If, as discussed above, fusion and fission occur normally at equal rates, one may simply have to inhibit fusions to promote the breakup of a large mitochondrion into smaller units. The phosphorylation or dephosphorylation of an essential docking protein may be all that is required, but identifying it is another problem.

Before meiosis can occur in yeast cells, one must have a diploid cell formed from the fusion of two haploid cells. When respiration-deficient yeast mutants were first investigated by means of mating experiments, and observations were made on the stability of phenotypes during mitotic divisions of the zygote, a puzzling observation was made. Zygotes containing heteroplasmic mtDNA populations generated homoplasmic progeny within approximately 20 generations, much faster than expected on the basis of stochastic models assuming random segregation (see references 81 and 82 for a recent review). This observation is even more puzzling in light of the discussion about the dynamic nature of the large mitochondrial structure and the implied constant mixing of mtDNAs. A study by Nunnari et al. (35) has reinvestigated this problem with a most elegant and ingenious approach. Mitochondria of one haploid cell were labeled with GFP; mitochondrial proteins of the other mating type were labeled with the vital dye TMR-CH_2Cl, which not only localized in mitochondria because of the membrane potential, but eventually becomes covalently attached to mitochondrial proteins. When such cells were allowed to mate, the fusion and complete mixing of the two mitochondrial protein populations could be directly observed by wide-field fluorescence microscopy: These observations were fully consistent with observations described above on the dynamic behavior of yeast mitochondria. The more provocative result was obtained when mtDNA from one mating partner was labeled selectively with bromodeoxyuridine (by a not totally trivial procedure). Its distribution in the zygote could subsequently be determined immunochemically after fixation and permeabilization of the cells. The results showed unequivocally that while mitochondrial proteins from the two mitochondrial populations mixed readily, the corresponding mtDNA populations remained segregated in the two lobes of the zygote derived from the haploid cells. This result was a visual confirmation of previous genetic experiments which had inferred the unequal distribution of mtDNA in the zygote from an analysis of mtDNA in zygotic progeny. In such progeny the type of mtDNA (genetically marked) depended on the position of the bud emergence from the zygote (83). To interpret these findings, many questions remain to be answered about the nature of the attachment of the mtDNA to either (a) structures within the matrix that restrict diffusion or (b) attachment sites that are connected to cytoskeletal structures on the outside.

As illustrated by the above discussion, genetic approaches are a powerful means of discovering genes in yeast encoding functions related to mitochondrial morphology, segregation, and stability. Mutants such as *mgm1* (mitochondrial genome maintenance (84, 85)) and the series of *yme* mutants (yeast mitochondrial DNA escape (86)) represent additional examples that still pose major challenges for the future. No other organism is comparably advanced and genetically more tractable than yeast. However, while yeast has been the model system *par excellence* for understanding many fundamental aspects of the biology of mitochondria, other future problems related to cell differentiation and development may not have counterparts in yeast biology.

3.3.3 During Cell Differentiation

Developmentally regulated fusion of mitochondria has been described in a limited number of mammalian tissues, where its functional significance is obscure. It is observed in many lower eukaryotes (protists, fungi, algae; reviewed by in reference 74) and in insect spermatogenesis, with much emphasis on *Drosophila* (24, 87, 88). Cytological and electron microscopic descriptions (e.g., reference 88) have documented dramatic morphological changes associated with spermatid differentiation after the completion of meiosis, and the behavior of mitochondria during this period has attracted special attention. All mitochondria first aggregate next to the haploid nucleus, and then appear to fuse into two giant mitochondria. These two giant mitochondria remain intimately associated in a spherical structure called a Nebenkern. The mitochondrial membranes form a series of concentric layers resembling an onion in cross section. When the axoneme is formed and flagellar growth procedes, the two mitochondria separate from each other and elongate along the axoneme. The juxtaposition of mitochondria and axoneme along the long axis assures an optimum transfer of ATP to the dynein motors of the axoneme to support flagellar movement. The hydrodynamics of the sperm tail motion is optimized by a decreasing thickness along its length, and this is achieved by a gradual decrease in mitochondrial size (diameter) in the sheath, while the diameter of the axoneme remains constant. The entire process has many facets for which molecular details remain to be unraveled.

The use of *Drosophila* as a model organism in genetics and developmental biology has led to the accumulation of a large collection of mutants. Among the male sterility mutants, a mutant termed *fzo* (fuzzy onions) was recognized to have a defect in the mitochondrial fusions leading to Nebenkern formation. As a consequence, postmeiotic sperm differentiation is arrested. In a typical illustration of the power of the "reverse genetic approach," Hales and Fuller (89) have mapped, cloned, and sequenced the gene, derived a protein sequence, and from it predicted the existence of a transmembrane GTPase. This protein is present on spermatid mitochondria during the relevant period when fusion takes place, as determined by indirect immunofluorescene microscopy. Homologues of the gene have been detected in mammals, nematodes, and yeast, and they constitute a novel family of multidomain GTPases. The biological function of this protein is of special interest, because Rab GTPases have been well-characterized as the enzymes that control the assembly of the SNARE complex required for vesicle docking in the secretory pathway (90). As described above, the original identification of the Fzo gene from *Drosophila* has stimulated much research in yeast, where genetic, biochemical, and cytological experiments are technically easier than with *Drosophila* gonads.

A comprehensive review under the provocative title "Sexuality of Mitochondria" by Kawano et al. (74) contains an expertly written account of mitochondrial fusions and recombination between mitochondrial genomes in the life cycle of the true slime mold *Physarum polycephalum*. In this organism, mitochondrial fusions are not serving in morphological rearrangements as required in spermatogenesis. The authors emphasize recombination ("mitochondrial meiosis") and view the process in the broader context of the evolution and manifestations of sex. This view is reinforced by the finding that only strains of *P. polycephalum* harboring a unique mitochondrial plasmid can display mitochondrial fusion, and the mitochondrial fusion-promoting (mF) plasmid may be representative of parasitic, selfish genes on plasmids that have been proposed on theoretical grounds to have played a role in the early stages in the evolution of sex.

The review contains a broad introduction to mitochondrial fusion, and the formation of giant mitochondria and reemphasizes the distinction between random fusions of mitochondria as observed in animal cells and periodic fusions as observed in protists, algae, and fungi, where it may have a much greater genetic significance. It is relevant that fungi such as *Saccharomyces cerevisiae* and *Physarum polycephalum* have mitochondrial nucleods (referred to as mt-nuclei in reference 74), and therefore it becomes an issue whether these mt-nuclei also "fuse." From the discussion in an earlier section, it appears that the mitochondrial genomes of two mating partners in a yeast mating remain segregated for some time after fusion, although recombination has indeed been observed in yeast. The true slime mold has a complicated life cycle during which programmed mitochondrial fusion occurs during two phases (see reference 74 for details). A germinating haploid spore gives rise to myxamoeba, which can fuse to form a zygote from which a plasmodium develops. During the earliest stages of plasmodium formation, mitochondrial fusion and fusion of mt-nuclei also occur in mF^+ strains. Another series of mF plasmid-dependent mitochondrial fusions occur during sporulation which can be induced in the mature plasmodium. The large amount of mtDNA in *P. polycephalum* mitochondria facilitiates the observation of mt-nuclei by DAPI staining, and the process of sporulation can be synchronized by an illumination cycle. A detailed analysis of mtDNA replication and of mt-nuclei behavior thus becomes possible, and its description can be found in the original papers by Kawano and colleagues. Many intricate details require attention for understanding the mating-type system of *P. polycephalum* and the genetic control of mtDNA transmission. Kawano and colleagues take advantage of restriction enzyme polymorphisms (RFLPs) in the mtDNA of various strains to do mitochondrial genetics.

The Mif^+ and Mif^- phenotypes with respect to mitochondrial fusion discovered first were ascribed to the existence of a single transmissible factor that eventually was identified as an ~16-kb mitochondrial plasmid. Plasmid sequences can also be found integrated into the mtDNA, and even mif^- strains were found to have certain integrated plasmid sequences. It was concluded that the missing sequences were responsible for the Mif^+ phenotype. A further analysis of the mitochondrial fusion frequencies in mif^+ strains also established the presence of nuclear alleles controlling the frequency of mitochondrial fusion, but their precise nature remains to be elucidated.

The linear mF plasmid has been sequenced. Ten potential ORFs have been identified, but deduced amino acid sequences do not show significant homology with any amino acid sequences in the SWISS-PROT database, except for one which has been tentatively identified as a DNA polymerase (91). Closely related DNA polymerase sequences have been found on other linear mitochondrial plasmids.

Several important biological questions related to the mF plasmid can be raised, but answers are not available or incomplete. Some of these have been elaborated in the section on mitochondrial plasmids in Chapter 4 (Section 4.1.6). To restate them briefly here:

1. The replication of a linear plasmid DNA poses special problems that can be overcome by the existence of terminal inverted repeat sequences (TIRs) and mechanisms for replication initiation and priming.
2. The nature of the gene products encoded by the plasmid is of interest to answer several questions.
3. How is the plasmid maintained in the mitochondria?
4. What functions are necessary to promote mitochondrial fusions?
5. How do plasmid sequences become integrated into mtDNA, and what are the biological consequences of such a recombination?

Space limitations preclude a pursuit of progress in this subject over the last decade. The interested reader can find an entry into this literature from a limited number of recent publications (92).

In summary, the discussion of developmentally regulated mitochondrial fusions is not intended to be exhaustive, but examples have been chosen to illustrate the potential rationale for the evolution of mechanisms of fusion. It is conceivable that mitochondrial fusions were originally part of the overall process of sexual reproduction, including genetic recombination of all genomes in the organisms. In organisms such as mammals, where recombination between mt genomes is no longer observed at any significant frequency, the rationale for fusion of mitochondria in fibroblasts is less obvious, unless one views it as the simple reverse of fission, where fission is required for mitochondrial multiplication. The cell-specific and tissue-specific regulated fusion seen in these organisms may represent special adaptations of the more primitive mechanisms observed in lower eukaryotes.

3.3.4 Turnover and Degradation

In Chapter 4 a summary of our current understanding of regulated protein degradation in mitochondria is presented. The emphasis there is on selective protein degradation by intra-mitochondrial ATP-dependent proteases (93–95). In the present section, only a brief account of what is known about turnover of the entire organelle will be presented. In general, the bulk of mitochondrial proteins are quite stable over prolonged periods of time. In mammalian cells, half-lives are measured in days, if not weeks. In yeast, a recent study found that mitochondria labeled with imported GFP (mito-GFP) did not become significantly altered over a period of 7 hours even when the yeast cells were shifted from aerobic conditions to anaerobic, glucose-repressed conditions.

Indirect evidence for some mechanism related to turnover can be deduced from the age-dependent onset of neuropathies and myopathies and the associated increase in the fraction of mutated mitochondrial genomes in subpopulations of muscle or neuronal cells. The time scale for such a change in the genetic and biochemical properties of mitochondrial populations is still not well-defined, and it may be very difficult to measure precisely *in vivo* over a period of years.

There are some physiological and perhaps pathological conditions where cells are forced to resort to extreme measures to stay alive. Under prolonged and severe starvation, many cells initiate a process referred to as autophagy (96–98). Autophagy can also be induced by mitochondrial dysfunction. Subcellular organelles can then find themselves engulfed by lysosomes (in animal cells) or by the vacuole (in yeast and plant cells), and this internal degradation is presumably initiated to mobilize resources for the most essential functions for survival. Alternatively, autophagy is one mechanism for quality control in which the entire damaged organelle, rather than individual proteins and other components, is removed from the cell. Mitophagy, specifically involving mitochondria, is triggered by damage due to oxidative stress or from mutations in mtDNA (99–101). Understanding these mechanisms, and perhaps their failure, will undoubtedly contribute much to our understanding of apoptosis, aging, and neurodegeneration.

3.3.5 Mitochondrial Alterations in Apoptosis

A special problem involving mitochondria, morphological alterations in their structure, and ultimately their degradation arises in the discussion of apoptosis and certain pathological conditions. This subject will be discussed in more detail in a later chapter. Earlier studies had placed the emphasis on permeability changes of the outer membrane and the release of proteins from the IMS. It is now becoming apparent that morphological changes and fragmentation of mitochondria are intimately associated with the apoptotic pathway at relatively early stages. The perturbation (e.g., by RNA interference or mutations) of the fission and fusion complexes can have pronounced effects on the activity of pro-apoptotic or anti-apoptotic factors (59).

3.3.6 Unsolved Problems for the Future

There are several broad areas in which more information is needed, and future progress can be expected. Such progress is likely to be more rapid in organisms in which genetic approaches are possible such that mitochondrial morphology and behavior can be manipulated or modulated. Yeast is the organism of choice here. It is anticipated that interesting new proteins and their genes will have homologues in other organisms, allowing the information to be generalized or extended in other systems. The fruit fly *Drosophila* may be a second experimental genetic system in which to study mitochondria in the context of development and differentiation of specialized cells. Other problems may be unique to vertebrates, for example, and even to specific cell types and tissues.

Among the major problems, one could list the following:

1. What controls the total surface area and hence the elaboration of cristae in a mitochondrion in a given tissue? Why do the cristae have such unusual morphologies in some mitochondria?
2. Is the matrix of mitochondria organized in a way that controls the position of the mtDNA, the location of protein synthesis, and perhaps even the shape of mitochondria by means of internal structural elements?
3. What are the structures (protein domains) on the surface of mitochondria capable of interacting with diverse cytoskeletal elements via molecular motors? How do such interactions contribute to the changing shape of mitochondria, to their movement within the cell, and their accumulation in subcellular regions?
4. What mechanisms control division and fusion of mitochondria? The fact that two membranes are involved adds complexity. What controls the specificity and the timing, with reference to cell division and differentiation?

REFERENCES

1. Yost, H. J. and Lindquist, S. (1986) *Cell* **45**, 185–193.
2. Matsuyama, S., Llopis, J., Dveraux, Q. L., Tsien, R. Y., and Reed, J. C. (2000) *Nature Cell Biol.* **2**, 318–325.
3. Goldstein, J. C., Munoz-Pinedo, C., Ricci, J. E., Adams, S. R., Kelekar, A., Schuler, M., Tsien, R. Y., and Green, D. R. (2005) *Cell Death Differ.* **12**, 453–462.
4. Steinert, M. and Steinert, G. (1960) *Exp. Cell Res* **19**, 421–424.
5. Perkins, G., Renken, C., Martone, M. E., Young, S. J., Ellisman, M., and Frey, T. (1997) *J. Struct. Biol.* **119**, 260–272.
6. Mannella, C. A., Marko, M., and Buttle, K. (1997) *Trends Biochem. Sci.* **22**, 37–38.
7. Mannella, C. A., Pfeiffer, D. R., Bradshaw, P. C., Moraru, I., Slepchenko, B., Loew, L. M., Hsieh, C. E., Buttle, K., and Marko, M. (2001) *IUBMB Life* **52**, 93–100.
8. Mannella, C. A. (2000) *J. Bioenerg. Biomembr.* **32**, 1–4.
9. Frey, T. G. and Mannella, C. A. (2000) *Trends Biochem. Sci.* **25**, 319–324.
10. Frey, T. G., Renken, C. W., and Perkins, G. A. (2002) *Biochim. Biophys. Acta* **1555**, 196–203.
11. Perkins, G. A. and Frey, T. G. (2000) *Micron* **31**, 97–111.
12. Mannella, C. A. (2006) *Biochim. Biophys. Acta* **1762**, 140–147.
13. Renken, C., Siragusa, G., Perkins, G., Washington, L., Nulton, J., Salamon, P., and Frey, T. (2002) *J. Struct. Biol.* **138**, 137.
14. Ponnuswamy, A., Nulton, J., Mahaffy, J. M., Salamon, P., Frey, T. G., and Baljon, A. R. (2005) *Phys. Biol.* **2**, 73–79.
15. John, G. B., Shang, Y., Li, L., Renken, C., Mannella, C. A., Selker, J. M., Rangell, L., Bennett, M. J., and Zha, J. (2005) *Mol. Biol. Cell* **16**, 1543–1554.
16. Bustos, D. M. and Velours, J. (2005) *J. Biol. Chem.* **280**, 29004–29010.
17. Arselin, G., Vaillier, J., Salin, B., Schaeffer, J., Giraud, M. F., Dautant, A., Brethes, D., and Velours, J. (2004) *J. Biol. Chem.* **279**, 40392–40399.
18. Paumard, P., Vaillier, J., Coulary, B., Schaeffer, J., Soubannier, V., Mueller, D. M., Brethes, D., Di Rago, J. P., and Velours, J. (2002) *EMBO J.* **21**, 221–230.
19. Scorrano, L. (2005) *J. Bioenerg. Biomembr.* **37**, 165–170.
20. Perfettini, J. L., Roumier, T., and Kroemer, G. (2005) *Trends Cell Biol.* **15**, 179–183.
21. Vogel, F., Bornhovd, C., Neupert, W., and Reichert, A. S. (2006) *J. Cell Biol.* **175**, 237–247.
22. Hoffman, H. P. and Avers, C. J. (1973) *Science* **181**, 749–751.
23. Stevens, B. (1981) Mitochondrial structure. In Strathern, J. N., Jones, E. W., and Broach, J. R., editors. *The Molecular Biology of the Yeast Saccharomyces, Life Cycle and Inheritance*, Cold Spring Harbor Laboratory Press, Cold Spring Harbor, NY. pp. 471–504.
24. Fawcett, D. W. (1981) Mitochondria. In *The Cell*, W. B. Saunders, Philadelphia.
25. Hackenbrock, C. R., Chazotte, B., and Gupte, S. S. (1986) *J. Bioenerg. Biomembr.* **18**, 331–368.
26. Chazotte, B. and Hackenbrock, C. R. (1991) *J. Biol. Chem.* **266**, 5973–5979.
27. Lipowsky, R. (1991) *Nature* **349**, 475–481.

28. Heggeness, M. H., Simon, M., and Singer, S. J. (1978) *Proc. Natl. Acad. Sci. USA* **75**, 3863–3866.
29. Ball, E. H. and Singer, S. J. (1982) *Proc. Natl. Acad. Sci. USA* **79**, 123–126.
30. Johnson, L. V., Walsh, M. L., and Chen, L. B. (1980) *Proc. Natl. Acad. Sci. USA* **77**, 990–994.
31. Chen, L. B. (1988) *Annu. Rev. Cell Biol.* **4**, 155–181.
32. Reers, M., Smiley, S. T., Mottola-Hartshorn, C., Chen, A., Lin, M., and Chen, L. B. (1995) *Methods Enzymol.* **260**, 406–417.
33. Bereiter-Hahn, J. (1990) *Int. Rev. Cytol.* **122**, 1–64.
34. Bereiter-Hahn, J. and Voth, M. (1994) *Microsc. Res. Tech.* **27**, 198–219.
35. Nunnari, J., Marshall, W. F., Straight, A., Murray, A., Sedat, J. W., and Walter, P. (1997) *Mol. Biol. Cell* **8**, 1233–1242.
36. Rizzuto, R., Pinton, P., Carrington, W., Fay, F. S., Fogarty, K. E., Lifshitz, L. M., Tuft, R. A., and Pozzan, T. (1998) *Science* **280**, 1763–1766.
37. Yadava, N. and Scheffler, I. E. (2004) *Mitochondrion* **4**, 1–12.
38. Szabadkai, G., Simoni, A. M., Bianchi, K., De Stefani, D., Leo, S., Wieckowski, M. R., and Rizzuto, R. (2006) *Biochim. Biophys. Acta.* **1763**, 442–429.
39. Goetz, J. G. and Nabi, I. R. (2006) *Biochem Soc. Trans.* **34**, 370–373.
40. Chang, D. T., Honick, A. S., and Reynolds, I. J. (2006) *J. Neurosci.* **26**, 7035–7045.
41. Chen, H. and Chan, D. C. (2006) *Curr. Opin. Cell Biol.* **18**, 453–459.
42. Overly, C. C., Rieff, H. I., and Hollenbeck, P. J. (1996) *J. Cell Sci.* **109**, 971–980.
43. Morris, R. L. and Hollenbeck, P. J. (1995) *J. Cell Biol.* **131**, 1315–1326.
44. Lee, K. D. and Hollenbeck, P. J. (1995) *J. Biol. Chem.* **270**, 5600–5605.
45. Morris, R. L. and Hollenbeck, P. J. (1993) *J. Cell Sci.* **104**, 917–927.
46. Hollenbeck, P. J. (2005) *Neuron* **47**, 331–333.
47. Hollenbeck, P. J. and Saxton, W. M. (2005) *J. Cell Sci.* **118**, 5411–5419.
48. Chada, S. R. and Hollenbeck, P. J. (2003) *J. Exp. Biol.* **206**, 1985–1992.
49. Miller, K. E. and Sheetz, M. P. (2004) *J. Cell Sci.* **117**, 2791–2804.
50. Berger, K. H. and Yaffe, M. P. (1996) *Experientia* **52**, 1111–1116.
51. Yaffe, M. P. (1999) *Science* **283**, 1493–1497.
52. Jensen, R. E. (2005) *Curr. Opin. Cell Biol.* **17**, 384–388.
53. Meeusen, S. L. and Nunnari, J. (2005) *Curr. Opin. Cell Biol.* **17**, 389–394.
54. Osteryoung, K. W. and Nunnari, J. (2003) *Science* **302**, 1698–1704.
55. Shaw, J. M. and Nunnari, J. (2002) *Trends Cell Biol.* **12**, 178.
56. Altmann, K. and Westermann, B. (2005) *Mol. Biol. Cell* **16**, 5410–5417.
57. Rube, D. A. and Van der Bliek, A. M. (2004) *Mol. Cell Biochem.* **256**–257, 331–339.
58. Anesti, V. and Scorrano, L. (2006) *Biochim. Biophys. Acta* **1757**, 692–699.
59. Okamoto, K. and Shaw, J. M. (2005) *Annu. Rev. Genet.* **39**, 503–536.

59a. Hoppins, S., Lackner, L., and Nunnari, J. (2007) *Annu. Rev. Biochem.* **76**, 751–780.

60. Schrader, M. (2006) *Biochim. Biophys. Acta* **1763**, 531–541.
61. Kuroiwa, T., Kuroiwa, H., Sakai, A., Takahashi, H., Toda, K., and Itoh, R. (1998) *Int. Rev. Cytol.* **181**, 1–41.

62. Song, B. D. and Schmid, S. L. (2003) *Biochemistry* **42**, 1369–1376.
63. Bleazard, W., McCaffery, J. M., King, E. J., Bale, S., Mozdy, A., Tieu, Q., Nunnari, J., and Shaw, J. M. (1999) *Nat. Cell Biol.* **1**, 298–304.
64. Kelly, R. B. (1995) *Nature* **374**, 116–117.
65. Olichon, A., Guillou, E., Delettre, C., Landes, T., Arnaune-Pelloquin, L., Emorine, L. J., Mils, V., Daloyau, M., Hamel, C., Amati-Bonneau, P., Bonneau, D., Reynier, P., Lenaers, G., and Belenguer, P. (2006) *Biochim. Biophys. Acta* **1763**, 500–509.
66. Sesaki, H., Dunn, C. D., Iijima, M., Shepard, K. A., Yaffe, M. P., Machamer, C. E., and Jensen, R. E. (2006) *J. Cell Biol.* **173**, 651–658.
67. Meeusen, S., McCaffery, J. M., and Nunnari, J. (2004) *Science* **305**, 1747–1752.
68. Koshiba, T., Detmer, S. A., Kaiser, J. T., Chen, H., McCaffery, J. M., and Chan, D. C. (2004) *Science* **305**, 858–862.
69. Amutha, B., Gu, Y., and Pain, D. (2004) *Biochem. J.* **381**, 19–23.
70. Olichon, A., Baricault, L., Gas, N., Guillou, E., Valette, A., Belenguer, P., and Lenaers, G. (2003) *J. Biol. Chem.* **278**, 7743–7746.
71. Ishihara, N., Fujita, Y., Oka, T., and Mihara, K. (2006) *EMBO J.* **25**, 2966–2977.
72. Jeyaraju, D. V., Xu, L., Letellier, M. C., Bandaru, S., Zunino.R., Berg, E. A., McBride, H. M., and Pellegrini, L. (2006) *Proc. Natl. Acad. Sci. USA* **103**, 18562–18567.
73. Escobar-Henriques, M., Westermann, B., and Langer, T. (2006) *J. Cell Biol.* **173**, 645–650.
74. Kawano, S., Takano, H., and Kuroiwa, T. (1995) *Int. Rev. Cytol.* **161**, 49–110.
75. Yaffe, M. P. (1995) *Methods Enzymol.* **260**, 447–453.
76. Burgess, S. M., Delannoy, M., and Jensen, R. E. (1994) *J. Cell. Biol.* **126**, 1361–1373.
77. Dimmer, K. S., Jakobs, S., Vogel, F., Altmann, K., and Westermann, B. (2005) *J. Cell Biol.* **168**, 103–115.
78. Dimmer, K. S., Fritz, S., Fuchs, F., Messerschmitt, M., Weinbach, N., Neupert, W., and Westermann, B. (2002) *Mol. Biol. Cell* **13**, 847–853.
79. Stewart, L. C. and Yaffe, M. P. (1991) *J. Cell. Biol.* **115**, 1249–1257.
80. Miyakawa, I., Aoi, H., Sando, N., and Kuroiwa, T. (1984) *J. Cell. Sci.* **66**, 21–38.
81. Birky, C. W., Jr. (1994) *J. Hered.* **85**, 355–365.
82. Birky, C. W., Jr. (1995) *Proc. Natl. Acad. Sci. USA* **92**, 11331–11338.
83. Azpiroz, R. and Butow, R. A. (1995) *Methods Enzymol.* **260**, 453–465.
84. Jones, B. A. and Fangman, W. L. (1992) *Genes Dev.* **6**, 380–389.
85. Chen, X.-J., Guan, M.-X., and Clark-Walker, G. D. (1993) *Nucleic Acids Res.* **21**, 3473–3477.
86. Thorsness, P. E. and Weber, E. R. (1996) *Int. Rev. Cytol.* **165**, 207–234.
87. Fuller, M. T. (1993) Spermatogenesis. In Bate, M. and Martinez-Arias, A., editors. *The Development of Drosophila*, Cold Spring Harbor Press, Cold Spring Harbor, NY, pp. 71–147.
88. Tokuyasu, K. T. (1974) *Exp. Cell Res.* **84**, 239–250.
89. Hales, K. G. and Fuller, M. T. (1997) *Cell* **90**, 121–129.
90. Pfeffer, S. R. (1996) *Annu. Rev. Cell Dev. Biol.* **12**, 2351–2359.
91. Takano, H., Kawano, S., and Kuroiwa, T. (1994) *Curr. Genet.* **25**, 252–257.

92. Sakurai, R., Nomura, H., Moriyam, Y., and Kawano, S. (2004) *Curr. Genet.* **46**, 103–114.
93. Augustin, S., Nolden, M., Muller, S., Hardt, O., Arnold, I., and Langer, T. (2004) *J. Biol. Chem.* **280**, 2691–2699.
94. Bota, D. A., Ngo, J. K., and Davies, K. J. (2005) *Free Radic. Biol. Med.* **38**, 665–677.
95. Bota, D. A. and Davies, K. J. (2001) *Mitochondrion* **1**, 33–49.
96. Seglen, P. O., Berg, T. O., Blankson, H., Fengsrud, M., Holen, I., and Stromhaug, P. E. (1996) *Adv. Exp. Med. Biol.* **389**, 103–111.
97. Aubert, S., Gout, E., Bligny, R., Marty-Mazars, D., Barrieu, F., Alabouvette, J., Marty, F., and Douce, R. (1996) *J. Cell Biol.* **133**, 1251–1263.
98. Scott, S. V., Hefner-Gravink, A., Morano, K. A., Noda, T., Ohsumi, Y., and Klionsky, D. J. (1996) *Proc. Natl. Acad. Sci. USA* **93**, 12304–12308.
99. Priault, M., Salin, B., Schaeffer, J., Vallette, F. M., Di Rago, J. P., and Martinou, J. C. (2005) *Cell Death Differ.* **12**, 1613–1621.
100. Gu, Y., Wang, C., and Cohen, A. (2004) *FEBS Lett.* **577**, 357–360.
101. Terman, A. and Brunk, U. T. (2004) *Exp. Gerontol.* **39**, 701–705.

4

BIOGENESIS OF MITOCHONDRIA

Summary

An estimated 800–1500 proteins are found in mitochondria at a relative abundance that varies from tissue to tissue. Of those, only 13 are encoded by the mitochondrial genome, but each is essential for oxidative phosphorylation. The biogenesis of mitochondria therefore includes several major themes. First, mtDNA has to be characterized and a mechanism for its replication has to be defined. Second, the mtDNA has to be transcribed, and the long primary transcripts have to be processed to yield two rRNAs, 22 tRNAs, and 13 mRNAs (in mammalian mitochondria). Third, a translational machinery for initiation and elongation of proteins exists in the mitochondrial matrix. The resulting proteins have to be inserted into the inner mitochondrial membrane, probably coupled to the assembly of the complexes of the oxidative phosphorylation system. Fourth, the vast majority of mitochondrial proteins are made in the cytosol. They must contain a signal for import into mitochondria, but there must be additional mechanisms assuring that each protein reaches its final, distinct localization: in the outer membrane, in the intermembrane space, in the inner membrane, and in the matrix. After the initial recognition by the TOM complex in the outer membrane, proteins are shuttled to complexes in the inner membrane (TIM) for distribution to their final destination. The import requires a membrane potential. Some proteins are subject to post-import processing by proteases. Turnover of proteins in mitochondria, or turnover of whole mitochondria by autophagy are processes under active investigation.

In many organisms (e.g., plants and fungi), some or all tRNAs have to be imported into the mitochondrial matrix from the cytosol. In many organisms RNA editing occurs: The original transcript is modified before it is translated into protein. This may involve a simple deamination (C to U) at selected sites, or it may include very substantial insertions or deletions of uridine tracts into the mRNA (e.g., in trypanosomes).

4.1 THE MITOCHONDRIAL GENOME

4.1.1 Introduction

The discovery of DNA in a compartment other than the eukaryotic nucleus was truly a discovery with far-reaching possibilities and consequences. It not only shed new light on theories concerned with the origin of this organelle, and ultimately provided overwhelming support in favor of the endosymbiont hypothesis (see Chapter 2), but also opened totally new vistas in the study of the biogenesis of mitochondria. At once the possibility was recognized that this subcellular structure was likely to contain genetic information required for its own assembly, and an immediate follow-up question was: How much? The earliest estimates came from studies with the electron microscope, after procedures were refined to isolate mtDNA molecules cleanly and in sufficient quantities (Figure 4.1) (1–4)).

MtDNA molecules derived from diverse multicellular animals (metazoans) were found to be circular structures, and occasionally catenated dimers or higher oligomers, with contour lengths corresponding to ~16,000 nucleotides. It was obvious that only a limited number of genes could be present. During the past three decades, hundreds of mitochondrial genomes from animals, plants, and fungi have been characterized genetically and even sequenced. A few milestones deserve special mention. The first complete mitochondrial genome to be sequenced from any organism was that of humans (5), and soon thereafter the bovine sequence was completed (6). Other than the genomes of bacteriophages and animal viruses, these were the first biologically well-defined and distinct DNA molecules (one might even say chromosomes) to

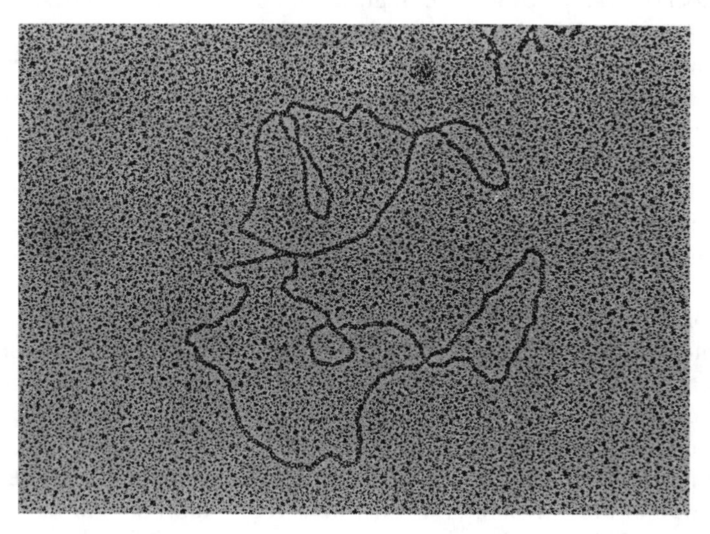

Figure 4.1 Mitochondrial DNA from an oocyte of *Xenopus laevis*. (Original photograph of I. Dawid; from *The Cell*, by D. W. Fawcett, 1981, W. B. Saunders Co.)

become known in their entirety. With the technology of the day, it was a major achievement (see also reference 6a). In contrast, the first complete mitochondrial genome from a plant (*Marchantia polymorpha*, liverwort) was not published until 1992 (7, 8), and an even larger sequence from *Arabidopsis thaliana* (mustard) appeared several years later (9). The latter was the first complete sequence from a flowering plant. The animal mtDNA was "only" 16,569 nucleotides long, while the plant mtDNAs contained 186,608 and 366,924 nucleotides, respectively, more than 20 times as many in the flowering plant compared to vertebrates. As will be illustrated in the following sections, sizes of mtDNAs vary considerably between metazoa, plants, and fungi, yet this difference is not necessarily paralleled by a proportionate difference in the number of genes retained by these genomes. The examples chosen will illustrate and support a number of generalizations, but at the same time they are selected to impress the reader with the tremendous variations found in different organisms. Clearly, one would like to understand the biological significance of any observed differences. One point of view is that the transfer of genes from the proto-mitochondria to the nucleus has progressed to different extents in different organisms purely by chance, and that further transfer of genes will occur in the future. Does the observation of a limited number of genes in mitochondria represent a snapshot in evolutionary time, or is it possible to interpret this finding in terms of structure and function? In metazoans the selection appears to have forced the reduction of the mt genome to its absolute minimum in terms of nucleotides—that is, the highest possible density of genes per unit length. It is not absolutely clear whether this represents a limit to the genes which can be transferred to the nucleus. At one time it was possible to argue that rRNAs and tRNAs for translation must be transcribed from genes inside the mitochondria, because the highly charged polynucleotides were not expected to be able to be imported across two membranes. However, in the meantime the import of tRNAs has been observed in plant mitochondria, in protozoa, and in yeasts. The proteins encoded by vertebrate mtDNA are very hydrophobic, prompting arguments that they also could not be imported from the cytosol. Such a rationalization fails to explain why many integral membrane proteins with multiple transmembrane domains (e.g., of complex II, many transporters) are imported from the cytosol in most organisms. In plants and fungi there is similar solid evidence that a very extensive reduction in the number of genes has occurred, but at the same time many other sequences mostly of no functional significance have been acquired, either by transfer and scavenging from other sources, or by some poorly understood mechanism of amplification of intergenic sequences, or by multiplication of transposon-related elements. It is particularly noteworthy that many non-metazoan mt genes have introns, and some of these introns contain ORFs.

When intergenic sequences are ignored, it becomes clear that the genetic information in the mitochondrial genome is remarkably similar in a majority of organisms. This could be because the loss of genes from the original symbiont was essentially complete before extensive branching of the phylogenetic tree occurred, or, alternatively, that this limited number was achieved independently, leaving only those genes inside the mitochondria whose products could not be imported at all. As mentioned above, this argument is no longer sustainable for the tRNAs. At the same time, modern molecular genetic techniques have made it possible to knock out mitochondrial genes and complement them with appropriately engineered versions in the nucleus (codon usage, import signals, see below) (10). Such experiments argue against the absolute necessity of retaining certain genes in the mitochondria.

While the majority of organisms examined have been found to have circular mtDNAs, an occasional claim of discovery of linear mitochondrial DNAs was made. Initially dismissed as artifact, or as a very rare exception to the circular genomes expected from a prokaryotic ancestor, the number of such examples has grown over the past few decades. Diverse organisms from among the ciliata (e.g., *Paramecium aurelia*), the algae (*Chlamydomonas reinhadtii*), the fungi (*Candida, Pichia*, and *Williopsis* species), and oomycetes are represented in a recent review of this subject (11). An even more radical view gaining acceptance is that at least in plants and fungi (including *Saccharomyces cerevisiae*) the major form of mtDNA *in vivo* is linear (11). The physical size of these linear mtDNAs is in the 30- to 60-kb range. Several criteria have been defined by investigators to distinguish true linear mtDNAs from linear concatomeric molecules arising from rolling circle mechanisms of DNA replication and other mechanisms. The most intriguing characteristic is a homogeneous terminal structure functionally analogous to telomeres. However, mitochondrial telomeres from diverse organisms do not share consensus sequences or sequence motifs indicative of a uniform solution to the problem of replicating 5′ ends.

It has been argued that the existence of linear mtDNA does indicate an independent evolutionary origin, since even closely related fungi may have either linear or circular mtDNAs. Rather, the linear form may have arisen by accident and become stabilized by the fortuitous existence of a replication machinery capable of dealing with linear ends (11).

The number of mitochondrial genomes sequenced as of September 2006 exceeds several hundred from different species. In the following, the salient characteristics of the mtDNAs of several important representatives of metazoans, plants, fungi, and protozoa will be described to impress the reader with the diversity of size, gene content, organization, and, ultimately, replication and gene expression. A more exhaustive coverage of different organisms is beyond the scope of this book, and the reader is referred to the appropriate databases on the Internet.

4.1.2 The Mitochondrial Genome in Metazoans

The size range of mtDNAs found in multicellular animals is relatively narrow (~16.5 kb), with some exceptions varying from 14 kb in the nematode, *Caenorhabditis elegans* (12), to 42 kb in the scallop, *Placeopecten megallanicus*

(13), and all are single, circular DNAs. A curious exception is the mitochondrial genome of the cnidarian *Hydra attenuata* which consists of two unique, linear DNA molecules of 8 kb (13a). The exceptionally large scallop mtDNA may represent an extreme; it has not been sequenced, and may even consist of a duplication. Complete sequences are now available from various mammals, chicken (*Gallus domesticus*), toads (*Xenopus laevis*), sea urchin (*Strongylocentrotus purpuratus*), fruit fly (*Drosophila Yakuba*), nematode (*Caenorhabditis elegans*), and the sea anemone (*Metridium senile*). This list is not exhaustive, but it is striking that deviations in length from the published human sequence (16,569 kb) are generally less than 1 kb (see Table 4.1). A metazoan mtDNA database is being maintained by Dr. M. Attimonelli at the University of Bari, Italy (see references 14 and 15). Updates on variations in human mitochondrial genomes can be found on the Internet (http://www.mitomap.org) (16).

TABLE 4.1 Representative Sequenced Metazoan Mitochondrial Genomes

Class	Organism	Genome Size (kb)
Nematoda	*Ascaris suum*	14284
	Caenorhabditis elegans	13794
	Meloidogyne javanica	20500
Insecta	*Drosophila yacuba*	16019
	Anopheles quadrimaculatus	15455
	Anopheles gambia	15363
	Apis mellifera	
Crustacea	*Artemia franciscana*	15770
Echinodermata	*Stronylocentrotus purpuratus*	15650
	Paracentrotus lividus	15700
	Asterina pectinifera	16260
	Arbacia lixula	15722
Cnidaria	*Metridium senile*	17443
Amphibia	*Xenopus laevis*	17443
Aves	*Gallus domesticus*	16775
Osteichthyes	*Crossostoma lacustre*	16558
	Ciprinus carpio	16364
Mammalia	*Bos taurus*	16338
	Mus musculus	16303
	Rattus norvegicus	16298
	Balaenoptera physalus	16398
	Balaenoptera musculus	16402
	Phoca vitulina	16826
	Halichoerus grypus	16797
	Equus caballus	16660
	Didelphis virginiana	17084
	Homo sapiens	16569

Metazoan mtDNA Database (MmtDB; Attimonelli, M., University of Bari, Italy; see reference 16.

When the human mtDNA sequence was first published in 1981, not all genes were immediately identifiable, but a complete characterization of the remaining URFs (unidentified reading frames) was achieved in 1986 (17, 18). All the other metazoan mtDNAs were subsequently found to contain the same genes with very few exceptions. On the other hand, the order of the genes is not always the same, suggesting that some rearrangements have occurred.

Metazoan mtDNAs encode two ribosomal RNAs (s-rRNA, l-rRNA), 22 tRNAs (with exceptions), and 13 peptides which become constituents (ND1–6, ND4L) of complex I (NADH-coenzyme Q oxidoreductase), complex III (cyt b), cytochrome oxidase (COI, COII, COIII of complex IV), and ATP synthase or complex V (ATPase 6, 8). The structure and function of these complexes in electron transport and oxidative phosphorylation will be covered in detail in another chapter. It is relevant to describe here that each complex consists of multiple peptides, and among those one can distinguish peptides deeply embedded in the membrane (integral membrane proteins, IMPs) and distinguish others that are firmly associated with domains of the IMPs extending into the mitochondrial matrix. The IMPs contain a large proportion of very hydrophobic amino acids, and many have multiple membrane-spanning segments based on hydropathy plots. It is often argued that such very hydrophobic peptides cannot be imported into the mitochondria, and thus their genes have been retained inside the mitochondrial matrix. Following or even during translation in the matrix, the peptides can be directly inserted into the inner membrane.

With the set of 13 peptides identified in many diverse species, the identification of open reading frames in new mtDNA sequences from additional organisms can be accomplished relatively easily from sequence alignments and hydropathy profiles. It should be noted, however, that DNA sequences for individual peptides from different organisms, even closely related species such as mammals, have diverged quite significantly to the point where probes from one species cannot generally be used to detect DNA or transcripts from another. The extraordinarily high rate of base changes in vertebrate mtDNA will be the subject of a separate discussion (Chapter 7). And even though most metazoan mtDNAs encode 13 peptides, the size of the corresponding reading frame can vary considerably. For example, the ND6 peptide in mouse has 172 amino acids, while the same peptide in *C. elegans* has 145. The ATPase6 peptide has 226 residues in the mouse and only 199 in *C. elegans*. Other peptides—for example, the cyt b peptide of complex III, or the peptides of cytochrome oxidase—exhibit less length variation from species to species. A more extensive compilation of such data can be found in the review by Wolstenholme (4).

The genetic information on the mt genome in metazoans is highly compacted on the circular mtDNA. The human mtDNA will serve as an example (Figure 4.2) to illustrate some of the general principles of organization of genes, but it should be stressed that there is variability, and some particularly striking exceptions will be noted to alert the reader of the deviations that can be expected in other organisms.

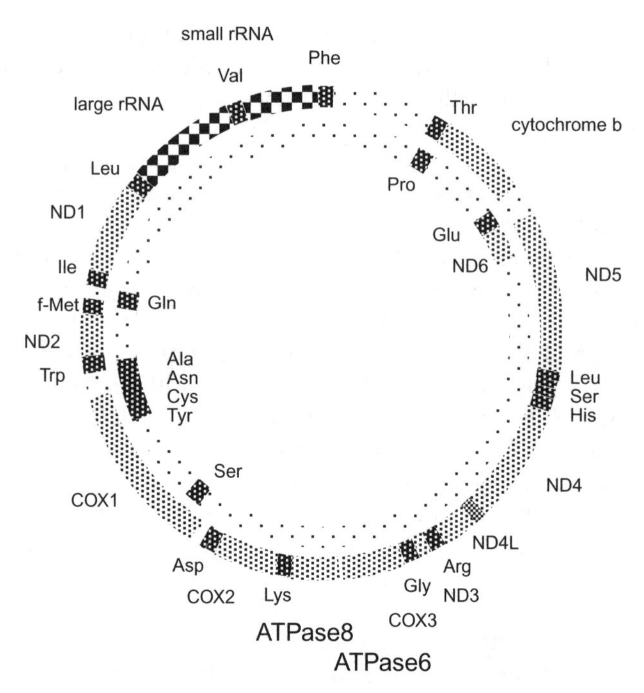

Figure 4.2 Schematic representation of the human mitochondrial genome map. The outer circle represents the heavy strand, encoding most of the peptides, the two rRNAs, and a large number of the tRNAs. ND1–ND6 genes encode peptides of complex I; COX genes encode peptides of complex IV.

Most, but not all, open reading frames for the peptides and the rRNA genes are separated by one or more tRNA genes—with few, if any, extra nucleotides in between. Small-rRNA and large-rRNA genes are separated by a single tRNA in vertebrates and *Drosophila*, but by multiple tRNA genes in other invertebrates. It is interesting to be reminded that the dispersion of tRNA genes around the mt genome was already discovered prior to the arrival of cloning techniques. By covalently labeling 4S RNA (tRNA) or rRNA with ferritin and hybridizing such "probes" to mtDNA, it was possible to map in 1976 the adjacent 16S and 12S rRNA genes and 19 tRNA genes distributed around the circular DNA (19). The transcription of the genome occurs in both directions, and hence open reading frames can be on either strand, although they are usually quite unevenly distributed between the heavy and light strands of mtDNA. (The distinction is based on an asymmetrical distribution of bases, giving rise to a difference in density on alkaline CsCl gradients.) Transcription, as will be discussed in more detail elsewhere, yields polycistronic RNA molecules. When the interspersed tRNAs are spliced out, individual mRNAs or rRNAs result. The mRNAs have extremely short or nonexisting 5′ and 3′ untranslated regions, and in some instances the first A from the polyadenylation step completes the terminal stop codon. Special problems are created for the initiation of translation, which will be elaborated in a later section. In HeLa cell and other vertebrate mitochondria the ATPase8 and ATPase6 and also the ND4L and ND4 reading frames overlap by a few nucleotides without a separating tRNA. The mature mRNA is bicistronic, and translation initiation must occur internally. A detailed consideration of mechanisms will be deferred, but in the current context these observations emphasize the extreme economy of packaging genetic information into a small genome. A discussion of plant and fungal mtDNA will present a distinct contrast.

There is one region in metazoan mtDNAs which deserves special mention here. Flanked by two tRNA genes ($tRNA^{pro}$ and $tRNA^{phe}$ in vertebrates), a noncoding region of variable length has been found. It is 1122 nt long in humans, 879 nt long in the mouse, and even larger in *Xenopus leavis*. Its sequence and function will be discussed in more detail under the appropriate headings. Here it suffices to state that this noncoding region contains a replication origin and the promoter regions for transcription in opposite directions. For these reasons it is being referred to as the "control region." In spite of the functional identity, the control region is not very well conserved between species, and it is polymorphic even within a single species such as humans. That is to say, there are short, functionally significant and conserved segments within the control region, while the remainder of the sequence is variable.

There are exceptions to the gene content and organization described for metazoans so far. They will not be catalogued exhaustively here, but a few examples will be presented to alert the reader to the possibilities that might be encountered with as-yet-unknown mt genomes. The genes for the rRNAs are generally encoded by the same DNA strand, separated by one or more tRNAs, but in the sea star they are encoded in opposite strands. This possibility should be considered when the analysis of a new mtDNA sequence fails to reveal one of the two rRNAs arranged in tandem. The creation of such a novel arrangement clearly requires an inversion of a portion of the mtDNA molecule. Other inversions and rearrangements must be responsible for the fact that the gene content is generally conserved in metazoans, while the gene order is conserved in fish, amphibians, and mammals, but not in birds, and even less so in invertebrates. Since processing of the polycistronic transcripts yields individual tRNAs, mRNAs, and rRNAs, the gene order may not be a matter of concern. Gene order is, however, of interest to the evolutionary biologists in their attempts to trace lineages over time.

As examples of variations in structural genes, one can mention the following. In nematodes the ATPase8 gene is missing. The ND3 and ND4L subunits of mitochondrial complex I are nucleus-encoded in *Chlamydomonas reinhardtii* (20). In the sea anemone *Metridium senile* the COI and the ND5 genes contain a group I intron, and one of these (the COI intron) has an open reading frame potentially capable of encoding either an RNA splicase or an endonuclease. The ND1 and ND3 genes are contained within the intron of the ND5 gene.

Most surprisingly, the sea anemone mtDNA has only two tRNA genes, for tryptophan and formyl-methionine. Since protein synthesis would obviously be impossible with only those two tRNAs, one has to confront the problem of how to import tRNAs into these mitochondria. Clearly, the import of such highly charged polynucleotides across two mitochondrial membranes presents a special challenge.

4.1.3 The Mitochondrial Genome in Plants

The mitochondrial genome in higher plants has many characteristics that probably make it one of the most interesting genomes to the molecular biologist. It has a potentially large coding capacity, is constantly reorganized by recombination, has captured genes encoded in the nucleus in other organisms, suffers mutations that can affect growth and sterility, and yields transcripts that have to be edited, or trans-spliced, and yet it lacks a subset of tRNAs that have to be imported from the cytoplasm. As can be expected, not all of these aspects are found all the time, and the functional/biological consequences of mitochondrial gene variations will be discussed at a later time.

It was mentioned earlier in this chapter that the first complete plant mtDNAs to be completely sequenced were those from *Marchantia polymorpha* (7) and *Arabidopsis thaliana* (9). Sufficient information is therefore available now to permit several generalizations (see references 8, 21, and 22 for critical reviews and exhaustive listings of primary references). The majority of data were derived from flowering plants (Angiospermophyta), only one of ten phyla within the plant kingdom. For this still narrow representation of plants, the following statements can be made:

1. The mitochondrial genome is much larger than that in metazoans, ranging from 200 to 2400 kb. It can be noted that 2400 kb is within the range of some prokaryotic genomes. Original estimates were based on renaturation kinetics (C_0t curve analysis), while others were based on summation of restriction fragments. Examinations by the electron microscope were inconclusive: Heterodisperse linear and circular molecules were seen, and the possibility of breakage could not be discounted. Similarly, pulsed field electrophoresis of either whole or digested molecules gave results that were either confusing or inconsistent with data derived by other approaches. Large circular DNA molecules are known to remain trapped in the well of the gel. Another possible source of confusion is the presence of small linear and circular DNA molecules in the mitochondria of several plant species which appear to be mitochondrial plasmids that can replicate autonomously and have open reading frames. One such plasmid, in all maize lines examined, carries the only functional $tRNA^{trp}$ gene (23). A consensus view from restriction mapping and other approaches (cosmid walking) is that the mitochondrial genome can be described as a large "master circle" containing all the mapped sequences. Actual structures found are derived from this master circle by mechanisms described next.

2. The arrangement of genes on these large genomes is extremely variable, even between closely related genera. This fluidity is evidenced by altered restriction maps and by absence of conserved linkages between genes. As an explanation for this observation, it has been proposed that plant mitochondria have an active recombination system (24, 25); and in order to account for the structural rearrangements resulting from homologous recombination, the presence and participation of repeated sequences in both direct or inverted orientation has been noted. Repeats of the order of 1–10 kb have been found in the majority of plant mtDNAs, and a role for a larger number of small repeats in the larger mt genomes has been proposed to make recombination responsible for the scrambling of plant mtDNA sequences (8, 26, 27).

In considering plausible models, one has to remember that these genomes are circular, that the repeats may be direct or inverted, and that the distance between the repeats is also variable. A simple "master" circle with two direct repeats at some distance from each other will yield two subgenomic circles if recombination is intramolecular; and if recombination is relatively active, one expects to find a dynamic equilibrium between the master circle and the two subgenomic circles, a situation that has been observed in plants such as *Brassica sp*., which have the smallest mtDNAs. Clearly, recombination between the master circle and one of the subgenomic circles can give rise to larger circles with duplications. If multiple small direct and inverted repeats are present, the possibilities are too numerous to count. A more stringent definition of "recombination repeats" has been proposed by Stern and Palmer (28). In the simplest case, such repeats occur in two copies relative to other sequences in the genome, but a probe directed against the repeats may detect four distinct restriction fragments, reflecting four different genomic environments for the repeat which result from recombinational events. If more repeats are present, more such genomic environments are predicted and have in fact been found in petunia line 3704 and maize CMS-T (21).

On the other hand, not all repeats found are found in multiple environments; that is, there are repeats that appear not to have been involved in recombination. It should be emphasized that the postulated recombination has not been formally demonstrated, but the consistency of observations makes recombination the mechanism of choice for restructuring these genomes. Specifically, it is not yet clear whether recombination is an active process at the present, responsible for the polymorphisms detected by restriction mapping, or for the different cosmid clones in a library. An alternative explanation is that recombination occurred in the evolutionary past, but the molecules generated then are replicated and continuously maintained in the mtDNA population of a given species. Evidence in favor of an active recombination system may be found from observations made in the production of somatic cell hybrids by protoplast fusion and the subsequent generation of a hybrid plant. In a few cases, the protoplast parents had distinct mt genomes based on restriction analysis, and hybrid plants were found to be homoplasmic (have a genetically stable mt genome) with recombinant mtDNA molecules (21). Curiously, recombination between mt genomes in somatic cell hybrids occurred at loci that were not normally considered to be recombination substrates (21, 29). A further discussion of mitochondrial inheritance in sexual reproduction and in mammalian somatic cell hybrids will be found in Chapter 7.

Another point worth noting is that in several cases, known genes flank or even extend into the recombination repeat. This allows genes to occur in more than one genomic context—that is, to have potentially more than a single promoter. Whether this has any biological and hence selective function remains to be seen. Among the genes found to be associated with repeats in some plants are various rrn genes, the atp6 gene, and the coxII gene, but no evidence for the presence of recombinase genes within such repeats has been found. Recombination repeats, and hence recombination involving repeats, are not essential, since at least one plant mt genome (*Brassica hirta*) has been found to be devoid of repeats. At the same time, recombination involving sites that are not repeats have been observed, both during tissue culture (see above) and in plants cultivated normally. In nature, such recombination events are rare, and they give rise to recombinant DNA molecules present at very low abundance relative to the majority mtDNA. Rare mtDNA molecules found at these substoichiometric levels have been termed "sublimons."

3. A third characteristic distinguishing the mitochondrial genes of higher plants from those of animals is that sequence divergence is strikingly slow. The plant mitochondrial genome is the most slowly evolving cellular genome so far characterized (8, 30). Thus, while large chunks of plant mtDNA have been moved around in multiple ways during evolution, sequences of individual genes have remained remarkably constant. The implication is that either mtDNA replication is exceptionally error-free or that plant mitochondria have a very efficient mismatch repair system.

In view of the widespread use of *Arabidopsis thaliana* as the model plant species to study plant molecular genetics and development, an analysis of the mt genome of this plant will serve here to illustrate and support generalizations; and again, prominent or instructive exceptions will be noted at the end of this section.

With 366,924 basepairs the *Arabidopsis thaliana* mt genome is at a medium position in the range found for plants. In light of its size, the gene content is of obvious interest. Because of previous results with related studies in plants, it was not a total surprise to find only 57 genes on this genome, including three rRNA genes, 22 tRNA genes, and the "standard" set of genes for the mitochondrial respiratory chain complexes. For someone exclusively focused on animal mtDNAs up to now, it will be news that plant mtDNAs generally contain genes for ribosomal proteins, along with a 5S rRNA gene in addition to the rRNA genes for the small and large ribosomal subunits. A relatively unique set of genes found only in the mustard and the liverwort so far are five genes encoding proteins required for cytochrome c biogenesis (ccb). Three

ORFs have not yet been identified from comparisons or homologies with genes in other organisms (plants or bacteria).

Among the "standard genes" for complexes I–V, some contrasting observations between plants and animals can be made. In the mustard, liverwort, and a green alga, *Prototheca wickerhami*, there are two additional mitochondrial genes for complex I (nad7 and nad9); plant mtDNAs encode three complex V genes (atp1, atp6, atp9), compared to two in human mtDNA (atp6, atp8), and in the liverwort but not the mustard the genes for the two membrane anchor proteins of complex II (succinate dehydrogenase) are found on mtDNA. The comparison of mustard, liverwort, and green alga suggests that plant mtDNAs encode ribosomal proteins, but their number may be different in different plants. Genes are encoded by both strands with no extreme strand bias.

Although the finding of 22 tRNAs may not raise any eyebrows at first, a more detailed examination has established that these 22 tRNAs are not capable of decoding all the codons found in the ORFs and that tRNA genes for six amino acids are missing. Previous results from the liverwort and the potato had alerted the field that the import of select tRNAs into plant mitochondria is required. Again, it is not possible to give a fixed number for plants, because it varies from two in the liverwort to 11 in the potato.

The summation of the sizes of the identified genes in *Arabidopsis thaliana* mtDNA leaves a large amount of sequence unaccounted for. There are 23 identified introns; eighteen of those are cis-splicing and comprise 30,764 nucleotides, or 8.4% of the genome. Their size ranges from 485 nucleotides to 4000 nucleotides. Five trans-splicing introns cover at least another 4500 nucleotides of the genome. One group II intron appears to contain an ORF that may encode a maturase gene. Contrasting these data with those from the liverwort, one finds a total of 8 ORFs in type II introns (out of a total of 25) and 2 ORFs in type I introns (out of a total of 7). The intron locations are different between the two plants, and this is likely to be another highly variable situation differentiating higher flowering plants from lower plants, and possibly even flowering plants from each other.

The availability of several cDNA clones for protein genes in *Arabidopsis* mitochondria provides evidence for RNA editing, but the analyses are incomplete to verify all editing sites precisely.

There are numerous regions where short fragments from the chloroplast DNA have been integrated, amounting to approximately 1% of the mt genome. These fragments (30–930 nt) are unlikely to be functional, and the mechanism of their transfer from one organelle to the other is still obscure. Similarly, about 4% of the mt genome is of nuclear origin and identifiable as fragments of retrotransposons.

A general feature of most plant mtDNAs, the presence of repeats involved in recombination, was also found in the *Arabidopsis thaliana* mtDNA, and from previous physical mapping and cosmid clones there is evidence for circular molecules of 233 kb and 134 kb, which represent subgenomic circles arising from recombination. The finding and localization of two large repeated sequences of 6.5 kb and 4.2 kb is fully consistent with their participation in recombination to generate the smaller circles. The repeats have sequences conserved over their entire lengths. One of them extends into the atp6 gene. One hundred forty-four duplications of repeats in the 30- to 560-nucleotide size range were found, contributing 14 kb to the genome. Repeats are >90% identical, and they are not implicated in any recombinational events so far.

Finally, 62% of the *Arabidopsis* mt genome is comprised of DNA sequences with no obvious informational content. In contrast to what is found in fungal mtDNAs, this spacer DNA exhibits no particular nucleotide bias, and its origin and biological relevance are completely obscure at this time.

4.1.4 The Mitochondrial Genome in Fungi

An extensive compilation of data for different fungi is compiled in an authoritative review by Clark-Walker (31). The reader is referred to this article for details on many individual representatives of this large group of organisms. At this point, one should not be surprised to find again significant variability, but there are also some features that distinguish this group from the metazoa and higher plants. The size of the mitochondrial genomes is typically between 40 and 60 kb, with extremes of ~20 kb for *Schizosaccharomyces pombe* (an ascomycetous fission yeast) and ~170 kb for *Agaricus bitorquis* (a filamentous basidiomycete). The average reader will be familiar with *Asperigillus nidulans* (33–40 kb), *Neurospora crassa* (60–73 kb), *Saccharomyces cerevisiae* (~85 kb), *Ustilago cynodontis* (a rust) (76.5 kb), and others. The size of these genomes is, on average, three to four times larger than that of animals, but significantly smaller than that of plants.

The evolution and possible polyphyletic origins of fungi have been described and interpreted from morphological and biochemical studies, and a detailed analysis of their mt genomes, their sequence organization, and nucleotide sequence comparisons of individual genes such as rRNA genes can shed much additional light on this subject and confirm or refine the evolutionary relationships. The intent here, however, is to focus on common characteristics, as well as on novel aspects of genome organization and fluidity unique to fungi. These organisms form a large group with many taxonomic groups and subgroups, but at this level they must be lumped together to avoid drowning the reader in molecular taxonomy. The budding yeast *Saccharomyces cerevisiae* mtDNA will serve as a useful reference point, since this organism has also become one of the best-known model systems (32–34).

Among the expected genes present are the genes for the small and large ribosomal subunits, 24 tRNAs, cytochrome b in complex III, and three subunits of complex V (atp6, atp8, atp9). Notably absent are the genes (nad) for complex I, and it appears that *Saccharomyces cerevisiae* mitochondria do not even have a typical complex I with the extraordinary number of >40 subunits, that is found in most other organisms. On the other hand, the lack of nad genes is not characteristic of all fungi. A single ribosomal protein encoded by a gene called VAR1 is present, and there are several unidentified or open reading frames whose significance is still unclear. Strains of *Saccharomyces cerevisiae* exist in which one or the other of these ORFs have been deleted, without any effect on growth on nonfermentable carbon sources, and they may not be expressed at all. There are also "optional introns" in *S. cerevisiae* mtDNA within which ORFs have been found. Examples are the COXI and the CYB genes, and it appears that the product encoded by these ORFs functions in the production of mature mRNAs by intron excision and splicing. In other words, the intron encodes an activity for its own removal, but when the intron is absent, the activity is not required for anything else. Optional introns account for a substantial part of the size variations in mtDNAs seen even within a single strain of yeast—for example, lengths of 68–85 kb found in *S. cerevisiae* (Figure 4.3). The complete sequence for Saccharomyces cerevisiae mtDNA can be found at http://www.genome.ad.jp/dbget-bin/www_bget?refseq+NC_001224. The corresponding genetic map has been published by Foury et al. (35)

The existence of introns, as well as the existence of "optional" introns, raises several questions. The mechanism for removal of introns from RNA transcripts is appropriately discussed in the chapter on transcription and RNA processing in mitochondria. The acquisition and mobility of introns within the genome is a subject for the present context (see reference 36 for a review).

Saccharomyces cerevisiae strains exist which differ in the presence or absence of a 1132-bp group I intron in the large rRNA gene at the so-called omega (ω^+) locus of mtDNA. The ω genetic system was discovered independently of the discovery of introns, but pioneering studies very rapidly related

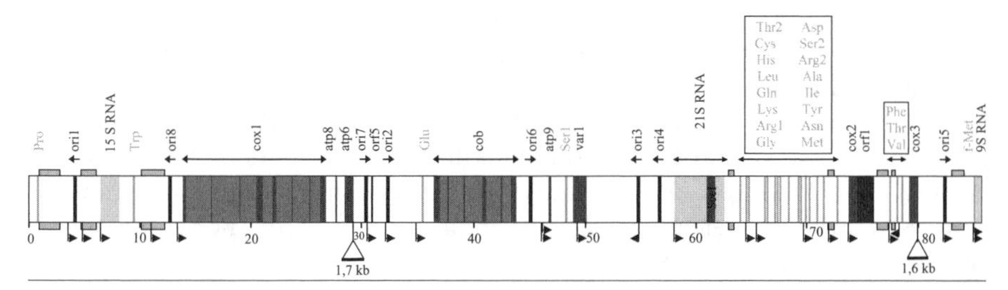

Figure 4.3 Schematic representation of the yeast mitochondrial genome map, strain FY1679, presented as a linearized map (35). Exons of protein coding genes are indicated in red (cox1, cox2, cox3, cob, atp9, var1), introns are gray; tRNA genes (green) are labeled with the corresponding amino acid, 9S, 15S, and 21S RNA (yellow) are labeled. Two significant deletions are found in this strain, and flags indicate transcription initiation sites. For details the original paper (35) should be consulted. (Reproduced from reference 35, with permission.) See color plates.

it to introns, and understanding its behavior has been a major guide in the understanding of group I intron mobility in general. When crosses between ω^+ and ω^- strains are made, the intron can move into the genome in which it was previously absent (ω^-), since in the progeny mt genomes with this intron are preferentially recovered. There is evidence that the intron has an ORF (235 nt) encoding the ω transposase later characterized as a site-specific, double-stranded endonuclease that is targeted to a specific sequence within the intron-free allele. It is speculated that site-specific cleavage initiates DNA recombination and conversion (see references 37 and 38, for reviews). The ω^+ intron in yeast was initially considered a rare oddity, but eventually similar introns were found in other organisms and at other loci in the yeast mt genome (36). The a14α intron in the COXI subunit gene can serve as another example (39, 40). It emerged that each intron-associated endonuclease was unique, with a different recognition site sufficiently complex to make it relatively infrequent in the genome. A plausible hypothesis proposed by Lambowitz (36) is that introns of group I (and group II) may initially have been self-splicing, with the involvement of a specific secondary structure. They can function only when inserted at sites where flanking exon sequences are compatible with splicing. ORFs were acquired later in evolution and adapted as endonucleases to facilitate transposition, or as maturases that assist in splicing by stabilizing the RNA in the catalytically active secondary structure. Endonuclease and maturase functions may even be combined in a single protein as suggested for the a14α intron.

The mechanism may be different for group II introns, and here speculation is driven by the finding that such introns may have ORFs encoding a protein related to retroviral reverse transcriptases. Mobility of the intron is therefore related to the RNA splicing reaction, where the RNA excision product could be reverse-transcribed to yield DNA that in turn is postulated to recombine with the intron-less gene (41). Alternative mechanisms proposed include a reversal of the splicing reaction into a compatible location of a different RNA, or even integration into a single-stranded DNA at a replication fork (36).

If introns can be mobilized during intraspecies crosses, the problem of whether and how introns may spread between species is still in the realm of speculation, fired by observations that some introns in different species show higher sequence similarity than the flanking sequences. As in all discussions about intron evolution, one may have to distinguish between early and late introns and establish clear phylogenetic relationships before horizontal intron transmission can be proved.

A second source of interspecific and intergeneric size variation has been determined from sequencing the regions between genes of several mt genomes of fungi including *S. cerevisiae* (32–34). They account for >60% of the *S. cerevisiae* genome and consist of sequences extremely rich in A + T (>90%), but there are also ~200 clusters of 20–50 bp which are rich in G + C. Circumstantial evidence (e.g., two nucleotide duplications in flanking positions) has been interpreted to mean that these G + C clusters are mobile by an as yet unknown mechanism. Transposition of such elements would also be consistent with the observed polymorphism of mt genomes with respect to these G+C clusters (42). Similar data from other fungi indicate that the phenomenon is widespread. A second element in *S. cerevisiae* mtDNA intergenic regions has been termed *ori* or *rep*, and there are still conflicting interpretations about its relationship to the mobile elements as opposed to an origin from mitochondrial promoters.

The expansion of (A+T)-rich intergenic sequences can be explained by several plausible mechanisms. Slippage during replication or DNA repair could create tandem repeats, and subsequently unequal crossing over could lead to further expansion or contraction. From experiments in the laboratory of Clark-Walker (reviewed in reference 31) it has been shown that relatively large deletions (3–5 kb) in intergenic regions have no consequences for mitochondrial gene expression and the ability of such strains to grow on nonfermentable substrates, and such regions may indeed be dispensable "junk" DNA.

An observation that was puzzling for a long time was that strains of *S. cerevisiae* produce spontaneous petite mutants at the high rate of approximately 1% per generation. The phenotype "petite" is caused by respiration deficiency that was shown to be due to deletions in the yeast mtDNA. It appears that these internal deletions are created by excisions occurring at directly repeated G+C clusters. This inference was made primarily from the comparison of the spontaneous generation of petites in *S. cerevisiae* and in *Candida glabrata*. The latter has no such G+C clusters to destabilize the mt genome, and as a result the spontaneous rate of petite formation in *C. glabrata* is four orders of magnitude lower (31).

Linkages between genes (e.g., COXI–atp8–atp6) have been preserved in many species, but evidence for rearrangements is abundant, and, most surprisingly, translocations have been found between interbreeding strains, calling into question the use of gene order in taxonomic classifications. Again direct repeats and recombination involving these repeats at distal sites may be responsible for the overall mechanism, and evidence for this "subgenomic pathway" has been accumulated by Clark-Walker (31). In these experiments, rearranged but complete genomes were deliberately produced, and no obvious deleterious effects on growth rates were observed. Such results could be explained if individual genes were expressed from individual promoters that are scrambled with the genes. As will be discussed later, there are multiple promoters on yeast mtDNA, and there are no recognized transcription termination sites, allowing constant levels of transcripts from rearranged genomes.

4.1.5 The Mitochondrial Genome in Kinetoplastid Protozoa

In organisms of the Protozoan order Kinetoplastida a particularly unusual structure of mtDNA is found (43). These "mitochondria" were detected by cytologists of the last century, but their unique morphology, cellular location, and assumed function caused them to be given a distinct name, kinetoplasts. The discovery of DNA in these organelles preceded the recognition that all mitochondria contained DNA, because kDNA constitutes about 7% of the total cellular DNA and hence is easily detectable by specific staining. Subsequently kinetoplasts were recognized to be highly specialized regions within the mitochondria which were colocalized with the basal body of the flagella of these highly motile unicellular organisms. Representatives of this group of organisms are the well-known human pathogens *Trypanosoma cruzi, Trypanosoma brucei*, and a related organism, *Leishmania tarentolae*, isolated from the gecko (Figure 4.4).

The kDNA exists of a network of ~40–50 maxicircles and 5000–10,000 minicircles that are topologically linked by catenation. There appears to be some hierarchical order in the catenation of clusters of minicircles, the catenation of these clusters, and the catenation of the maxicircles into the entire network. Basket-shaped structures have been recognized, and the packing within the mitochondrion is by stacking and alignment of DNA fibrils, most likely with the help of specific proteins bound at regular sites. A combination of the activities of topoisomerases and restriction enzymes is employed to maintain order and to segregate these structures after DNA replication. Numerous reviews on the topology, packaging, and fractionation of these DNA circles are available (44, 45).

The size range of the maxicircles extends from less than 20 kb in the African trypanosomes to about 35 kb in Leishmania, while the minicircles range from 645 pb in *T. lewisi* to 2500 bp in *Crithidia fasciculata*. Analysis of renaturation kinetics (C_0T analysis) with *T. brucei* DNA has indicated that kDNA has a total complexity of ~350 kb (the highest complexity observed among kinetoplastids), and this result immediately suggested that these circles must be quite heterogeneous. However, the gene content and even the gene order in the maxicircles in several species was shown to be identical. Size variation among species can be largely accounted for by the change in repeat number of a simple repeat in one variable region (45). Thus, heterogeneity must be sought among the minicircles, and indeed there are an estimated 400 different minicircles in *T. brucei* kDNA. The analysis is complicated by the fact that they are not found in equal copy numbers, and for two examples the copy numbers were found to be about 60 and 500, respectively. In *Leishmania tarentolae* the complexity is lower and 17 minicircles make up all of the total amount of minicircle DNA in the UC strain, whereas approximately 80 minicircle sequence classes are found in the LEM125 strain. Recombination may contribute to the generation of minicircles, and at this time it seems impossible to give precise descriptions of their distribution in *T. brucei*. Another complication in the interpretation of renaturation kinetics is that all the minicircles have one conserved sequence of ~120 bp and three conserved pairs of 18-bp inverted repeats.

The network of catenated maxi- and minicircles is highly sensitive to perturbations by intercalating molecules such as ethidium bromide or acridines.

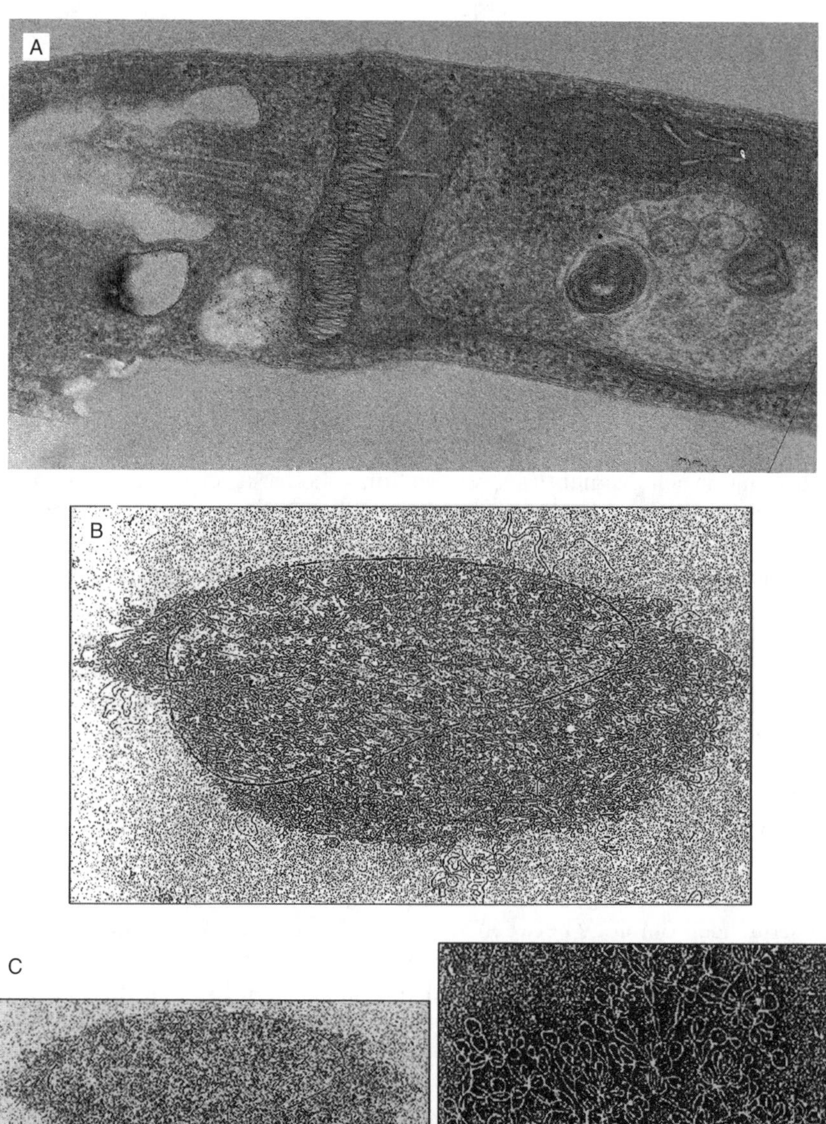

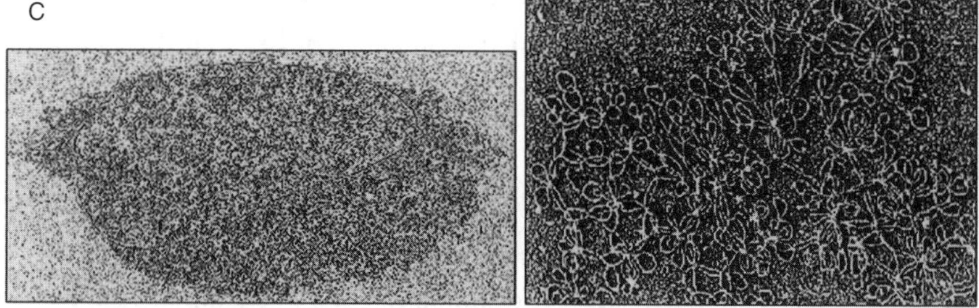

Figure 4.4 (A) Electron micrograph showing a thin section through *Leishmania tarentolae*. The basal body and axoneme extend to the left. (B) Kinetoplast DNA. The maxi circle and many minicircles appear associated in an aggregate. (C) Higher magnification view of many minicircles in various toplogical linkages. (Photographs provided by L. Simpson, UCLA.)

Treatment with such drugs causes loss of kDNA and ultimately the loss of the capability to carry out mitochondrial respiration. Since respiration is required at least during part of the life cycle of these parasites—for example, in the insect host for the African trypanosome—the usefulness of such drugs in treatment and prevention of these tropical diseases has been explored extensively. It is also obvious that the replication of this network of DNA is intimidating to contemplate.

By now the complete sequences of the maxicircles of several species have been determined, and the genes present have been identified by homology with genes on mtDNA, some nuclear genes, and even chloroplast DNA. The challenge was not trivial, because there is extensive RNA editing in these species. The discovery of this process, not as a relatively rare oddity but as a widespread phenomenon in the kinetoplasts, was one of the great surprises in molecular genetics in the 1980s, and the elucidation of its mechanism represents a major achievement (46; see Section 4.4.6 for a further discussion).

What genes were found? Not unexpectedly, kDNA maxicircles encode five subunits of NADH-CoQ reductase (complex I), cyt b of CoQ-cytochrome c reductase (complex III), three subunits of cytochrome oxidase (complex IV), one presumed subunit of ATPsynthase (complex V), and seven or eight unidentified reading frames. Two ribosomal RNA genes yield rRNAs of 1150 and 610 nt (12S and 9S, respectively), which are exceptionally small but still capable of forming secondary structures resembling the characteristic structures found in *E. coli* rRNAs. One would also expect to find tRNA genes, but RNA editing may obscure their identification by routine genome sequencing.

The maxicircles of some species have clearly been shown to encode some guide RNA (gRNA) genes, but many more gRNA genes can be found on the minicircles (see below).

Gene order is preserved between species where the analysis is complete, but sequence homology between species is recognizable (by computer and by cross-hybridization) only for genes whose transcripts are not heavily edited, while genes encoding heavily edited RNAs show little or no homology. A fertile field exists here for molecular taxonomists and evolutionary biologists, but the discussion exceeds the scope of this treatise.

The variable regions on the maxicircle between the ND5 gene and the 12S rRNA gene differ in size and sequence between species and even between stocks of *T. brucei*. It is unlikely to encode a functional protein, and speculation about its function centers around a role as replication origin.

The information or gene content of minicircles can be summarized only in very general terms, since their size, organization, and complexity are all variables among species. However, all minicircles have a conserved sequence (at about the 90% level) of ~120 nt. Within this sequence a 13-nt sequence is recognized as a small gap in replicating minicircles, possibly in association with a small RNA. Hence its function in DNA replication is plausible. The presence of several stop codons and the lack of a start codon argue against any coding function within this region. A second common feature of minicircles is a number of runs of adenines at regular intervals, which give rise to a "bend" or "kink" in the DNA and cause such a molecule to have an abnormal electrophoretic mobility compared to a normal DNA molecule of identical size. The position can only be related to the position of the conserved 120-nt sequence, and where it has been examined, it appears to be at no particular distance from this landmark in different species. The binding of a protein at this location has been hypothesized to be related to the stacking of the circles within the kinetoplast.

The major class of transcripts from the minicircles are the small (<100 nt) guide RNAs. In some species (e.g., African trypanosomes) the gRNAs are from within an ~100-bp region flanked by 18-nt inverted repeats, a so-called coding cassette, but such cassettes are not recognizable in *T. cruzi* or *Leishmania tarentolae*. The role played by these gRNA in RNA editing will be discussed in Section 4.4.6.

4.1.6 Mitochondrial Plasmids

In a previous section, reference has been made to mobile introns in the yeast mt genome. Additional genetic elements found in the mitochondria of numerous fungi, protozoa, and plants are small circular or linear plasmids. They have origins of DNA replication, replicate independently of the much larger mtDNA, and may even encode some of their own replication machinery such as DNA polymerase. Some of these may become integrated into the mt genome, disrupting genes and making their presence felt by causing a recognizable phenotype. A listing of over 50 such plasmids in various fungi and higher plants can be found in a recent review on sexuality in mitochondria by Kawano et al. (47). In some organisms such as *Neurospora* they may be the rule rather than the exception (48).

A number of these plasmids have been sequenced, but no significant insights about their biological function have become apparent. They range in size from ~1.5 to ~10 kb, with a few exceptions, the largest being the linear mF plasmid of *Physarum polycephalum* (~16 kb). They may be true molecular parasites adapted for autonomous replication and transmission with the mitochondrial genome. Thus, ORFs encoding DNA polymerase are commonly found, and RNA polymerases are also encoded by many plasmid genomes. When biological effects are observed, they may be incidental in most cases, since loss of the plasmid does not affect the organism's viability. The mF plasmid in *P. polycephalum* has been shown to encode a function required for developmentally regulated mitochondrial fusions during the life cycle of the slime mold, but strains missing the mF plasmid appear to be equally capable of sporulating or of forming a plasmodium.

The linear plasmids share several properties related to the mechanism of their replication, most notably the presence of terminal inverted repeats (TIRs) with proteins bound to their 5′ ends. A replication mechanism resembling that of linear DNA viruses has been proposed, and further speculation centers around their origin as bacteriophages that have degenerated in the course of evolution. Since they are stable in the absence of selective pressure and have a high copy number, their use as vectors in yeasts has been proposed (49). Transformation protocols using microprojectiles have been developed yielding transformants with useful efficiencies (50).

The replication of the circular plasmids has also been mostly speculated on, using information from bacterial systems as a guide. Replication intermediates described as θ and σ (51) have been looked for in a few plants, and a rolling circle type of mechanism was found for the mp1 plasmid in the higher plant *Chenopodium album* (L.) (52). In its most simplified formulation, the rolling circle mechanism of replication of circular genomes requires the following steps, first elucidated for bacteriophages such as X RFI (51): (1) Site-specific nicking within the origin creates a free 3′-OH group; (2) displacement of the 5′ end from the circular template by a helicase, and complex formation with single-strand binding proteins; (3) extension of the 3′-OH end by DNA polymerase displaces more 5′ single strand, and at this stage the molecule has the appearance of a "σ" when viewed by electron microscopy; (4) when continuous synthesis around the circular template has reached the origin, another cleavage of the displaced single strand creates a linear molecule which has to be circularized by ligation; and (5) discontinuous synthesis converts the single-stranded circle into another double-stranded circle. The cycle can repeat itself. In contrast, two replication forks moving in opposite directions on a circular DNA will create an intermediate with the appearance of a "θ" in the electron microscope.

Three significant biological effects have been associated with the presence of plasmids in mitochondria of fungi or plants:

4.1.6.1 Fungal Senescence Among the natural Hawaiian isolates of *Neurospora intermedia* the phenomenon of senescence was discovered, in contrast to the immortal nature of most laboratory and field-derived strains of *Neurospora* (see references 48 and 53 for reviews). A mitochondrial linear plasmid named kalilo (~9kb) was found to be responsible. In juvenile strains, kalilo DNA is harmless even at high copy number, but insertion into mtDNA appears to initiate the senescence process, and eventually almost all mtDNAs contain inserts at death. A systematic search uncovered a *Neurospora crassa* strain in India carrying a linear plasmid (maranhar) which also causes senescence by inserting into mtDNA, but kalilo DNA and maranhar DNA do not share any sequence homology. The investigation of 171 natural isolates of *N. crassa* and *N. intermedia* yielded another 28 senescing strains. Most strains, senescent or nonsenescent, carried plasmids; the plasmids were linear or circular; some strains had more than one plasmid, even mixtures of linear and crcular plasmids. A cross-homology study established sequence relatedness between plasmids from geographically distant origins, but a generalizable relationship of plasmids or sequences to senescence has not yet emerged. Thus, the current idea is that the plasmids act as insertional mutagens causing an abnormal mitochondrial physiology. However, the defect is not the result of an accumulation of random insertions at different sites; instead, a single mtDNA with a specific insertion predominates in a senescent culture.

The mechanism for the insertion may depend on the plasmid. Frequently the plasmid suffers short deletions, or base changes when inserted, and host sequences (mtDNA) may be duplicated in the form of long inverted repeats (48).

It should be pointed out that fungal senescence can also occur in the absence of any recognizable exogenous plasmid DNA as an inherent property of mitochondrial DNA itself (53). This phenomenon will be discussed in Section 7.6.

4.1.6.2 Phytopathogenicity As briefly summarized by Meinhardt et al. (49), the phytopathogenicity of fungi has been examined with the hope of correlating virulence and host range with the presence of specific plasmids. Some promising results have appeared, but far-ranging conclusions are not yet possible.

4.1.6.3 Cytoplasmic Male Sterility (CMS) Different wild and cultivars of maize (*Zea mays*) mitochondria harbor a variety of plasmids: A linear 2.3-kb plasmid is found in all maize plants, while linear S-plasmids (S_1, S_2) and R-plasmids (R_1, R_2) distinguish different maize lines. Cytoplasmic male sterility (CMS) appears to be the consequence of the integration of the S-plasmids into the mitochondrial genome, linearizing it in the process. The presence of the free plasmids (S or R) is not associated with CMS.

CMS is a common trait of higher plants characterized by abnormal pollen development. Maternal inheritance suggests in general a mitochondrial mutation. Such mutations may be created by plasmid integration, or they may result from aberrant recombination events involving the multiple circular mt genomes present in plant mitochondria (54). The situation is complicated by the possibility of suppressing male sterility and restoring normal pollen development when specific nuclear genes (*Rf*—restorers of fertility) are present. Molecular mechanisms for these phenomena are still elusive. The analysis is made especially challenging by the large size of plant mt genomes. CMS as a result of chimeric genes created by rare recombination events will be taken up again in Chapter 7 (Section 7.7).

4.2 NUCLEAR GENES ENCODING MITOCHONDRIAL PROTEINS

The number of different polypeptides in each mitochondrion had been estimated to be greater than one thousand (55), and more recent proteomic analyses have identified 500–800 proteins, with more suspected, but beyond the limit of the analysis (56–61). It has also become very clear that the composition and relative abundance is highly tissue specific in mammals. This means that at least 1000 nuclear genes contribute to the biogenesis and function of mitochondria. Many of these are known, but many remain to be identified and characterized. Attempts have been made to identify mitochondrial proteins from the sequences of complete genomes, based on predicted presequences for import into the organelle (http://www.123genomics.com/files/analysis.html). Success has been limited, because there is no good consensus N-terminal sequence for targeting, and many proteins have internal targeting signals that are poorly characterized. Logically, one can subdivide this large group of genes and their proteins into two major groups: (1) genes necessary for the biogenesis of normal, fully functional mitochondria and (2) genes encoding proteins that play a role in the various specialized biochemical activities of mitochondria. It is not the intent to list these exhaustively here, but the major categories and a few examples will exemplify the two groupings. They will also provide a logical starting point for the discussion of several aspects of the biogenesis of mitochondria which have not yet been presented, and in a later chapter they will lead into the discussion of the biochemical reactions localized in these organelles.

4.2.1 Enzymes Required for Maintenance and Expression of the Mitochondrial Genome

In past chapters many references have already been made to enzymes required for the replication of the mitochondrial genome, its transcription, the processing of the primary transcripts, and the translation of mature mRNAs. The major DNA polymerase belongs to the polymerase γ family (62–64). Primers for replication are made by an RNA primase that may also function as a helicase (65), and one can anticipate the need for a ligase to make covalently closed, supercoiled circles. Mitochondrial topoisomerases have also been described (66, 67). A list of enzymes involved in DNA repair and recombination has been accumulating (68–71), and the relative importance of these activities is variable in different organisms. In this context one might mention that there are no histones in mitochondria, although mitochondrial DNA appears to be complexed to a protein that resembles the bacterial HU protein or the eukaryotic nuclear HMG1 proteins (72, 73). For transcription, an RNA polymerase and at least one of two transcription factors or activators is required (65, 74, 75). Many of these proteins are associated with mtDNA forming structures described as nucleods (75–77). The number of activities associated with RNA processing, splicing, and editing depends on the species, although several types of nucleases for the cleavage of polycistronic RNA and the maturation of tRNAs are likely to be found in all mitochondria. Polyadenylation of RNA transcripts in metazoan mitochondria requires at least one enzyme.

The translation of mitochondrial mRNAs will be discussed more extensively in Section 4.5. Ribosomal RNAs are universally encoded by mtDNAs, as are some, but not all, of the tRNAs required. Similarly, in metazoan mitochondria all the ribosomal proteins are imported (and hence nuclear-encoded), while some organisms (notably plants and fungi) have some of their r-proteins still encoded by mtDNA. Specific as well as nonspecific translation initiation factors and presumably nonspecific elongation factors will be discussed in Section 4.5.5; all are encoded by nuclear genes. The aminoacyl-tRNA activating enzymes are also all imported.

The complicated nature of mitochondrial gene expression may be illustrated with the finding that 18 nuclear genes have been shown to be required for the expression and maturation of the yeast mitochondrial cytochrome c oxidase subunit 1 (78). In general, one would expect a mutation in any of these nuclear genes, especially a deletion or null mutation, to have serious consequences for the biological function of mitochondria. Such mitochondria would be incapable of respiration, but may continue to perform other important functions such as heme synthesis, lipid synthesis, amino acid metabolism, and biosynthesis of Fe–S clusters.

DNA replication is not absolutely required for the biogenesis, since mtDNA-less (ρ^0) mitochondria have been described for yeast and mammalian cells. Similarly, mitochondrial protein synthesis can be completely abolished by mutations, without affecting the continued assembly of this organelle in mammalian cells (79, 80). Therefore, replication, transcription, and translation seem to serve exclusively the purpose of maintaining and assembling a functional electron transport chain and oxidative phosphorylation mechanism, a vital but by no means the only function of mitochondria.

The contribution of large number of genes (and proteins) is required for the import of the vast majority of mitochondrial proteins into the matrix, into the two membranes, and into the intermembrane space. This subject will be dealt with in Section 4.6, where the function of the various proteins will be explored in detail. In broad categories one can list the proteins of the import machinery in the outer and inner membranes (receptors, transporters/channels), proteases/peptidases responsible for protein maturation, and mitochondrial heat-shock-like proteins (chaperones) that assist in the folding and assembly of the imported peptides. Cytosolic chaperones are equally essential, although they do not belong exclusively to mitochondria. It is easily understood that a serious defect in any of these major factors required for protein import will not only eliminate respiration and oxidative phosphorylation, but will affect mitochondrial biogenesis in general and may therefore be lethal.

Many, if not most, of these proteins have been discovered by a genetic approach, and for obvious reasons the majority of the corresponding genes have been identified in the yeast *Saccharomyces cerevisiae*. This yeast is a facultative anaerobe and hence is capable of growth without respiration in the presence of abundant glucose for glycolysis. A variety of clever and elegant screens have been developed for the identification of yeast mutants with defects in mitochondrial biogenesis. Historically, the first mutants were the so-called nuclear *petite* or *pet* mutants. *Petite* (small) colonies arise when a respiration-deficient yeast mutant is grown on glucose: Normal colonies will continue to grow even after the glucose is exhausted, because they are capable of metabolizing the ethanol accumulated during the first phase of growth. The first *petites* discovered had "cytoplasmic" mutations, later identified as large deletions or complete loss of mtDNA. Nuclear petites have the same phenotype due to a nuclear mutation affecting respiration. They can be most readily distinguished with the help of a tester strain lacking mtDNA (ρ^0). After fusion and formation of diploid cells, the diploid cells will grow in the absence of a fermentable carbon source if the mutation is nuclear and recessive.

The selection of mutants after mutagenesis makes use of replica-plating techniques. It can also be combined with selections at permissive and nonpermissive temperatures to obtain conditional mutants. Additional wrinkles in the design of selections have been described, but the reader is referred to the original literature for details. Once a *pet* mutant has been identified, a number of methods are available to establish some basic facts: Is it a new mutation, and what kind of mutation is it? Complementation analysis with existing, well-characterized mutant panels is one option, but it may be quite laborious, since several hundred complementation groups have already been identified. Cloning of the complementing gene from a yeast DNA library is probably a more efficient means of obtaining a sequence which can be checked against the existing yeast genome data base (81). An expert and comprehensive review of the *PET* genes of *Saccharomyces cerevisiae* has been written by Tzagoloff and Dieckmann (82). Not surprisingly, the collection of pet mutants includes many mutations in nuclear genes encoding subunits of the mitochondrial electron transport chain. The activity of specific complexes is affected. Less expected was the finding of mutants with defects in a specific complex, but the defective gene was not a structural gene for a known subunit. Instead, further analysis has revealed that the assembly of the various complexes requires additional proteins. Only a few examples will be mentioned here, with additional details to be presented in the discussion of the assembly, structure and function of each complex in Chapter 5. *COX14* is a gene whose product, Cox14p, is required for the assembly of cytochrome oxidase (complex IV) (83). A newly discovered member of a family of ATPases is required for the assembly of several complexes including complex V (84).

Many *pet* mutants have a pleiotropic phenotype, and are often found to be defective in mitochondrial protein synthesis. Others are defective in heme synthesis, lipoic acid synthesis, coenzyme Q synthesis, or in enzymes of the Krebs cycle.

The first mutants specifically isolated to be defective in mitochondrial protein import were described by Yaffe and Schatz (85), and many more have been identified in the intervening years (86–88). A detailed discussion is presented in Section 4.6.

Finally, proteins and their nuclear genes required for the many biochemical reactions in mitochondria are too numerous to mention. Every student of Biochemistry is introduced to the enzymes of the tricarboxylic acid cycle (Krebs cycle), and a defect in one of these will also lead to a respiration-deficient phenotype. The Krebs cycle reactions are also responsible for the interconversion of various four-carbon and five-carbon molecules which by trans-amination yield amino acids such as aspartate and glutamate. The urea cycle, fatty acid metabolism, and a number of other biochemical pathways require many enzymes that have to be imported into mitochondria. Defects in any of the latter enzymes may have serious consequences at the organismic level, but may not be recognizable as mitochondrial defects in the sense of perturbing the biosynthesis and morphology of this organelle.

There will undoubtedly be many more genes and gene products not mentioned in the brief survey so far. They may be associated with processes already mentioned, or may be relevant for the generally less familiar pathways in mitochondria—for example, the biosynthesis of cardiolipin. A number of the more interesting ones will be discussed under specific headings such as the mitochondrial permeability transition and apoptosis.

4.2.2 Nucleo-mitochondrial Interactions

4.2.2.1 Introduction The existence of two spatially separated genomes, each contributing, however asymmetrically, to the biogenesis of mitochondria, has raised for a long time already questions about "interactions" between the two genomes. The focus of these questions can vary. One may wonder how the number of mitochondria in a cell is maintained over many cell divisions, or how it is possibly reduced or expanded, as appears to be the case in oogenesis. Is mitochondrial DNA replication in any way coupled to nuclear DNA replication to assure a more or less constant ratio between the two genomes? How is the composition in different cell types altered? Most obviously, there are differences in the morphology of the cristae or in the relative surface area of the inner mitochondrial membrane, which may reflect differences in respiratory activity in different tissues. It may be conceptually easy to understand that one can express the enzymes for the urea cycle at different levels in different tissues and hence have differing specific activities in mitochondria. On the other hand, to down-regulate or up-regulate the activity of, for example, cytochrome oxidase requires not only a coordinate regulation of at least 13 structural genes in mammals, but three of these are on the mt genome, while the other 10 are in the nucleus. If one wishes to regulate oxidative phosphorylation and electron transport, should all complexes be up- or down-regulated; that is, are the nuclear genes encoding subunits for complexes I–V coordinately expressed? Or is there a rate-limiting complex (a "bottleneck" in electron transport) whose level determines the rate for the overall pathway? The mitochondrial genome is transcribed into polycistronic transcripts, but the translation of individual processed mRNAs may still be controlled separately to provide subunits at different levels. It is also conceivable that there is no stringent coordination of the synthesis of different subunits of a complex, but one critical subunit is made in controlled, limiting amounts, assembly takes place, and all other excess subunits are turned over rapidly.

It has already been mentioned that a crucial factor in the replication and expression of the mtDNA in mammals is the mitochondrial transcription factor, mtTFA (89), encoded by a nuclear gene (see Section 4.4.1). Regulating the availability of this factor might be sufficient to control the expression of mitochondrial genes, and if the expression of mtTFA were coordinately regulated with the expression of other nuclear "mitochondrial" genes, a simple mechanism suggests itself.

A number of different approaches to answer questions related to nucleo-mitochondrial interactions have been taken. Again, yeast has led the way with the exploitation of powerful genetic techniques. In mammals and plants the number of relevant mutants is quite limited, even with cell cultures, and alternate means must be sought. The two systems will therefore be discussed separately.

***4.2.2.2 In Yeast,* Saccharomyces cerevisiae** Yeast is a good model system not only because of the advanced state of genetics of this system, but also because yeast cells can be experimentally manipulated in a way that directly addresses the regulation of the biogenesis of mitochondria. When grown in a medium in which glucose is abundant (YPD (dextrose) medium), the synthesis of the electron transport chain is almost completely suppressed, and the cells use fermentation (glucose → ethanol) exclusively to satisfy their energy needs. The morphology of the residual mitochondria (with normal levels of mtDNA) is quite abnormal, making the mitochondria barely recognizable by the non-expert. Almost immediately after a switch to a nonfermentable carbon source such as glycerol, ethanol, or acetate, there is a rapid induction of the transcription of many genes, including those encoding subunits of the electron transport chain complexes. The phenomenon is referred to as "glucose repression." In general, it also includes the repression of nonmitochondrial enzymes involved in the uptake and utilization of carbon sources, of which invertase (*SUC2*)[1] is a prominent example, since this gene has been used extensively in the genetic analysis of glucose repression, and results obtained with this gene have been extrapolated to other genes. A number of expert reviews have been written on this subject (90–93). A possible physiological explanation is that it is more economical for the yeast cell to "waste" ethanol and not invest extensively into the biogenesis of the inner mitochondrial membrane. It may also be that growth in the absence of respiration is preferred under these conditions because it saves the cells from the potential onslaught of reactive oxygen species. However, in the course of evolution, yeast has learned to adapt rapidly to changing conditions; and when transferred from rotting fruit to a much more dilute aqueous environment, a switch to the more efficient use of a limiting carbon source is expected to be advantageous. Another example of chang-

[1] Following the convention in yeast molecular genetics, genes will e designated by italicized, capital letters (e.g., *SDH2*), while their respective peptides or protein will carry the same name with only the first letter capitalized, followed by a p (eg. Sdh2p).

ing environmental conditions encountered by such microorganisms is a change in oxygen availability. Thus, oxygen deprivation also results in changes in gene expression including those encoding mitochondrial complexes (see reference 94 for a review).

Until recently, all responses of yeast to glucose levels or oxygen pressure were interpreted in terms of transcriptional regulation. A signaling pathway for glucose repression can be decomposed into the following major subdivisions: (1) Glucose must be taken up by the cell, and it or a metabolite must be sensed by a key protein in the pathway; (2) the signal must be transmitted through the cytoplasm via a series of protein–protein interactions (which may involve protein kinases and phosphatases), or by second messengers such as cAMP, Ca^{2+}, or inositol phosphate derivatives; and (3) final targets of the signal transduction pathway are nuclear transcriptional activators or repressors that regulate the transcription of specific genes. Therefore, among the major goals of research in this area have been the characterization of the postulated specific transcription factors and the identification of the corresponding cis-acting sequences [upstream activating sequence (UAS) and upstream repressor sequence (URS)] of the genes in question. There may be more than one UAS, and they may be distinguished as UAS1A and UAS1B, for example. A more explicit description of the transcriptional activators or repressors and their interaction with the basic transcriptional machinery (RNA polymerase, TATA-binding protein, TFIID, and initiation factor TFIIB) will not be attempted here. Broadly speaking, an activator/repressor protein has (a) a DNA-binding domain for specific interactions with target DNA sequences and (b) another domain for interaction with the transcriptional initiation complex. In some cases these domains are on the same peptide, while in other cases different subunits of a complex contribute these distinct domains. There may also be another domain serving a regulatory function (see below). Pioneering research on the regulation of the cytochrome c gene in yeast (*CYC1*) was carried out by the Guarente's group (95, 96), and the cumulative conclusions of the Guarente laboratory can illustrate some general principles. First, the *CYC1* promoter region was found to contain several CCAAT boxes that were identified to be the target of the Hap2/3/4 complex. This heterotrimeric complex is a transcriptional activator whose activity is low in glucose-rich medium and high in glycerol-containing medium. It is one of the targets in the signal transduction pathway initiated by the uptake and phosphorylation of glucose. Glucose not only controls its activity (via phosphorylation or dephosphorylation), but also controls the transcription of some of the *HAP* genes themselves. Not all CCAAT boxes are equal; and in the *CYC1* promoter one of the distal, upstream CCAAT boxes is the target of the Hap1 transcription factor whose activity is controlled by oxygen, via heme as an intracellular oxygen sensor (94, 96–98).

It is not yet clear whether all genes encoding subunits of respiratory complexes are regulated by Hap1p and or the Hap2/3/4 complex, but the required CCAAT boxes have been identified in the *SDH2* gene of complex II (99), and the *SDH3* gene was cloned specifically on the expectation that it was transcriptionally regulated by the Hap2/3/4 complex and glucose (100). Cyb2p is a lactate cytochrome c oxido-reductase localized in the intermembrane space. Expression of *CYB2* is repressed by glucose and dependent on heme. Its promoter has been characterized by Lodi and Guiard (101) and shown to have two positively acting cis elements and one negatively acting cis element. One of these binds the Hap1 (Cyp1) activator, and the Hap2/3/4 complex is another regulatory factor for this gene. Not all the genes studied to date are regulated by both Hap1p and the Hap2/3/4 complex. Among those subject to both are *CYC1, CYB2, CYT1, QCR2*, and *COX6*, all genes encoding peptides in the electron transport chain complexes and therefore useful in the presence of oxygen and required with a nonfermentable substrate. Several genes for enzymes/peptides of the Krebs cycle are regulated by the Hap2/3/4 complex alone (the target of glucose signaling), while another subset of enzymes needed for heme, sterol, or fatty acid biosynthesis are regulated by Hap1p alone (see reference 98 for specific references).

Hap1p, in addition to DNA-binding and transactivation domains, also has a domain for heme or metal binding. It is speculated that the redox state of the cell, depending on the level of oxygen, is sensed via the heme, and heme binding unmasks the DNA-binding domain. The activator is therefore active only in the presence of oxygen. The activity of Hap1p may be further modulated by the specific DNA sequence to which it binds; that is, its activity is found to be promoter-dependent. One hypothesis to explain the behavior of Hap1p postulates allosteric interactions between the various domains. Since Hap1p also controls heme biosynthesis, the possibility exists for the establishment of complex feedback loops and very intricate fine-tuning of gene expression to optimize activity to a given environment.

The Hap2, Hap3, and Hap4 peptides form a heterotrimer in which Hap2 and Hap3 domains control assembly and interaction with the 5′-ACCAATNA-3′ sequence, and Hap4p contains the transactivation domain. The assembly and/or activity may be controlled by a kinase (Snf1p), but the transcription of the *HAP2* and *HAP4* genes is repressed by glucose. Thus both the level and the activity of the complex are under control by the carbon source, and perhaps even by oxygen, since it has been suggested that heme may influence the translation of the *HAP4* mRNA.

Another gene with a prominent role in the control of the expression of respiratory chain peptides and related functions is the *ROX1* gene. The transcription of this gene is dependent on heme, and the Rox1 protein acts as a transcriptional repressor of a variety of genes for mitochondrial proteins. A challenging problem in yeast has been the finding of isoforms of proteins encoded by gene pairs. Examples are *COX5A* and *COX5B, CYC1* and *CYC7*, and others (98). Their expression is alternate; that is, when one is expressed under aerobic conditions, the other is expressed under anaerobic or hypoxic conditions. The preferential expression one or the other cytochrome c or cytochrome oxidase subunit may again reflect an optimization and adaptation to a variety of conditions, to transient conditions in shifting from respiration to fermentation, and to different levels of availability of heme. *HAP1* and *ROX1* may therefore play antagonistic roles in the precise control of the ratios of these isoforms.

Although the number of genes for mitochondrial proteins which have been explicitly examined is limited, one could probably draw the conclusion with some confidence that many, if not most, of these will have promoter elements recognized by either the Hap1 peptide or the Hap2/3/4 complex, and therefore a basis for the coordinate expression of many genes under conditions of glucose deprivation is emerging. Additional factors and their DNA targets, such as Rox1p, may be discovered, and they may be required for fine-tuning of gene expression and precise adjustments to the available carbon source, oxygen pressure, and other conditions. By regulating Hap1p and Hap2/3/4p activities, mitochondrial biogenesis can be controlled as a whole. When the availability of glucose is the only external factor, it remains to describe the cascade of signals initiated by this metabolite and interpreted eventually by nuclear factors. A large volume of publications have addressed this problem, and only a bare outline of the upstream signaling pathway can be presented in the present context.

There is agreement that glucose has to be phosphorylated, but uncertainty begins already in the distinction between a signal from glucose-6-phosphate or one of its metabolites, or a signal via an allosteric mechanism involving the hexokinase. Glucose addition to yeast has also been shown to activate the RAS-adenylate cyclase pathway (e.g., reference 102), but current thinking does not include this pathway in the phenomenon of glucose repression. Genetic analysis has quite definitively implicated the product of the *REG1* gene in the pathway, and Reg1p has recently been identified as one of possibly several regulatory proteins for the major protein phosphatase in yeast encoded by the *GLC7* gene (103). A protein kinase definitely implicated in the pathway is the Snf1 kinase (92, 104). It is not yet clear, however, whether the Snf1 kinase phosphorylates transcriptional activators such as Hap2/3/4p directly, or whether there are intervening steps. Similarly, the target in this pathway of the Glc7 phosphatase has not been determined. Recent reviews on glucose repression should be consulted on the latest developments in this active field (90–92, 105).

Until recently, most attention was devoted to understanding how gene expression is regulated by the carbon source at the transcriptional level. A new dimension to the analysis of this phenomenon was added by studies on the steady-state levels of the mRNAs for the iron–sulfur subunit (Ip) and the flavoprotein subunit (Fp) of complex II in yeast. Transcriptional regulation plays a role, but an equally important mechanism was found to be the control of the half-life of these transcripts by the carbon source (99, 105, 106). In medium with a nonfermentable carbon source the mRNAs have long half-lives, contributing to the high steady-state levels of these mRNAs and therefore to the rapid synthesis of the SDH peptides. In the presence of abundant glucose the half-life of these mRNAs is very short (<5 min), and the addition of glucose to an induced culture causes an extremely rapid turnover of these specific mRNAs. As a result, steady-state levels are maintained at a very low level in glucose by a combination of repression of transcription and rapid turnover of the transcripts. From another point of view, a three- to fourfold induction of transcription and a stabilization of the message in glycerol leads to a 12- to 20-fold induction at steady state.

The mechanism of this very specific destabilization of a subset of mRNAs in glucose has been partially elucidated (105, 106). The important cis-acting element in the Ip-mRNA is the 5′ untranslated region. The addition of glucose triggers a very rapid decapping of this mRNA, followed by an equally rapid degradation by the 5′–3′ exonuclease (Xrn1p). Since these activities take place in the cytosol, it is not surprising that various nuclear factors such as Hap2/3/4p are not required. More surprising was the finding that the Snf1 kinase was also not required for this glucose-triggered degradation mechanism. On the other hand, a phosphorylation/dephosphorylation step or cascade is indicated by the requirement for the Reg1 protein, a phosphatase regulator.

Theoretical considerations and observations with a temperature-sensitive eIF3 mutant have focused attention on a competition between different events at the 5′ end of the mRNA (106). On the one hand, initiation factors (eIFs), the 40S small ribosome, and F-met-tRNA will bind the 5′ UTR at the cap and then translocate to the start codon where the 60S ribosomal subunit is added and peptide elongation begins. At the same time a decapping activity is present (107), aiming for the same 5′ end of the transcript. Once decapped, the mRNA is not only no longer a good target for the initiation factors, but is also a good substrate for the constitutively active 5′ exonuclease. A working hypothesis therefore proposes that glucose influences the outcome of this competition in favor of the degradative pathway. These observations fit into a more general picture of mRNA turnover developed over the past few years (108–110). Constitutively short-lived mRNAs are susceptible to a rapid shortening of their poly(A) tails, which precedes the decapping. The rate-controlling step appears to be deadenylation, which in turn is controlled by a specific nuclease (111). The activity of this nuclease is regulated by its interaction with other factors binding to sequences within the mRNA, which are responsible for the specificity of this process. Recent experiments have emphasized that the 5′ end and the 3′ end (poly(A) tail) of mRNAs interact, most likely indirectly via proteins such as the poly(A)-binding protein, eIFs, ribosomal proteins, and other factors yet to be characterized (112–114). This interaction may have a major influence on either (a) the stability of an mRNA or (b) the efficiency with which it is translated.

The signal generated by glucose could therefore have as its cytoplasmic target a factor influencing the deadenylation reaction. Alternatively, it may directly modify one of the initiation factors and hence affect initiation, and hence decapping, without an obligatory deadenylation. A precedent for mRNA turnover in the absence of deadenylation is the turnover of mRNAs with premature stop codons by the UPF-mediated pathway (114).

Post-transcriptional mechanisms as well as transcriptional regulation must be considered in yeast when the regulation of respiratory capacity and assembly of the oxidative phosphorylation system in yeast mitochondria is under discussion. One can speculate whether such a multitude of mechanisms is required for fine-tuning, for establishing the overall range of the activities involved, or for a rapid response to rapidly changing environmental conditions. It will be important to keep these considerations in mind when similar questions are raised about various mammalian cell types in intact organisms.

4.2.2.3 Regulation of Nuclear Respiratory Genes in Mammalian Cells The general considerations and the observations made with yeast have encouraged hypotheses and experiments aimed at similar goals for mammalian cells: (a) characterization of transcription factors and their targets on the relevant promoters, which might explain the coordinate expression of the large number of nuclear genes encoding mitochondrial proteins, and (b) the nuclear–mitochondrial interactions, which are assumed to take place to make nuclear gene expression match mitochondrial gene expression or vice versa. A genetic approach is technically and logistically difficult, if not impossible, with diploid mammalian cells, and no conditions are known in tissue culture which would induce or repress the biogenesis of mitochondria to permit willful manipulation of the experimental system. *A priori* one would not expect mammalian cells in culture to be able to respond like yeast to changing carbon sources, since they have become adapted in evolution to a very stable environment *in vivo* and hence have no need regulate their mitochondria over a short term. Experiments with Chinese hamster fibroblasts have shown that a large fraction >50% of the energy needs of these cells is satisfied by glycolysis (115); and even when glucose is deliberately limited in culture, there is no induction of transcripts of complex II, for example (80), and only a modest increase in respiration. On the other hand, it has become apparent that different tissues and cell types may have widely differing energy needs and hence may have to adjust their level of mitochondrial respiration and oxidative phosphorylation to these needs as part of the process of differentiation. From the study of mitochondrial diseases and the discovery of the differential sensitivity of different tissues to mutations in mtDNA, neurons and muscle cells were deduced to be the most active tissues, as already anticipated from their physiological functions. Cardiac muscle is a specific example of a tissue deriving most of its ATP from respiration; and not surprisingly, cardiac myopathies are prominent among the clinical manifestations associated with mutations in mtDNA.

To repeat, one may not expect to find rapid, short-term fluctuations in mitochondrial biogenesis in a given cell type, but one can anticipate that the expression of a large number of genes encoding respiratory and other mitochondrial proteins may be expressed at a higher level in some cell types compared to others, and thus the existence of common regulatory molecules may be the simplest mechanism for a coordinate expression.

The number of mammalian genes investigated fully is still small. For practical reasons, cDNA clones are the first to be isolated; and the step to the full genomic clone, or at least the promoter region, is nowadays made possible by the availability of whole genome sequences. From the examples investigated, one may tentatively conclude that there is no single universal cis-acting sequence element with its associated transcriptional activator which is common to all such genes. However, at least two factors have been characterized which play key roles in the regulation of entire subsets of genes. Which genes belong to the set, and why, is not yet clear.

Evans and Scarpulla (116, 117) were the first to define a factor, designated as NRF-1 (nuclear respiratory factor 1), from their analysis of the promoters of the mammalian (rat) cytochrome c and cytochrome oxidase subunit Vb (COXVb). Sequence analysis of the cytochrome c promoter revealed the presence of several consensus sites for well-known transcription factors such as Sp1, CCAAT, and the cAMP response element binding protein CREB. In addition, a new cis-acting element was defined as a palindromic motif consisting of tandemly repeated GC sequences over a single helical turn that interacts with the NRF-1. The case was strengthened when over the years several other nuclear genes encoding mitochondrial proteins were found to have the same element (COXVb, COXVIIaL, COXVIc, mtTFA, a ubiquinone-binding protein, a mtRNA processing enzyme, aminolevulinate synthase) (118), prompting Scarpulla and his colleagues to purify the human protein to near homogeneity from cultured cells. Partial peptide sequences were used to synthesize degenerate primers, which in turn led to the isolation of full-length cDNA for NRF-1 and the production of authenticated recombinant protein in bacteria (119, 120). NRF-1 is a single peptide with a novel DNA-binding domain (related to DNA-binding domains in two developmental regulatory proteins found in *Drosophila* and the sea urchin) and a transactivating domain. The finding of an NRF-1 site in the promoter for the mitochondrial transcriptional activator (mtTFA) is of special interest. Since mtTFA is required for both replication and transcription of mtDNA (see Section 4.4.1), one can begin to build a case for a role of NRF-1 in the nuclear control of mtDNA copy number and gene expression. Aminolevulinate synthase is also not an arbitrary, irrelevant enzyme for this discussion. It is the first and rate-limiting enzyme in the pathway of heme synthesis in the matrix. Although a role for heme in the control of expression of mammalian cytochromes is not yet as well-defined as in yeast, the connection is tantalizing.

When several other promoters of respiratory genes (e.g., COXIV) either failed to show an NRF-1-binding site or could not be footprinted with pure NRF-1, the search for additional recognition sites and factors by Virbasius and Scarpulla succeeded in the identification of a second factor termed NRF-2 (121, 122). The core GGAA motif was somewhat familiar from the previous characterization of the ETS-domain family of transcription factors, but when the NRF-2 site from COXIV was used in an affinity column, a novel DNA binding activity could be purified (123). The DNA-binding activity co-purified with five distinct polypeptides (α, β_1, β_2, γ_1, γ_2). The α subunit was subsequently identified as the DNA-binding peptide, but it is found as a heteromeric complex with one of the other peptides ($\alpha\beta_1$, $\alpha\beta_2$, $\alpha\gamma_1$, $\alpha\gamma_2$). These complexes differ in their affinity for the NRF-2 site and to tandem NRF-2 sites, a characteristic that had already been observed for the related ETS-domain transcription factor GABP of the mouse. Further cloning and sequencing of all five human NRF-2 subunits established NRF-2 as the homologue of the mouse GABP, a widely expressed transcription factor first defined by the study of viral promoters. Consensus sequences for the NRFs are the following: NRF-1: . . . YGCGC**A**Y**GCG**C**R** . . . ; NRF-2: . . . ACC**GGAA**GAG. . . .

Reminiscent of the Hap2/3/4 complex in yeast, NRF-2 has DNA-binding domains and transactivation domains on separate peptides. In addition, the affinity of some of the heterodimers for a promoter can be further enhanced by cooperative binding to tandem NRF-2 motifs, and overall a mechanism is created to modulate promoter strength for different genes.

It was also first found by Scarpulla and his co-workers that some genes have both NRF-1 and NRF-2 motifs that contribute to the control of expression of these genes (124). Another example was found in the promoter for the iron–sulfur subunit of succinate dehydrogenase (and complex II) (125, 126). In this promoter the two sites are within 200 nucleotides upstream from the transcription start site, and mutagenesis of either site reduced the expression of a reporter gene from this promoter 5- to 10-fold. Both NRF-1 and NRF-2 were shown to footprint the respective sites independently. Other promoter elements were found further upstream (Sp1, CCAAT), but deleting them had no significant effect on the expression of a reporter gene in fibroblasts, myoblasts, or differentiated myotubes. For a more detailed and expert review the reader should consult (127, 128). A highly reduced summary of many studies in mammals is that environmental conditions influence the expression of PGC-1 family coactivators (PGC-1alpha, PGC-1beta, and PRC), which, in turn, target specific transcription factors (NRF-1, NRF-2, and ERR alpha) in the expression of respiratory genes (128a). The notion of having a limited, select number of transcription factors and promoter sequences exclusively devoted to the transcription of respiratory genes and mitochondrial proteins is apparently too simple. Several such genes have been found to lack NRF-1 or NRF-2 sites altogether (based on sequence inspection), while other genes unrelated to mitochondrial function contain such sites in their promoter region.

Mammalian cytochrome oxidase (COX, complex IV) consists of 13 subunits, 10 of which are encoded in the nucleus. It has been intriguing that three of the nuclear-encoded subunits (VIa, VIIa, and VIII) exist as multiple isoforms that are differentially expressed in different tissues. The L-form is named after its identification in liver, and this form is also found in many other tissues. The H-form is found predominantly in skeletal and cardiac muscle (heart). A detailed analysis of the COXVIaH promoter region with a reporter construct in transgenic mice as well as in tissue culture has verified the highly specific elevation of expression in heart and muscle, with some activity in the brain but virtually none in other tissues. Expression of this gene was suppressed in myoblasts, but dramatically induced when these myoblasts were induced to differentiate into myotubes. A 300-bp promoter segment was entirely sufficient to mimic the expression of this gene and its dependence on muscle differentiation. No NRF-1, NRF-2, or other element previously associated with respiratory genes (see above) was found in this region. Instead, an MEF2 consensus sequence and an E-box element were shown to be essential by mutational analysis (129). The muscle enhancer factor (MEF) comprises a family of four genes induced by myogenic HLH proteins (myo-D, myogenin, Myf5, and MRF4), and MEF2 has been demonstrated to be significantly involved in muscle differentiation (see references 130 and 131 for summaries and examples). Thus, the activation of the COXVIaH gene during muscle cell differentiation is understandable from a consideration of its promoter, but the promoter does not betray that the protein functions in mitochondria.

A very similar analysis was made with the SDH2 promoter (iron–sulfur subunit of complex II) (126). Transcripts of this gene are also significantly elevated in adult skeletal and cardiac muscle. In contrast to COXVIaH, however, the expression of the SDH2 gene was not up-regulated upon myotube formation, consistent with the absence of promoter elements responding to myogenic transcription factors. If there is no general, coordinate induction of respiratory chain peptides, one may wonder about the purpose of inducing COXVIaH mRNA by at least an order of magnitude in myotubes compared to myoblasts. It has been proposed from experiments with the homologous subunit in yeast that this subunit is responsible for sensing ATP and phosphate levels—that is, for the modulation of enzymatic activity of complex IV under varying metabolic conditions (132). For this factor to participate in the regulation of cytochrome oxidase activity requires one of two possible mechanisms: Either the subunit VIaH replaces the preexisting subunit VIaL directly, or there is turnover of the entire complex, with newly made complex including the muscle-specific VIaH subunit.

Thus, at first sight, the well-controlled and already thoroughly investigated rat or mouse myoblast/myotube system in tissue culture may serve as a good model system in which to study mitochondrial biogenesis and the control of respiratory gene expression. Much previous attention has been focused on muscle specific proteins such as creatine kinase and the various cytoskeletal proteins of the sarcomere. On the other hand, the preliminary lessons are that more subtle mechanisms for modulating mitochondrial respiration may be at work, requiring only a limited number of changes in the composition of the electron transport complexes. A second and very interesting preliminary conclusion is that myotube differentiation in culture is incomplete, in the sense that maturation, stimulation, and contractile activity of a myofibril may be required for further changes in respiratory gene expression and mitochondrial capacity.

Two other characterizations of genes had once raised expectations about finding more universal promoter elements that might be unique or specific for mitochondrial proteins. The analysis of the promoters for the ATP/ADP translocator and for the β subunit of the F1-ATP synthase had identified so-called OXBOX and REBOX elements that were postulated to respond to nuclear factors present at variable levels in myoblasts, myotubes, and HeLa cells (133, 134). A generalization of these conclusions awaits the characterization of more genes in this category, but at this time one can state that many other relevant genes already analyzed do not contain these particular elements as a recognizable consensus sequence.

4.2.2.4 Co-evolution of Nuclear and Mitochondrial Genomes From the above description it is clear that several dozen nuclear-coded proteins interact either with the mitochondrial genome (replication, transcription, etc.) or directly with proteins encoded by the mt genome. The mt genome suffers sequence substitutions at a significantly faster rate than the nuclear genome. If altered proteins result from mtDNA substitutions, compensatory changes may have to occur in the nuclear genome to assure proper strong, functional interactions between the gene products, in forming a complex of the electron transport chain, for example. Thus the two genomes are expected to co-evolve, and strong evidence for this concept has been deduced from studies with interspecific hybrids and cybrids of mammalian cells. The first experiments addressing this issue were reported by Clayton et al. (135) from their analysis of mtDNAs in human-mouse interspecies hybrids. Further pioneering studies of Wallace and Eisenstadt (136) indicated quite convincingly that the nuclear and mitochondrial genome must be compatible; that is, there is species specificity, and a human mtDNA cannot coexist for long in a rodent cell, or vice versa. More specifically, in the Scheffler laboratory, it was found that nuclear mutations in a hamster cell mutant causing a defect in complex I could be complemented by a gene on a hamster or mouse X chromosome, but not by a human X-linked gene (137). This result was also interpreted in terms of the incompatibility of heterologous gene products.

A very informative test of genome compatibility was recently reported by Kenyon and Moraes (138). By fusing enucleated cells from the hominoid apes (chimpanzee, pigmy chimpanzee, gorilla, and orangutan) with ρ^0 (mtDNA-less) human cells these authors created "xenomitochondrial cybrids" with ape mtDNA and human nuclear DNA. Only the combinations of human nuclei with mitochondria from the most closely related species (chipanzee, gorilla) yielded cells capable of oxidative phosphorylation. Orangutan mitochondria in such cells did not express a functional OXPHOS system. An even more specific example of incompatibility between species was reported by Yadava et al. (139). The MWFE subunit of complex I is highly conserved in mammalian species, and yet a human protein fails completely to complement a deficiency in a hamster cell. A change of two amino acids (out of 70) in the human protein restores the function of this protein in hamster cells.

4.3 REPLICATION AND MAINTENANCE OF MITOCHONDRIAL DNA

4.3.1 DNA Replication in Mammalian Mitochondria

Transcription and replication of mtDNA are basic processes which are closely related to both the function of the organelle and its biogenesis (75). The present section will deal with mtDNA replication, with an emphasis on animal mitochondria. When mtDNAs were discovered almost 50 years ago and the first estimates of their size became available, it also became clear that the relatively small size of animal mtDNAs made them attractive candidates for study. Expectations were not disappointed, and our understanding of DNA replication and transcription in animal mitochondria is relatively far advanced. Insights from such studies will be presented first. An update on studies in other organisms will summarize the progress, provide some contrasting or unique characteristics, and emphasize some of the issues for which answers still have to be sought.

Mitochondrial DNA replication can be investigated conveniently in tissue culture, where intracellular mitochondrial proliferation is loosely coupled to cellular division and multiplication. The total number of mtDNAs has to double in each cell cycle. Replication intermediates can be found readily. Before entering into the details of this process, one should be reminded that the situation may be more complicated in an intact animal, and several aspects of the process need to be discussed separately. It is necessary to distinguish between the number of mitochondria per cell and the number of mtDNA molecules per cell (estimated to be a few thousand). *A priori* it makes sense for mitochondrial DNA replication to be controlled to maintain a roughly constant number of mitochondria per cell, and thus there has to be coordination with DNA replication in the nucleus. However, during embryonic development and differentiation in mammals the number of mitochondria per cell may be altered, depending on the energy requirements of particular tissues. It is possible that in adult tissue such as muscle and brain there is also turnover of mitochondria and hence mtDNA replication which is not linked to nuclear DNA replication and cell division. A particularly fascinating situation may occur during oogenesis, where only a small subset of the total population of mtDNA molecules may be involved in mtDNA replication to generate the large number of mitochondria found in the mammalian egg. While there is as yet no proof of this hypothesis, such a "bottleneck" has been invoked to explain observations on mitochondrial inheritance and the phenotypic expression of mitochondrial mutations in heteroplasmic cells (140). Alternatively, a bottleneck can also be created if there is no replication of mtDNA during the first 10–15 zygotic divisions following fertilization. A more complete discussion of this topic will be found in Chapter 7.

In cells in tissue culture, mtDNA replication seems to be loosely coupled to nuclear DNA replication, with the precise mechanism of the control of copy numbers yet to be elucidated. On the other hand, the process of replication itself is quite well understood (141–143). Mammalian mtDNAs, and probably most vertebrate mtDNAs, have two origins of DNA replication, one responsible for each strand. The asymmetric distribution of nucleotides (G + C) in mammalian mtDNA allows one to distinguish a "heavy" and "light" strand on alkaline CsCl gradients, and hence one can also speak of O_H as the origin of H-strand synthesis and O_L as the origin of L-strand synthesis. O_H is located within the region devoid of genes referred to as the displacement loop, or D-loop region, and at other times as the control region. Initiation and elongation of the H-strand is the first event, and it proceeds a considerable distance around the circle; only after the second origin (O_L) is displaced as a single-stranded template will DNA replication of the other strand and in the opposite direction begin. O_H is therefore dominant, and O_L is not an independent origin, since DNA replication beginning at this origin without the prior activation of O_H has never been observed. It is significant that O_L is located within a cluster of five tRNA genes. When this region has been displaced and is found in single-stranded form, it is likely that a secondary structure is formed which contributes to the functioning of the origin. Support for such a hypothesis comes from the observation that in chickens the O_L sequence itself (found in mammals) is absent, but a similar cluster of five tRNAs is still found at this location (144).

Strand elongation by a mitochondrial DNA polymerase (polymerase γ) proceeds in the usual 5′–3′ direction. The nature of this polymerase and its relationship to the various bacterial and eukaryotic DNA polymerases is an interesting one (see reference 51 for a more explicit discussion). Its low abundance has made biochemical studies a challenge, but the enzyme has now been characterized from several organisms including humans (63, 64, 145–148). Pol γ prefers ribohomopolymer templates, and therefore it was initially thought to represent a cellular reverse transcriptase resembling the reverse transcriptase of tumor viruses. Later it was recognized that nuclear preparations had been contaminated with mitochondria. It is antigenically completely different from the viral enzymes, and it is further distinguished by its inability to use natural RNAs as templates. Other distinguishing characteristics include stimulation by salt, resistance to aphidocolin, inhibition by N-ethylmaleimide, and by dideoxynucleotide triphosphates (51).

The most highly purified preparations suggested that it consists of a holoenzyme of ~125–140 kDa ($POL_{\gamma}A$) that has both catalytic and exonuclease activity, and it is associated with a smaller subunit of 35–54 kDa ($POL_{\gamma}B$) that functions as an essential processivity factor. When pol γ genes were cloned (e.g., MIP1 from *S. cerevisiae*), a 140-kDa polypeptide was found to include both polymerase and exonuclease domains with recognizable homology to prokaryotic, A-type DNA polymerases—for example, *E. coli* DNA polymerase I. The cloning of *X. laevis* and *D. melanogaster* pol γ genes confirmed the conclusion that the polymerization and 3′ to 5′ exonuclease functions are combined in one polypeptide (143). The reconstitution of a minimal mtDNA replisome *in vitro* has recently been achieved in the laboratory of M.

Falkenberg (149a). Three proteins were defined to act in concert at the replication fork by a mechanism that resembles T4 or T7 replisomes: Polymerase γ, mtSSB (a single-strand binding protein), and TWINKLE, a mitochondrial DNA helicase with 5′ to 3′ directionality. Single-stranded DNA molecules could be generated at a rate of 180 bp/min, close to the rate reported for *in vivo* DNA replication.

While there are no novel or unusual aspects of the mechanism of DNA strand elongation in mitochondria, the same cannot be said about priming and initiation of DNA replication. It became clear some time ago that DNA replication and transcription of mtDNA were directly and intimately linked. The first indication was that the two known promoters for transcription are both located in the control region. They are ~150 bp apart in mouse and human mtDNA and control transcription in opposite directions; hence they can be referred to as the light-strand promoter (LSP) and the heavy-strand promoter (HSP). Their precise identification became possible with the establishment of a faithful *in vitro* system for transcription which will be described in further detail in the section on transcription. Three key findings were made when detailed characterizations were made of the major nascent DNA chains at the earliest stages in DNA replications, and of all the transcripts from the LSP. First, all newly initiated DNA chains had a 5′ end that corresponded to the sequence ~200 nt downstream from the LSP. Second, LSP transcripts were of two types: (a) the expected long transcripts that would later be processed to yield eight tRNAs and one mRNA and (b) short transcripts with 3′ termini precisely or closely adjacent to the 5′ termini of the nascent DNA chains. It was a very plausible jump to the conclusion that the RNA transcripts might serve as primers for the start of DNA replication from the O_H origin. A satisfying confirmation of this idea was provided by the third finding: LSP transcripts still covalently attached to the nascent DNA (in the mouse). Transcriptional activation is therefore closely linked to, and a prerequisite for, leading-strand DNA replication. A current model incorporating many findings is presented in Figures 4.5 and 4.6 (143).

Figure 4.5 shows the control region of the vertebrate mt genome between the Phe-tRNA on the left and the Thr-tRNA on the right. Initiation of transcription at the L-strand promoter generates an RNA that remains firmly associated with the DNA in the region defined by the conserved sequence blocks (CSBs), displacing the DNA strand to form a loop. Subsequent processing on either side releases a long downstream transcript and a short upstream fragment. The RNA segment remaining in the R-loop can then serve as primer for DNA synthesis. The D-loop is formed when the extension of the newly synthesized H-strand is arrested (see below). Initiation of mtDNA replication in other vertebrates is likely to be quite similar, with some variations in the size of the control region, the nature of the CSB elements (with CSB1 apparently the most essential), and the spacing between them (143).

Recent studies *in vitro* have provided additional support to this model (150). RNA of defined length made *in vitro* was capable of annealing to

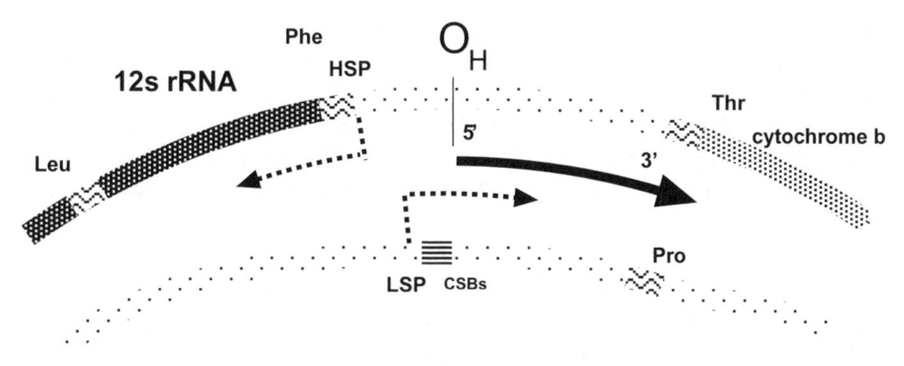

Figure 4.5 The D-loop region of mammalian mitochondrial DNA and the flanking genes. HSP and LSP are the heavy-strand and light-strand promoters. A processed transcript from the LSP serves as the primer for DNA replication from the O_H origin. (Modified after D. Clayton.)

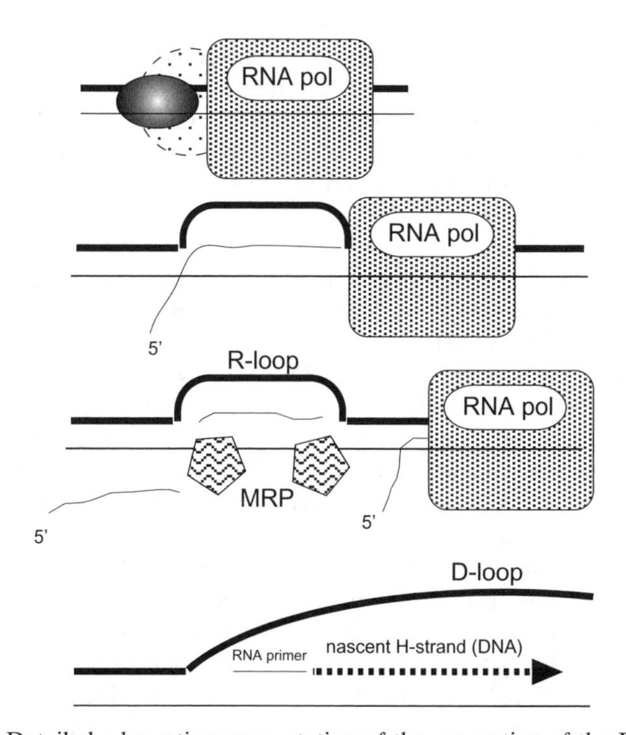

Figure 4.6 Detailed schematic representation of the generation of the RNA primer for DNA replication. RNA polymerase and factor(s) generate an RNA transcript. This transcript is modified by a mitochondrial RNA processing enzyme (MRP), leaving a short RNA tightly annealed to the DNA and creating an RNA loop. Priming from this RNA of DNA replication leads to the generation of the larger D-loop. (Modified after D. Clayton.)

supercoiled mouse mtDNA under defined conditions, creating an "R-loop." This R-loop had been previously observed in *in vitro* transcription experiments with purified mitochondrial RNA polymerase activity, and it had been shown to consist of a "persistent" RNA–DNA hybrid. It was also possible to create the R-loop by *in vitro* transcription with SP6 polymerase from an SP6 promoter inserted into mouse mtDNA control region. Its formation was not dependent on the presence of a particular RNA polymerase or additional protein, but did require negative superhelical turns in the mtDNA.

The R-loop with the RNA–DNA hybrid must have unusual stability, since it is stable to manipulation and purification without branch migration, and even linear restriction fragments containing the R-loop are stable for considerable time at 37°C without loss of one of the polynucleotide strands. The RNA in the R-loop is not base paired with the complementary DNA strand along its entire length. From a detailed examination and comparison of RNase H and RNaseT1 cleavage sites in the R-loop with those of a double-stranded hybrid made from the same RNA and its cDNA, it was concluded that a unique structure must be formed. A model of the R-loop has emerged in which intramolecular RNA–RNA interactions stabilize and protect distinct regions against single-strand-specific nucleases (stem-loops?), while other segments interact with unique sites on the DNA. An examination of the relevant DNA sequence has revealed three conserved sequence blocks (CSB I, CSB II, CSB III), and both CSB II and CSB III contain poly G tracts and are absolutely required for R-loop formation. From the results of the probing with single-strand-specific nucleases and RNase H, the whole structure cannot be represented by conventional Watson–Crick basepairing schemes. The CSBs sequences attracted attention as potential cis-acting regulatory regions as early as 1981, because they were short, highly conserved regions within an otherwise highly divergent region (141, 143).

Endonucleolytic processing of the RNA in this hybrid structure is postulated to create a primer which is positioned in the correct position for extension by deoxyribonucleotides with DNA polymerase. A candidate enzyme performing this function was isolated by Clayton's laboratory in 1987 (142, 151, 152). It was shown to be a site-specific endonuclease capable of cleaving RNA transcripts from the D-loop region between CSB II and CSB III; and most intriguingly, this RNase MRP (mitochondrial RNA processing) contains an RNA component encoded by nuclear genes in mouse and humans. Base-pairing interactions between this RNA and the substrate RNA for this nuclease are therefore likely to guide substrate selection and cleavage site selection. Undoubtedly, the secondary structure of the LSP transcript within the R-loop is also a strong determinant of specificity and activity. Identical ribonucleoproteins have been isolated from several mammals, the frog *Xenopus laevis*, and even yeast. Cloning of these genes as well as the genes encoding the RNA component has confirmed a high degree of evolutionary conservation (143). As discussed in the review by Shadel and Clayton (143), it was a challenge to show that the MRP is indeed found in mitochondria, since a very similar activity is also found in the nucleus, where it is involved in RNA processing. Two types of experiments now support the conclusion that there are two pools of MRP: (1) MRP RNA was localized in mitochondria by *in situ* hybridization experiments, and (2) mutations in the yeast MRP RNA gene have been shown to cause a "mitochondrial phenotype."

Finally, one must consider that the stability of the R-loop creates a new problem at the completion of the heavy DNA strand synthesis. Before the ends can be ligated to form a covalently closed circle, the R-loop structure has to be removed. The existence of RNase H-like activities found in submitochondrial protein fractions suggests candidates for the removal of the primer, since mitochondrial DNA polymerase does not have an associated RNase H activity (51).

L-strand synthesis beginning at O_L is also primed by an RNA primer complementary to a T-rich segment of O_L flanked by tRNA genes. The T-rich segment has the potential for stem loop formation, once it is dissociated as a single-stranded DNA. At a location close to the base of the stem a switch occurs from ribonucleotide incorporation to deoxyribonucleotide incorporation by DNA polymerase. The priming RNA is thought to be synthesized by an mtDNA primase that recognizes O_L. An activity with such a capacity has been described (153), but much needs to be learned about the recognition of the origin, the limited synthesis of the primer, and the ultimate cleavage of the primer from the L-strand. Very recently, Twinkle homologues have been proposed to have both helicase and primase activity in a broad range of eukaryotes. Mutations in Twinkle have been shown to cause mtDNA deletions that in humans result in the autosomal dominant mitochondrial disease adPEO (progressive external ophthalmoplegia) (65). Since the transition from RNA to DNA is localized within a sequence that is normally transcribed as a complete tRNA within a polycistronic transcript, it is possible that the activity/mechanism for splicing out the tRNA performs a dual role in removing the RNA primer from the L-strand.

The model presented above is referred to as the strand-asymmetric model of mtDNA replication, and it continues to be favored. However, it has been challenged by experiments from the Holt laboratory (154–157). An examination of replication intermediates by 2-D gel electrophoresis was interpreted in favor of a coupled leading- and lagging-strand DNA synthesis. A modified strand-displacement model has recently been proposed that includes a new appreciation of alternative light strand origins and a consideration of potential brand migration in replicating mtDNA molecules (158, 159). It claims to reconcile most of the controversial data with the original hypothesis.

A curious and still unresolved problem is related to the existence and persistence of the D-loop and its function (143). The D-loop is a bubble in which one strand of the control region has been copied and the other has been displaced. It includes the ~1-kb control region of mammalian mtDNA. The exact size of the D-loop is species-specific. Near the 3′ end of the nascent (arrested) strand, conserved termination-associated sequences (TASs) have been identified in vertebrates, and the existence of a specific DNA-binding activity has been proposed from experiments with bovine mitochondrial extracts (160). It is not yet known whether D-loop strands are simply elongated when replication proceeds, or whether there is a distinct cycle of initiation through the D-loop region. The discrete size of the D-loop indicates that there is a pause in strand synthesis, and overcoming this arrest may be another mode of control. It has also been suggested that it plays a role in transcriptional regulation or mtDNA segregation. A novel twist in this story has been introduced by the report from Attardi's laboratory (161) of another major replication origin in the D-loop. When originating from this position, the nascent chains did not terminate at the 3′ end of the D-loop but proceeded well beyond this control point. Fish et al. (161) interpret these as the true replicating strands made during DNA replication under steady-state conditions, in contrast to DNA synthesis originating at previously identified D-loop origins. The latter is suggested to occur during recovery from mtDNA depletion, or when increased mtDNA replication is stimulated by novel physiological demands. This point is indeed worth noting. What controls the copy number of mtDNA per cell, and under what conditions can a cell be induced to change this number?

It is generally stated that mtDNA replication is not tightly coupled to nuclear DNA replication during the S-phase of the cell cycle. Some control is undoubtedly exercised through the availability of deoxyribonucleotides in the cell, and hence through the activity of ribonucleotide reductase. How the import of such deoxyribonucleotides into mitochondria is achieved and whether pool size in the matrix is different from cytosolic/nuclear pools are unanswered questions at this time. Since the copy number of mtDNA per cell remains constant in tissue culture, there must be some check on mtDNA replication, and likely sites of control involve initiation at the origin in the D-loop (O_H). In the current model, transcription from the LSP is a prerequisite, but a rate-limiting step could still be at the level of the RNase MRP which creates the RNA primer. In mouse L cells lacking cytoplasmic thymidine kinase, but having a mitochondrial thymidine kinase, mtDNA replication is conveniently studied by incorporation of radioactive thymidine, or of bromo-deoxyuridine. Such studies have revealed that mtDNAs are apparently selected at random, with some being replicated twice during the cell cycle while others are not replicated at all. Organelle genomes like mtDNA (and chloroplast DNA) have been referred to as "relaxed genomes," and both their replication and their segregation can be considered to be relaxed (162). This behavior leads to their general failure to obey Mendel's Laws: Alleles can segregate during mitosis as well as during meiosis, inheritance is often uniparental, and there is the possibiltiy of intracellular selection of advantageous alleles among hundreds, if not thousands, of copies of the genome.

A very interesting study by Davis and Clayton (163) has revisited the issue of mtDNA replication in mammalian cells, taking advantage of laser-scanning confocal microscopy to detect newly incorporated BrdU into mtDNA. The total population of mitochondria within a cell was viewed by dye-labeling (using Mitotracker™), and BrdU incorporation was detected by immunocytochemistry. When short labeling periods were used (1–2 hours), BrdU incorporation occurred preferentially into perinuclear mitochondria, and distal mitochondria could be stained with antibody only after prolonged labeling. Since mtDNA was present in all mitochondria (determined by ethidium bromide staining), mtDNA replication must occur preferentially in the vicinity of the nucleus. The dependence on the nucleus was further demonstrated by experiments with enucleated HeLa cells. In such cytoplasts, mitochondrial DNA replication was completely arrested (even though mitochondria looked normal and retained their membrane potential). Similarly, in human platelets no BrdU was incorporated into mtDNA. Thus, not only is a nucleus required, but factors provided by the nucleus are apparently not freely diffusible throughout the cell. On the other hand, it is not necessary to have ongoing DNA replication in the nucleus. This was shown most clearly with PC12 cells induced by NGF to differentiate into neuron-like cells with a cell body and long extensions (neurites) and axons. Mitochondria in the cell body (perinuclear) incorporated BrdU within a period of a few hours, while mitochondria in the neurites and growth cones were not labeled after such relatively short labeling periods. With time, labeled mitochondria appeared in the periphery, and the authors speculated that they had been transported there by axonal transport. Inhibiting DNA replication with the chain terminator dideoxy cytidine (ddC) while still allowing some BrdU incorporation suggested that such mitochondria (with unfinished replication intermediates) remain in the cell body—that is, are not subject to axonal transport toward the periphery.

The interpretation of these provokative results is highly speculative at this time. The authors suggest that either (a) some type of tethering to a cytoskeletal perinuclear element is required or (b) an essential component of the mtDNA replication machinery must be restricted to the perinuclear region of the cell. The nature of the required factor is not further specified. A nuclear-encoded protein would have to be synthesized within this restricted domain of the cell, or concentrated there before import into the mitochondria. It is also possible that the synthesis of deoxynucleoside triphosphates is limited to this region of the cells, although one would expect such small molecules to diffuse rapidly throughout the cell.

The RNA and DNA polymerases as well as all of the nucleases, maturases, transcription factors, repair enzymes, and various DNA-binding proteins are encoded by nuclear genes, and ultimate control of their expression is exerted in the nucleus. A full discussion of the nuclear genes relevant for mitochondrial biogenesis and their expression has been presented earlier in this chapter (Section 4.2). Whether or not the transcriptional activator mtTFA acts as the master switch (in the earlier literature it had been referred to as MtTF1 and is now referred to as TFAM)) was an open question (143), but much new information now suggests that TFAM plays critical roles in multiple aspects to maintain the integrity of mitochondrial DNA: transcription, replication, nucleoid formation, damage sensing, and DNA repair (164). Curiously, it does not bind exclusively at the two promoters (LSP and HSP); it also binds within the region of the CSBs, where it may interact with the RNase MRP and control primer production.

The initiation of transcription will be discussed in a later section. The basic machinery for initiation is now known to consist of a mitochondrial RNA polymerase and two transcription factors, mtTFA and mtTFB1 or mtTFB2. Experiments by Ekstrand et al. (165) have investigated the consequences of overexpressing the human TFA in transgenic mice. In this heterologous system it is known that hu-TFA is not very effective in promoting transcriptional initiation, but curiously the mtDNA copy number was elevated significantly. Thus, the role of TFA is not restricted to transcription. There are reports that the amount of TFA is far in excess of the amount needed for binding to the promoters, and in fact it may bind in a more nonspecific manner to the entire mtDNA. How it influences the copy number remains a challenge for the future.

It has been noted that whereas gene order in animal mtDNA is quite stable over long evolutionary periods, the nucleotide sequences of individual genes differ markedly between species, and a significant sequence heterogeneity (polymorphisms) is found even among human populations, a fact that is of great interest to evolutionary anthropology and in forensic applications (see Chapter 8). Of even greater potential interest is the current hypothesis that an accumulation of mutations in mtDNA is contributing to senescence in general, and perhaps specifically to the symptoms of patients suffering from Parkinsonism and other neurodegenerative diseases (see Chapter 7). A standard interpretation is that DNA replication in animals is exceptionally error-prone, and/or there is no or only a limited capacity for DNA repair such as mismatch repair or removal of thymine dimers. An error rate near 1 per million nucleotides incorporated is relatively low, thanks to the proofreading by the 3′ to 5′ exonuclease (51).

4.3.2 mtDNA Repair in Mammalian Mitochondria

The capacity for mtDNA repair has been examined by a number of authors, with damage induced with a variety of agents. Not too long ago the prevalent view was that mitochondria have only a very limited capacity for repairing damage in mtDNA, but the past decade has seen considerable progress in the elucidation of a variety of repair mechanisms, depending on the nature of the damage. Prominent lesions induced by oxidative stress are removed by a base excision repair pathway (166–168). There is evidence that the excision repair may be induced during oxidative stress (169). The various enzymes strongly resemble (or are identical to) those operating in the nucleus, and in the mitochondria they appear to be associated with the inner membrane by electrostatic interactions that are sensitive to 150–300 mM NaCl; this association is observed even in the absence of mtDNA (in rho-zero mutants) (170).

Evidence for mismatch repair has been obtained only recently from a rat liver mitochondrial lysate (70).

4.3.3 Recombination in Mammalian Mitochondria

There is only limited evidence for recombination between mammalian mtDNAs, in part because the process is difficult to study due to the strictly maternal inheritance of mtDNA (see Chapter 7). More recently, some generalizations about the absence of recombination and repair mechanisms have been challenged (171). Since a genetic approach is still out of reach for the study of mammalian mitochondria, biochemical assays were made with mitochondrial extracts to look for activities that could carry out recombinational repair on genetically engineered substrates *in vitro*. Since the presence of a "robust" homologous repair activity was demonstrable in liver mitochondria, the authors argued that this mitochondrial mechanism was exclusively used for DNA repair through gene conversion. The previous failures to observe genetic recombination were due to the failure to detect crossovers.

4.3.4 mtDNA Maintenance and Replication in Other Organisms

Mitochondrial DNA replication in nonvertebrate organisms and plants will not be addressed here in detail, and much remains to be explored. The yeast, *S. cerevisiae*, has, as expected, served as another model system with the advantage of the exploitation of genetic approaches. Thus, the pol γ gene (*MIP1*) was first isolated and characterized from yeast (62), two mitochondrial transcription factors (sc-mtTFA and sc-mtTFB) have been defined, and the gene for the RNA component of the RNase MRP gene (*NME1*) has been cloned. However, the size of the yeast mtDNA has made the characterization of replication intermediates and the identification of replication origins difficult. The same problem is even more acute with the still larger plant mtDNAs.

Now that it is established that many plant and fungal mtDNAs are linear molecules (11, 172) and that terminal sequences may have a telomere-like function, understanding mtDNA replication includes another challenge: How are linear 5′ ends dealt with by the replication machinery. The search for proteins active at the ends is on, and some initial success in *Candida parapsilosis* has been reported (11).

4.4 TRANSCRIPTION OF MITOCHONDRIAL DNA–RNA METABOLISM

4.4.1 Transcription in Mammalian Mitochondria

In the previous section on mtDNA replication reference has already been made to transription and the role it plays in providing the primers for DNA replication in mammalian mitochondria. The discussion will focus first on the mammalian system because of its relative simplicity and the advanced level of understanding that has been reached in this system (19, 74, 75, 142, 173).

One needs to be reminded first of the absolute compactness with which genes are arranged on the mammalian mt genome. From the first complete sequences it became apparent that there was virtually no room for promoter elements between genes, and the choice was between (a) having very few promoters from which large segments of the mtDNA were transcribed or (b) postulating promoters that overlapped the transcribed and coding sequences of the various genes. Pioneering experiments in the laboratories of Clayton and Attardi and others came to converging conclusions and interpretations.

In a chronological order that was dependent on available technology, the first relevant experiments addressed the question, Is the mitochondrial genome transcribed? Labeling with [^{3}H]uridine and cell fractionation studies gave the earliest indications that rapidly labeled RNA was associated with a membrane fraction that included mitochondria (19). MtDNA had played a crucial part in (a) the development of methodology employing CsCl gradients in the ultracentrifuge and (b) the subsequent discovery of the phenomenon of DNA supercoiling with the help of the intercalating dye ethidium bromide. As a result, relatively pure preparations of mtDNA were available when RNA–DNA hybridizations were first exploited to study gene content, some years before sequencing and cloning became the tools for high-resolution analysis. Such studies gave the first indications that most sequences of mtDNA in HeLa cells were transcribed (19). Another question raised immediately was, How many peptides are made in mitochondria? The capacity for autonomous protein synthesis was strongly suggested by the discovery of ribosomes in mitochondria, although their properties differed quite significantly from bacterial and eukaryotic cytoplasmic ribosomes. Using isolated mitochondria, or whole cells in which cytoplasmic protein synthesis was inhibited by cycloheximide, it could be shown by labeling with [^{35}S] methionine and electrophoretic fractionation that a small number of peptides was synthesized in mitochondria, and eventually all peptides predicted from the primary sequence could be accounted for on gels. Later, as cloned probes for subregions of mtDNA became available, Northern analyses and S1 protection analyses were used to establish that each peptide was encoded by its own distinct mRNA, but very large molecules were also seen which could be identified as presumptive precursors by pulse-chase experiments. From such studies the consensus emerged that a few relatively large transcripts from both strands must be made which are subsequently spliced and processed to create individual mRNAs, rRNAs, and tRNAs. It is now accepted that the transcription units include the entire H-strand and the entire L-strand. Having the entire mtDNA sequence from humans and cows was clearly of enormous help in the interpretation. In this regard it is worth noting that mtDNA sequencing and the characterization of mature mtRNAs were carried out simultaneously in Cambridge and at Cal Tech, with an ongoing exchange of complementary information. The papers on the human mtDNA sequence and on the precise mapping of the transcripts were published in the same issue of *Nature* (5, 174, 175). A very broad and authoritative review from a historical perspective has been written by one of the pioneers and major contributors to this accumulated knowledge, G. Attardi (19). Not the least of the technical problems to be solved in the course of these studies was to culture an estimated kilogram of HeLa cells.

The existence of the D-loop identified by electron microscopy and by the finding of an ~1-kb region free of coding information between two tRNA genes drew attention to this region as a possible control region, but in the early 1980s there were no criteria for the identification of a mitochondrial promoter, and no procedures existed then or now to test promoters with reporter genes *in vivo* in mitochondria. A breakthrough was achieved in 1983 when the first *in vitro* system for initiating transcription from mtDNA in a human mitochondrial extract was described (176). Transcription starting specifically from the D-loop region was observed, leading in follow-up experiments to the identification of the L-strand and H-strand promoter elements (LSP and HSP) in the D-loop region. For clarification it may be necessary to state that the D-loop identifies the control region (D-loop region), but the D-loop itself (the bubble created by L-strand displacement by a short DNA strand) is somewhat shorter than the entire region. A systematic destruction of these elements by the insertion of small linkers (linker-scan analysis across the entire region) confirmed their importance and their limited extension within the control region. The two promoters are approximately 150 bp apart, but they do not overlap and they function independently. Upon further analysis, it was confirmed that L-strand transcripts are giant polycistronic transcripts containing the mRNA sequence for the NAD6 subunit and eight tRNAs. However, transcription in the opposite direction from HSP starts at two closely spaced initiation sites with differing activity (HSP1 and HSP2 (75)). One transcript contains the $tRNA^{Phe}$ and $tRNA^{Val}$ genes and the rRNA genes, terminating at the 3′ end of the 16S rRNA. It is most likely responsible for the much higher steady-state levels of rRNAs compared to the mRNAs. The second, overlapping transcript is another giant polycistronic RNA that is processed to yield the other mRNAs and most of the tRNAs encoded by the H-strand. The higher rate of initiation and the termination at the 3′ end of the 16S rRNA assures that the rRNA species are synthesized in much greater abundance (15- to 50-fold) than the mRNAs.

The D-loop region is functionally conserved in all vertebrates, but the nucleotide sequences of the different elements themselves are not conserved. Because of this lack of sequence conservation, it is also less clear whether all vertebrates (mammals?) contain all of the elements that are found in the human D-loop. However, it is flanked in almost all cases by the $tRNA^{Phe}$ downstream from the HSP (Figure 4.5) and by a $tRNA^{Pro}$ downstream from the LSP (Figure 4.5). An immediately adjacent gene is the 12S rRNA gene. Since the tRNA and rRNA sequences are highly conserved in vertebrates, these flanking genes can provide (degenerate) primer sequences for use in

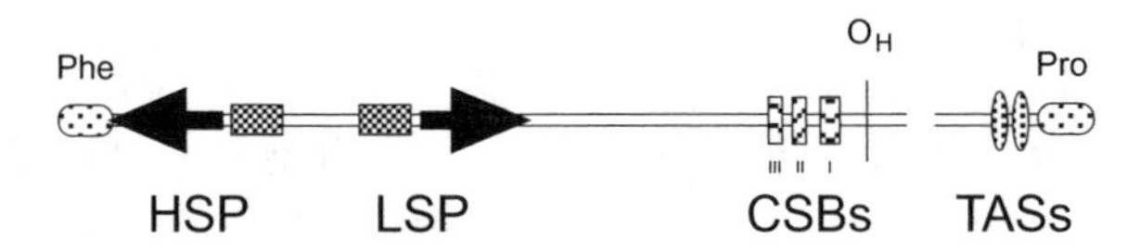

Figure 4.7 A detailed view of the promoter regions of mammalian mitochondrial DNA. (After D. Clayton.)

PCR amplification, and D-loop regions from other vertebrates can be conveniently amplified and cloned for further analysis (177). The detailed description of the human and mouse D-loops (Figure 4.7) will serve as examples, illustrating common features, but it should emphasized that D-loops in different vertebrates vary in size, in sequence, and even in the number of distinct promoters they contain. In chicken and frog (*X. laevis*), D-loops are small and promoters may overlap, making them appear to be bidirectional promoters. There may be multiple transcription initiation sites in some vertebrates, and some promoters (human, mouse) have additional cis-acting elements for tuning promoter activity.

The stage was set for a further fractionation of mammalian mitochondrial extracts, which yielded two significant protein components: a mtRNA polymerase and the mitochondrial transcription factor now referred to as mtTFA (89). A dedicated mitochondrial RNA polymerase was first identified from yeast and subsequently from humans (hu-POLRMT) (178). These polymerases are related to phage T3 and T7 RNA polymerases, but distinguished by their inability to bind to promoters in the absence of additional factors. Stimulated by results obtained from transcriptional studies in yeast, a second mammalian mitochondrial transcription factor, mtTFB, was found more recently (179). It is required in combination with mtTFA to stimulate transcription from the (L)-strand promoter (LSP). Subsequently a second mtTFB with ~25% sequence identity was discovered (180), and it appears that these two factors, mtTFB1 and mtTFB2, do not represent a simple redundancy. A more surprising finding was that these factors belong to a family of rRNA methyltransferases that are capable of binding *S*-adenosylmethionine (SAM) and are most likely responsible for methylating certain target residues in the small rRNA of mitochondrial ribosomes (173). This methylating activity of mtTFB1 is required for ribosome biogenesis, but is not required for its function as a transcription factor in *in vitro* experiments. Three proteins, POLRMT, mtTFA, and either mtTFB1 or mtTFB2, constitute the basal mitochondrial transcription machinery *in vitro* in combination with a DNA fragment containing the promoter (180). MtTFB2 appears to be ~100 times more active than mtTFB1 (see reference 74 for a more detailed discussion). An expert, authoritative review of the human mitochondrial transcription machinery has recently been published by Bonawitz et al. (75).

The cDNA and genes for mtTFA have been characterized, and the basic structural features of the 25-kDa protein have been deduced from protein sequence comparisons (181, 182). A mitochondrial targeting sequence of 42 amino acids is presumably cleaved off, leaving a mature protein with 204 amino acids. The most notable features are two segments of ~70 amino acids recognized to be present also in high-mobility group proteins (HMG) of the nucleus. These segments are likely to play a role in DNA binding, although the mechanism of transcriptional activation is not obvious. There are 31 residues separating HMG Box 1 from HMG Box 2, along with very short flanking regions at the N-terminal and C-terminal, respectively.

Binding of mtTFA to the D-loop DNA could be demonstrated in the absence of mtRNA polymerase, confirming a direct interaction of the factor with the DNA. This conclusion was corroborated and extended by footprinting and additional linker-scan analyses. MtTFA bound upstream of each of the two transcription start sites (defined by nucleotides from positions −10 to −40); but surprisingly, it also bound to a region between the conserved sequence boxes I and II (CSB I and CSB II), which play a role in the generation of the RNA primer for H-strand synthesis. The affinity of each promoter for mtTFA is different: The LSP is favored, consistent with its greater activity *in vitro*. It is probably also significant that the LSP controls primer synthesis for DNA replication. When the two sequences at each promoter identified by the footprinting are compared, homology emerged only when one sequence was inverted with respect to the orientation of the promoter. In other words, if this homology is significant, it implies that mtTFA can function in a bidirectional manner. In the most recent model for transcription initiation in human mitochondria the binding of mtTFA creates a "uniquely bent" or contorted DNA conformation at the HSP and LSP. The C-terminal tail of mtTFA then becomes exposed for interaction with mtTFB protein(s), forming a bridge for interaction with the mtRNA polymerase (173). It becomes clear that these factors are important for initiation. However, questions still remain about the precise role contributed by each factor to the precise promoter recognition and specific transcription initiation at each site.

Again, generalizations from the discoveries in human or mouse mitochondria to other vertebrates must be made with caution, since promoters differ significantly in nucleotide sequence and in the number and spacing of critical elements. There are indications that mtTFA binds to mtDNA at other sites, raising the possibility of other functions being performed by this factor. Suggestions have been made that it is very abundant, even to the level of coating the entire mtDNA. At the same time, overexpression of hu-mtTFA in the mouse leads to an increase in the mtDNA copy number without increasing transcription or respiratory chain capacity. Thus, transcriptional activation by mtTFA may be dissociated from its other function(s) as a regulator of mtDNA copy number (165). The multiple roles of mtTFA have also been expertly reviewed by Kang and Hamasaki (164).

A comparison of the human and mouse control region and experiments in both systems had confirmed the functional significance of this region in DNA replication and transcription and had revealed a similar arrangement of the relevant sequence elements. Unexpectedly, there was a notable divergence in nucleotide sequence, explaining, perhaps, why human mitochondrial extracts failed to transcribe mouse mtDNA and vice versa. When the mtRNA polymerase and the mtTFA from each species were sufficiently well purified, it became possible to set up mixed assays to determine whether the species specificity was entirely attributable to the recognition of the mtDNA by mtTFA. Human mtTFA was found to be a poor activator of mouse mtDNA transcription (74).

In the past few years a major new insight into the mechanism and machinery of transcription has been gained, primarily from studies in yeast but most likely applicable to mammalian mitochondria as well (75, 173, 183). A short version of this still-emerging story places one or more mtDNAs into a nucleod that is associated with the inner mitochondrial membrane. Nucleods can be visualized by staining for DNA, but also by fusing nucleod-associated proteins with GFP. The number of such proteins is still increasing. Among the expected proteins, one finds mtTFA, the mitochondrial ssDNA-binding proteins, and components of the transcriptional machinery. More intriguing was the finding of a variety of proteins required for translation of the mRNAs. These have been defined mostly from molecular genetics experiments in yeast. The relevant hypothesis for the present context is that translation is coupled to transcription (173), but it is not far-fetched to consider that nascent peptide strands are co-translationally inserted into the inner mitochondrial membrane (see below). A completely unexpected protein found in nucleods was the TCA-cycle enzyme aconitase (77, 183). This protein is essential for mtDNA maintenance, even when mutated to abolish the formation of the [4Fe–4S] cluster. The authoritative review by Chen and Butow (183) should be consulted for details on several other components of mitochondrial nucleods and their role in a variety of phenomena. These include: (1) metabolic remodeling of nucleods in response to amino acid starvation, glucose repression, and retrograde (RTG) signaling; (2) nucleod division and segregation; and (3) mtDNA recombination (in yeast). Finally, nucleod stability is linked to mitochondrial dynamics, and specifically fusion. A comprehensive model by these authors proposes an apparatus comprised of proteins in the outer membrane, in the inner membrane, and the nucleod. It is still in the speculative phase, but it should greatly stimulate investigations into the coordination of mitochondrial fusions and fissions, nucleod replication and segregation, the expression of mitochondrial genes, and the integration of the encoded proteins into the OXPHOS complexes. Last but not least, these processes must be regulated in response to bioenergetic and metabolic requirements in the cell. Here it is worth pointing out that a yeast cell is probably much more versatile and responsive as a result of its exposure to a variety of natural environments, while mammalian cells of a given tissue may not see such extremes under most normal conditions.

4.4.2 Transcription of mtDNA in the Yeast *Saccharomyces cerevisiae*

A comparison of the structure of the yeast mt genome with that of animals immediately suggests that many more transcriptional start sites must exist, since the genome is about five times larger, the genes are dispersed, and AT-rich, noncoding sequences are inserted between them. At least 13 transcription initiation sites have been mapped, and there are an additional four sites for the synthesis of primers to be used in DNA replication (184, 185). The simple promoters all have a 9-bp consensus sequence: 5′-ATATAAGTA(+1)-3′. The last A also corresponds to the 5′ end of the transcript. Like in animal cells, polycistronic RNAs (but smaller) are still produced which must be cleaved and processed to yield mRNAs encoding peptides, tRNAs and rRNAs. A total of eight peptides are encoded by the mRNAs: cytochrome oxidase subunits I, II, III (COXI, COXII, COXIII), apo-cytochrome b (COB), subunits 6, 8, and 9 of the F_0 complex of the mitochondrial ATP synthase, and a ribosomal protein VAR1. There are some nonessential peptides encoded by introns of COXI, COB, and the 21S rRNA gene which are required when the introns are present (maturases), or can be active in the transposition of the introns, but in intronless strains they are absent and not missed. Three open reading frames have so far been identified only from sequencing, not from isolation of transcripts or from the discovery of mutations within these sequences. It has been speculated that such as yet unknown functions may be required for DNA replication or general recombination (186).

Transcription itself is accomplished by a core RNA polymerase (145 kDa, encoded by the nuclear *RPO41* gene), and a specificity factor (43 kDa, encoded by the nuclear *MTF1* gene). Both genes have been cloned and the proteins have been characterized (187). Like the mammalian enzymes, the polymerase bears a relationship to polymerases of the bacteriophages T3 and T7, while the specificity factor resembles the bacterial sigma (σ) factors, but it has also been compared to the mtTFA protein in mammalian mitochondria. The Mtf1 protein is released from the transcriptional complex after the first few nucleotides have been polymerized and is then recycled. Its function is primarily to assure specificity, stimulating polymerase activity only a few-fold. When the MTF1 gene is deleted, the maintenance of the yeast mt genome is no longer assured.

4.4.3 Transcription of mtDNA in Plant Mitochondria

An up-to-date and expert review of the present state of knowledge has recently been written by Binder, Marchfelder, and Brennicke (188), and references to much of the primary literature can be found in this review. The area is in an explosive state of growth, and many examples in the literature on the one hand permit increasingly reliable generalizations on certain aspects, while on the other hand plant-specific differences are also apparent. For example, monocot and dicot plants share many broad features but may differ in detail. A distinction for all plants is the simultaneous presence of chloroplasts. Mitochondria and chloroplasts are often discussed together, not only because of obvious similarities—for example, the observation of RNA editing in both plastids—but because it is likely that factors may be shared between these two organelles.

The variable size of plant mtDNAs has already been emphasized. At the same time, a large fraction of this DNA does not have any obvious coding or regulatory function, and generally the same limited number of proteins, rRNAs and tRNAs, is encoded by plant mtDNA as in the mtDNA of animals and fungi. Only a limited number of multicistronic transcription units have been described so far, and these may be more common in plant mitochondria having relatively small genomes with a denser spacing of genes. As described earlier, the intergenic regions of plant mtDNA can be expanded or contracted by frequent recombination events, which can also rearrange the position of genes relative to each other.

Many examples of monocistronic transcripts encoding proteins (atp6, atp9, cytb, cox1, cox2, and cox3) have been described. Such transcripts include extended untranslated regions at the 5′ and 3′ end, frequently hundreds of nucleotides in length. In contrast to other species, plant mitochondrial transcripts are not polyadenylated. Since the primary 5′ ends of such transcripts are not capped, they can be capped by an *in vitro* reaction, which subsequently permits a distinction between a 5′ end of a primary transcript and the 5′ end of a mature mRNA generated from RNA processing. By this method, at least 15 primary transcripts (and promoters) have been identified in *Oenothera* (188), with the promoters distributed over the entire genome. The analysis of primary transcripts can be complicated by the existence of multiple promoters, spaced over several hundred nucleotides upstream of the coding sequence. Multiple transcripts of different size may therefore be found. It has not yet been resolved whether such arrangements have a regulatory significance in differentiation and development. Another potential complication is the possibility of multiple copies of a gene in the genome which may yield transcripts with differing 5′ and/or 3′ UTRs.

A detailed analysis of several promoters has been greatly facilitated by the development of *in vitro* transcription systems, permitting the examination of promoters and derived variants with deletions, substitutions, and linker-insertions. From such an approach a promoter region of about 17 nucleotides has been identified in monocot plant mitochondria (wheat, maize) and includes (a) a conserved tetranucleotide CRTA at the transcription initiation site and (b) a conserved purine-rich stretch about 15 nt upstream of the transcription initiation site. In dicots (e.g., pea), the CRTA motif is also conserved, and an A-rich region is found upstream which is important as demonstrated by *in vitro* mutagenesis experiments. *In vitro* studies also permit the testing of a promoter from one species with extracts from another species, and by such means functioning promoter sequences can be confirmed, but differences in efficiencies are indicative of differences in the interaction of promoters with species-specific factors. Such differences cannot be deduced from simple inspection of primary sequences.

Not all promoters are identical on a specific mtDNA. Ribosomal RNAs are apparently transcribed from promoters with a distinct sequence in both monocots and dicots, based on sequence inspections or on *in vitro* experiments with homologous or heterologous mitochondrial lysates. Such observations have given rise to speculations about different transcriptional activators, and even about the possibility of more than one RNA polymerase in plant mitochondria, a situation encountered in chloroplasts of higher plants (189). Differences in RNA polymerases, factors, and promoter sequence suggest that activity at different promoters may variable and regulated. Such a conclusion is supported by run-on experiments in isolated organelles or lysates—that is, elongation of pre-initiated transcripts with added nucleotides. Ribosomal RNAs and ribosomal protein genes were found to be most actively transcribed, and distinctions were found in one case between the other transcripts. Comparison of such data with the observed steady-state levels of transcripts *in vivo* is clearly indicative of the existence of extensive post-transcriptional regulatory mechanisms, most likely at the level of mRNA stability.

4.4.4 Transcriptional Termination

Transcription of an entire circular template in animal cells raises the question of whether true transcriptional termination occurs. This has been a difficult question to answer, but a region at the 3′ terminus of the rRNA gene and immediately adjacent to the $tRNA^{Leu(UUR)}$ gene has attracted attention based on (a) *in vitro* experiments showing transcription termination at this site (190) and (b) a very provocative finding of an ~34-kDa protein that can footprint this sequence. The same sequence is mutated in a patient with MELAS (myopathy, encephalopathy, lactic acidosis, and stroke-like symptoms; see Chapter 7) (191), and the mutated form has a lowered affinity for this presumed termination protein (192). A factor named mTERF (mitochondrial transcription termination factor) has recently been purified by DNA affinity chromatography using the boundary sequence between the 16S rRNA and the $tRNA^{Leu(UUR)}$ genes (193). The important role of this protein is most likely related to the need for making more ribosomes than mRNAs (see above). It is not clear whether it can directly influence the rates at which the two closely spaced initiation sites are used, but by terminating the majority of the transcripts at the boundary site the overproduction of rRNAs compared to mRNAs is assured.

4.4.5 RNA Processing in Mitochondria

An equally challenging problem has been the elucidation of the precise mechanism by which the primary transcripts are processed to form individual

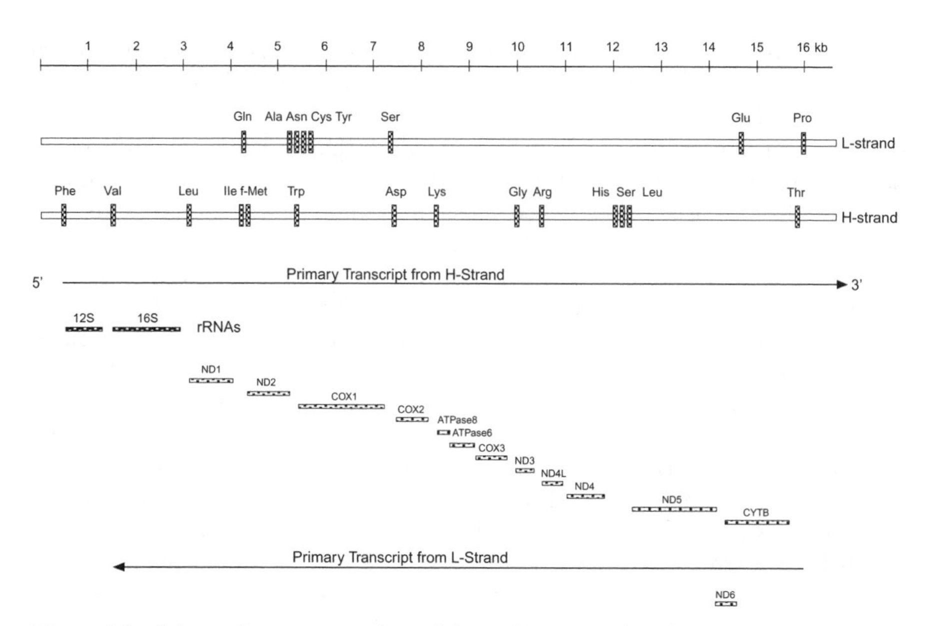

Figure 4.8 Schematic representation of the major transcripts from the H-strand, and the subsequent processing to form mature rRNAs, tRNAs (not shown), and mRNAs. Only a single mRNA (ND6) and several tRNAs are derived from the transcript of the L-strand.

mRNAs, rRNAs, and tRNAs. It is clear that the process must recognize the interspersed tRNAs (Figure 4.8), because they are joined directly to rRNA or coding sequences. In the original publication describing the localization of the tRNA genes, Attardi and his colleagues made reference to the "tRNA punctuation model" for mt RNA processing (175), an apt description of how the compact genome is organized and interpreted. In attempts to reproduce this processing *in vitro*, the primary transcript has been produced with prokaryotic RNA polymerases, but efforts to achieve processing with mitochondrial extracts have been quite unsuccessful. Thus it is possible that most of the processing occurs already with nascent transcripts.

Splicing out a tRNA requires two endonucleolytic cleavages: at the 5′ end and at the 3′ end. Cleavage at the 5′ end could occur as soon as the entire tRNA sequence has been transcribed and has assumed the appropriate secondary structure; in mitochondria this cleavage would produce the 3′ end of an mRNA, for example. From the study of tRNA precursors and their processing in bacteria and other organisms, a nuclease, RNase P, has been characterized which performs the 5′ cleavage, prompting a search for a corresponding activity in mitochondria of yeast and mammals which only recently has met with success (194). Most significantly, it was found that the yeast enzyme was a ribonucleoprotein that consisted of a peptide encoded in the nucleus and an RNA encoded by the mt genome in yeast, but presumably by the nuclear genome in other organisms. The RNA has two regions highly conserved in RNAs found in other RNase Ps. They are believed to participate in basepairing interactions leading to pseudoknot formation and the formation of the catalytic core of the enzyme.

The other steps in RNA processing are even less well understood. In general, the activities of interest are present at low abundance, making purification difficult, even when a specific substrate for the assay was available from *in vitro* transcription of an engineered template. A genetic approach in yeast is a possibility, but the specific mutant would presumably have to be distinguished from a large number of other respiration-deficient mutants (petites).

The following types of activities remain to be purified and characterized in detail:

1. The nucleases complete the processing of the polycistronic transcripts—for example, the cleavage at the 3′ end of the tRNAs.
2. Mature mRNAs and even rRNAs are polyadenylated. Symptomatic for the extreme economy in the genome is the finding that some mRNAs do not have complete stop codons until the third adenine nucleotide is added as part of the polyadenylation reaction.
3. Fungi and plant mtDNAs encode ORFs interrupted by introns. These introns can be of type I or type II. Some of these introns include ORFs encoding maturases or reverse transcriptase activities that play a role in intron splicing and mobility. Some of the diversity and variability of these introns in different species and even within a species have been discussed in the section on gene organization. In *Saccharomyces cerevisiae* introns have been found in the mt COXI gene and the mt COB gene. They have been characterized as self-splicing group II introns (195–197).

Processing of transcripts in plant mitochondria is certainly necessary when the transcript is polycistronic—for example, the 18S–5S–nad5 transcript in *Oenothera* mitochondria. Little is known at this time about the details of the enzymatic reactions and about the processing signals recognized by the trans-acting factors. 5′ and 3′ processing is required, depending on the transcript. For example, the 18S rRNA is transcribed from an upstream promoter and requires the removal of 25–120 nt, depending on the plant. Various mRNA have been found to have a double stem loop at the 3′ end, and it has been hypothesized that the transcript initially extends beyond the stem loop, but is trimmed back by a 3′ exonuclease until the stabilizing secondary structure prevents further hydrolysis (188).

For the processing of tRNAs, processing systems have been established *in vitro* from mitochondrial extracts, and such extracts have been shown to contain an RNase-P-like enzyme with an RNA moiety necessary for activity. As discussed above, this enzyme is distinct from the necessary 3′ endonuclease.

In higher plant mitochondria a number of transcripts have been found to contain group II introns that are similar to this class of introns found in yeast mitochondria. A relatively unique feature is that some introns are physically disrupted by a large intervening portion of the genome. In virtually all flowering plants, such disrupted introns have been discovered in the nad1, nad2, and nad5 genes. mRNA maturation requires trans-splicing to transcripts with exons that are independently transcribed. One specific example will illustrate the surprises and unpredictability of transcription and splicing implants. The penultimate intron of the nad1 gene appears to contain an ORF potentially encoding the only maturase identified so far in plant mitochondria (198). Trans-splicing of this intron occurs in petunia and wheat, but in the first example the maturase ORF is in the upstream half of the intron, while in wheat it is found in the downstream portion of the intron. In the broad bean this intron is not trans-spliced at all, and the maturase ORF is located between the two exons (188). Small wonder that a review on the subject is entitled "The Mitochondrial Genome: So Simple Yet So Complex" (199).

4.4.6 RNA Editing in Kinetoplastid Protozoa

One of the most bizarre and unexpected observations was the discovery of RNA editing in mitochondria of kinetoplastid protozoa. After its discovery in these organisms, editing was subsequently found in a few other organellar or nuclear transcripts in a wide range of organisms, but the extent to which it is being used in kinetoplastids is astonishing. The authors of one review point out that it appears to be completely redundant, because there is no obvious reason why the same polypeptides could not be encoded directly by the genes and the resulting transcripts (200). What is editing? In the most general sense it means that an RNA, after it has been transcribed from the corresponding gene, is altered by the insertion or deletion of nucleotides that are not encoded by the gene. The unedited transcript could not be translated because there is no sensible ORF. An ORF is created only after a series of nucleotides have been inserted or deleted (or altered). As pointed out by one of the pioneers in the field (201), the term RNA "editing" has been used rather indiscriminately to describe many types of RNA modifications occurring post-transcriptionally; and in many of these, site-specificity is achieved with the help of small complementary RNAs. RNA interference and siRNA-mediated degradation of mRNAs may be included under this broader definition. The present discussion will be concerned with phenomena defined by the more limited definition above. A discussion of RNA editing from a perspective of an evolutionary biologist can be found in the review by Covello and Gray (200).

The major effort has been devoted to the study of the trypanosomatid genera *Trypanosoma and Leishmania*, since they are the pathogens responsible for several widespread diseases of humans and animals in tropical Africa and South America. Their interesting life cycle includes stages in both a mammalian host and insect host which serves as the vector in transmitting the pathogen.

The structure of the kDNA in the kinetoplast within the single mitochondrion has been described in an earlier chapter. It remains to define and explain how mature, translatable mRNAs are produced from transcripts of the "cryptogenes" on the maxicircle. RNA editing in these organisms requires the insertion, and less frequently the deletion, of uridylate residues, not as a rare event but in a process involving more than half of the maxicircle transcripts. Not only are the majority of the transcripts affected, but the addition of Us is extensive, such that in some cases the transcript is doubled in size, and essential, because the addition of Us creates start and/or stop codons and the desired open reading frame. Unedited transcripts are believed not to be recognized by the translational initiation mechanism. This amazing violation of orthodoxy has been and continues to be met with questions about What? How? Why? and When? (200, 202, 203). The first two can now be answered with some confidence and in detail (201, 204, 205). The last is a question about the evolution of this phenomenon, and answers are coming in as more and more examples become available to construct reliable phylogenetic trees (200, 202). Why? is the question most difficult to answer, and a speculative answer will be deferred until it has been made clear what is happening to the transcripts.

RNA editing is not confined to trypanosomes (206, 207). However, in mammals less than a handful of nuclear gene transcripts and some viral transcripts have been found to be edited, usually at a single position, where a C is converted to a U, or the change is U-to-C. In marsupial mitochondria a C-to-U conversion in the anticodon loop converts a $tRNA^{Gly}$ to $tRNA^{Asp}$ (208). The COX1 mRNA in mitochondria of the slime mold *Physarum polycephalum* is edited by a C-to-U conversion. In plants, more extensive editing occurs in mitochondria and in chloroplasts (C to U). In distinction to the editing observed in trypanosomes, editing in the other examples involves a base change—for example, a C-to-U conversion in apolipoprotein B, or an A-to-I conversion in the AMPA receptor—by hydrolytic deamination, while for editing in trypanosomes the change entails insertions and deletions of uridines. One should also distinguish clearly between editing occurring within an RNA sequence and the "editing" that is exemplified by the polyadenylation of mRNA, or the addition of the sequence CCA at the 3′ end of tRNAs. The addition of a 3′ oligo(U) tail to the kinetoplastid mRNAs and gRNAs (see below) also belongs to a distinct category, requiring the activity of a mitochondrial terminal uridylyl transferase.

The editing proceeds in a systematic manner from the 3′ end to the 5′ end, and it requires the participation of so-called guide RNAs (gRNAs). As one portion of the mRNA becomes edited, it can form a basepair with another gRNA, and the editing step can be repeated further upstream in the mRNA. The gRNAs are encoded in both maxicircles and minicircles, with their abundance being assured by the abundance of minicircles (45). The complexity of the minicircles and the total number of gRNAs varies from species to species;

T. brucei has over 300 different minicircle classes, each encoding three gRNAs. The gRNA contain sequences that are antisense to portions of the mRNA to be edited, and short "anchor" duplexes are formed between the two RNAs just 3′ of the sequence to be edited. The subsequent steps can so far be modeled by several schemes involving either transesterifications or cleavage–ligation ("cut, insert, and paste," also described as the enzyme cascade model), with recent *in vitro* studies favoring a cleavage–ligation model. Each scheme still includes alternatives for the terminal reactions (207, 209, 210).

A model based on successive transesterification reactions is shown in Figure 4.9A. The initial reaction is a nucleophilic attack at a site specified by the guide RNA, upstream from the anchor sequence. The resulting 3′ hydroxyl group on the upstream portion of the pre-edited mRNA can then make another nucleophilic attack leading to the insertion of only one uridine, if the initial nucleophile is UTP (scheme on the right), and may lead to the insertion of one or more uridines if the nucleophile is the 3′ end of the oligo (U) tail of the guide RNA. Thus, in this model the gRNA has two functions: It serves to specify the site of insertion by the specificity of the anchor sequence, and it may provide the uridines inserted into the mRNA. While this model was initially appealing in its resemblance to RNA-catalyzed splicing reactions, and hence indicative of a potential, common evolutionary origin, recent evidence leads to a rejection of this model in favor of the cascade model shown in Figure 4.9B. The anchor sequence of the guide RNA again serves to specify a point of cleavage by an endonuclease, and cleavage occurs precisely at the first mismatched nucleotide upstream of the duplex formed between the gRNA and the pre-edited mRNA. Uridylate residues can be added to the 3′ end by a terminal transferase (addition, middle, and right scheme). 3′–5′ exonuclease trimming may be necessary, and may be responsible for the guided deletion of U's (left scheme) (211), and finally an RNA ligase restores the continuity in the mRNA. At this point the gRNA must be released (presumably with the help of a helicase), and the steps can be repeated at a new position.

Evidence favoring the second model is derived from significant advances in reproducing some of the required steps *in vitro* (see Simpson (201) for an up-to-date review and a listing of many references). In addition to a model system demonstrating the gRNA-dependent insertion of Us directly from UTP (and a requirement for exogenous UTP), it could also be shown that the required ATP is hydrolyzed between the α–β bond, as expected for an RNA ligation reaction in which AMP becomes covalently linked to an intermediate. Chemical blockage of the 3′ end of the gRNA by periodation did not prevent U-insertion, eliminating this end as a potential nucleophile. As discussed in more detail by Alfonzo et al. (209), a major modification of the original enzyme-cascade model includes the addition of multiple U's to the 5′ cleavage fragment, followed by trimming by a 3′ exonuclease. This model could thus accommodate additions, deletions, and even misediting in a single mechanism. In subsequent years an additional insight was that separate but interconnected enzymatic pathways exist for U-insertion and U-deletion sites (212).

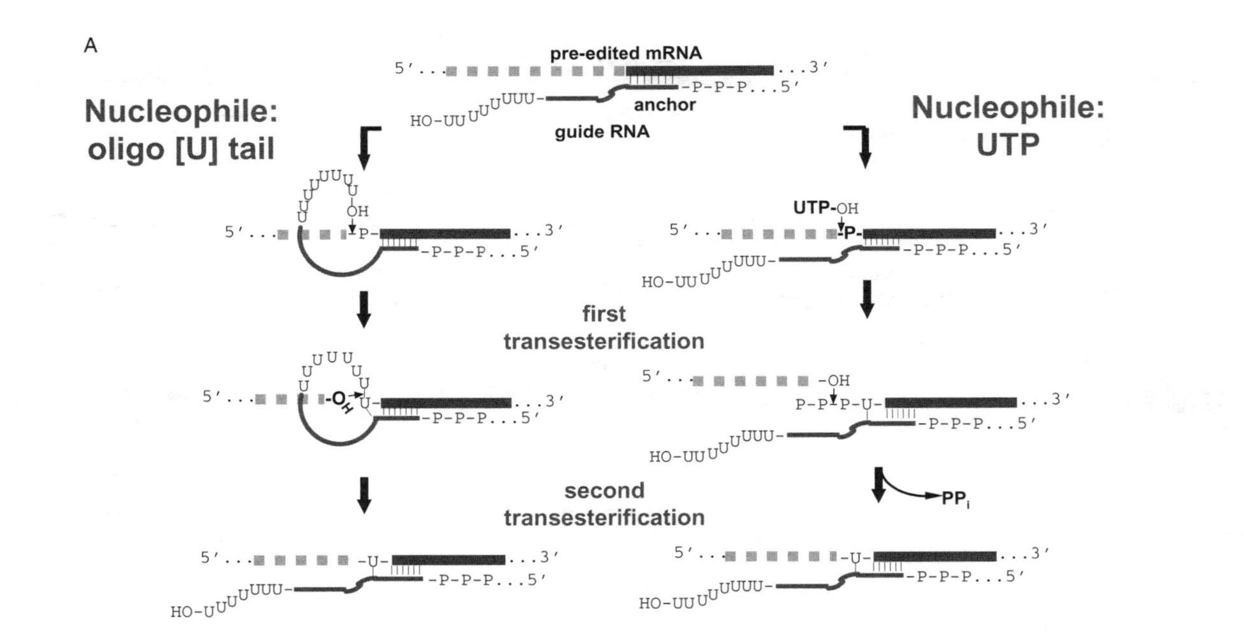

Figure 4.9 RNA editing in Trypanosomes (see reference 201 for details). (A) Double transesterification models, showing only the U-insertions. (B) The modified enzyme cascade model showing U-deletion, or U-addition or misedited U-addition as alternatives. In the simplest version the U-tail of the guide RNA is shown as a free overhang, but it is also possible that it may pair with a purine-rich sequence of the pre-edited RNA. The number of Us added to the 5′ fragment in the U-addition pathway (middle) is show as 13, but this number is actually variable. If the trimming is precise, correct guided products are obtained. If trimming is incomplete or excessive, a misedited product may result. (The figures were provided by Dr. L. Simpson. With permission from Oxford University Press.) See color plates.

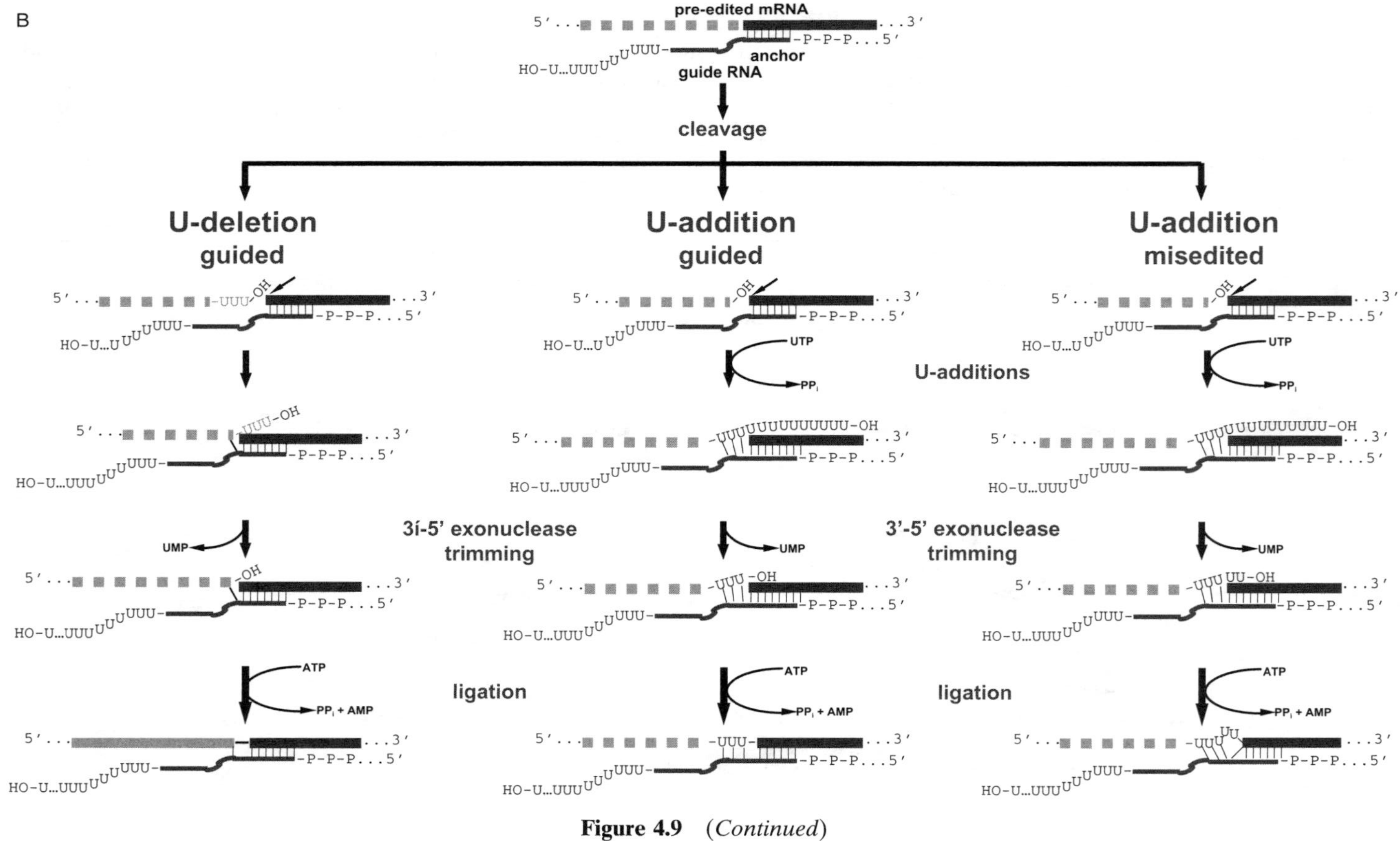

Figure 4.9 *(Continued)*

Significant progress has been made in fractionations of mitochondrial extracts from *T. brucei* and *L. tarentolae*, and the reconstitution of *in vitro* systems performing progressive editing at multiple sites (for example, see reference 205). Many of the relevant enzyme activities have been characterized (terminal uridyl transferase, endoribonuclease, RNA ligase) and obtained as recombinant proteins. These activities together with mRNA and gRNA are assembled in a multiprotein complex; glycerol gradient sedimentation has yielded a series of often heterodisperse complexes, depending on the species examined. Many details, references, and a model for the editing RNP supercomplex can be found in the review by Simpson (201).

The evolutionary history and the biological significance of the phenomenon should perhaps not be discussed as dissociated topics. The extensive "pan-editing" described here has been found so far only in kinetoplastid protozoa, which are representative of the earliest divergence of eukaryotic cells containing mitochondria. When trans-esterification was still a viable model, speculations that "editing and splicing have a common origin in an RNA world" were defensible, but the currently favored model exhibits little resemblance to RNA splicing, and pan-editing may have an independent origin. Editing may become superfluous in the evolution of species when edited RNAs can replace the cryptogenes by a mechanisms termed retroposition (202), unless there is continuous selective pressure to maintain it. In this context it is interesting that continuous culturing of a *Leishmania tarentolae* strain in the laboratory was observed to have lost minicircles and hence gRNA complexity when compared to a newly isolated strain. The laboratory lifestyle is clearly different from the lifestyle in the wild, with passages through insect and vertebrate hosts and corresponding adaptive changes in energy metabolism.

4.4.7 Editing in Plant Mitochondria

RNA editing in plants is not nearly as dramatic as in trypanosomes, but it is found both in mitochondria and in chloroplasts. A comprehensive review by Maier et al. (213) can serve as an introduction. A very important distinction is that in this case, editing does not involve the insertion or deletion of nucleotides as specified by guide RNAs. Rather, most of the editing reactions lead to specific C-to-U conversions, but a few U-to-C conversions have also been observed. While the detailed mechanism is not understood, there are indications that at least in mitochondria, RNA editing may be a simple deamination ($C \rightarrow U$). From a broader perspective, the variety of complex post-transcriptional processes of plant mitochondria such as 5′ and 3′ RNA processing, intron splicing, RNA editing, and controlled RNA stability has been expertly reviewed by Binder and Brennicke (214).

It is obvious that until RNA editing in plant plastids was discovered, there was potential confusion in identifying or translating open reading frames. Some apparent differences between the plant mitochondrial genetic code and the universal genetic code disappeared when editing was considered, and sequence differences between homologous plant mitochondrial genes also disappeared after editing was discovered independently in three laboratories in 1989 (215–217). Most significantly, editing creates the AUG initiation codon in the nad1 transcript of wheat mitochondria (188) and thus defines the ORF. Similarly, in potato mitochondria editing of the rps10 transcript creates the correct start codon and a new stop codon. Therefore, editing must be considered in all analyses of newly sequenced plant mtDNAs.

The basic approach to identify editing is to clone cDNA sequences from organelle mRNAs and then compare the sequences with the corresponding sequences on the mt genome. The eight genes and their transcripts for subunits of complex I of wheat mitochondria (nad1-7,9) can serve to illustrate the extent of editing observed in mitochondria (see reference 213 for a summary of the results and the primary references). Editing sites appear to be distributed at random in most transcripts, and the density of sites per unit length of RNA is variable, ranging from 5.5 per 1000 nt in nad5 to 59.3 per 1000 nt in nad3. Exon 2 of nad4 and exons 1, 2, and 3 of nad5 are not edited at all, whereas the flanking exons are edited. An interpretation of this curious fact is still highly speculative, and it includes the possibility that a portion of the gene represents the incorporation by homologous recombination of a cDNA obtained from reverse transcription of the edited mRNA (213). There is to date only one plant mitochondrial mRNA which is not edited at all, but it may be a special case, since it is the transcript of a chimeric gene composed of 3′ regions of the 26S rRNA gene, an unknown sequence, and a part of the 26S rRNA gene. Structural RNA (e.g., rRNA) genes in plant mitochondria are subject to very limited editing. The unedited transcript was discovered because it had been found to be responsible for the Texas male sterility of maize (218).

What makes a particular cytidine a substrate for the presumed deaminase activity? It appears to depend on local sequences. This conclusion was reached from the observed editing in partial gene sequences in chimeric transcripts which were made from chimeric genes created by the frequent recombination events in plant mt genomes. For example, a 193-nt fragment of exon 1 of cox1 is edited at exactly the same positions as found in the intact cox1 mRNA (213). A large number of plant mitochondrial gene sequences have been accumulated, but the information cannot be properly interpreted if sites to be edited cannot be predicted accurately. Thus, information on known editing sites has been incorporated into computer programs that can predict editing sites in novel sequences (219, 220).

Editing sites have been found at lower frequencies in rRNAs and tRNAs, in intergenic regions of multicistronic transcripts, and in the cis-spliced or trans-spliced group II introns mentioned above. The C-to-U conversions can be found in stems of secondary structures where a perfect A–U basepair would presumably be stabilizing relative to an A–C basepair. Editing may therefore contribute to the formation of the secondary structure necessary for the splicing or processing mechanism.

It is clear that editing is a post-transcriptional event, since partially edited transcripts can be routinely identified in plant mitochondria. Furthermore, when such partially edited transcripts are analyzed further, it is found that the editing process is random; that is, it does not proceed with any apparent polarity, in contrast to the editing process in trypanosome mRNAs. This observation is consistent with a recognition of the site to be edited based on local sequences by the enzyme catalyzing the deamination (or other, more complex reactions such as base or nucleotide exchange). Since at any time a population of partially edited mRNAs exists in mitochondria, one can speculate whether such mRNAs are translated, presumably into peptides with amino acid sequences differing from the normal peptide derived from fully edited mRNA. Mechanisms may exist which prevent partially edited mRNAs from associating with the translational machinery. In one test case the atp9 subunit was produced from an unedited transcript in the nucleus of transgenic tobacco as well as from the endogenous mitochondrial atp9 gene (and edited transcript). The nuclear atp9 transcript also encoded a signal sequence for mitochondrial import, and thus both the normal and the abnormal atp9 peptide (sequence differences at 7 positions) were available for assembly of complex V. The large fraction of plants with cytoplasmic male sterility were interpreted to result from the assembly of inactive ATPsynthases (221).

Efforts to establish *in vitro* systems for RNA editing from mitochondrial extracts are showing promise (213, 222), and with the help of such *in vitro* experiments it will be possible to answer one of the more intriguing questions: How are cytidines selected for deamination? So far an analysis of neighboring regions has not shown any sequence conservation (consensus sequence) and has not given indications of localized secondary structures.

4.4.8 Control of mRNA Levels by Turnover

The integration of mitochondrial gene expression with the synthesis and import of peptides encoded by the nuclear genome has long been recognized as an important issue. One need not belabor the point that there must be a coordinate expression of many genes to make a functional electron transport chain, and a unique kind of complexity derives from the need for coordinating the expression from two genomes in separate subcellular compartments. In this chapter the focus will be on the mt genome, which can probably be considered to play a subordinate role, since nuclear gene products are required for all essential processes: replication, transcription, and translation. An interesting and important question is whether there is any signal from the mitochondria which directly influences gene expression in the nucleus.

A priori there are several levels at which control could be exerted: (1) transcription of the genome, (2) stability of the mRNA, and (3) translational control—that is, efficiency of initiation or elongation. The consequences of the joint activity of the first two steps can be measured by the steady-state levels of the mRNA(s) under consideration. It is also quite apparent that the rate of processing of the mitochondrial polycistronic transcripts can be an important factor, especially evident in the case of vertebrate mitochondria.

The structure of the promoters on mt genomes is quite simple, as described in an earlier section, and it is apparent that these promoters are not differentially active in combination with a wide variety of transcriptional activators or repressors. Instead, the presence of the RNA polymerase in combination with two major accessory/transcription factors (mtTFA and mtTFB1 or mtTFB2 in mammals) is likely to determine a steady rate of transcription at all promoters. The levels of the mitochondrial RNA polymerase and the MtFA are most likely controlled by nuclear factors, which will be the subject of a separate section.

It has already been discussed how transcriptional termination in mammalian mitochondria is responsible for the synthesis of unequal amounts of rRNAs relative to many of the mRNAs. Obviously, all mRNAs encoded on polycistronic transcripts are made in equal amounts, and any observed differences in their steady-state levels must be accounted for by differential turnover. Is there any evidence for differential turnover? Mitochondrial mRNAs have been described to have relatively short half-lives, but the whole subject has not been explored in great detail, either in different tissues of a given organism or in different organisms (however, see reference 223). A series of recent reviewers appear to have found an insufficient number of studies on mitochondria, and the subject is discussed together with the corresponding situation in chloroplasts (224–226). Chloroplasts development is beyond the scope of this book, meriting its own detailed treatment. Nevertheless, there may be lessons derived from the studies of chloroplasts. There is control in the context of the development of the whole plant, and there are significant responses to environmental cues, notably light. The only comparable system studied extensively with respect to mitochondria is the adaptation of the yeast *Saccharomyces cerevisiae* to changes in external carbon sources.

The earliest studies on mRNA turnover in HeLa cells were conducted by measuring the kinetics of labeling with [5-^{3}H] uridine and the decay of labeled mt mRNAs after blocking transcription with 3′-deoxyadenosine (cordycepin). It was shown that because of turnover of mRNAs, rRNA species were significantly more abundant (~50-fold) than mRNAs (223). Half-lives in the range of 25–90 minutes were reported. Subsequent studies gave conflicting results—in particular, studies showing the destruction of mtDNA (labeled with 5′ bromouracil and irradiated to stop transcription) (227). It was suggested that turnover is coupled to transcription and/or processing and that the stability increased after inhibition of transcription. The true situation is likely even more complicated, since rapid mt mRNA decay in fibroblasts can also be observed when protein synthesis is inhibited, either by a nuclear mutation affecting the translational machinery or by the addition of chloramphenicol

(80). It is noteworthy that in these studies, not all mRNAs were affected equally. At this time, much remains to be learned in mammalian cells, with the impetus derived in part from the study of human patients with various mitochondrial mutations, including mutations that affect mitochondrial gene expression (see Chapter 7). A review by Gagliardi et al. (228) focuses on the role of polyadenylation of mitochondrial transcripts. In the cytosol of eukaryotes, polyadenylation tends to stabilize mRNAs, while polyadenylation mediated mRNA decay in prokaryotes (and in chloroplasts). In view of the origin of mitochondria, it is tempting to speculate that polyadenylation in mitochondria should also promote turnover. However, these authors report, on the basis of a comprehensive survey of post-transcriptional processes in yeast, plant and mammalian mitochondria, that contrasting situations exist in plants and animals: poly(A) stabilizes mt mRNAs in animal cells, but promotes turnover in higher plant cells. The review by Gagliardi et al. (228) should serve as a valuable interim report on post-transcriptional processes in mitochondria; but as the authors emphasize, much needs to be learned about the regulation of mRNA turnover, and they anticipate some "important evolutionary surprises" in the years to come.

In yeast there have been a number of interesting and intriguing observations of altered stability of specific mitochondrial transcripts that remain to be fully understood, but it appears that these observations also establish a link between mRNA stability and the translation of the particular mRNA. Since this subject will be covered fully in a following section, only a brief description of the phenomena will be given here.

The yeast mitochondrial cytochrome b gene is transcribed as a discistronic precursor RNA requiring extensive processing. Two introns have to be removed to form the COB mRNA itself, and a $tRNA^{Glu}$ has to be released from the 5′ end to create a 5′ UTR for translational initiation. Most relevant for the present discussion is the finding that translation of the COB mRNA requires two translational activators encoded by the nuclear genes *CBS1* and *CBS2*. In *cbs1* mutant strains the transcript is degraded rapidly after the tRNA has been cleaved off, and it has been postulated that the Cbp1 protein stabilizes the COB mRNA by an interaction with its 5′UTR and, further, that it may either negatively regulate a nuclease or induce a processing event which makes the COB mRNA resistant to nucleases. (see reference 228a). CBP1 and CBP2 are not the only mRNA-specific translational activators in yeast mitochondria. Translation of the COX3 mRNA requires three proteins (nuclear genes *PET54, PET494*, and *PET122*). Similarly, the OLI1 mRNA is dependent on the nuclear genes products *AEP1* and *AEP2*. Translation and stability are affected by the Pet122 or Aep2 proteins, respectively, again interacting with the 5′ UTR of these mRNAs.

The primary role of these proteins is most likely to be in translation; but as in the mammalian example described above, mitochondrial mRNAs appear to be protected from degradation when they are being translated.

4.5 TRANSLATION OF MITOCHONDRIAL mRNAs

4.5.1 Introduction

The mitochondrial translation system in the matrix is an independent molecular machinery composed of components encoded by the nuclear and mitochondrial genomes. It is not essential in the sense that many cells can grow quite well under conditions where glucose is abundantly available for glycolysis. The main function of the mitochondrial translation machinery is to synthesize 13 proteins for the assembly of the oxidative phosphorylation system. The potential redundancy of mitochondrial protein synthesis was recognized early in the history of yeast genetics. Somewhat later it was discovered, however, that in the absence of mitochondrial protein synthesis the mitochondrial genome is lost for still unexplained reasons (229). In mammalian fibroblasts, mitochondrial protein synthesis can be inhibited more than 95% by a still undefined nuclear mutation without the loss or even decrease in the amount of mtDNA per cell (80).

A statement encountered frequently in the earlier literature is that mitochondrial ribosomes and general translation factors are similar to their prokaryotic counterparts, but a closer look can identify numerous aspects of translation in mitochondria which differ significantly. Further distinctions have to be made in comparisons of different organisms. Broadly speaking, the observed differences can be assigned to following major levels:

1. Changes in tRNA structure and use of altered codons
2. Changes in the fine structure of ribosomes
3. Cis-acting elements (5′ and 3′UTRs) of mRNA
4. Initiation factors

They will be discussed in turn.

4.5.2 Codon Usage and tRNA Structure

One of the major surprises was discovered when the human and bovine mtDNA sequences were completed and compared. The genetic code was not universal after all (see Table 4.2). The termination codons, UAA and UAG, defined for both prokaryotes and the eukaryotic cytoplasmic translation machinery did not function as termination codons in mitochondria, and a different set of termination codons was used. Also, additional codons could be used as initiation codons. The data cannot be summarized in simple statements for all species, because there are species-specific differences as well. TAA is found as a termination codon in most species examined, but frequently the ORF of a specific mRNA ends with either a T or a TA, and the termination codon is completed by the polyadenylation step. TAG and AGA have been found as termination codons in some species (see Table V in reference 4).

TABLE 4.2 Codon Usage in Mitochondria in Various Organisms

CODON	Standard Code	Mammals	*Drosophila*	*Neurospora*	Yeasts	Plants
UGA	STOP	Trp	Trp	Trp	Trp	STOP
AGA, AGG	Arg	STOP	Ser	Arg	Arg	Arg
AUA	Ile	Met	Met	Ile	Met	Ile
AUU	Ile	Met	Met	Met	Met	Ile
CUU, CUC, CUA, CUG	Leu	Leu	Leu	Leu	Thr	Leu

Mammalian mtDNA encodes 22 tRNAS, far fewer than the number available in the cytosol. Therefore, a single tRNA must be able to read all codons of a four-codon family, but it is not yet clear whether in some cases two nucleotide pairs are sufficient, or whether a uridine in the anticodon wobble position can pair with all four third-position nucleotides. A uridine in the wobble position (first position of the anticodon) is frequently modified to pseudouridine in some tRNAs that recognize two-codon families ending in G or A, leading to speculation that this structure prevents misreading the two-codon family ending in C or U. A similar rationale for accurate codon anticodon recognition has been proposed to explain specific modifications found in the nucleotide immediately following the anticodon (4).

Metazoan mtDNAs encode only one $tRNA^{f\text{-}Met}$ having the anticodon CAT, and it is the only tRNA for methionine. Not only internal AUG, AUA and all AUN codons have to be recognized by this tRNA, but the other start codons TTG, GTG, and GTT must be recognized by this tRNA as well, if it is assumed that all peptides are initiated with formyl-methionine. The universal stop codon UGA is translated as tryptophan in mitochondria of many species, and the universal UAA and UAG stop codons are translated as glutamine in some species such as *Acetabularia, Tetrahymena*, and *Paramecium*.

It is appropriate to mention here that only a small number of proteins made in mitochondria have been sequenced directly. Many other protein sequences are deduced from homology with bacterial or other eukaryotic mitochondrial peptides, and codon assignments are often deduced but not proved by experiments. In fact, it has so far been impossible to translate mitochondrial mRNAs with a mitochondrial set of ribosomes, tRNAs, and factors *in vitro*. The earliest attempts failed because the altered codon usage was not recognized, and "mixed" systems were tested—for example, cytoplasmic mRNAs with crude mitochondrial extracts. However, even if the abnormal codon usage is taken into account in such experiments, it has not been possible to translate mt mRNAs in a homologous system. A possible reason for this failure will be discussed below.

The complete sequencing of a tRNA was one of the early triumphs of nucleotide sequencing before the modern era of DNA sequencing by the

Sanger or Maxam–Gilbert techniques. From this first sequence the familiar clover-leaf secondary structure was proposed, and it has proved to be a universal representation of the structure of all cytoplasmic tRNAs and even chloroplast tRNAs. There are three major loops—referred to as the anticodon loop, the D-loop, and the pseudo-uridylate loop—and the amino acid acceptor stem. When more and more mitochondrial tRNA sequences were elucidated from a large variety of organisms, variations in size and structure of the loops and arms were encountered, although generally most vertebrate and invertebrate mt tRNAs can still be represented by a model resembling the standard tRNA. In extreme cases either the arm with the D-loop or the arm with the pseudo-U loop can be almost completely absent (Figure 3 in reference 4). In some instances the total length of the presumed tRNA and the secondary structure predicted deviated so much from the norm that additional data had to be interpreted for confirmation. For example, in the nematode *C. elegans* there were 22 tRNA genes, as in other vertebrate and invertebrate mtDNAs, and the anticodon in each of the postulated anticodon arms was compatible with codon usage in these organisms. To show that such unusual tRNA structures are indeed produced, oligonucleotide probes were designed and successfully hybridized to a fraction of small (<150 nt) RNAs from mitochondria of *C. elegans* (4), and the possibility of the creation of "standard" tRNAs by a trans-splicing or RNA editing mechanism was specifically ruled out.

The characteristic 3′CCA end serving as the aminoacyl acceptor site is not encoded in the mt genome of metazoans and other species, and it constitutes one of several post-transcriptional modifications. While this modification might have been expected, many other base modifications encountered upon direct sequencing of tRNAs are more surprising, because they require additional enzymes and substrates to be imported into the mitochondria. For example, pseudouridine has already been mentioned; other modified bases include 1-methyladenosine, 1-methylguanosine, N6-isopentenyladenosine, 5-methylcytidine, and N2-methylguanosine (4).

4.5.3 Mitochondrial Ribosomes

When ribosomes in mitochondria were discovered, their "similarity" to bacterial ribosomes became one of the aspects feeding early speculations on the evolutionary origin of mitochondria. However, this similarity was deduced primarily from their sensitivity to chloramphenicol and insensitivity to cycloheximide, in contrast to eukaryotic cytoplasmic ribosomes. Their hydrodynamic properties were also quite different from those of cytoplasmic ribosomes with a sedimentation coefficient of 55S compared to 80S for cytoplasmic ribosomes, but a comparison with bacterial ribosomes (70S) already gave an indication that the similarity was limited. When the size of the ribosomal RNAs was compared, the discrepancy became even more apparent. A summary of relevant parameters is presented in Table 4.3. The numbers given for the mitochondrial rRNAs are approximate numbers for metazoans, since variations

TABLE 4.3 Ribosomal RNAs

	Prokaryotes	Eukaryotes (mammalian)	Mitochondria
Large rRNA	2900 nt / 23 S	4800 nt / 28 S	~1600 nt / 16 S
Small rRNA	1540 nt / 16 S	1900 nt / 18 S	~950 nt / 12 S
5.8 S	—	160 nt / 5.8 S	—
5 S	120 nt / 5 S	120 nt / 5 S	120 nt / 5 S

are observed between species from as low as 953 nt for the l-rRNA of *C. elegans* to as high as 1640 nt for the l-rRNA of *X. laevis*; the nematode also has the smallest s-rRNA (697 nt), but the mouse and chicken s-rRNAs contain over a hundred more nucleotides than that of the frog (819 nt). There is no equivalent to a 5.8S RNA in mitochondria, and the 5S RNA is found mainly in plant mitochondria, but not in metazoans.

Bacterial and other ribosomal RNAs have been subject to very considerable and fruitful attempts to model secondary structure with the goal of mapping protein/RNA interactions and deriving a structure for the entire ribosome. Sequences from diverse organisms, when compared not as primary sequences but on the basis of secondary structural elements and conserved ribosomal subunit interactions, have revealed sequence blocks that are universally conserved. When mitochondrial rRNAs are included in such comparisons, they fit the same models with some secondary structure elements deleted. Their reduced size arises from deletions of specific internal segments rather than from the deletion of one or few nucleotides throughout their sequence. In other words, the "active sites" in this ribonucleoprotein complex have been highly conserved to catalyze the universal steps in peptide synthesis. For example, the formation of a new peptide bond between the nascent peptidyl-tRNA in the P site and the amino acyl-tRNA in the A site is catalyzed by the "peptidyl transferase," but this is not a conventional enzyme, but rather an activity intrinsic to the large ribosomal subunit. The active site is referred to as the peptidyl transferase center (PTC), and its structure is determined by a combination of domains (IV and V) of the 23S rRNA and several r-proteins, L2, L3, L4, L15, L16, L27. These relationships have been worked out primarily for *E. coli* ribosomes, and the reader is referred to recent comprehensive reviews on this very extensive topic (230–233).

Domain V of the 23S rRNA of *E. coli* contains highly conserved nucleotides, three of which are methylated on the 2′-O of ribose. Yeast mitochondrial 21S rRNA is significantly less modified globally compared to *E. coli 23S* rRNA, but the functionally and structurally homologous region has two ribose methylations and one pseudouridine (see reference 234 for references). The conservation of the modified nucleotides is already suggestive of their potential importance in the formation of the PTC, but even more weight can be placed on this interpretation from the finding that the *pet56* mutation in a nuclear gene eliminates a rRNA ribose methyltransferase—that is, an activity which is responsible for the formation of 2′-*O*-methylguanosine at G2270 in yeast 21S rRNA. The same activity can methylate *in vitro* the corresponding G2251 in 23S rRNA of *E. coli* (234). In other words, failure to methylate the G2270 causes a defect in mitochondrial protein synthesis and a respiration-deficient phenotype. Additional implications of the *pet56* mutation and other mutations in the PET56 gene are discussed by Mason et al. (234), who also present another tour de force example of the power of present-day yeast molecular genetics: T7 RNA polymerase was expressed from a nuclear plasmid, but with a signal sequence for import into mitochondria. The *E. coli* 23S rRNA gene with a T7 promoter was transformed into yeast mitochondria. Thus, *E. coli* 23S rRNA made in yeast mitochondria could be examined for the formation of the three expected nucleotide modifications, and at least two have been confirmed so far (234). This heterologous system obviously has potential for many more interesting comparisons and analyses.

From a determination of buoyant densities in CsCl gradients, it can be deduced that mitochondrial ribosomes (1.43 g/cm^3) have a higher protein: nucleic acid ratio than cytoplasmic ribosomes (1.58 g/cm^3) (235). About half of the ribosomal proteins are homologous to their prokaryotic counterparts, whereas the others are unique to mitochondrial ribosomes. A procedure for the large-scale isolation of relatively pure mitochondrial ribosomes from bovine liver has been worked out (235), but an explicit account of all the peptides present is still technically quite challenging. Chances are much better in yeast. It is estimated that there are a total of ~80 ribosomal proteins encoded in the nucleus.

To the extent that homology is a guide, the whole yeast genome can be scanned for genes encoding ribosomal proteins that are common to bacteria and yeast, but this approach may have its limitations if sequence divergence is extensive. At least 40 proteins are found in the 54S large subunit, 30 of which have been identified from the current Yeast Protein Database (236) as confirmed or likely constituents. Seventeen of those have clearly recognizable homology or similarity to ribosomal proteins of *E. coli.* A listing of the yeast r-proteins together with their bacterial homologues can be found in a recent review by Mason and colleagues (234). Unfortunately, a standard nomenclature has not yet been adopted for the yeast r-proteins, making cross-references to bacterial r-proteins somewhat confusing. The identification and cloning of yeast mitochondrial ribosomal protein genes has made it possible to examine the function of each of these proteins and even domains within each protein by the powerful techniques of molecular genetics. Constructs made *in vitro* can be substituted for the wild-type gene, and the effect of specific mutations on respiratory capacity can be examined *in vivo*. Guided by the insights derived for the peptidyl transferase center (PTC) in *E. coli*, the precise role of the presumed homologues in yeast mitochondria can be investigated, as exemplified by the work of Mason et al. (234). Many of the detailed conclusions are beyond the scope of this treatise, but a few interesting findings provide provocative contrasts or additional confirmations of structural relationships deduced for one system. Among the examples, one can cite the surprising fact that while L27 is not essential in *E. coli*, the homologous yeast mitochondrial r-protein Rml27p is absolutely essential; moreover, conserved as well as non-conserved domains must be present. It was also speculated that the additional mass of the yeast mitochondrial r-protein Rml27p may, in combination with other peptides, provide a "surrogate for the 5S ribonucleoprotein complex" found in prokaryotic and cytoplasmic ribosomes (234).

Other proteins found exclusively in mitochondrial ribosomes have been identified by chance through genetic experiments. For example, suppressor genes of a mutation in a mitochondrial translational activator protein *PET122* have been characterized and identified as nuclear genes for ribosomal proteins that are not found in other prokaryotic or eukaryotic ribosomes (see below (237, 238)).

A straightforward and definite classification of mitochondrial ribosomal proteins is possible when their genes are actually encoded by the mtDNA. No such genes are found in metazoan mitochondrial genomes, but in yeast there is one such gene (VAR1), and in plant mitochondria a variable number is found, depending on the species. Significantly, the VAR1 gene is not an r-protein gene with homologues in prokaryotic ribosomes which failed to be translocated to the nucleus in the course of evolution. It is rather unique, and as the name implies, it encodes a variant protein, with the most curious variations in different strains of yeast. The observed variation in size is the result of changes in coding sequences at one or more of four different locations: At two of these a 46 basepair cluster of GC can be inserted; at the two other positions, expansions or contractions (in phase) of AAT repeats are observed (codons for Asn). Except for the GC clusters, most of the gene consists of As and Ts. All versions of the protein are active; and, when missing, mitochondrial protein synthesis is defunct—that is, the gene is essential. Since mutations in the VAR1 gene or nuclear mutations affecting its expression would block mitochondrial protein synthesis and cause a deletion of mtDNA, such mutations are rare. The functional role played by Var1p is still subject to speculation. Proposed functions include an involvement of the peptide in small subunit assembly or a participation in translational initiation requiring message-specific translational activators (see below) and the small ribosomal subunit. In ongoing experimental approaches to study this r-protein, the gene has been reengineered for expression from a nuclear plasmid and promoter, and import into mitochondria in which the endogenous gene has been knocked out (234, 239). Because the gene functions perfectly well when expressed from a nuclear location, arguments cannot be supported that its retention in yeast mtDNA has a selective advantage or functional significance.

It has already been mentioned in another context that mt ribosomes, like prokaryotic ribosomes, are sensitive to chloramphenicol but insensitive to cycloheximide. This property makes it possible to study mitochondrial protein synthesis in intact mammalian cells in tissue culture, for example, or to eliminate residual contamination of protein synthesis in isolated mitochondria by the cytoplasmic translation machinery. The binding site for chloramphenicol can be modified by a mutation in the rRNA gene, resulting in chloramphenicol resistant mammalian cell mutants in tissue culture. Such mutants were the first cytoplasmic mutants of mammalian cells to be isolated (136, 240). Somewhat later, other phenotypes resulting from mutations in mammalian mtDNA were described: Oligomycin and antimycin resistance can be due to mutations in peptides of complex V or complex III, respectively (241–243). The isolation of such mutants also constituted one of the first indications that these peptides were encoded by mtDNA. A further discussion of mitochondrial mutations in mammalian cells will be presented in Chapter 7.

4.5.4 Cis-Acting Elements

The extreme economy of spacing genes on mammalian (and most metazoan) mt genomes has been confirmed by the analysis of the polycistronic transcripts and their processing into rRNAs, tRNAs, and mRNAs. The process yields mRNAs with no 5′ UTR, and the terminal stop codon of some is completed by the post-transcriptional polyadenylation reaction. There is also no 5′ capping on mitochondrial mRNAs. In contrast, almost all cytoplasmic mRNAs have caps, have 5′ UTRs varying in length from 20 to hundreds of nucleotides, and have 3′ UTRs of a bewildering variety and size, sometimes exceeding the entire coding sequence. The importance of the 5′ and 3′ UTRs has become recognized only during the last decade when it became quite clear that these structural elements of an mRNA can serve very crucial functions in determining the intracellular localization of the mRNA, the stability (half-life) of the mRNA, or the efficiency with which an mRNA is translated (224, 244). In each case, the primary sequence or the secondary structure resulting from the sequence can act as the cis-acting elements that, in conjunction with transacting RNA-binding proteins, govern the behavior of the mRNA.

One has to wonder whether similar mechanisms are applicable to mammalian mitochondrial mRNAs; and, if so, the relevant cis-acting elements must be contained entirely within the coding sequences. The same dilemma is not encountered for mRNAs in mitochondria of yeast and plants, for example. The genomes are much larger, lengthy intergenic sequences exist, and about 20 transcription start sites have been mapped on the yeast, *S. cerevisiae*, mt genome. Polycistronic transcripts containing only a small number of ORFs are processed to leave significant stretches of (A + U)-rich 5′UTRs and 3′UTRs in the mature mRNAs. Similarly, plant monocistronic mt mRNAs have long sequences flanking the open reading frame. In contrast, in *S. pombe* the mt mRNAs are also short and lacking 5′ and 3′ UTRs.

The importance of the 5′ UTR of yeast mt mRNAs has been convincingly demonstrated by genetic experiments (245). It is the target for imported proteins controlling the turnover and translation of these mRNAs, as will be elaborated below. At this time, only one specific sequence 5′-UUUAUA-3′ has been demonstrated to have significance, since it is found in all yeast mitochondrial mRNAs and constitutes part of a binding site for a 40-kDa protein. The binding site appears to be made up of a helical region containing this sequence and an upstream segment of unpaired nucleotides of undefined sequence. This protein, it is to be noted, is found to interact with all mitochondrial mRNAs of *S. cerevisiae*, and it therefore must be distinguished from the mRNA-specific factors that are required for stabilization and translation of the individual mRNAs (see below). Other structural characteristics (primary sequence, stem loops) responsible for protein recognition have not yet been further defined. In particular, there is no evidence for the presence of sequences which might be involved in a Shine–Dalgarno type of interaction characteristic of the association of the small ribosomal subunit with mRNAs in prokaryotes.

All the 3′ termini of yeast mitochondrial mRNAs have the conserved dodecamer sequence 5′-AAUAAUAUUCUU-3′. It is created directly by endonucleolytic processing of polycistronic transcripts, while the resulting 5′ end has to be further processed to yield the mature downstream mRNA. The conserved motif may not be responsible exclusively for specificity of the endonuclease, but may have an additional function in the stabilization and/or translation of the mRNA (186).

4.5.5 Translation Factors

It may be quite astonishing to learn that decades after the discovery of genes, mRNA, and ribosomes in mitochondria, there still is no *in vitro* system for carrying out mitochondrial protein synthesis with components derived exclusively from mitochondria. Only some isolated steps of the entire process have been successfully reproduced *in vitro*, frequently with artificial mRNAs. As a result, it has been difficult to devise an assay for the fractionation and purification of all the required initiation and elongation factors, and our understanding of translational initiation in mitochondria is still incomplete. Two approaches have been partially successful in identification of at least some of the factors. A biochemical approach has been based on assays for factors capable of supporting a limited number of steps of the reaction (e.g., binding of fMet-tRNA to mitochondrial ribosomes), and a genetic approach has been successful primarily in the yeast *S. cerevisiae*, because of the large number of mutations that have already been selected, screened, and categorized.

Before the mitochondrial factors are described, a brief recall of the factors active in normal cytosolic translation is in order. In prokaryotes there are three factors required for initiation: IF1, IF2, IF3. In eukaryotes there are five "factors" involved in initiation: eIF1A, eIF2, eIF3, eIF4, and eIF5. There are two factors needed for elongation, termed Tu and Ts in prokaryotes and called EF1 and EF1β in eukaryotes. Finally there are termination factors (TF). A flurry of activity and the powerful combination of biochemistry and genetics has led to the elucidation of both the structure and individual functions of these factors in prokaryotes and in the cytosol of eukaryotes. In prokaryotes the three IFs bind to the small ribosomal subunit. F-Met tRNA, the ribosomal subunit, and mRNA then form a ternary complex from which IF3 has been released. The complex includes the Shine–Dalgarno sequence and the initiation codon on the mRNA. Finally, the large subunit is assembled, displacing IF1 and IF2 to complete the 70S initiation complex, and elongation is ready to start. In eukaryotes, in contrast, a complex including eIF1A, eIF3, eIF2, the 40S ribosomal subunit, and f-Met tRNA is assembled, while the mRNA is separately bound by eIF4. The two are combined at the 5′ cap site, and a scanning mechanism moves the 40S ribosome to the first AUG, where finally the other factors are released, and the large 60S ribosomal subunit is assembled with the help of eIF5.

The "factors" are somewhat misnamed, because each eIF consists of multiple subunits whose precise function at each step is still being refined (e.g., references 246 and 247). For example, eIF4 consists of several peptides: one for binding to the 5′ cap of a eukaryotic mRNA, another performing the function of a helicase, presumably involved in the scanning of the 5′UTR and the removal of any secondary structure (stem-loops). Many of these peptides are now recognized as substrates for one or more serine–threonine kinases which can modulate the activity of the factor and thus control the rate of initiation of translation, either globally or more selectively with mRNAs having the appropriate cis-acting sequences.

In contrasting the prokaryotic and eukaryotic systems, one should note specifically that in prokaryotes the small ribosomal subunit appears to be positioned directly over the start codon with the help of an interaction between the Shine–Dalgarno sequence four to seven nucleotides upstream from the start codon in the mRNA and a complementary nucleotide sequence on the 3′ end of the 16S ribosomal RNA. In eukaryotes the initial assembly occurs at the extreme 5′ end (cap), followed by a scanning mechanism to find the start codon. Secondary structure can interfere and must be overcome. The stability of such obstructions, and mechanisms to overcome them, can be additional control points for translational control.

Does mitochondrial protein synthesis follow the prokaryotic or the eukaryotic model? There are no caps and no Shine–Dalgarno sequences in mitochondrial mRNAs. Mammalian mitochondrial mRNAs have no 5′UTR, and hence no need for scanning, but certainly a need exists for proper positioning of the ribosomes over the first codon. In other organisms, relatively long 5′ UTRs exist, with significant secondary structure. What factors do we need? Which of these are common? Which of these are unique to one or a small subset of organism?

During the last decade, two initiation factors (IF-2_{mt}, IF-3_{mt}) and three elongation factors (EF-Tu_{mt}, EF-Ts_{mt}, and EF-G_{mt}) have been purified and characterized from animal mitochondria (see (248) for an update and original references). Typical starting material is fresh liver from male Angus cattle, and about 30 g of mitochondria are used for making the initial crude extract (249). The activity of the initiation factor IF-2_{mt} can be assayed by its stimulation of the binding of f-Met-tRNA to mitochondrial ribosomes (or 28S subunits) and poly (A, U, G) as a template. The 85-kDa protein appears to function as a monomer similar to the bacterial IF-2. The subsequent isolation of the corresponding genes from various bacteria and eukaryotes has permitted a detailed comparison of several domains of this protein, and such studies have revealed a high degree of functional conservation. Interactions with Met-tRNA and fMet-tRNA as well as with the ribosome (and specific ribosomal proteins) have been characterized. The mammalian initiation factor IF-3_{mt} was difficult to isolate by biochemical approaches, but was eventually identified by "homology cloning/cybersearching" using IF-3 sequences from *Mycoplasma* and *Euglena gracilis*. The protein, expressed in *E. coli*, can promote certain initiation steps *in vitro*, suggesting that it promotes the dissociation of the ribosome into subunits (248), but a more detailed definition of its role requires further experiments. So far no orthologue for the bacterial IF-1 has been found in mitochondria. Some molecular modeling studies indicate that an insertion in the mammalian IF-2_{mt} can assume the role of IF-1 and make a separate mitochondrial IF-1 obsolete (248).

The EF-Tu_{mt} activity was shown to be able to replace the corresponding factor from *E. coli* in the synthesis of polyphenylalanine in the presence of bacterial ribosomes and a poly(U) "message." This is probably one of the clearest indications of the relationship of the mitochondrial factor to the prokaryotic factor. In a similar vain, EF-Ts_{mt} can stimulate the exchange of guanine nucleotides bound to *E. coli* EF-Tu, and EF-G_{mt} can substitute for *E. coli* EF-G in poly(U)-directed polymerization of phenylalanine with bacterial ribosomes. By now the properties of EF-Tu_{mt} and EF-Ts_{mt} have been characterized in considerable detail. Many primary sequences from a variety of organisms are available, and secondary and tertiary structures have been determined or predicted. As pointed out in an authoritative review (248), a remaining challenge is to understand how these factors interact with the unusual tRNAs in mammalian mitochondria. And, while EF-Ts_{mt} has been purified, its characterization lags behind. The problem of elongation in mammalian mitochondria may therefore be close to a solution, to the extent that we understand the same problem in *E. coli*. Initiation remains a problem awaiting further exploration.

There is also the genetic approach. Although a mammalian mutant cell line with a defect in mitochondrial protein synthesis (and presumably in initiation) has been described (79, 80, 250), it is quite apparent that the mammalian cell in tissue culture is not the optimal system for isolating such mutants. In the most elaborate attempt to date, about 50 respiration-deficient mutant cell lines were isolated and grouped into seven complementation groups, one of which had the phenotype of interest here (251). This is to be contrasted to the very large collection of yeast, *S. cerevisiae, pet* mutants, so-called because they make only small (petite) colonies on media with nonfermentable carbon sources such as glycerol, but grow normally on glucose (82). Because of the many nuclear genes required for making respiration competent and oxidative phosphorylation-competent mitochondria, many different complementation groups have been identified, and many of these mutants are still frozen away in private collections waiting to be further characterized. Quite a few have been examined in detail, and among them a few have been shown to have defects that have shed light on aspects under discussion in this chapter. It is difficult to select yeast mutants specifically defective in mitochondrial protein synthesis (possibly as suppressors of other mutations?), but the ease of isolation of *pet* mutants makes it possible to obtain a large sample of mutants including those of potential interest for any aspect of mitochondrial biogenesis. Some specific strategies for isolating *S. cerevisiae* mutants defective in mitochondrial translation factors have been proposed by Fox (237).

A complete deficiency of a factor essential for mitochondrial protein synthesis leads for unknown reasons to a total loss of mitochondrial DNA. Only leaky mutations could be identified with sufficient activity to maintain the mitochondrial genome, but below the level required to sustain respiration and growth on nonfermentable carbon sources. The difference in growth rate in the presence or absence of glucose could be exploited to isolate the corresponding genes by complementation, and frequently their identity could be deduced by sequence homology with known genes. Among others, nuclear genes encoding homologues of the bacterial initiation factor 2 (IF2), the bacterial elongation factor G (EF-G), a release (termination) factor, and several amino acyl tRNA synthetases have been cloned and characterized (237) by this approach. The yeast mitochondrial elongation factor EF-Tu has been cloned with the help of the corresponding bacterial gene as a probe in a screen of a yeast DNA library.

The reader is referred to the excellent and authoritative essay by Fox (237) for the other potential strategies and the many intricate aspects of yeast genetics which can be exploited and/or must be considered in the design of selection schemes.

The most unexpected initial discovery was that nuclear mutations could prevent the assembly of an active cytochrome oxidase because of the failure to synthesize a single mitochondrially coded subunit. In other words, the deficiency did not affect transcription of mtDNA, or have a global affect on mitochondrial protein synthesis, but instead it blocked the translation of a specific mitochondrial mRNA. The first such genes to be identified by such means were *PET494* and *PET111*, which are required for the translation of *COX3* and *COX2* mRNAs, respectively (238). Since then, the need for specific translation factors has been demonstrated for five of the seven membrane proteins encoded by the yeast mtDNA (252, 253). An eighth mitochondrial mRNA in yeast encodes a hydrophilic ribosomal protein. The translation of *COX2, COX3*, and *COB* mRNAs has received most of the attention, and therefore our understanding of the potential role of these factors is the most advanced. Again, the lack of a faithful *in vitro* system has slowed the analysis and has made it dependent on the fortuitous discovery of nuclear mutations and rearrangements of the yeast mitochondrial genome. A summary of present insights will serve to make cautious generalizations where possible, but also to emphasize where information and understanding is still lacking. Some of the conclusions derived from the study of translation in yeast mitochondria may also apply to the translation of at least some chloroplast mRNAs; similarly, studies with chloroplasts may have relevance and serve as guidance in mitochondrial investigations.

The translation of the COX3 mRNA requires three nuclear genes found so far: *PET494, PET54*, and *PET122*. The gene products are normally present at very small and undetectable levels, but overproduction from multicopy plasmids allows immunological detection and localization of these gene products exclusively in mitochondria. They are all found associated with the membrane, except that ~50% of Pet54p was found soluble in the matrix (where its additional function in splicing an optional intron in *COX1*–pre-mRNA may be required). Membrane extractions with increasingly harsher conditions indicate that Pet494p and Pet122p are strongly membrane bound (integral membrane proteins? Have transmembrane regions been identified in the peptide sequence?), while Pet54p acts as a typical peripheral membrane protein. A variety of indirect experiments strongly suggest the formation of a complex including all three peptides. In a yeast two-hybrid system, pairwise interactions were detected between Pet54p and Pet122p and between Pet54p and Pet494p, but not between Pet122p and Pet494p, the two integral membrane proteins. One can speculate that their matrix domains bind and are bridged by the Pet54 peptide. An interaction is also strongly indicated by allele-specific pet45 missense suppression by a missense substitution in PET122 (238).

PET111 required for *COX2* mRNA translation is so far the only nuclear gene identified. In overproducers it can also be detected as an integral membrane protein in mitochondria. For *COB* mRNA translation the nuclear gene products Cbs1p and Cbs2p are necessary. Cbs1 is integrated into the mitochondrial inner membrane, while Cbs2 has properties of a peripheral membrane protein. It is not clear whether they interact directly, or only when called into action in the initiation of translation of *COB* mRNA.

Experimentally, there are serious challenges to demonstrate the specificity of the factors for the 5′UTR of a single mRNA, and even more so in the identification of target sequences within the 5′UTRs (254). Standard molecular genetic techniques can make the constructs of interest for testing, but the problem is to get the genes into the mitochondria. Some early studies took advantage of rearranged or deleted mitochondrial genomes from which chimeric transcripts are made with the 5′UTRs from one gene fused to coding and 3′UTRs of other genes. (255). Such genomes can be maintained as suppressor rho^- genomes in the presence of wild-type (rho^+) genomes. A full discussion of the genetics of such systems is beyond the scope of the current discussion. It is sufficient to say that many yeast mitochondrial genomes with deletions and rearrangements have been discovered, and strains carrying such genomes can be constructed and maintained.

A new and potentially very powerful approach is to make the constructs to be tested *in vitro* and then introduce them into yeast mitochondria by microprojectile bombardment. Such plasmids can stably transform yeast mitochondria lacking endogenous mtDNA (256). Mating of appropriate yeast strains can combine the plasmid with an otherwise wild-type mitochondrial genome, and recombination can lead to the creation of specific changes in the 5′UTR of the COB gene. The more complex aspects of this procedure are also not to be discussed here in detail. An explicit description and the relevant background can be found in a recent paper by Mittelmeier and Dieckmann (257).

A single or a combination of approaches outlined above has confirmed that the translation of a chimeric mRNA required a translational activator that was specified by the 5′UTR, and not by any other downstream elements. Such a conclusion strongly implies that specific cis-acting sequences within the 5′UTR must be responsible for recognition and binding. Simple inspection of sequences has not been fruitful. The regions are often but not always quite long, they are (A + U)-rich, and they have the capacity for secondary structure formation as predicted by computer modeling. Since each transcript is unique and requires its own factor(s), a search for consensus sequences is potentially productive only in comparisons of the 5′UTRs of the same mRNA from several related yeast strains. Attempts to define functional elements require deletions in the 5′UTRs. Those may be found serendipitously, or induced deliberately by a procedure outlined above.

The *COX3* 5′UTR is 613 nt long. Deletion analysis has indicated that a 150-nt region between −480 and −330 relative to the start codon is required for translational activation (238). A more systematic analysis of sequences required for translation has been made with the COB mRNA, which has a 5′UTR of 954 nucleotides (257). Two distinct regions were found to be important: The sequence elements between −232 and −4 are required for translation; observations with additional overlapping deletions confine the required regions to two: between −170 and −104 and between −60 and −4. A further refinement suggested that proximal sequence between −33 and −4 is responsible for the selection of the correct start codon. The latter conclusion was based on the interesting result that the deletion of this proximal sequence did not arrest translation of COB mRNA, but caused the appearance of a novel protein that is a truncated form of cytochrome b, possibly by initiation at the first in frame AUG codon at +96. It was still not possible to distinguish clearly between loss of a binding site for Cbs1p or Cbs2p or even ribosomes, or a change in the three-dimensional structure of the RNA necessary for creating a crucial binding site.

In contrast, the *COX2* 5′UTR is only 54 nucleotides long, the shortest of all the mitochondrial mRNAs in yeast. In principle, it would therefore be easier to define a target for an activator by site-directed mutagenesis, but the task of getting such mutations into mitochondria is not reduced.

One view of translational activators is by analogy with transcriptional activators. A site on the 5′ end of mRNA positions them such that they can interact with and stabilize the basic machinery for translation, in this case one or more initiation factors, a ribosomal subunit or the entire ribosome, and probably the f-Met-tRNA. Interestingly, a Pet122 peptide with a truncated C-terminus has been shown to be inactive, but the mutation can be suppressed by mutations at three other nuclear loci which on further examination were found to encode peptides of the small ribosomal subunits. These ribosomal peptides are unique to mitochondrial ribosomes and have no homology to ribosomal peptides from other systems (238). Other partial or complete deletions or point mutations in the *PET122* gene are not suppressed by these mutated ribosomal proteins, indicating that suppression requires protein–protein interactions. Their distinction from initiation factors may seem almost semantic, but their specificity for individual mRNAs is an important distinction. A second fact to be kept in mind is that so far they have been found only in yeast, primarily because of the attention which has been devoted to this organism. It is probably a good guess that other organisms in which mitochondrial mRNAs have substantial 5′UTRs will have similar nuclear encoded factors; the case for such factors in mammalian mitochondria is more difficult to make. There is no 5′UTR and there is effectively no space to position such an activator upstream of the start codon. The alternative would be to position such a protein downstream in a transient mode; that is, it would assist in the assembly of the translational initiation complex and then be displaced by the process of elongation. Much luck would be required to define such a gene in mammals by a genetic approach.

A second property that makes the yeast translational activators unique is their firm association with the mitochondrial inner membrane (245, 258). Furthermore, the COX1, COX2, and COX3 mRNA-specific translational activator proteins interact with each other and with other factors on the membrane as shown by co-precipitations and two-hybrid systems (245). Overexpression of Pet111p (the COX2 translational activator) prevents the translation of COX1 mRNA (259). Thus it appears that the synthesis of these three cytochrome oxidase subunits is coordinated at the membrane and presumably tightly linked to their assembly. From this perspective, another important function immediately comes to mind. The limited number of peptides made in the mitochondrial matrix have from the beginning been recognized to constitute the most hydrophobic, integral membrane proteins of the various complexes of the electron transport chain and the ATPsynthase (complex V). Their insertion into the membrane may be one of the first and crucial steps in the assembly of complexes I–V. It is therefore possible to consider the membrane-associated translational activators somewhat analogous to the components of the membrane complex which attracts polysomes to the endoplasmic reticulum for the co-translational insertion of membrane proteins and secretory proteins into the ER. However, this analogy cannot be pushed too far or made universal. The mechanism of membrane insertion may be different in mammalian (metazoan) mitochondria where the absence of a 5′UTR makes the binding of such factors to the mRNA impossible (see above). However, it is still feasible to have membrane complexes which interact with the ribosome directly without the need for a site on the mRNA. It is also instructive to consider that two of the four peptides of complex II are integral membrane proteins with three postulated transmembrane segments, but they are made in the cytosol and inserted into the inner membrane during import. It is not obligatory to synthesize all integral membrane proteins of the inner membrane in the matrix. Only very recently have organisms been found where genes for these anchor proteins have been retained on the mitochondrial genome (260–262), and therefore they must be synthesized in the matrix and integrated into the membrane like the other hydrophobic peptides.

As pointed out by Fox (238), the Var1 peptide is a mitochondrial translation product with mostly hydrophilic domains which becomes associated with a small ribosomal subunit. It is argued that since no nuclear mutations have been found affecting the expression of the VAR1 gene, there may be no translational activators for the VAR1 mRNA. On the other hand, genetic arguments can be made that such mutations would not yield viable cells in the genetic screens used so far, and Fox makes the suggestion that the presence of a VAR1 gene on a plasmid in the nucleus, engineered to have an ORF for translation in the cytosol and import into mitochondria, could provide a parental strain with which mutations blocking the expression of the mitochondrial *VAR1* gene might be screened for.

Finally, it has been noted that the absolute number of Pet494p molecules is extremely small, between 2 and 60 per cell, which can be compared with the estimate of ~100 copies of the *COX3* gene per diploid cell (238). Similarly, the abundance of the other translational activators is in the same range. They therefore appear to be the limiting factors for the assembly of complex IV (cytochrome oxidase) in mitochondria. Not surprisingly, the steady-state levels of *PET494* mRNA is subject to regulation depending on the growth conditions, being up-regulated when yeast is grown in media with nonfermentable carbon sources and down-regulated in the presence of glucose (263). A more detailed discussion of the coordinate regulation of gene expression and of the assembly of the complexes of the mitochondrial electron transport chain will be deferred to a later chapter.

In the discussion of transcription (Section 4.4.1) it has already been pointed out that transcription and translation are likely to be coupled by the association of various known and unknown factors in the nucleoids. The mtRNA polymerase of *S. cerevisiae* has an amino terminal domain that is a binding site for the Nam1 protein. The function of Nam1p has been proposed to facilitate the association of newly synthesized mRNA (or active transcription complexes) with the mitochondrial inner membrane (173, 245). The corresponding N-terminal extension of the human mtRNA polymerase exhibits no homology to that of yeast, and so far no factors have been characterized that might target this sequence in mammalian mitochondria.

4.6 PROTEIN IMPORT INTO MITOCHONDRIA

4.6.1 Mitochondrial Targeting of Proteins

Mitochondria are distinct subcellular organelles. The existence of an outer and an inner membrane therefore defines two compartments topologically separated from the cytosol and from each other, with the matrix on the inside and the intermembrane space between the two membranes. The identification of specific proteins in the intermembrane space occurred relatively more recently, but from the time of the understanding of the morphology of mitochondria the following question naturally arose: How do proteins get into the matrix? This question was soon broadened to include the proteins found to be associated specifically with the inner and outer mitochondrial membranes.

The problem also became more urgent as soon as it was recognized that the coding capacity of the mitochondrial genome was extremely limited and that therefore the vast majority of the mitochondrial proteins were encoded by the nuclear genome and synthesized in the cytosol. Pioneering and elegant studies initiated in the laboratories of Schatz and Neupert with yeast, *Saccharomyces cerevisiae*, and many contributions by the disciples trained in these laboratories and others have provided answers to increasingly complex and detailed questions. Many complementary studies were also conducted with the mold *Neurospora crassa*. Whenever it has been possible to test key features or mechanisms with mammalian mitochondria, the similarities have been striking, and it appears that the principal insights into mitochondrial protein import derived from yeast will be applicable to mitochondria from many if not all organisms. A question not often raised is the question about the evolutionary origin of the import machinery (86). With some exceptions, the proteins involved are unique and unrelated to other proteins of eukaryotic cells. From the point of view of the endosymbiont theory of the origin of mitochondria, it is clear that the development of a protein import system must have preceded the significant transfer of genes from the endosymbiont cell to the nucleus; that is, it constitutes an early event in the evolution of mitochondria. There was no obvious need for protein import in the prokaryotic ancestors of mitochondria, unless scavenging for "loose" proteins constituted a possible source of nutrients. One might therefore expect closely related and homologous proteins in phylogenetically diverse organisms dedicated to this task. The explosion of genomic sequence information from many organisms in the past decade has provided a wealth of supportive data for the specialist (see Chapter 2). Plants present the additional challenge of understanding the biogenesis of chloroplasts, where similar problems with protein import have to be solved. In chloroplasts there is one additional compartment formed by the thylakoid membranes. Comparative studies on these two organelles promise to be revealing.

As always, the development of experimental approaches and model systems was crucial for progress in this field, and two breakthroughs can be singled out. First, Schatz and his collaborators succeeded in using isolated mitochondria from yeast in their import studies, with peptides to be imported also made *in vitro*, using the reticulocyte lysate or wheat germ systems. This approach became feasible when it was discovered that at least *in vitro* protein import was post-translational, in contrast to the situation encountered with protein synthesis and translocation into the endoplasmic reticulum. The second methodological advance is becoming a refrain, or in more colorful language, "déjà vu all over again": The combined approach of molecular genetics and cell biology/biochemistry in yeast has facilitated the analysis in unprecedented detail, because of (a) the ingenuity and power of selective methods and (b) the efficient way in which mutants can be analyzed *in vivo* and by *in vitro* model systems.

There is still some uncertainty whether protein import is post-translational or co-translational *in vivo* (86). The absence of a prominent coating of mitochondria by polysomes (as in the rough endoplasmic reticulum) argues against co-translational import, but it can occur, since ribosomes have been found to be associated with mitochondria and specific cDNAs have been cloned from mRNAs attached to mitochondria (264). Polysome attachment was specifically promoted in these experiments by slowing down protein synthesis with cycloheximide. A problem that was solved in broad outline quite early in this field was concerned with the distinction between (a) proteins made in the cytosol destined for import into mitochondria and (b) those which were unrelated to any mitochondrial function and had to be discriminated against by the import machinery. In the early, pioneering studies all proteins to be imported were found to have a signal sequence (or targeting sequence) at the N-terminal. It consists of 20–50 amino acid residues and is usually cleaved off upon import on the matrix side of the inner membrane. The functional significance of such a signal sequence was demonstrated most convincingly by showing that almost any protein—for example, the cytoplasmic protein dihydrofolate reductase (DHFR)—could be imported into mitochondria when it was equipped with such a signal at the N-terminal (265, 266). Since most of the early studies were performed with soluble proteins targeted to the matrix, the N-terminal sequence in the present context will be referred to as the "matrix targeting signal." Many such sequences have been characterized to date in the hope of defining criteria that would allow one to predict simply from inspection whether a protein, known from its cloned and sequenced gene, was a mitochondrial protein (267–269). The number of such sequences was surprisingly diverse in length and composition. The major distinguishing feature appears to be the ability to form an amphiphilic α-helix with positive charges on one side when viewed down the long axis (268). No obvious distinctions are found between yeast and mammalian cells, for example. As the number of examples increased, attempts were made to define consensus sequences (or patterns), and computer programs can now be found on the Internet that will identify mitochondrial proteins from genomic sequence information (www.123genomics.com/files/analysis.html). While useful, all of these programs fail to identify a substantial fraction of mitochondrial proteins. There are several reasons. A number of proteins are imported into mitochondria without further processing by a peptidase, making it more problematical to define the leader sequence. Other proteins, especially the large number of proteins found as integral membrane proteins in the inner membrane, have internal targeting sequences, and such sequences are particularly difficult to identify and categorize (see below).

The targeting sequence interacts with receptors on the mitochondrial surface, but its removal on the matrix side is appropriately discussed here also, since sequence recognition within or near the matrix targeting signal by one or more highly specific endopeptidases is required. An exhaustive survey of the literature is impossible, but a few examples will illustrate some major discoveries and generalities. Human patients diagnosed with citrullinemia and a defect in ornithine transcarbamylase (an enzyme in the urea cycle) prompted efforts to understand this mitochondrial enzyme, leading Rosenberg and his colleagues to perform some path breaking studies on mitochondrial protein import in mammalian cells. It was discovered that the matrix targeting signal of OTC was processed in two steps (270, 271). A survey of proteolytic cleavage sites identified amino acid motifs recognized by the matrix proteases (269). In general, such studies have been helpful in identifying cleavage sites in the maturation of mitochondrial proteins (the "ARg2 rule"), but an absolutely certain prediction is still not possible in all cases.

Finally, the cDNA for the β subunit of the mitochondrial processing peptidase was cloned from rat liver, and sequence comparisons established relationships with a proposed family of metallopeptidases (272). A few years ago, several of the prominent contributors to this field established a uniform nomenclature for the mitochondrial peptidases responsible for protein maturation after these laboratories had independently purified proteins, and they characterized genes from rat liver, *Saccharomyces cerevisiae, Neurospora crassa*, and potato (273, 274). Three major activities are recognized at this time: α-MPP and β-MPP are the subunits of the mitochondrial processing peptidase responsible for the initial cleavage of the targeting sequence. MIP (mitochondrial intermediate peptidase) carries out the second cleavage required for maturation of some matrix proteins, removing an additional eight amino acids from the N-terminal. For proteins sorted into the intermembrane space (see below), sorting signals have to be removed as well, and at least two integral membrane proteins with peptidase activity have been identified. They are designated IMP I, and so on, and it appears that they are heterodimers. An alternate approach to cloning the corresponding peptidases from yeast (originally designated MAS1 and MAS2 (275)) will be discussed below.

The import of a protein into the matrix requires the traverse of two membranes and hence the existence of a mechanism that allows a large, water-soluble polypeptide to pass through the lipid bilayers. Before discussing the proteins involved, and how they were identified and assigned their specific roles, it should be stated that this discussion will use the nomenclature adopted by the leaders in the field (276). It therefore becomes possible to talk about the "TOM and TIM complexes" (87) and remove at least one potential type of confusion. Tom proteins are integral (or peripheral) membrane proteins of the outer membrane participating in protein import. Tim proteins are integral membrane proteins located in the inner membrane and participating in the same process. Additional soluble proteins with distinct names are carrying out important functions on the cytosolic side of the outer membrane; similarly, a variety of soluble proteins on the matrix side have equally essential roles in the overall process. More recently, a variety of Tim proteins in the intermembrane space have attracted attention. Tom and Tim proteins are distinguished further by numbers corresponding to their molecular mass in kDa. For example, Tom20, Tom22, Tom37, Tom70, and so on, represent outer membrane proteins of the import machine.

It is difficult to do justice to the elegance and ingenious design of the many experiments that have made the system less and less of a black box, or a few rectangles symbolizing a pore through the membranes. But before losing ourselves in details, an outline and some highlights will provide perspective and allow the formulation of guiding questions.

A globular, soluble, cytoplasmic protein (for example, aquorin or the green fluorescent protein), when provided with a matrix targeting signal at the N-terminal, was found to be imported into yeast and mammalian mitochondria *in vitro* and *in vivo*. Therefore, the targeting signal had to be recognized by a receptor on the mitochondrial outer surface. Simple notions of polypeptide tails "diving" into lipid bilayers were entertained only temporarily. Since the process was post-translational, one had to assume that the protein was in a folded form; in fact the test protein, DHFR, was shown to be active before import and after import. A simple but clever experiment provided evidence, however, that the protein had to be unfolded to be imported: When the tertiary structure of the protein was stabilized by the binding of a ligand (methotrexate), import was prevented. From this and related experiments the notion was generated that mitochondrial proteins are unfolded in the process of import and threaded through an import pore as extended polypeptides. This reduces the required pore size, but creates a problem at the beginning and at the end: The protein has to be unfolded first, and the protein has to be refolded once it has arrived on the inside. More will be said about this later.

Since two membranes are involved, one could *a priori* imagine the import process to occur in two stages, with the intermembrane space being an intermediate, temporary location. However, matrix proteins were never found there, even transiently, and no conditions were found which would have arrested the process at such an intermediate stage. It could be shown that proteins had their targeting signal removed before they were even completely transferred to the matrix. For example, at lower temperatures, in *vitro* import/translocation can be initiated, but the peptide remains "stuck" in the membrane. The interpretation was that the N-terminal protruding into the matrix was subject to proteolytic processing by the matrix peptidases while the remainder of the peptide was still in transit. Another observation made with the electron microscope was that the inner and outer membrane were frequently in close contact, and it was hypothesized that these were locations where a complex of proteins in the outer and inner membrane created a channel traversing both membranes for protein import. Such a structure may indeed exist, but only transiently (see below).

A key observation made relatively early in the course of these investigations was that the isolated mitochondria used in the *in vitro* experiments had to be in a state where a membrane potential was maintained. The addition of ionophores or uncouplers destroyed the membrane potential and also made these mitochondria incapable of importing a peptide. This was and continues to be a puzzling observation, although attempts to rationalize it have been made. In general, current thinking attributes to the membrane potential a driving force that pulls the positively charged matrix targeting signal across the inner membrane, since the inside is negative relative to the outside (277). This problem is part of a larger question about the energetics of the process. What are the thermodynamic parameters that make this process feasible, efficient, and essentially unidirectional. Electrophoresis of the targeting signal across the inner membrane is part of, but not the complete, answer.

Mitochondrial protein import is more akin to active transport, first because of the recognized need for a membrane potential, and later because of the involvement of the chaperones and their associated ATPase activities. The conditions for manipulating external and/or internal ATP concentrations *in vitro* in order to study import requirements have been described (278), and some general conclusions can be stated from such experiments. An obvious one is that no internal ATP is required if a protein does not cross the inner membrane. If matrix ATP is required, then the peptide or a segment of it will have crossed the inner membrane on its path to its final destination.

4.6.2 The Protein Import Machinery of Mitochondria

It now remains to understand each of these processes more detail and to explain how the molecular architecture of the TOM and TIM complexes is designed to execute translocation of a polypeptide across two membranes, or to deliver them to other compartments. A genetic approach in yeast proved invaluable. In pioneering studies, Yaffe and Schatz (85) were the first to isolate and characterize yeast mutants defective in general protein import. Recognizing that mutations affecting all mitochondrial functions might be lethal, a collection of temperature sensitive mutants was screened for the inability to import the β subunit of the F1-ATPase at the nonpermissive temperature, 37°C. The assay was based on a clear distinction in electrophoretic mobility between the precursor in the cytosol and the mature (processed) peptide after import. More than 1000 mutagenized strains were screened; and in the first publication, two nonallelic, recessive mas mutants (*mas1* and *mas2*, mitochondrial assembly) were characterized in more detail. As described expertly and in some detail by Yaffe (279), such mutants can be analyzed for import and mitochondrial function *in vivo*, and isolated mitochondria can be further investigated *in vitro*. Specificity, observed processing, dependence on a membrane potential, and protection of the imported protein against exogenous proteases were among the criteria defined for import experiments. A large collection of reviews of methods can also be found in volume 260 of *Methods of Enzymology* (1995). By one definition, protein import is defined as the process that leads to the protection of the imported peptide against an exogenously added protease. A simple test of this type will not distinguish between import into the matrix, insertion into the inner membrane, or import into the intermembrane space. Insertion into the outer membrane may also leave a significant domain of a protein susceptible to proteolytic attack from the outside. A further distinction can be achieved by a variety of schemes. Dilution of mitochondria in hypotonic medium causes swelling and rupturing of the outer membrane, leading to the formation of mitoplasts. Outer membrane vesicles can be made and purified for functional analysis and biochemical/immunochemical assays of composition. Well-characterized antibodies, especially monoclonals, directed against specific epitopes can also give useful information about the location and orientation of a protein of interest. Nowadays the focus is often less on the mechanism of import, but on a determination of the precise location of a mitochondrial protein that co-purifies with mitochondria.

Over the years, many more *mas* mutants were characterized and the corresponding genes cloned. As described above, they were renamed Toms and Tims, but can be found in the original literature under their original "mas" designations. Not all Tom and Tim peptides were discovered from mutant screens and cloning. A number of peptides, especially in the outer membrane, were first defined by cross-linking studies to already known components. Progress reports from this field have been published on an almost annual basis from the laboratories of Neupert, Pfanner, and others; only the most recent reviews are cited here (280–286a).

Multiple subunits of one kind may be found in the complexes being formed. One of the major and uncontroversial conclusions about these complexes is that they are not stable over time, but instead are transiently assembled from the interaction of subcomplexes. An attractive and rational hypothesis is that the assembly of one may promote or stabilize the assembly of the other in the juxtaposed portion of the membrane, leading to the temporary alignment of the two channels through the outer and inner membranes (Figure 4.10).

A logical start is with the accessory proteins in the cytosol. It was first discovered by Schekman's laboratory (287) that a subset of stress proteins was required for (a) the translocation of secretory proteins across the endoplasmic reticulum membrane and (b) the translocation of mitochondrial proteins into the matrix. It is now recognized that proteins such as those belonging to the hsp70 heat-shock protein family bind to hydrophobic domains of denatured proteins, thus stabilizing an unfolded state, or they may bind already to the

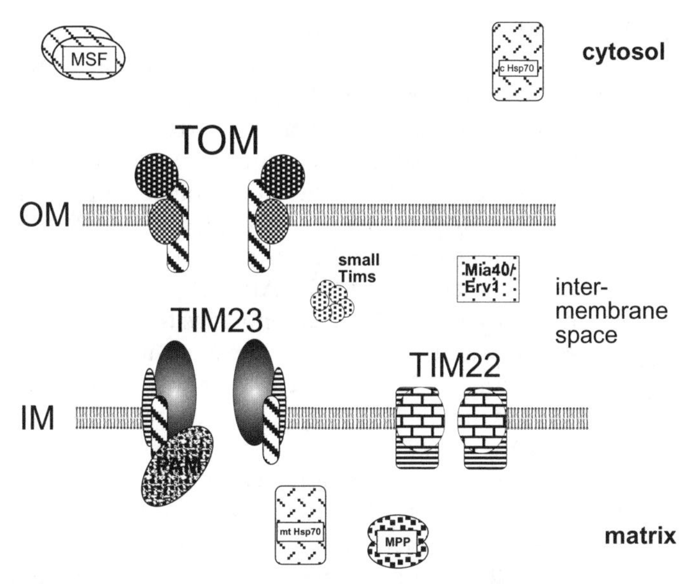

Figure 4.10 Highly schematized representation of the mitochondrial protein translocation machinery. Not all peptides are shown. In this view, the complex in the outer membrane (TOM) and the complex in the inner membrane (TIM23) are aligned to form a continuous channel through both lipid bilayers for import of proteins into the matrix. This association may be transient, as explained in the text. Proteins destined for the inner membrane also pass thorough TOM, but they are guided to TIM22; they may also interact with the small Tims in the IMS. Soluble factors on the cytosol side (MSF and cHsp70), in the IMS (Mia40/Erv1), and on the matrix side (MPP and mtHsp70) are also shown. Both TOM and TIM are made up of integral membrane proteins and peripheral membrane proteins. See text for detailed discussion.

nascent peptide emerging from the ribosome during synthesis (288). Matrix proteins are also bound by these "chaperones," so called because they are believed to chaperone a peptide to its eventual destination, or to assist it in folding into its functional tertiary structure. A mitochondrial protein is kept in an unfolded state by association with hsp70 proteins and is thus able to present the matrix targeting sequence to the receptor on the outer surface. When import has been initiated, the chaperone proteins are postulated to be sequentially released as more and more of the peptide is transferred into the mitochondrion. The dissociation of the chaperone from the imported peptide requires ATP hydrolysis. While hsp70 family members are participants in protein translocation into several subcellular organelles, a factor that is exclusively engaged in mitochondrial import has also been discovered. It is also an ATP-dependent chaperone consisting of two subunits (30 and 32 kDa, respectively) and is referred to as MSF (mitochondrial import stimulating factor). The two chaperones, hsp70 and MSP, appear to be involved in distinct import pathways that differ in the nature of the receptor on the outer surface and in the final destination of the imported protein (288, 289) (see below). Their behavior has also been contrasted in a review by Mihara and Omura (290).

All precursors destined for the mitochondria are received by the outer-membrane translocase (TOM complex); an important constituent of this complex must be a receptor capable of recognizing the targeting signal. The TOM complex consists of seven different subunits. Tom20, Tom22, and Tom70 are receptor subunits, Tom40 is the channel-forming subunit (291), and Tom5, Tom6, and Tom7 are small subunits of uncertain function, but appear to be required for stabilizing the complex (292). After the initial recognition the protein is translocated through the β-barrel topology/structure of Tom 40, and then the pathway of the imported protein depends on its structure and final destination: outer membrane (OM), intermembrane space (IMS), inner membrane (IM), and matrix. One authoritative review characterizes six different classes of proteins (281).

The simplest OM proteins have a membrane anchor in the OM, with a large C-terminal domain extending into the cytosol. The TOM complex assists in membrane insertion and then releases the protein into the OM. It is still an open question whether all OM proteins with a single transmembrane domain follow this pathway. Another class of outer-membrane proteins has a more complex topology, exemplified by the β-barrel structure of porin (VDAC) and of Tom40 itself. Such proteins enter into the IMS through the TOM complex, where they become transiently associated with some "small Tim" proteins to be delivered to the SAM complex (Sorting and Assembly Machinery) in the outer membrane. The SAM complex consists of three subunits (Sam37 alias Mas37, Sam50 (Omp85), and Sam35 alias Tom38). The need for the SAM complex was demonstrated by genetic experiments; *sam* mutants fail to incorporate β-barrel proteins properly into the outer membrane. Sam50 is homologous to the bacterial Omp85 protein involved in β-barrel protein export from the periplasm into the outer membrane and thus represents additional evidence for the prokaryotic origin of mitochondria.

From genetic experiments in yeast (293) and with an impetus from the study of the X-linked deafness–dystonia–optic atrophy syndrome in humans (294), a group of "small Tims" (Tim8, Tim9, Tim10, . . . Tim13) was discovered; these are small proteins localized in the IMS (295–297). They form specific aggregates (e.g., Tim9/Tim10) and act as chaperones for unfolded proteins in transit through the IMS (see below). More recently, two other components of this IMS import machinery have been identified: Mia40 and Erv1. All of these IMS proteins contain characteristic cysteine motifs, and it is speculated that the oxidation of sulfhydryl groups with the formation of disulfide bonds is an essential part of the mechanism of action of these proteins (295). Erv1 functions as a sulfhydryl oxidase with Mia40 as its substrate. Proteins are imported in their unfolded form, and upon folding and disulfide bond formation they are trapped in the IMS (298, 299). Alternatively, it had been proposed that the sulfhydryl groups form ligands for a metal ion (Zn^{2+}) stabilizing the folded structure of these proteins (296). A subset of small Tims associates with β-barrel proteins destined for the outer membrane via the SAM complex, while another subset (Tim9/Tim10) chaperones multi-pass integral membrane proteins (carriers) destined for the inner membrane.

The TIM complex has since differentiated into two distinct complexes in the inner membrane. The first one (TIM23) is identified by the Tim23 subunit in association with Tim50, Tim17, Tim21, and Tim44. It receives unfolded proteins with an N-terminal targeting sequence (pre-sequence) that are to be localized in the matrix. This ensemble of integral membrane proteins in the IM has been referred to as the translocation unit (280). Tim50 makes the initial contact with the precursor emerging from TOM. Inserting the pre-sequence through the membrane channel formed by the Tim23 subunits requires a membrane potential ($\Delta\Psi$). Tim 23 has a most unusual topology. Fifty amino acids at the N-terminal actually cross the outer membrane, followed by a linker, and a C-terminal domain with four transmembrane helices in the inner membrane forms the actual channel. The protein thus "cross-links" the OM and IM, but it does not appear to interact specifically and strongly with the TOM complex. The linker may serve as an additional receptor for the presequence. The C-terminal domain of Tim17 is homologous to that of Tim23; it is required for the function of the translocation unit but cannot replace Tim23.

A second essential component of TIM23 is the import motor (280). Other reviewers refer to this motor as the PAM complex (281). Its constituents are peripheral membrane proteins. In its fully assembled form, Tim44 acts as the organizer binding to Tim17 and Tim23, it binds the Tim14-Tim16 subcomplex, and it recruits the mtHsp70 chaperone from the mitochondrial matrix. As the unfolded precursor emerges from the Tim23 channel, it immediately becomes engaged by mtHsp70, a well-studied protein that binds and releases unfolded peptides as a result of conformational changes coupled to ATP hydrolysis. For the operation of the motor, two models have been proposed. A Brownian ratchet has been invoked to create a biased diffusion of the precursor into the matrix. In this model a protein in transit is considered to diffuse back and forth (oscillate), with chaperones on the outside and on the inside competing for segments to bind to in a dynamic equilibrium. If the affinity of the chaperones on the inside is slightly higher than that of the outside chaperones, the diffusion through the channel will be biased (300). Alternatively, the ATPase of mtHsp70 could execute a power stroke and actively pull the precursor through the channel to which it is attached. The analogy with motors operating on cytoskeletal elements (microfilaments) has been proposed; but in that case, polymeric filamentous structures are involved with regularly spaced binding sites, while in the case of mitochondrial import a single peptide chain with almost any sequence can be imported. Can a chaperone creep along a peptide chain? As noted by the authoritative review by Mokranjac and Neupert (280), many more mysteries of the TIM23 translocase remain to be uncovered.

The presequence is cleaved by a matrix metallo-protease before the entire protein has traversed the TIM23 complex. The mature protein is, however, not released directly into the matrix, but is delivered to a large complex made of hsp60 and cpn10 proteins and referred to as chaperonin. These proteins are related to the bacterial chaperones GroEL and GroES. Further ATP hydrolysis leads to the eventual release into the matrix of a mature, biologically functional, soluble enzyme. This is the simplest scenario.

Increasingly more complex situations arise when the protein, after import into the matrix, becomes associated with other, heterologous subunits and integral membrane protein domains on the inner membrane. A relatively simple example is the assembly of the flavoprotein and the iron–sulfur subunit of succinate dehydrogenase with each other and with the two membrane anchor proteins to form complex II, and a more complex problem is the assembling of peptides of the F_1 oligomeric complex ($\alpha_3\beta_3\gamma\delta\varepsilon$) and the F_0 oligomeric complex of the ATP synthase (complex V, see Section 5.5). Much needs to be learned about the pathway of assembly of these complexes (see Chapter 5). It is not surprising that the hsp60/cpn10 chaperonin complex is not the only activity assisting with protein folding. For example, in some complexes of the electron transport chain proteins have to become equipped with iron–sulfur centers (see Section 6.8).

The second TIM complex is now referred to as TIM22, with subunits Tim22, Tim54, Tim18, and a subset of small Tims. It is primarily responsible for the insertion of polytopic integral membrane proteins into the IM. Early studies focused on the ATP/ADP carrier (adenine nucleotide translocase, ANT), but nowadays a very large number of metabolite carriers, shuttles, and uncoupling proteins are recognized to have a similar topology in the IM and to follow a common import mechanism. After traversing the TOM complex, such proteins interact with an aggregate of small Tims that keeps them unfolded and soluble in the IMS until delivered to the TIM22 complex. This complex then guides these carrier proteins into the IM with the required topology for function (see Section 6.9). A characteristic feature of these carrier proteins is the lack of an N-terminal presequence. Instead, internal and still poorly defined targeting signals serve initially in the recognition by the TOM complex, and later in their interaction with TIM22.

Some special examples can also be considered. Cytochrome b2 is found in the IMS. The precursor made in the cytosol has a matrix targeting sequence at the N-terminal followed almost immediately by another short peptide sequence for further targeting. Neither of them can be found on the mature protein. Models for import differ in their view of the precise role of the second targeting sequence. In the first, the precursor is translocated into the matrix following the path of a normal matrix peptide. Removal of the matrix targeting signal exposes the second targeting sequence whose function is to initiate a second translocation of the peptide in the opposite direction from the matrix into the IMS. Presumably, a receptor and pore exist for this purpose. In the alternative model the second peptide signal serves as a stop-transfer signal similar to such signals found in protein translocation across the endoplasmic reticulum. In this case, the matrix targeting signal initiates translocation and reaches the matrix for processing, but the imported peptide is trapped by the second signal inside the pore of the Tim machinery, while further translocation across the Tom channel takes place until the entire peptide is found in the IMS. A proteolytic cleavage close to the outer surface of the inner membrane would liberate the protein as a soluble constituent in the intermembrane space. Integral membrane proteins with peptidase activity have been identified (see above) which could perform this cleavage.

As another example one can consider the two membrane anchor proteins of complex II, which interact with the succinate dehydrogenase heterodimer composed of the iron–sulfur subunit and the flavoprotein. These anchor proteins (C_{II-3}, C_{II-4}) are synthesized in the cytosol in most organisms, although a few exceptions have been found where one or more of these peptides are encoded by mtDNA and made in the matrix (260–262). The peptides are relatively small (15, 12–13 kDa, respectively), and hydropathy plots have suggested three transmembrane segments, with the relatively large N-terminal domain exposed to the matrix (now confirmed by X-ray crystallography). They also have N-terminal presequences that are removed in the mature proteins. A plausible mechanism for import includes a combination of N-terminal and internal-membrane targeting signals and stop transfer signals, and hence an insertion of the peptide into the membrane in a sequential, loop-by-loop fashion. The precursor is imported via TOM and TIM23 like a matrix protein, but is then prevented from entering the matrix by a stop-transfer sequence (the first transmembrane helix). Further transfer through TOM locates the C-terminal domain in the IMS. It is now necessary to postulate the formation of a hairpin loop and the insertion of this loop (with or without the help from TIM22) into the IM to form the second and third transmembrane helices of the mature protein. A radically different mechanism is suggested by the existence of organisms in which the C_{II-3} anchor protein is made in the mitochondrial matrix. Here the peptide clearly has to be inserted into the membrane from the inside. In this case the mechanism of membrane insertion is likely to resemble the mechanism for topogenesis of the other 13 mitochondrial proteins encoded by the mtDNA.

The membrane insertion of proteins synthesized in the matrix is likely to occur by a co-translational mechanism that includes mRNA-specific factors (see Section 4.5.4) as well as the protein Oxa1. It remains to be seen whether it is also dependent on the presence of other select subunits of the appropriate complex in which they function. Oxa1 is an integral membrane protein belonging to the conserved protein family Oxa1/YidC/Alb3 that is also represented in prokaryotes. It has a C-terminal domain on the matrix side that has been shown to interact with mitochondrial ribosomes (301, 302). Oxa1 not only is required for the insertion of mitochondrially encoded proteins, but also plays a role in the insertion of nuclear-encoded inner-membrane proteins as well (303). In a study of its own synthesis and localization, it was found to involve TIM23, but it may be then inserted into the IM in a mtHsp70-independent manner (304). Some further discussion of the membrane insertion of mitochondrially encoded proteins will be deferred until the assembly of individual complexes of the electron transport chain is discussed.

The discussion so far has rested heavily on experiments conducted with *Saccharomyces cerevisiae* and *Neurospora crassa* mutants and mitochondria. Will there be dramatic differences or novel concepts to be discovered in other organisms? Some experience with mammalian mitochondria and with plant mitochondria would lead to the conclusion that the basic mechanisms have been conserved. Tissue-specific and developmentally regulated differences in efficiency may be found, but the use of matrix targeting sequences and the proteolytic processing are a universal aspect of protein import, and the Tom and Tim machinery may show differences only in detail.

4.7 IMPORT OF TRANSFER RNA INTO MITOCHONDRIA

Up till now it has been tacitly assumed that proteins are the only macromolecule imported into mitochondria. A highly charged polyelectrolyte such as an RNA would be expected to have difficulty crossing two membranes; and for some time, RNA import into mitochondria was considered irrelevant. An exception, the import of the RNA associated with the mitochondrial RNA processing RNase, has already been mentioned in an earlier section (Section 4.3). From sequencing the mitochondrial genomes of diverse organisms, it is now becoming apparent that many such genomes do not encode all of the tRNAs required for translation of mitochondrial mRNAs, and thus tRNAs have to be imported. In one of the most extreme cases, mitochondrial tRNAs of *L. tarentolae* (305) and of *Trypanosoma brucei* (306) are all synthesized in the nucleus and imported. This was established first by showing that the purified mitochondrial tRNAs from *L. tarentolae* hybridized exclusively with nuclear DNA, but not with minicircle or maxicircle DNA. A large fraction of mitochondrial tRNAs in plants are imported, and there are significant differences among different species and even between relatively closely related plants. It has therefore been suggested that the ability to import specific tRNAs has been acquired at different times during the evolution of higher plants (307). However, it has to be kept in mind that a complete set of such tRNAs is required at all times for translation, and therefore one must imagine that import of a nuclear-encoded tRNA has to precede the loss of the isoaccepting tRNA transcribed from the mitochondrial genome. Illustrating this idea, in *Saccharomyces cerevisiae* there are three lysine isoacceptors: One is made and found exclusively in mitochondria, a second is found exclusively in the cytosol, and a third is found in the cytosol (95%) and in mitochondria (5%) (308, 309).

The specific mechanism operating in the import of tRNAs is starting to come into view (306, 307, 309–311). Plant tRNA genes have been identified and used to construct transgenic plants in order to demonstrate that a heterologous tRNA is also imported into the mitochondria of the transgenic plant—

that is, to demonstrate that some universal signal/sequence must be recognized (310, 312). An emerging consensus is that the tRNA is associated with the precursor of the cognate mitochondrial tRNA synthetase, and it may be imported in its aminoacylated form, but this conclusion may not apply to all organisms. In *L. tarentolae*, mutated $tRNA^{Gln}$ that cannot be charged by the synthetase can still be imported into mitochondria as long as an import signal in the D-loop is intact (313). Sequence and structural changes in the imported yeast $tRNA^{Lys}$ have been investigated by *in vivo* and *in vitro* experiments to show that binding to the synthetase, and the capacity to be aminoacylated are well-correlated with the import efficiency (308). The protein translocation apparatus (Tom and Tim machine) is clearly involved (311), but this raises a problem if the RNA/protein complex is to remain intact. Peptides are thought to be translocated in an unfolded form. Since the known mechanism for aminoacylation of tRNAs does not include any covalently joined intermediate between tRNA and the enzyme, it is likely that the enzyme and the tRNA would have to preserve their secondary and tertiary structure during import to remain noncovalently bound to each other. In support of this idea, Entelis et al. (308) have demonstrated that when tRNA is nicked, the 3′ end can still direct mitochondrial import of the 5′ end. The latest extension of the studies of the import of tRNALys in yeast has identified the glycolytic enzyme enolase as an essential factor in the formation of the complex between the tRNA and lysyl-tRNA synthetase (314). Thus, the glycolytic enzyme can execute the function of a molecular chaperone. It remains to be seen whether this is a peculiarity of the yeast enolase, where only a limited number of tRNAs have to be imported. What are the factors and what is the specificity in the case of organisms where many, if not all, tRNAs have to be imported?

4.8 REGULATED PROTEIN DEGRADATION IN MITOCHONDRIA

When yeast and mammalian mitochondria were analyzed biochemically, proteases represented a prominent enzymatic activity. Further examination revealed that proteolysis was greatly stimulated by ATP and in fact was associated with ATP hydrolysis. This protease activity was distinct from the very limited activity required for the removal of the signal sequence from imported proteins, as described above. However, the target of these proteases and the real biological significance of their presence has become clearer only in recent years.

Fractionation of mammalian mitochondrial matrix proteins yielded an ATP-dependent (vanadate-sensitive) protease with properties resembling those of the *E. coli* Lon protease (315). Cloning of the corresponding genes in yeast and humans has been accomplished within the past few years, and sequence analysis confirmed a relatedness to the *E. coli* enzyme, although the eukaryotic peptides appeared to be longer with sequences extensions having an as yet unknown function. The gene was designated *PIM1* (proteolysis in mitochondria). A striking aspect of these serine proteases is their high molecular mass, due to the formation of an oligomeric (hexamer) functional form.

With the gene identified, the yeast system offered powerful means of establishing the function of the PIM1 protease, since the gene could be knocked out and the physiological consequences could be examined. It may have come as a mild surprise that *pim1* mutants are respiration-deficient and affected in mitochondrial biogenesis (316). More puzzling is the observation that *pim1* mutants accumulate deletions in mitochondrial DNA. It is not clear whether this is a direct effect—for example, overproduction of a Pim1p-sensitive protein with a destabilizing effect on mtDNA. Alternatively, mtDNA stability in yeast is dependent on respiration, and its integrity is compromised in other respiration-deficient mutants. The *pim1* phenotype could be rescued by the expression of the *E. coli* Lon protein in yeast, provided that it was expressed with a mitochondrial targeting sequence (317).

What is the target of the PIM1 protease? First of all, it appears to degrade specific peptides such as the β-subunit of the mitochondrial processing peptidase and the β-subunit of the F1-ATPase when cytoplasmic protein synthesis is inhibited. One can consider this an example of regulated protein degradation, and it may be part of the overall response of a yeast cell to amino acid starvation. Furthermore, mutated, misfolded polypeptides imported into the matrix are prevented from accumulating and aggregating by the activity of Pim1p. In the latter case, misfolded proteins have to be "presented" to the PIM1 protease by the mitochondrial Hsp70 system, in conjunction with the activity of two other components, Mdj1p and Mge1p (318). In the authoritative review by Langer and Neupert (319), another interesting issue is now raised: The mt-Hsp70 system is intimately involved in the normal import of proteins from the cytoplasm, aiding in their folding to their native conformation after translocation through the Tom and Tim machine (see Section 4.6). This process also requires ATP. The multiple roles played by the mt-Hsp70 system therefore require that a distinction be made between normal proteins transiently associated wth Hsp70 and abnormal proteins to be degraded by the PIM1 protease. Langer and Neupert make the suggestion that a normal protein would be released to the mtHsp60 system for its final folding, while an abnormal protein may also be released, but failing to fold properly, it may be recaptured by the mtHsp70, and after prolonged dynamic association with the mtHsp70 system the PIM1 protease may finally gain access.

Taking a cue from the existence of a second class of ATP-dependent proteases in bacteria (referred to as Clp proteases), a search for a eukaryotic equivalent function has been successful in humans (320), but a *Saccharomyces* homologue could not be identified from sequence comparisons with the entire yeast genome (319).

When cytoplasmic protein synthesis is stopped, and proteins made inside the mitochondria now fail to be assembled into complexes in the inner membrane, they are also rapidly degraded. A further examination of this process has revealed the existence of two additional ATP-dependent metallo peptidases responsible for the degradation of nonassembled mitochondrial translation products. These proteases are integral membrane proteins in the inner membrane. From amino acid sequence data the proteins can be assigned to the AAA family of proteins (ATPases associated with a variety of cellular activities) characterized by an ATPase domain comprised of 230 conserved amino acids, and as proteases they represent a disitinct subgroup of this family. Another conserved motif is the HEXXH sequence shown to be responsible for divalent metal ion binding. The two peptidases have been named m-AAA-protease and i-AAA-protease, since it has been shown that the active site of one of them is on the matrix side of the inner membrane (m-AAA-protease), while the other has its active site in the intermembrane space (i-AAA-protease).

The m-AAA-protease is an 850-kDa complex made up of heterologous but closely related peptides encoded by the YTA10 and YTA12 genes (321). Each peptide has two transmembrane helices; a short N-terminal domain and a large C-terminal domain are exposed to the matrix and ATP, and metal-ion-binding sites can be found in these domains. The Yta10p and Yta12p constituents of the complex may have different substrate specificities. A knockout of either gene leads to a failure of cells to grow on nonfermentable carbon sources; that is, the m-AAA proteases on the inner membrane appears to be essential for the assembly or maintenance of the electron transport chain and the capacity for oxidative phosphorylation. In fact, a chaperone-like activity of the m-AAA protease for the assembly of the F_0 complex of the ATP synthase has been suggested based on genetic studies. The proteolytic and chaperone-like activities appear to be independent of each other(319).

The i-AAA-protease is a similarly large complex (850 kDa), but it is made up of a single type of subunits encoded by the *YME1* gene, alias *YTA11*, alias *OSD1*. Multiple independent approaches had identified this gene (e.g., reference 322), the first time from an analyses of mutants with an increased rate of escape of mtDNA to the nucleus (323). The Yme1p has been given a broad role in the maintenance of mitochondrial morphology under various conditions, since its absence causes not only a respiratory deficiency at elevated temperatures, but also morphologically abnormal mitochondria. Destruction of the proteolytic activity by a specific point mutation causes a similar phenotype, and this has been interpreted to mean that the proteolytic activity is the most essential function (324).

The arrangement of the m-AAA- and i-AAA-proteases in the inner mitochondrial membrane with active sites on opposite sides equips the mitochondria with the proteolytic machinery for degrading integral membrane proteins with domains on either side. It has been pointed out that the complete degradation of such targets ultimately requires that the transmembrane region(s) also become exposed to the proteases. The chaperone-like activity of the m-AAA protease may be an essential function for this process (319).

Protein turnover in the cytosol was recognized several decades ago. Detailed mechanisms have become apparent only more recently. Numerous functions can be attributed to the process: (1) quality control by removal of defective peptides, (2) control of gene expression in response to diverse signals, and (3) regulation of cell cycle progression, to name a few. As distinct organelles with topologically separate compartments, mitochondria require their own set of proteases to achieve some of the same objectives.

REFERENCES

1. Dawid, I. B. and Wolstenholme, D. R. (1967) *J. Mol. Biol.* **28**, 233–245.
2. Hudson, B., Clayton, D. A., and Vinograd, J. (1968) *Cold Spring Harbor Symp. Quant. Biol.* **33**, 435–442.
3. Hudson, B., Upholt, W. B., Devinny, J., and Vinograd, J. (1969) *Proc. Natl. Acad. Sci. USA* **62**, 813–820.
4. Wolstenholme, D. R. (1992) *Int. Rev. Cytol.* **141**, 173–216.
5. Anderson, S., Bankier, A. T., Barrell, B. G., de Bruijn, M. H., Coulson, A. R., Drouin, J., Eperon, I. C., Nierlich, D. P., Roe, B. A., Sanger, F., Schreier, P. H., Smith, A. J., Staden, R., and Young, I. G. (1981) *Nature* **290**, 457–465.
6. Anderson, S., de Bruin, M. H. L., Coulson, A. R., Eperon, I. C., and Sanger, F. (1982) *J. Mol. Biol.* **156**, 683–717.

6a. Anderson, S., Bankier, A. T., Barrell, B. G., de Bruijn, M. H., Coulson, A. R., Drouin, J., Eperon, I. C., Nierlich, D. P., Roe, B. A., Sanger, F., Schreier, P. H., Smith, A. J., Staden, R., and Young, I. G. (1981) *Nature* **90**, 457–465.

7. Oda, K., Yamato, K., Ohta, E., Nakamura, Y., Takemura, M., Nozato, N., Akashi, K., Kanegae, T., Ogura, Y., Kohchi, T., and Ohyama, K. (1992) *J. Mol. Biol.* **223**, 1–7.
8. Gray, M. W. (1992) *Int. Rev. Cytol.* **141**, 233–357.
9. Unseld, M., Marienfeld, J. R., Brandt, P., and Brennicke, A. (1997) *Nature Genet.* **15**, 57–61.
10. Guy, J., Qi, X., Pallotti, F., Schon, E. A., Manfredi, G., Carelli, V., Martinuzzi, A., Hauswirth, W. W., and Lewin, A. S. (2002) *Ann. Neurol.* **52**, 534–542.
11. Nosek, J., Tomaska, L. F. H., Suyama, Y., and Kovac, L. (1998) *Trends Genet.* **14**, 184–188.
12. Okimoto, R., Macfarlane, J. L., and Wolstenholme, D. R. (1992) *Genetics* **130**, 471–498.
13. LaRoche, J., Snyder, M., Cook, D. I., Fuller, K., and Zouros, E. (1990) *Mol. Biol. Evol.* **7**, 45–64.

13a. Warrior, R. and Gall, J. (1985) *Arch. Sci. Geneva* **38**, 439–445.

14. Saccone, C. (1994) *Curr. Opin. Genet. Dev.* **4**, 875–881.
15. Lanave, C., Licciulli, F., De Robertis, M., Marolla, A., and Attimonelli, M. (2002) *Nucleic Acids Res.* **30**, 174–175.
16. Kogelnik, A. M., Lott, M. T., Brown, M. D., Navathe, S. B., and Wallace, D. C. (1997) *Nucleic Acids Res.* **25**, 196–199.
17. Chomyn, A., Mariottini, P., Cleeter, M. W. J., Ragan, C. I., Matsuno-Yagi, A., Hatefi, Y., Doolittle, R. F., and Attardi, G. (1985) *Nature* **314**, 592–597.

18. Chomyn, A., Cleeter, M. W., Ragan, C. I., Riley, M., Doolittle, R. F., and Attardi, G. (1986) *Science* **234**, 614–618.
19. Attardi, G. (1986) *Bioessays* **5**, 34–39.
20. Cardol, P., Lapaille, M., Minet, P., Franck, F., Matagne, R. F., and Remacle, C. (2006) *Eukaryot. Cell* **5**, 1460–1467.
21. Hanson, M. R. and Folkerts, O. (1992) *Int. Rev. Cytol.* **141**, 129–172.
22. Wolstenholme, D. R. and Fauron, C. M. R. (1995) Mitochondrial genome organization. In Levings, C. S. I. and Vasil, I. K., editors. *The Molecular Biology of Plant Mitochondria*, Kluwer Academic Publishers, Dordrecht, pp. 1–59.
23. Leon, P., Walbot, V., and Bedinger, P. (1989) *Nucl. Acids Res.* **17**, 4089–4099.
24. Schardl, C. L., Pring, D. R., and Lonsdale, D. M. (1985) *Cell* **43**, 361–368.
25. Lonsdale, D. M., Hodge, T. P., and Fauron, C. M. (1984) *Nucleic Acids Res.* **12**, 9249–9261.
26. Fauron, C. M., Havlik, M., and Brettell, R. I. (1990) *Genetics* **124**, 423–428.
27. Small, I., Suffolk, R., and Leaver, C. J. (1989) *Cell* **58**, 69–76.
28. Stern, D. B. and Palmer, J. D. (1984) *Nucleic Acids Res.* **12**, 6141–6157.
29. Rothenberg, M. and Hanson, M. R. (1988) *Genetics* **118**, 155–161.
30. Wolfe, K. H., Li, W. H., and Sharp, P. M. (1987) *Proc. Natl. Acad. Sci. USA* **84**, 9054–9058.
31. Clark-Walker, G. D. (1992) *Int. Rev. Cytol.* **141**, 89–127.
32. Chen, J.-J., Throop, M. S., Gehrke, L., Kuo, I., Pal, J. K., Brodsky, M., and London, I. M. (1991) *Proc. Natl. Acad. Sci. USA* **88**, 7729–7733.
33. Koeller, D. M., Horowitz, J. A., Casey, J. L., Klausner, R. D., and Harford, J. B. (1991) *Proc. Natl. Acad. Sci. USA* **88**, 7778–7782.
34. Pavan, W. J. and Reeves, R. H. (1991) *Proc. Natl. Acad. Sci. USA* **88**, 7788–7791.
35. Foury, F., Roganti, T., Lecrenier, N., and Purnelle, B. (1998) *FEBS Lett.* **440**, 325–331.
36. Lambowitz, A. M. (1989) *Cell* **56**, 323–326.
37. Dujon, B., Colleaux, L., Jacquier, A., Michel, F., and Monteilhet, C. (1986) *Basic Life Sciences* **40**, 5–27.
38. Dujon, B. (1989) *Gene* **82**, 91–114.
39. Delahodde, A., Goguel, V., Becam, A. M., Creusot, F., Perea, J., Banroques, J., and Jacq, C. (1989) *Cell* **56**, 431–441.
40. Wenzlau, J. M., Saldanha, R. J., Butow, R. A., and Perlman, P. S. (1989) *Cell* **56**, 421–430.
41. Morl, M. and Schmelzer, C. (1990) *Cell* **60**, 629–636.
42. Weiller, G., Schueller, C. M., and Schweyen, R. J. (1989) *Mol. Gen. Genet.* **218**, 272–283.
43. Stuart, K. and Feagin, J. E. (1992) *Int. Rev. Cytol.* **141**, 65–88.
44. Englund, P. T., Hajduk, S. L., and Marini, J. C. (1982) *Annu. Rev. Biochem.* **51**, 695–726.
45. Simpson, L. (1987) *Annu. Rev. Microbiol.* **42**, 363–382.
46. Simpson, L. (1997) *Mol. Biochem. Parasitol.* **86**, 133–141.

47. Kawano, S., Takano, H., and Kuroiwa, T. (1995) *Int. Rev. Cytol.* **161**, 49–110.
48. Griffiths, A. J. F. (1995) *Microbiol. Rev.* **59**, 673–685.
49. Meinhardt, F., Kempken, F., Kamper, J., and Esser, K. (1990) *Curr. Genet.* **17**, 89–95.
50. Johnston, S. A., Anziano, P. Q., Shark, K., Sanford, J. C., and Butow, R. A. (1988) *Science* **240**, 1538–1541.
51. Kornberg, A. and Baker, T. (1992) *DNA Replication*, 2nd ed., W. H. Freeman, New York.
52. Backert, S., Meissner, K., and Börner, T. (1997) *Nucleic Acids Res.* **25**, 582–589.
53. Griffiths, A. J. F. (1992) *Annu. Rev. Genet.* **26**, 351–372.
54. L'Homme, Y., Stahl, R. J., Li, X.-Q., Hameed, A., and Brown, G. G. (1997) *Curr. Genet.* **31**, 325–335.
55. Schatz, G. (1995) *Biochim. Biophys. Acta Mol. Basis Dis.* **1271**, 123–126.
56. Forner, F., Foster, L. J., Campanaro, S., Valle, G., and Mann, M. (2006) *Mol. Cell Proteomics.* **5**, 608–619.
57. Johnson, D. T., Harris, R. A., Blair, P. V., and Balaban, R. S. (2006) *Am. J. Physiol. Cell Physiol.* **292**, C698–707.
58. Prokisch, H., andreoli, C., Ahting, U., Heiss, K., Ruepp, A., Scharfe, C., and Meitinger, T. (2006) *Nucleic Acids Res.* **34**, D705–D711.
59. Vo, T. D. and Palsson, B. (2006) *Am. J. Physiol. Cell Physiol.* **292**, C164–177.
60. Da Cruz, S., Parone, P. A., and Martinou, J. C. (2005) *Expert Rev. Proteomics.* **2**, 541–551.
61. Gabaldon, T. and Huynen, M. A. (2004) *Biochim. Biophys. Acta* **1659**, 212–220.
62. Lecrenier, N., Van Der Bruggen, P., and Foury, F. (1997) *Gene* **185**, 147–152.
63. Kaguni, L. S. (2004) *Annu. Rev. Biochem.* **73**, 293–320.
64. Yakubovskaya, E., Chen, Z., Carrodeguas, J. A., Kisker, C., and Bogenhagen, D. F. (2006) *J. Biol. Chem.* **281**, 374–382.
65. Shutt, T. E. and Gray, M. W. (2006) *J. Mol. Evol.* **62**, 588–599.
66. Tua, A., Wang, J., Kulpa, V., and Wernette, C. M. (1997) *Biochimie* **79**, 341–350.
67. Zhang, H., Barcelo, J. M., Lee, B., Kohlhagen, G., Zimonjic, D. B., Popescu, N. C., and Pommier, Y. (2001) *Proc. Natl. Acad. Sci. USA* **98**, 10608–10613.
68. Anson, R. M., Mason, P. A., and Bohr, V. A. (2006) *Methods Mol. Biol.* **314**, 155–181.
69. Stuart, J. A. and Brown, M. F. (2006) *Biochim. Biophys. Acta* **1757**, 79–89.
70. Mason, P. A., Matheson, E. C., Hall, A. G., and Lightowlers, R. N. (2003) *Nucleic Acids Res.* **31**, 1052–1058.
71. Croteau, D. L., Stierum, R. H., and Bohr, V. A. (1999) *Mutat. Res.* **434**, 137–148.
72. Megraw, T. L. and Chae, C.-B. (1993) *J. Biol. Chem.* **268**, 12758–12763.
73. Megraw, T. L., Kao, L. R., and Chae, C. B. (1994) *Biochimie* **76**, 909–916.
74. Gaspari, M., Larsson, N. G., and Gustafsson, C. M. (2004) *Biochim. Biophys. Acta* **1659**, 148–152.
75. Bonawitz, N. D., Clayton, D. A., and Shadel, G. S. (2006) *Mol. Cell* **24**, 813–825.
76. Bogenhagen, D. F., Wang, Y., Shen, E. L., and Kobayashi, R. (2003) *Mol. Cell Proteomics.* **2**, 1205–1216.

77. Shadel, G. S. (2005) *Trends Biochem. Sci.* **30**, 294–296.
78. Pel, H. J., Tzagoloff, A., and Grivell, L. A. (1992) *Curr. Genet.* **21**, 139–146.
79. Burnett, K. G. and Scheffler, I. E. (1981) *J. Cell Biol.* **90**, 108–115.
80. Au, H. C. and Scheffler, I. E. (1997) *Somat. Cell Mol. Genet.* **23**, 27–35.
81. Hieter, P., Bassett, D. E., Jr., and Valle, D. (1996) *Nature Genet.* **13**, 253–255.
82. Tzagoloff, A. and Dieckmann, C. L. (1990) *Microbiol. Rev.* **54**, 211–225.
83. Glerum, D. M., Koerner, T. J., and Tzagoloff, A. (1995) *J. Biol. Chem.* **270**, 15585–15590.
84. Tzagoloff, A., Yue, J., Jang, J., and Paul, M.-F. (1994) *J. Biol. Chem.* **269**, 26144–26151.
85. Yaffe, M. P. and Schatz, G. (1984) *Proc. Natl. Acad. Sci. USA* **81**, 4819–4823.
86. Schatz, G. (1996) *J. Biol. Chem.* **271**, 31763–31766.
87. Pfanner, N. and Meijer, M. (1997) *Curr. Biol.* **7**, R100–R103.
88. Stuart, R. A., Ono, H., Langer, T., and Neupert, W. (1996) *Cell Struct. Funct.* **21**, 403–406.
89. Shadel, G. S. and Clayton, D. A. (1996) *Methods Enzymol.* **264**, 149–158.
90. Trumbly, R. J. (1992) *Mol. Microbiol.* **6**, 15–21.
91. Ronne, H. (1995) *TIG* **11**, 12–17.
92. Johnston, M. and Carlson, M. (1993) Regulation of carbon and phosphate utilization. In Broach, J., Jones, E. W., and Pringle, J., editors. Cold Spring Harbor Press, Cold Spring Harbor, NY, pp. 193–281.
93. Entian, K.-D. and Barnett, J. A. (1992) *Trends Biochem. Sci.* **17**, 506–510.
94. Zitomer, R. S. and Lowry, C. V. (1992) *Microbiol. Rev.* **56**, 1–11.
95. Forsburg, S. L. and Guarente, L. (1989) *Annu. Rev. Cell Biol.* **5**, 153–180.
96. Schneider, J. C. and Guarente, L. (1991) *Mol. Cell. Biol.* **11**, 4934–4942.
97. Haldi, M. L. and Guarente, L. (1995) *Mol. Gen. Genet.* **248**, 229–235.
98. De Winde, J. H. and Grivell, L. A. (1993) *Prog. Nucleic Acids Res.* **46**, 51–91.
99. Lombardo, A., Cereghino, G. P., and Scheffler, I. E. (1992) *Mol. Cell. Biol.* **12**, 2941–2948.
100. Daignan-Fornier, B., Valens, M., Lemire, B. D., and Bolotin-Fukuhara, M. (1994) *J. Biol. Chem.* **269**, 15469–15472.
101. Lodi, T. and Guiard, B. (1991) *Mol. Cell. Biol.* **11**, 3762–3772.
102. Thevelein, J. M. (1991) *Mol. Microbiol.* **5**, 1301–1307.
103. Tu, J. G. and Carlson, M. (1995) *EMBO J.* **14**, 5939–5946.
104. Jiang, R. and Carlson, M. (1996) *Genes Dev.* **10**, 3105–3115.
105. Cereghino, G. P. and Scheffler, I. E. (1996) *EMBO J.* **15**, 363–374.
106. Cereghino, G. P., Atencio, D. P., Saghbini, M., Beiner, J., and Scheffler, I. E. (1995) *Mol. Biol. Cell* **6**, 1125–1143.
107. Beelman, C. A., Stevens, A., Caponigro, G., LaGrandeur, T. E., Hatfield, L., Fortner, D. M., and Parker, R. (1996) *Nature* **382**, 642–646.
108. Beelman, C. A. and Parker, R. (1995) *Cell* **81**, 179–183.
109. Sachs, A. B. (1993) *Cell* **74**, 413–421.
110. Decker, C. J. and Parker, R. (1994) *TIBS* **19**, 336–340.

111. Brown, C. E., Tarun, S. Z., Jr., Boeck, R., and Sachs, A. B. (1996) *Mol. Cell. Biol.* **16**, 5744–5753.
112. Brown, A. J. P. (1993) *Trends Cell Biol.* **3**, 180–183.
113. Jacobson, A. (1995) Poly(A) metabolism and translation: The closed loop model. In Hershey, J., Mathews, M., and Sonenberg, N., editors. *Translational Control*, Cold Spring Harbor Press, Cold Spring Harbor, NY, pp. 451–480.
114. Jacobson, A. and Peltz, S. W. (1996) *Annu. Rev. Biochem.* **65**, 693–739.
115. Donnelly, M. and Scheffler, I. E. (1976) *J. Cell. Physiol.* **89**, 39–52.
116. Popot, J.-L. and deVitry, C. (1990) *Annu. Rev. Biophys. Biophys. Chem.* **19**, 369–403.
117. Evans, M. J. and Scarpulla, R. C. (1990) *Genes Dev.* **4**, 1023–1034.
118. Scarpulla, R. C. (1996) *Trends Cardiovasc. Med.* **6**, 39–45.
119. Chau, C. A., Evans, M. J., and Scarpulla, R. C. (1992) *J. Biol. Chem.* **267**, 6999–7006.
120. Virbasius, C. A., Virbasius, J. V., and Scarpulla, R. C. (1993) *Genes Dev.* **7**, 2431–2445.
121. Virbasius, J. V. and Scarpulla, R. C. (1990) *Nucleic Acids Res.* **18**, 6581–6586.
122. Virbasius, J. V. and Scarpulla, R. C. (1991) *Mol. Cell. Biol.* **11**, 5631–5638.
123. Virbasius, J. V., Virbasius, C. A., and Scarpulla, R. C. (1993) *Genes Dev.* **7**, 380–392.
124. Seelan, R. S., Gopalakrishnan, L., Scarpulla, R. C., and Grossman, L. I. (1996) *J. Biol. Chem.* **271**, 2112–2120.
125. Au, H. C., Ream-Robinson, D., Bellew, L. A., Broomfield, P. L. E., Saghbini, M., and Scheffler, I. E. (1995) *Gene* **159**, 249–253.
126. Au, H. C. and Scheffler, I. E. (1998) *Eur. J. Biochem.* **251**, 164–174.
127. Scarpulla, R. C. (2006) *J Cell Biochem.* **97**, 673–683.
128. Scarpulla, R. C. (2002) *Biochim. Biophys. Acta* **1576**, 1–14.
128a. Ryan, M. T. and Hoogenraad, N. J. (2007) *Annu. Rev. Biochem.* **76**, 701–722.
129. Wan, B. and Moreadith, R. W. (1995) *J. Biol. Chem.* **270**, 26433–26440.
130. Molkentin, J. D. and Olson, E. N. (1996) *Proc. Natl. Acad. Sci. USA* **93**, 9366–9373.
131. Olson, E. N. and Klein, W. H. (1994) *Genes Dev.* **8**, 1–8.
132. Taanman, J.-W. and Capaldi, R. A. (1993) *J. Biol. Chem.* **268**, 18754–18761.
133. Li, K., Hodge, J. A., and Wallace, D. C. (1990) *J. Biol. Chem.* **265**, 20585–20588.
134. Chung, A. B., Stepien, G., Haraguchi, Y., Li, K., and Wallace, D. C. (1992) *J. Biol. Chem.* **267**, 21154–21161.
135. Clayton, D. A., Teplitz, R. L., Nabholz, M., Dovey, H., and Bodmer, W. (1971) *Nature* **234**, 560–562.
136. Wallace, D. C. and Eisenstadt, J. M. (1979) *Somat. Cell Genet.* **5**, 373–396.
137. Day, C. and Scheffler, I. E. (1982) *Somat. Cell Genet.* **8**, 691–707.
138. Kenyon, L. and Moraes, C. T. (1997) *Proc. Natl. Acad. Sci. USA* **94**, 9131–9135.
139. Yadava, N., Potluri, P., Smith, E., Bisevac, A., and Scheffler, I. E. (2002) *J. Biol. Chem.* **277**, 21221–21230.

140. DiMauro, S. and Wallace, D. C., editors (1993); Raven Press, New York, pp. 1–206.
141. Clayton, D. A. (1992) *Int. Rev. Cytol.* **141**, 217–232.
142. Clayton, D. A. (1991) *TIBS* **16**, 107–111.
143. Shadel, G. S. and Clayton, D. A. (1997) *Annu. Rev. Biochem.* **66**, 409–435.
144. Desjardins, P. and Morais, R. (1990) *J. Mol. Biol.* **212**, 599–634.
145. Carrodeguas, J. A., Pinz, K. G., and Bogenhagen, D. F. (2002) *J. Biol. Chem.* **277**, 50008–50014.
146. Ponamarev, M. V., Longley, M. J., Nguyen, D., Kunkel, T. A., and Copeland, W. C. (2002) *J. Biol. Chem.* **277**, 15225–15228.
147. Lim, S. E., Ponamarev, M. V., Longley, M. J., and Copeland, W. C. (2003) *J. Mol. Biol.* **329**, 45–57.
148. Longley, M. J., Graziewicz, M. A., Bienstock, R. J., and Copeland, W. C. (2005) *Gene* **354**, 125–131.
149. Korhonen, J. A., Pham, X. H., Pellegrini, M., and Falkenberg, M. (2004) *The EMBO J.* **23**, 2423–2429.
149a. Falkenberg, M. Larsson, N. G., and Gustafsson, C. M. (2007) *Annu. Rev. Biochem.* **76**, 679–699.
150. Lee, D. Y. and Clayton, D. A. (1996) *J. Biol. Chem.* **271**, 24262–24269.
151. Chang, D. D. and Clayton, D. A. (1987) *Science* **235**, 1178–1184.
152. Chang, D. D. and Clayton, D. A. (1987) *EMBO J.* **6**, 409–417.
153. Wong, T. W. and Clayton, D. A. (1986) *Cell* **45**, 817–825.
154. Holt, I. J., Lorimer, H. E., and Jacobs, H. T. (2000) *Cell* **100**, 515–524.
155. Bowmaker, M., Yang, M. Y., Yasukawa, T., Reyes, A., Jacobs, H. T., Huberman, J. A., and Holt, I. J. (2003) *J. Biol. Chem.* **278**, 50961–50969.
156. Holt, I. J. and Jacobs, H. T. (2003) *Trends Biochem. Sci.* **28**, 355–356.
157. Yasukawa, T., Yang, M. Y., Jacobs, H. T., and Holt, I. J. (2005) *Mol. Cell* **18**, 651–662.
158. Brown, T. A., Cecconi, C., Tkachuk, A. N., Bustamante, C., and Clayton, D. A. (2005) *Genes Dev.* **19**, 2466–2476.
159. Brown, T. A. and Clayton, D. A. (2006) *Cell Cycle* **5**, 917–921.
160. Madsen, C. S., Ghivizzani, S. C., and Hauswirth, W. W. (1993) *Mol. Cell. Biol.* **13**, 2162–2171.
161. Fish, J., Raule, N., and Attardi, G. (2004) *Science* **306**, 2098–2101.
162. Birky, C. W., Jr. (1994) *J. Hered.* **85**, 355–365.
163. Davis, A. F. and Clayton, D. A. (1996) *J. Cell Biol.* **135**, 883–893.
164. Kang, D. and Hamasaki, N. (2005) *Ann. NY. Acad. Sci.* **1042**, 101–108.
165. Ekstrand, M. I., Falkenberg, M., Rantanen, A., Park, C. B., Gaspari, M., Hultenby, K., Rustin, P., Gustafsson, C. M., and Larsson, N. G. (2004) *Hum. Mol. Genet.* **13**, 935–944.
166. LeDoux, S. P. and Wilson, G. L. (2001) *Prog. Nucleic Acid Res. Mol. Biol.* **68**, 273–284.
167. Mandavilli, B. S., Santos, J. H., and Van Houten, B. (2002) *Mutat. Res.* **509**, 127–151.

168. Stuart, J. A., Hashiguchi, K., Wilson, D. M., III, Copeland, W. C., Souza-Pinto, N. C., and Bohr, V. A. (2004) *Nucleic Acids Res.* **32**, 2181–2192.
169. Grishko, V. I., Rachek, L. I., Spitz, D. R., Wilson, G. L., and LeDoux, S. P. (2005) *J. Biol. Chem.* **280**, 8901–8905.
170. Stuart, J. A., Mayard, S., Hashiguchi, K., Souza-Pinto, N. C., and Bohr, V. A. (2005) *Nucleic Acids Res.* **33**, 3722–3732.
171. Thyagarajan, B., Padua, R. A., and Campbell, C. (1996) *J. Biol. Chem.* **271**, 27536–27543.
172. Williamson, D. (2002) *Nat. Rev Genet.* **3**, 475–481.
173. Shadel, G. S. (2004) *Trends Genet.* **20**, 513–519.
174. Montoya, J., Ojala, D., and Attardi, G. (1981) *Nature* **290**, 465–470.
175. Ojala, D., Montoya, J., and Attardi, G. (1981) *Nature* **290**, 470–474.
176. Walberg, M. W. and Clayton, D. A. (1983) *J. Biol. Chem.* **258**, 1268–1275.
177. Shadel, G. S. and Clayton, D. A. (1996) *Methods Enzymol.* **264**, 139–148.
178. Tiranti, V., Savoia, A., Forti, F., D'Apolito, M. F., Centra, M., Racchi, M., and Zeviani, M. (1997) *Hum. Mol. Genet.* **6**, 615–625.
179. McCulloch, V., Seidel-Rogol, B. L., and Shadel, G. S. (2002) *Mol. Cell Biol.* **22**, 1116–1125.
180. Falkenberg, M., Gaspari, M., Rantanen, A., Trifunovic, A., Larsson, N. G., and Gustafsson, C. M. (2002) *Nat. Genet.* **31**, 289–294.
181. Larsson, N. G., Barsh, G. S., and Clayton, D. A. (1997) *Mamm. Genome* **8**, 139–140.
182. Reyes, A., Mezzina, M., and Gadaleta, G. (2002) *Gene* **291**, 223–232.
183. Chen, X. J. and Butow, R. A. (2005) *Nat. Rev. Genet.* **6**, 815–825.
184. Attardi, G. and Schatz, G. (1988) *Annu. Rev. Cell Biol.* **4**, 289–333.
185. de Zamaroczy, M. and Bernardi, G. (1985) *Gene* **41**, 1–22.
186. Costanzo, M. C. and Fox, T. D. (1990) *Annu. Rev. Genet.* **24**, 91–113.
187. Mangus, D. A. and Jaehning, J. A. (1996) *Methods Enzymol.* **264**, 57–66.
188. Binder, S., Marchfelder, A., and Brennicke, A. (1996) *Plant Mol. Biol.* **32**, 303–314.
189. Gruissem, W., Barkan, A., Deng, X. W., and Stern, D. B. (1988) *Trends Genet.* **4**, 258–263.
190. Christianson, T. W. and Clayton, D. A. (1988) *Mol. Cell Biol.* **8**, 4502–4509.
191. King, M. P., Koga, Y., Davidson, M., and Schon, E. A. (1992) *Mol. Cell. Biol.* **12**, 480–490.
192. Hess, J. P., Parisi, M. A., Bennett, J. L., and Clayton, D. A. (1991) *Nature* **351**, 236–239.
193. Micol, V., Fernández-Silva, P., and Attardi, G. (1996) *Methods Enzymol.* **264**, 158–173.
194. Groom, K. R., Dang, Y. L., Gao, G.-J., Lou, Y. C., Martin, N. C., Wise, C., and Morales, M. J. (1996) *Methods Enzymol.* **264**, 86–99.
195. Belfort, M. and Perlman, P. S. (1995) *J. Biol. Chem.* **270**, 30237–30240.
196. Podar, M., Perlman, P. S., and Padgett, R. A. (1995) *Mol. Cell Biol.* **15**, 4466–4478.

197. Padgett, R. A., Podar, M., Boulanger, S. C., and Perlman, P. S. (1994) *Science* **266**, 1685–1688.
198. Chapdelaine, Y. and Bonen, L. (1991) *Cell* **65**, 465–472.
199. Bonen, L. (1991) *Curr. Opin. Genet. Devel.* **1**, 515–522.
200. Covello, P. S. and Gray, M. W. (1993) *Trends Genet.* **9**, 265–268.
201. Simpson, L. (2005); in *RNA World* (editors: R. Gesteland, Cech, T. R., and Atkins, J. F.). Cold Spring Harbor Press, Cold Spring Harbor, NY. Chapter 14, pp. 401–417.
202. Simpson, L. and Maslov, D. A. (1994) *Curr. Opin. Genet. Dev.* **4**, 887–894.
203. Simpson, L. and Thiemann, O. H. (1995) *Cell* **81**, 837–840.
204. Simpson, L., Sbicego, S., and Aphasizhev, R. (2003) *RNA* **9**, 265–276.
205. Kang, X., Gao, G., Rogers, K., Falick, A. M., Zhou, S., and Simpson, L. (2006) *Proc. Natl. Acad. Sci. USA* **103**, 13944–13949.
206. Scott, J. (1995) *Cell* **81**, 833–836.
207. Benne, R. (1996) *Curr. Opin. Genet. Dev.* **6**, 221–231.
208. Janke, A. and Paabo, S. (1993) *Nucl. Acids Res.* **21**, 1523–1525.
209. Alfonzo, J. D., Thiemann, O., and Simpson, L. (1997) *Nucl. Acids Res.* **25**, 3751–3759.
210. Byrne, E. M., Connell, G. J., and Simpson, L. (1996) *The EMBO J.* **15**, 6758–6765.
211. Seiwert, S. D., Heidmann, S., and Stuart, K. (1996) *Cell* **84**, 1–20.
212. Cruz-Reyes, J., Zhelonkina, A., Huang, C. E., and Sollner-Webb, B. (2002) *Mol. Cell. Biol.* **22**, 4652–4660.
213. Maier, R. M., Zeltz, P., Kössel, H., Bonnard, G., Gualberto, J. M., and Grienenberger, J. M. (1996) *Plant Mol. Biol.* **32**, 343–365.
214. Binder, S. and Brennicke, A. (2003) *Philos. Trans. R. Soc. Lond. B Biol. Sci.* **358**, 181–189.
215. Covello, P. S. and Gray, M. W. (1989) *Nature* **341**, 662–666.
216. Gualberto, J. M., Lamattina, L., Bonnard, G., Weil, J. H., and Grienenberger, J. M. (1989) *Nature* **341**, 660–662.
217. Hiesel, R., Wissinger, B., Schuster, W., and Brennicke, A. (1989) *Science* **246**, 1632–1634.
218. Dewey, R. E., Levings, C. S. I., and Timothy, D. H. (1986) *Cell* **44**, 439–449.
219. Cummings, M. P. and Myers, D. S. (2004) *BMC Bioinform.* **5**, 132.
220. Mower, J. P. (2005) *BMC Bioinform.* **6**, 96.
221. Hernould, M., Mouras, A., Litvak, S., and Araya, A. (1992) *Nucl. Acids Res.* **20**, 1809–1818.
222. Araya, A., Domec, C., Begu, D., and Litvak, S. (1992) *Proc. Natl. Acad. Sci. USA* **89**, 1040–1044.
223. Gelfand, R. and Attardi, G. (1981) *Mol. Cell. Biol.* **1**, 497–511.
224. Belasco, J. and Brawerman, G. (1993) *Control of Messenger RNA Stability*, Academic Press, San Diego.
225. Gruissem, W. and Schuster, G. (1993) Contol of mRNA degradation in organelles. In Belasco, J. and Brawerman, G., editors. *Control of Messenger RNA Stability*, Academic Press, San Diego. Chapter 14, pp. 329–365.

226. Gillham, N. W., Boynton, J. E., and Hauser, C. R. (1994) *Annu. Rev. Genet.* **28**, 71–93.

227. Lansman, R. A. and Clayton, D. A. (1975) *J. Mol. Biol.* **99**, 761–776.

228. Gagliardi, D., Stepien, P. P., Temperley, R. J., Lightowlers, R. N., and Chrzanowska-Lightowlers, Z. M. (2004) *Trends Genet.* **20**, 260–267.

228a. Dieckmann C. L. and Staples R. R. (1994) *Int. Rev. Cytol.* **152**, 145–182.

229. Myers, A. M., Pape, L. K., and Tzagoloff, A. (1985) *EMBO J.* **4087**, 2087–2092.

230. Noller, H. F. (1991) *Annu. Rev. Biochem.* **60**, 191–227.

231. Garrett, R. A. and Rodriguez-Fonseca, C. (1996) The peptidyl transferase center. In Zimmerman, R. A. and Dahlberg, A. E., editors. *Ribosomal RNA: Structure, Evolution, Processing, and Function in Protein Synthesis*, CRC Press, Boca Raton, FL, pp. 327–355.

232. Lieberman, K. R. and Dahlberg, A. E. (1995) *Prog. Nucl. Acids. Mol. Biol* **50**, 1–23.

233. Garrett, R. A. and Rodrivez-Fonseca, C. (1996) The Peptidyl Transferase Center. In *Ribosomal RNA, Structure, Evolution, Processing, and Function in Protein Biosynthesis*. Zimmermann, R. A. and Dahlberg, A. E., editors. CRC Press, Boca Raton, FL.

234. Mason, T. L., Pan, C., Sanchirico, M. E., and Sirum-Connolly, K. (1996) *Experientia* **52**, 1148–1157.

235. O'Brien, T. W. and Denslow, N. D. (1996) *Methods Enzymol.* **264**, 237–248.

236. Garrels, J. (1995) *Nucl. Acids. Res.* **24**, 46–49.

237. Fox, T. D. (1996) *Methods Enzymol.* **264**, 228–237.

238. Fox, T. D. (1996) *Experientia* **52**, 1130–1135.

239. Sanchirico, M. E., Tzellas, A., Fox, T. D., Conrad-Webb, H., Perlman, P. S., and Mason, T. L. (1995) *Biochem. Cell. Biol.* **73**, 987–995.

240. Bunn, C. L. D., Wallace, D. C., and Eisenstadt, J. M. (1974) *Proc. Natl. Acad. Sci. USA* **71**, 1681–1685.

241. Breen, G. A. M. and Scheffler, I. E. (1980) *J. Cell Biol.* **86**, 723–729.

242. Harris, M. (1978) *Proc. Natl. Acad. Sci. USA* **75**, 5604–5608.

243. Howell, N. and Sager, R. (1979) *Somat.Cell Genet.* **5**, 833–845.

244. Sachs, A. B., Sarnow, P., and Hentze, M. W. (1997) *Cell* **89**, 831–838.

245. Naithani, S., Saracco, S. A., Butler, C. A., and Fox, T. D. (2003) *Mol. Biol. Cell* **14**, 324–333.

246. Richter, J. D. and Sonenberg, N. (2005) *Nature* **433**, 477–480.

247. Sonenberg, N. and Dever, T. E. (2003) *Curr. Opin. Struct. Biol.* **13**, 56–63.

248. Spremulli, L. L., Coursey, A., Navratil, T., and Hunter, S. E. (2004) *Prog. Nucleic Acid Res. Mol. Biol.* **77**, 211–261.

249. Schwartzbach, C. J., Farwell, M., Liao, H.-X., and Spremulli, L. L. (1996) *Methods Enzymol.* **264**, 248–261.

250. Ditta, G. S., Soderberg, K., and Scheffler, I. E. (1977) *Nature* **268**, 64–67.

251. Scheffler, I. E. (1986) Biochemical genetics of respiration-deficient mutants of animal cells. In Morgan, M. J., editor. *Carbohydrate Metabolism in Cultured Cells*, Plenum, London, pp. 77–109.

252. Grivell, L. A. (1995) *Crit. Rev. Biochem. Mol. Biol.* **30**, 121–164.
253. Fox, T. D. (1996) Genetics of mitochondrial translation. In Hershey, J. W. B., Mathews, M. B., and Sonenberg, N., editors. *Translational Control*, Cold Spring Harbor Press, Cold Spring Harbor, NY, pp. 733–758.
254. Bonnefoy, N., Bsat, N., and Fox, T. D. (2001) *Mol. Cell. Biol.* **21**, 2359–2372.
255. Costanzo, M. C. and Fox, T. D. (1988) *Proc. Natl. Acad. Sci. USA* **85**, 2677–2681.
256. Fox, T. D., Sanford, J. C., and McMullin, T. W. (1988) *Proc. Natl. Acad. Sci. USA* **85**, 7288–7292.
257. Mittelmaier, T. M. and Dieckmann, C. L. (1995) *Mol. Cell. Biol.* **15**, 780–789.
258. Green-Willms, N. S., Butler, C. A., Dunstan, H. M., and Fox, T. D. (2001) *J. Biol. Chem.* **276**, 6392–6397.
259. Fiori, A., Perez-Martinez, X., and Fox, T. D. (2005) *Mol. Microbiol.* **56**, 1689–1704.
260. Viehmann, S., Richard, O., Boyen, C., and Zetsche, K. (1996) *Curr. Genet.* **29**, 199–201.
261. Leblanc, C., Boyen, C., Richard, O., Bonnard, G., Grienenberger, J.-M., and Kloareg, B. (1995) *J. Mol. Biol.* **250**, 484–495.
262. Burger, G., Lang, B. F., Reith, M., and Gray, M. W. (1996) *Proc. Natl. Acad. Sci. USA* **93**, 2328–2332.
263. Marykwas, D. L. and Fox, T. D. (1989) *Mol. Cell. Biol.* **9**, 484–491.
264. Suissa, M., Suda, K., and Schatz, G. (1984) *EMBO J.* **3**, 1773–1781.
265. Horwich, A. L., Kalousek, F., Mellman, I., and Rosenberg, L. E. (1985) *EMBO J.* **4**, 1129–1135.
266. Eilers, M. and Schatz, G. (1986) *Nature* **322**, 228–232.
267. Allison, D. S. and Schatz, G. (1986) *Proc. Natl. Acad. Sci. USA* **83**, 9011–9015.
268. Roise, D. and Schatz, G. (1988) *J. Biol. Chem.* **263**, 4509–4511.
269. Hendrick, J. P., Hodges, P. E., and Rosenberg, L. E. (1989) *Proc. Natl. Acad. Sci. USA* **86**, 4056–4060.
270. Sztul, E. S., Hendrick, J. P., Kraus, J. P., Wall, D., Kalousek, F., and Rosenberg, L. E. (1987) *J. Cell Biol.* **105**, 2631–2639.
271. Kalousek, F., Hendrick, J. P., and Rosenberg, L. E. (1988) *Proc. Natl. Acad. Sci. USA* **85**, 7536–7540.
272. Paces, V., Rosenberg, L. E., Fenton, W. A., and Kalousek, F. (1993) *Proc. Natl. Acad. Sci. USA* **90**, 5355–5358.
273. Kalousek, F., Neupert, W., Omura, T., Schatz, G., and Schmitz, U. K. (1993) *Trends Biochem. Sci.* **18**, 249.
274. Gakh, O., Cavadini, P., and Isaya, G. (2002) *Biochim. Biophys. Acta (BBA) Mol. Cell Res.* **1592**, 63–77.
275. Yang, M., Jensen, R. E., Yaffe, M. P., Oppliger, W., and Schatz, G. (1988) *EMBO J.* **7**, 3857–3862.
276. Pfanner, N., Douglas, M. G., Endo, T., Hoogenraad, N. J., Jensen, R. E., Meijer, M., Neupert, W., Schatz, G., Schmitz, U. K., and Shore, G. C. (1996) *Trends Biochem. Sci.* **21**, 51–52.
277. Martin, J., Mahlke, K., and Pfanner, N. (1991) *J. Biol. Chem.* **266**, 18051–18057.

278. Glick, B. S. (1995) *Methods Enzymol.* **260**, 224–231.

279. Yaffe, M. P. (1991) *Methods Enzymol.* **194**, 627–643.

280. Mokranjac, D. and Neupert, W. (2005) *Biochem. Soc. Trans.* **33**, 1019–1023.

281. Stojanovski, D., Rissler, M., Pfanner, N., and Meisinger, C. (2006) *Biochim. Biophys. Acta* **1763**, 414–421.

282. Wiedemann, N., Frazier, A. E., and Pfanner, N. (2004) *J Biol. Chem.* **279**, 14473–14476.

283. Dolezal, P., Likic, V., Tachezy, J., and Lithgow, T. (2006) *Science* **313**, 314–318.

284. Lister, R. and Whelan, J. (2006) *Curr. Biol.* **16**, R197–R199.

285. Perry, A. J., Hulett, J. M., Likic, V. A., Lithgow, T., and Gooley, P. R. (2006) *Curr. Biol.* **16**, 221–229.

286. Lister, R., Hulett, J. M., Lithgow, T., and Whelan, J. (2005) *Mol. Membr. Biol.* **22**, 87–100.

286a. Neupert, W. and Herrmann, J. M. (2007b) *Annu. Rev. Biochem.* **76**, 723–749.

287. Deshaies, R. J., Koch, B. D., Werner-Washburne, M., Craig, E. A., and Schekman, R. (1988) *Nature* **332**, 800–805.

288. Beddoe, T. and Lithgow, T. (2002) *Biochim. Biophys. Acta (BBA) Mol. Cell Res.* **1592**, 35–39.

289. Hoogenraad, N. J., Ward, L. A., and Ryan, M. T. (2002) *Biochim. Biophys. Acta (BBA) Mol. Cell Res.* **1592**, 97–105.

290. Mihara, K. and Omura, T. (1996) *Trends Cell Biol.* **6**, 104–108.

291. Rapaport, D. (2005) *J. Cell Biol.* **171**, 419–423.

292. Sherman, E. L., Go, N. E., and Nargang, F. E. (2005) *Mol. Biol. Cell.* **16**, 4172–4182.

293. Koehler, C. M., Merchant, S., Oppliger, W., Schmid, K., Jarosch, E., Dolfini, L., Junne, T., Schatz, G., and Tokatlidis, K. (1998) *EMBO J.* **17**, 6477–6486.

294. Tranebjaerg, L., Hamel, B. C., Gabreels, F. J., Renier, W. O., and Van Ghelue, M. (2000) *Eur. J. Hum. Genet.* **8**, 464–467.

295. Koehler, C. (2004) *Trends Biochem. Sci.* **29**, 1–4.

296. Lutz, T., Neupert, W., and Herrmann, J. M. (2003) *EMBO J.* **22**, 4400–4408.

297. Wiedemann, N., Pfanner, N., and Chacinska, A. (2006) *Mol. Cell* **21**, 145–148.

298. Mesecke, N., Terziyska, N., Kozany, C., Baumann, F., Neupert, W., Hell, K., and Herrmann, J. M. (2005) *Cell* **121**, 1059–1069.

299. Rissler, M., Wiedemann, N., Pfannschmidt, S., Gabriel, K., Guiard, B., Pfanner, N., and Chacinska, A. (2005) *J. Mol. Biol* **353**, 485–492.

300. Schatz, G. and Dobberstein, B. (1996) *Science* **271**, 1519–1526.

301. Jia, L., Dienhart, M., Schramp, M., McCauley, M., Hell, K., and Stuart, R. A. (2003) *EMBO J.* **22**, 6438–6447.

302. Szyrach, G., Ott, M., Bonnefoy, N., Neupert, W., and Herrmann, J. M. (2003) *EMBO J.* **22**, 6448–6457.

303. Hell, K., Neupert, W., and Stuart, R. A. (2001) *EMBO J.* **20**, 1281–1288.

304. Reif, S., Randelj, O., Doman, S. G., Dian, A., Krimmer, T., Motz, C., and Rassow, J. (2005) *J. Mol. Biol.* **354**, 520–528.

305. Simpson, A. M., Suyama, Y., Dewes, H., Campbell, D. A., and Simpson, L. (1989) *Nucleic. Acids. Res.* **17**, 5427–5445.

306. Hauser, R. and Schneider, A. (1995) *EMBO J.* **14**, 4212–4220.

307. Kumar, R., Maréchal-Drouard, L., Akama, K., Small, I., and Marechal-Drouard, L. (1996) *Mol. Gen. Genet.* **252**, 404–411.

308. Entelis, N. S., Kieffer, S., Kolesnikova, O. A., Martin, R. P., and Tarassov, I. A. (1998) *Proc. Natl. Acad. Sci. USA* **95**, 2838–2843.

309. Tarassov, I. A. and Martin, R. P. (1996) *Biochimie* **78**, 502–510.

310. Marechal-Drouard, L., Small, I., Weil, J. H., and Dietrich, A. (1995) *Methods Enzymol.* **260**, 310–327.

311. Tarassov, I., Entelis, N., and Martin, R. P. (1995) *J. Mol. Biol.* **245**, 315–323.

312. Dietrich, A., Marechal-Drouard, L., Carneiro, V., Cosset, A., and Small, I. (1996) *Plant J.* **10**, 913–918.

313. Nabholz, C. E., Hauser, R., and Schneider, A. (1997) *Proc. Natl. Acad. Sci. USA* **94**, 7903–7908.

314. Entelis, N., Brandina, I., Kamenski, P., Krasheninnikov, I. A., Martin, R. P., and Tarassov, I. (2006) *Genes Dev.* **20**, 1609–1620.

315. Desautels, M. and Goldberg, A. L. (1982) *J. Biol. Chem.* **257**, 11673–11679.

316. Suzuki, C. K., Suda, K., Wang, N., and Schatz, G. (1994) *Science* **264**, 273–276.

317. Teichmann, U., Van Dyck, L., Guiard, B., Fischer, H., Glockshuber, R., Neupert, W., and Langer, T. (1996) *J. Biol. Chem.* **271**, 10137–10142.

318. Wagner, I., Arlt, H., Van Dyck, L., Langer, T., and Neupert, W. (1994) *EMBO J.* **13**, 5135–5145.

319. Langer, T. and Neupert, W. (1996) *Experientia* **52**, 1069–1076.

320. Boss, P., Andresen, B. S., Knudsen, I., Kruse, T. A., and Gregersen, N. (1995) *FEBS Lett.* **377**, 249–252.

321. Arlt, H., Tauer, R., Feldmann, H., Neupert, W., and Langer, T. (1996) *Cell* **85**, 875–885.

322. Thorsness, P. E., White, K. H., and Fox, T. D. (1993) *Mol. Cell Biol.* **13**, 5418–5426.

323. Thorsness, P. E. and Weber, E. R. (1996) *Int. Rev. Cytol.* **165**, 207–234.

324. Weber, E. R., Hanekamp, T., and Thorsness, P. E. (1996) *Mol. Biol. Cell* **7**, 307–317.

5

MITOCHONDRIAL ELECTRON TRANSFER AND OXIDATIVE PHOSPHORYLATION

Summary

Bioenergetics can be said to have started with the experiments by Lavoisier showing that the amount of heat produced per volume of carbon dioxide is the same from the combustion of charcoal and from a living animal. This chapter reviews the highlights of the history of thermodynamics and bioenergetics in the nineteenth century leading to the major breakthroughs in the twentieth century. Among others, these include the discovery of cytochromes, the recognition of the role of ATP, and the elucidation of the role of mitochondria in respiration and oxidative phosphorylation. Decades passed between the idea of an "Atmungsferment" and the detailed understanding of the electron transport chain in mitochondria. A culmination of these studies was the formulation of P. Mitchell's chemiosmotic hypothesis that provided new insights into the interconversion and storage of chemical free energy in living organism. A final milestone was the meeting of bioenergetics and structural biology, when the structure of the ATP synthase was solved to explain how a proton and electrochemical gradient across a membrane could drive a molecular rotary engine and thus the synthesis of ATP. Turnover as understood in the context of enzyme kinetics became in this case directly associated with a rotational motion. Respiration and electron transport leads to a significant membrane potential associated with the inner mitochondrial membrane, and this energy source is utilized not only for ATP synthesis, but also for a large number of transport mechanisms for moving proteins, metabolites, and ions into or out of mitochondria.

Mitochondria, Second Edition, by Immo E. Scheffler

5.1 HISTORICAL INTRODUCTION

The earliest history of research on mitochondria has been presented in Chapter 1. Some far-sighted conclusions and very intelligent guesses made during the first half of the twentieth century turned out to be ahead of their time but could not be proved with the technology of the time, or were not pursued. From the biochemist's point of view, a major milestone was the development of methods for the reproducible isolation of intact mitochondria. Biochemical reactions unique to mitochondria could therefore be studied for the first time, and it could be established that the reactions of the Krebs cycle and fatty acid oxidation were major or even exclusive pathways in mitochondria. Thus, the pathways and cycles leading to the complete oxidation of carbon compounds to carbon dioxide were elucidated, adding substance and detail to the conjecture already expressed by Lavoisier in 1789 that living organisms were slow combustion engines burning carbon compounds and hydrogen to CO_2 and water.

> This fire stolen from heaven, this torch of Prometheus, does not only represent an ingenious and poetic idea, it is a faithful picture of the operations of nature, at least for animals that breathe; one may therefore say, with the ancients, that the torch of life lights itself at the moment the infant breathes for the first time, and it does not extinguish itself except at death.
>
> —Lavoisier, 1862, Vol. II, pp. 691–692, quoted by Fruton (2)

In a celebrated experiment with a few forgivable flaws, Lavoisier and Laplace demonstrated that the amount of heat released per volume of CO_2 produced was equal in the combustion of charcoal and in the respiration of a guinea pig. Combustion was believed then to take place in the lungs, and mitochondria were not yet on the horizon. By 1807, Lazzaro Spallanzani's findings had been translated and publicized that carbon dioxide can be formed in a large variety of animal tissues, but for some time afterwards the lungs were thought of as specialized combustion engines. Later on in the nineteenth century the blood became the favored site of respiration.

We should remind ourselves that thermodynamics was not fully developed until many decades later. What eventually became the first law of thermodynamics was first formulated by a physician, Robert Mayer, in 1842, reflecting on Lavoisier's theory about animal heat being derived from a combustion process, and blood entered the considerations because of the observations of color changes associated with heat production and oxidation. In a later article, Mayer correctly surmised that "blood corpuscles play the role of an oxygen carrier," but his ideas were not widely accepted until about 1875 when the role of hemoglobin in blood had been discovered and respiration in other tissues had been convincingly demonstrated. For example, one of the most dramatic demonstrations was to observe respiration in a frog whose blood had been replaced by a saline solution (Oertmann, 1877) (2). James Joule provided a firm experimental foundation for the first law of thermodynamics in his celebrated experiments during 1843–1849. These studies inspired Helmholtz, a physiologist studying fermentation, to become a physicist and to write his memorable and influential article "Uber die Erhaltung der Kraft," in which he clearly enunciated (a) the idea of chemical potential energy released from respiration/combustion and (b) the conservation of this energy in the form of heat and mechanical energy. Assuming that mechanical energy (by muscles) was negligible, and in ignorance of the energy consumed by biosynthetic reactions and many other subcellular processes, he concluded that combustion of nutritional components yields the same quantity of energy of heat as that released by the animals.

The history of physiology in the late nineteenth century is full of illustrious personalities and their struggles with understanding blood and oxygen and combustion in animals. Two extensive quotes will suffice to illustrate the issues. In a long battle with opposing views, E. Pfluger wrote in 1872:

> Here lies, and I want to declare this once and for all, the real secret of the regulation of the oxygen consumption by the intact organism, a quantity determined only by the cell, not by the oxygen content of the blood, not by the tension of the aortic system, not by the rate of blood flow, not by the mode of cardiac action, not by the mode of respiration.

At about the same time (1878–1879) the famous French physiologist C. Bernard wrote:

> . . . it is not in direct combustion that this gas (oxygen) is used. The usual formula repeated by all the physiologists that the role of oxygen is to support combustion is not correct. . . . It is quite certain that this gas is fixed in the organism and that it thereby becomes one of the elements of organic structure or creation. But it is not at all through its combination with the organic matter that it incites vital function. Upon contact with the tissues, it makes them excitable; they cannot exist without this contact. It is therefore as an agent of excitation that it would participate intimately in most of the phenomena of life.

A more detailed account of the history of intracellular respiration can be found in the fascinating book by J. S. Fruton (*Molecules and Life, Historical Essays on the Interplay of Chemistry and Biology*, Wiley-Interscience, New York, 1972). By the 1950s it was perfectly clear from a basic understanding of chemical thermodynamics that the combustion of sugars and fats to CO_2 was accompanied by the release of a substantial amount of energy, some of which was used to maintain our body temperature, but much of it interconverted into a chemical form of "free energy" to be used to drive biosyntheses and physiological reactions. The question was, How?

The early decades of this century led to the important recognition that iron played a central role in oxygen consumption and respiration, and already in 1924 O. Warburg was able to formulate the theory that

> . . . molecular oxygen reacts with divalent iron, whereby there results a higher oxidation state of iron. The higher oxidation state reacts with the organic substance with the regeneration of divalent iron. . . . Molecular oxygen never reacts with the organic substance.

In a remarkable paper he defined the respiratory enzyme (*Atmungsferment*) as ". . . the sum of all the catalytically active iron compounds present in the cell." The jump from the *Atmungsferment* to the definition of the mitochondrial electron transport complexes I–IV took almost another 40 years. Cytochromes were first identified by MacMunn as histohematins and myohematins in the 1880s. Their role in the respiratory chain was, however, not appreciated until Keilin in 1925 placed them in the appropriate context, although Warburg continued to have doubts about the intracellular function of cytochromes. The objection was sustained on the one hand by Warburg's attachment to the notion of the Atmungsferment as a single enzyme. Another good reason for Warburg's resistance was that he had found that oxygen uptake by yeast was inhibited by carbon monoxide, and therefore cytochrome could not be the Atmungsferment, since cytochromes had not been discovered to react with carbon monoxide (1927). It took another 12 years for Keilin and Hartree to demonstrate the existence of a CO-sensitive cytochrome absorbing near 600 nm, and its spectroscopic properties were identical to those of Warburg's *Atmungsferment* complexed to CO. A detailed personal history of the evolution of ideas about respiration and the role of the cytochromes has been written by Keilin (1), and the reader is referred to this highly readable account.

We should remind ourselves that all of these studies were carried out with whole-animal tissues by spectroscopic techniques, since mitochondria had not been isolated, and the solubilization of the cytochromes with the exception of cytochrome c had not been achieved.

There was a widespread assumption for some decades that all the oxygen consumed during respiration was accounted for in the carbon dioxide produced, and the hydrogens of carbohydrates appear to have been temporarily ignored by many. Again to quote C. Bernard (1876):

> Later, more precise studies having shown that all of the oxygen was not represented in the carbonic acid, it was supposed that the surplus was used to burn hydrogen and form water. But, as noted by M. M. Regnault and Reiset, that is a gratuitous hypothesis. It has no reason for existence except to explain a deficit, itself hypothetical.

Experiments started primarily by Wieland in around 1912 drew attention to the activation of hydrogen by palladium black, and biochemists were made aware that the conversion of ethanol to acetic acid (in *Acetobacter*) was also an oxidation which could, however, proceed anaerobically—that is, in the absence of oxygen. Similarly, observations on putrefaction of organic materials had shown that hydrogen-rich compounds could be formed (methane, hydrogen sulfide), and it began to dawn on some that organic substances can be changed and cleaved by the action of water. From a number of pathbreaking discoveries on the interconversion of organic substances in various cell extracts, enzymology was born. From a more limited perspective it also became clear that oxidative processes can also be considered dehydrogenations. This was shown elegantly by T. Thunberg (1917–1920), who prepared frog muscle as a suspension in an anaerobic environment and demonstrated that several organic acids (succinate, citric acid, glutamic acid, etc.) could be oxidized with the simultaneous reduction of methylene blue. These experiments and ideas culminated around 1924–1926 with the hypothesis that

> hydrogen is transported, not directly from primary donators to oxygen, but by stages. It would appear that the path may in a given case be smoothened, and so the velocity of transport increased, by the intervention of a substance which can alternately act as an intermediate acceptor and donator. Such a substance acts therefore catalytically, as a carrier of hydrogen (Hopkins, 1926; Szent-Gyorgyi, 1924) (2).

We can begin to see the glimmer of the idea of an electron transport chain, and the reader is invited to pursue the detailed lineage of ideas on this fascinating subject in the wonderful and scholarly account written by H. J. Fruton (2). Some updated information may be found in a more recent book by the same author (3). Some further milestones in the history of biochemistry and bioenergetics will be presented in Chapter 6. We have reached the stage where it is clearly understood that organic acids and ultimately NADH are the hydrogen donors and that a "carrier" exists which separates these substances to be oxidized from molecular oxygen.

The discussion so far has focused on only one aspect of respiration as the driving force behind biological work, namely, the nature of the redox reactions and the participation of oxygen as the ultimate electron acceptor. Thermodynamic theory predicted that part of the free energy change resulting from these reactions would be in the form of the release of heat, as in ordinary combustions, but it became clear that many biochemical phenomena and reactions are endergonic and therefore required an input of energy in some form. What was the nature of the chemical free energy driving the biosynthesis of macromolecules, muscular contraction, or conduction of signals in the nervous system?

"Energy-rich" compounds were first isolated from muscle, leading to the characterization of ATP by Lohman, and of creatine phosphate, but the universal significance of ATP was not immediately recognized. In 1934, ATP was still considered a "co-enzyme" in glycolysis. On the other hand, when in 1939 Engelhardt and Lyubimova identified myosin in muscle as an ATPase, and they and others linked the contractile process directly to the energy derived from ATP hydrolysis, the need for synthesizing ATP (or recycling it from ADP) became obvious. Many illustrious names in the history of biochemistry are linked to the clarification of ATP synthesis in the process of glycolysis, and the details of "substrate-level phosphorylation" were beginning to be elucidated, culminating in the most influential articles by Kalckar and Lipman in 1941. These authors defined the concept of "high-energy phosphate bonds" and clarified the thermodynamics of phosphoryl-transfer reactions. From this initial focus on muscle the general notion of ATP as the "energy currency" of all cells evolved, and within the next two decades the role of ATP in the synthesis of DNA, RNA, and proteins and other macromolecules was rationalized and explained. It took significantly longer to fully understand how ATP was essential for the conduction of nerve impulses. Transmission of electrical signals required a membrane potential, and how such a membrane potential could be created could be understood only after the role of ATP in primary active transport was defined. It will become obvious from the discussion of the chemiosmotic hypothesis in the section below how the problem of ATP synthesis in mitochondria is conceptually related to active transport. The role of ATP in force generation and motility is appreciated now not only for muscle, but also for the dynein-based movement of cilia and flagella. The discovery of whole families of molecular motors (myosin, dynein, kinesin) using ATP has expanded the horizon even further. ATP is now recognized to be essential for protein targeting to subcellular compartments, membrane fusions in the secretory pathway, protein degradation by the proteasome, and last but not least: What would signal transduction be without ATP, kinases, and phosphatases?

However, it is necessary to step back and dwell on another problem. Glycolysis and ATP production in the course of glycolysis are only part of the story, and every undergraduate is taught how much energy is "wasted" when lactate (or ethanol) is the end product of glycolysis (or fermentation) under anaerobic conditions. Converting pyruvate to CO_2, along with trapping the released chemical free energy in the form of ATP, became the next great challenge. It was decomposed into several problems: (1) the pathways of the breakdown of carbon compounds, solved brilliantly by H. Krebs with his formulation of the tricarboxylic acid cycle (Krebs cycle); (2) the oxidation of NADH and succinate by the mitochondrial electron transport chain; and (3) the coupling of electron transport to ATP synthesis (oxidative phosphorylation), culminating in the Chemiosmotic Hypothesis of P. Mitchell.

5.2 THE ELECTRON TRANSPORT CHAIN

5.2.1 The Biochemical Components

The only component of the electron transport chain which was successfully purified early on (1930s and 1940s) was cytochrome c, since it could be easily solubilized. The purified protein was further characterized as consisting of a

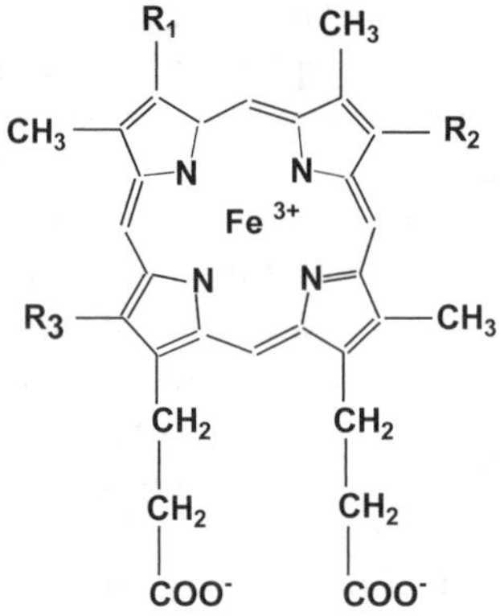

Figure 5.1 The basic structure of a heme prosthetic group.

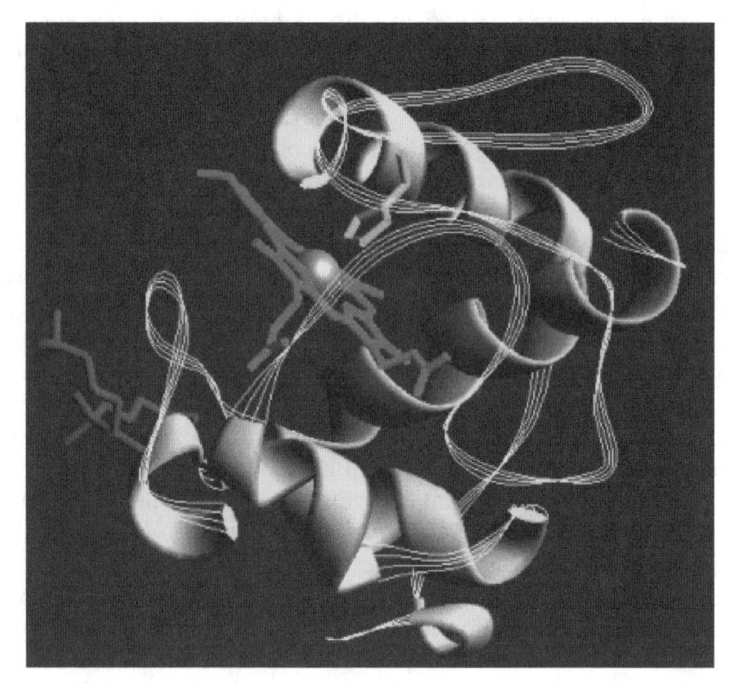

Figure 5.2 Structure of cytochrome c showing the association of the heme with the polypeptide chain. See color plates.

peptide (~13 kDa), the apo-protein, and a heme linked covalently to two cysteine side chains in the peptide (Figures 5.1 and 5.2). Because of the ease of its purification, it could be obtained readily from many different species. Furthermore, its small size invited efforts to determine the amino acid sequence long before cloning and DNA sequencing made it a routine. In turn, sequence comparisons and other characteristics, as well as its ubiquitous distribution in living organisms, made it a favorite for molecular evolutionists. It is sobering to learn that already in 1871 Lankester expressed the idea that "The chemical differences among the various species and genera of animals and plants are certainly as significant for the history of their origins as the differences in form reference" (2), p. 314). The first phylogenetic trees constructed from amino acid sequence comparisons were made by Margoliash and Dayhoff, and those of cytochrome c were very prominent among them.

When mitochondria were finally identified as the "powerhouse of the cell," it became easier to look for the "carrier" between organic hydrogen and oxygen, but almost another 15 years elapsed before the "carrier" was resolved into a series of complexes forming the electron transport chain, with ubiquinone (coenzyme Q) and cytochrome c acting as electron carriers between the complexes. It is difficult to do justice to all the discoveries and to define in a simple, linear progression the chain of ideas which contributed to the picture of the electron transport chain as presently found in elementary Biology textbooks. As noted earlier, cytochromes were characterized at first mainly by spectroscopic methods, and eventually their chromophore was established as a heme, bound to proteins, and thus localized in a microenvironment that determined its spectroscopic properties and its redox potential. Type c cytochromes have covalently bound heme, while type a and b cytochromes have hemes in a noncovalent attachment. Parallel studies of heme in soluble hemoglobin were, of course, most valuable.

Another major breakthrough belonging to the earlier phase was Warburg's discovery of the "old yellow enzyme" (1932–1933), the identification of the coferment (cofactor) as a flavin, and the concept of a flavoprotein. The structures of riboflavin phosphate and flavin-adenine dinucleotide (FAD) are now familiar to most students (at least at one time in their careers) (Figure 5.3).

The discovery of ubiquinone (coenzyme Q) in 1957 contributed a key, low-molecular-weight lipid constituent with a central role in electron transport (4, 5). The use of ubiquinone as an acceptor, and of ubiquinol as a donor of hydrogen, proved to be invaluable in the assays devised by Hatefi to fractionate the complexes of the electron transport chain.

Ingenious spectroscopic studies with relatively intact systems, in combinations with highly specific inhibitors (rotenone, antimycin, CO and cyanide to name a few), were able to define a pathway which can be outlined as follows:

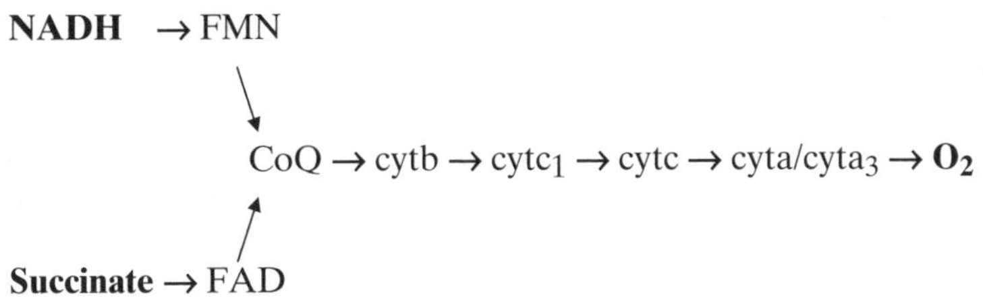

NADH → FMN
↘
CoQ → cytb → cytc$_1$ → cytc → cyta/cyta$_3$ → **O_2**
↗
Succinate → FAD

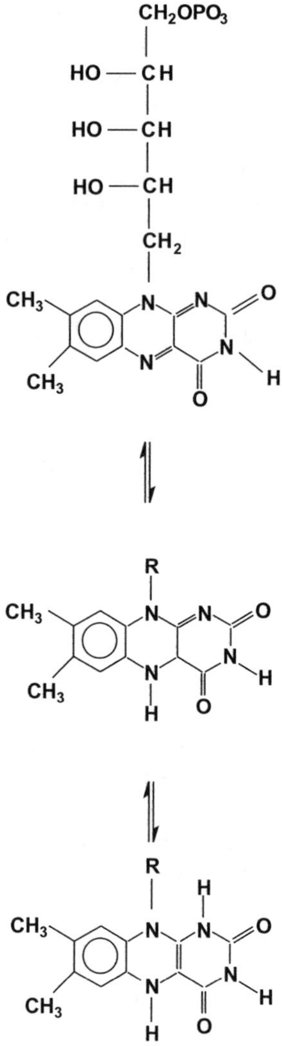

Figure 5.3 The structure of flavin mononucleotide (FMN) showing the flavin in the fully oxidized form (top) and in the fully reduced form (bottom).

Many individuals have contributed to the precise definition of this electron transport chain, and the reader is referred to the monograph of W. Wainio (6) for an authoritative and detailed review of the literature up to about 1970. Key contributions were many, but the experiments from the B. Chance laboratory using dual-beam, dual-wavelength spectrophotometers stand out. They contributed to a further definition of the spectroscopic properties of the cytochromes, and kinetic and inhibitor studies resolved the direction of electron flow, defined coupling sites, and further supported earlier thermodynamic calculations by Ball and Lipman showing that the free energy change associated with the oxidation of NADH is sufficient for the generation of three ATPs. Oxidation of succinate was predicted to yield only two ATPs. The enzyme succinate dehydrogenase (SDH) was found to be the only enzyme of the Krebs cycle which was insoluble and attached to the inner mitochondrial membrane, coupling the Krebs cycle to the electron transport chain. Thus, succinate can reduce the flavin in SDH directly. The recognition of several distinct flavoproteins was important, because the above simple scheme emphasizes the distinction between FMN and FAD.

The advent and application of more sophisticated spectroscopic techniques in the 1950s and 1960s brought a final major discovery about electron transport with proteins. Electron spin resonance (ESR), now referred to as electron paramagnetic resonance (EPR), which detects the spins of unpaired electrons in a transition metal ion, was applied to whole heart tissue or mitochondria in pioneering studies by H. Beinert (7), leading to the discovery of a novel kind of metalloprotein containing iron in a nonheme form. The iron is found in so-called iron–sulfur clusters, and their diversity and abundance in the ETC was at first quite a surprise (7). They had escaped detection because the optical properties, measurable by spectrophotometry, were obscured by the strong absorption of the cytochromes. The structures of many such clusters have since been characterized (Figure 5.4). In addition to their role in electron transfer, their structural versatility has been adapted in a large variety of proteins to “accept, donate, shift, and store electrons.” Cluster construction as well as cluster interconversions by ligand exchanges and oxidative degradation constitute interesting biological reactions (8). The past decade has seen a particularly dramatic progress in our understanding of the biosynthesis of iron–sulfur clusters (9–11), and this subject will be reviewed in a separate section (6.8).

The iron is coordinated with two to four cysteine side chains from the polypeptide, and in addition the cluster generally contains an equal number of sulfide ions (S^{2-}); these sulfides are acid-labile and released at low pH as hydrogen sulfide (H_2S), a biochemical diagnostic for the presence of such clusters. In most clusters, each iron is coordinated to a total of four S atoms in a roughly tetrahedral arrangement. The [Fe–S] cluster, with a single Fe linked to four cysteine residues (and no sulfide), has been found only in bacteria. Eukaryotic proteins have been found to contain the clusters [2Fe–2S], [3Fe–4S], and [4Fe–4S]. The iron atoms in each cluster form a conjugated system, and instead of a single iron forming a Fe^{2+}/Fe^{3+} redox couple, the entire cluster can lose or gain electrons. For example, the [4Fe–4S] cluster contains one Fe(II) and three Fe(III)’s in the oxidized form and two each of Fe(II) and Fe(III) in the reduced form. In general the cluster can be represented by $[m\text{Fe–}n\text{S}]^{x-}$ Each cluster has its characteristic redox potential, and with unpaired electrons each cluster also has a characteristic EPR spectrum (when $x = 1$ or $x = 3$). EPR can thus detect the gain or loss of electrons at each cluster.

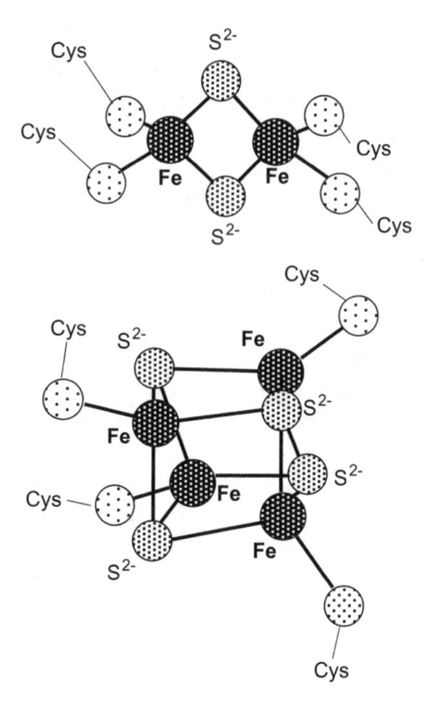

Figure 5.4 Structure of the [2Fe–2S] and [4Fe–4S] iron-sulfur clusters.

The presence of such clusters in the mitochondrial ETC provided another mechanism for the passage of an electron through a protein. It should be mentioned that Fe–S clusters were eventually also found in many proteins unrelated to the mitochondrial electron transport chain. Not surprisingly, they were found in proteins involved in photosynthesis and nitrogen fixation, but also in soluble proteins such as xanthine oxidase, ferredoxins and numerous hydrogenases in bacteria. A particularly interesting example is the protein binding to the iron-responsive element (IRE) in certain mRNAs. This protein has been identified as the cytosolic aconitase, and iron binding involves the formation of an Fe–S center accompanied by a conformational change in the protein. Disassembly of the center in the absence of iron leads to a conformation with high affinity for the IRE. Not all Fe–S centers are devoted to electron transfer, but they may also serve in the stabilization of the tertiary structure of proteins. The literature on the subject has become vast, with whole books on the subject appearing periodically. A very insightful and comprehensive review on recent developments has been written by Beinert et al. (8). In mitochondria, the next challenge was to localize them and place them in a position relative to the flavins, ubiquinones, and cytochromes. The original discoveries (reviewed by Beinert (7)) were made in the NADH and succinate dehydrogenases (see below), and a third such center was found in what is now known as Rieske's protein in the cytochrome b-c_1 complex (complex III), named after the original discoverer (12). More centers in mitochondria were discovered when EPR spectroscopy was carried out at very low temperatures with a significant gain in resolution, but their specific position within the overall ETC could not be determined without further purification.

The assembly of iron–sulfur centers in proteins was originally thought to be a simple process, but research within the last decade has shown that it depends on many proteins. These will be discussed in more detail in a separate section (13).

5.2.2 Physical Separation of the Complexes of the ETC

5.2.2.1 Biochemical Fractionations A hotbed of biochemical research on the electron transport chain was the laboratory of D. Green at the Enzyme Institute in Madison, from which many of the pioneers in the field emerged in the late 1950s and early 60s. The stage was set for the pioneering breakthroughs achieved by Y. Hatefi and his collaborators in the late 1960s (14–16). Starting with enormous quantities of beef heart, mitochondria were isolated and purified in correspondingly large amounts and subjected to systematic solubilizations and fractionations (Figure 5.5). It should be recalled that our concepts of membrane structure and membrane proteins were rudimentary at that time. While other laboratories had attempted to isolate individual cytochromes, Hatefi sought to isolate proteins or protein complexes which would be functionally intact and capable of carrying out individual redox reactions that had been associated with respiration. He exploited the observation that coenzyme Q (ubiquinone; Figure 5.6) and cytochrome c could be easily and reversibly removed from mitochondria by an appropriate solvent or salt extraction, respectively, and the suggestion that these two compounds were mobile electron carriers: Ubiquinone was found to act between complexes I or II and III, and cytochrome c was found to act between complexes III and IV. Based on these insights, assays for the function of each individual complex could be developed. Complex I oxidized NADH with ubiquinone as electron acceptor; the same electron acceptor was used in the oxidation of succinate by complex II; a reduced complex III was able to donate electrons to cytochrome c; reduced cytochrome c could be oxidized by complex IV with oxygen as electron acceptor. Finally, complex V was assayed as an ATPase. The devised scheme allowed the isolation of all five complexes from the same batch of mitochondria. For the first time it became apparent that each complex consisted of multiple peptides and required the presence of lipids for solubilization and stabilization.

These isolation procedures provided relatively pure complexes with which more detailed biochemical and biophysical analyses could be performed. The earliest data were confirmed and refined over the years, and there is no reason to dwell on some of the uncertainties that were resolved with time. The most challenging problem was to establish very precise stoichiometries in the various complexes with respect to flavin content (complexes I and II), iron content,

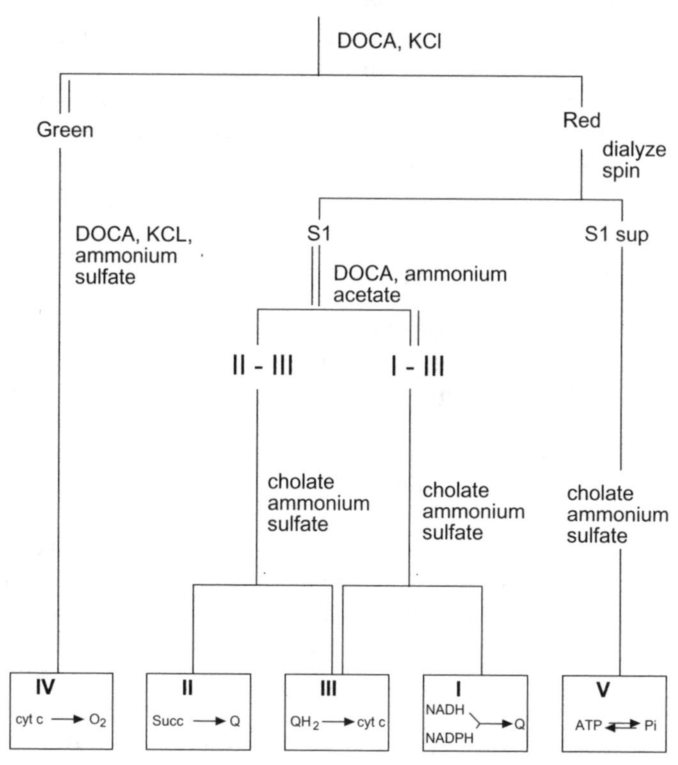

Figure 5.5 An outline of the purification scheme devised by Hatefi to obtain all five complexes from the inner mitochondrial membrane of bovine heart. The complexes are assayed and defined by the reactions shown in the boxes at the bottom.

and acid-labile sulfide (complexes I, II, and III). Another issue was to determine the number of polypeptides that were specifically associated with each complex, and while this number was settled quite quickly for complexes II to V, the peptide composition of complex I continued to be a significant challenge over the years. Initial estimates of over 10 polypeptides (14) were revised to approximately 25 polypeptides based on two-dimensional gel electrophoresis (17), and the most recent analysis by high-resolution chromatography and sequencing yielded at least 45 peptides (18, 19). As discussed elsewhere in this book, all complexes except complex II were shown to contain peptides made in the mitochondrial matrix, and their number and identity were settled after the complete human mtDNA sequence had been published and analyzed (20, 21).

In the early 1970s the mitochondrial electron transport chain in mammalian mitochondria had taken on its present form which is schematically represented

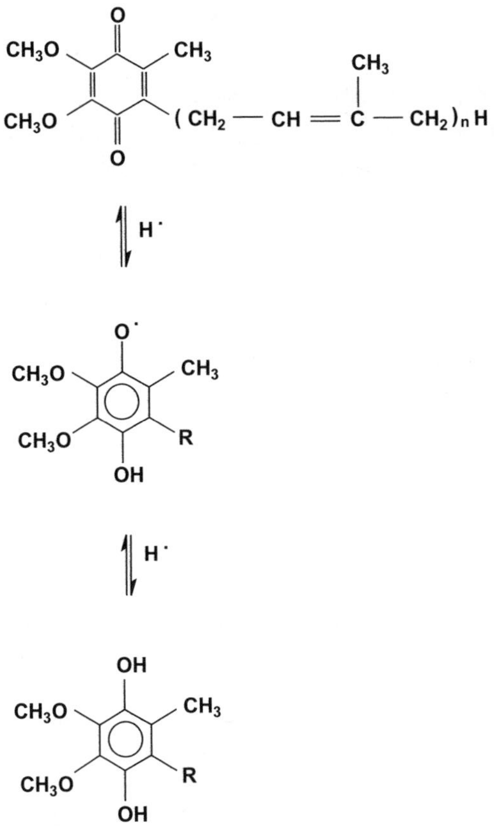

Figure 5.6 Structure of ubiquinone (coenzyme Q) in the oxidized, partly reduced (free radical) and fully reduced form. The number of isoprenoid units in the long hydrophobic tail may vary slightly in different organisms.

in Figure 5.7. Indeed, in 1975 each complex could be presented only as a box, or a by some other more or less rounded, asymmetrical shape. For the biochemist, structural biochemist, and biophysicist the challenge was to (a) determine the interaction of the various subunits within each complex and (b) orient these peptides relative to the inner and outer surface of the inner mitochondrial membrane. The final goal was to determine the crystal structure of each complex and to understand (1) electron transport through the complex, (2) proton translocation across the inner membrane which is coupled to electron transport, (3) the site and mechanism of action of specific inhibitors, and (4) most recently, the effect of specific missense mutations associated with human mitochondrial diseases (Chapter 7). The inner mitochondrial membrane, although exceptionally protein-rich, is nevertheless still a fluid membrane (22, 23), and the assumption of each complex being distinct but connected functionally (for electron transport) to the next via mobile carriers

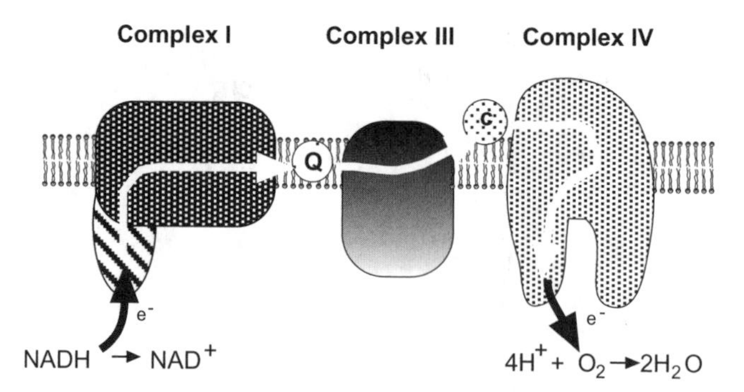

Figure 5.7 Schematic representation of the electron transport chain from NADH to oxygen. Complexes I, III, and IV are shown, as well as the mobile carriers, Q and cytochrome c.

(ubiquinone or cytochrome c) was assumed for a long time. It appears that in bovine heart mitochondria each complex is present in a stoichiometric ratio to the other complexes: Original estimates (1:2:3:6) have been reported by Hatefi, but no such precise quantitative determinations have been made for other tissues. Furthermore, as discussed in Chapter 3, there is probably a developmentally regulated and tissue-specific expression of the genes encoding the electron transport chain, which adapts the capacity of the mitochondrial oxidative phosphorylation system to the requirements of specific tissues and cells. If complexes were to function completely in isolation, connected by mobile carriers, there would *a priori* be no compelling reason to maintain precise stoichimetric ratios. On the other hand, based on results from novel technologies, there has been a radical rethinking in the field.

5.2.2.2 Supercomplexes The invention of Blue-Native gel electrophoresis by Schagger and colleagues (24–27) has had a profound impact on the study of the complexes of the ETC. In principle, the method achieves in one gel the fractionation of all five complexes in their native form. They can be visualized by staining, by histochemical reactions, or by conventional Western blotting. For example, the entire complex I of molecular mass >950 kDa can be separated cleanly from isolated mitochondria after solubilization in a suitable detergent; for details on the methodology the reader is referred to the specialized literature (26). Thus, the method is particularly suitable for the study of the biogenesis of the individual complexes, for the determination of assembly intermediates under normal conditions, or for the case where mutations in one subunit cause assembly defects. Only a few examples of a voluminous literature can be cited here (28–31).

The use of BN gels also revealed a novel property: In addition to their expected positions in the gel (the "monomers") there were also bands at positions corresponding to higher aggregates. Those of complex V (ATP synthase) will be discussed briefly in a later section. Complexes I, III, and IV in particular were found to be associated into super- or supracomplexes of an apparently fixed stoichiometry (24). In yeast, with no complex I, complexes III and IV exist as a larger complex of ~1000 kDa (32). These recent findings appear contradictory to the earlier studies of Hackenbrock (22, 23), suggesting that the inner mitochondrial membrane acts as a fluid mosaic in which complexes diffuse freely and the mobile carriers shuttle electrons between the complexes. In the revised and more "solid-state" model for the ETC complex I is found to be firmly associated with a dimer of complex III, and a variable number of complexes IV are added to form a structure named a "respirasome." The formation of such a respirasome can be rationalized in a number of ways, but firm experimental support for some of these ideas is just beginning to emerge. One simple proposal is that (1) substrate channeling of quinones can occur to speed up the electron transfer reaction and (2) the reactive, intermediate semiquinones can be trapped to avoid side reactions. Experimental evidence for this interpretation is still missing (33). Another very plausible possibility is that individual complexes are stabilized by supermolecular assembly. A finding that supports the second possibility comes from the investigation of respirasome formation in human patients with a primary genetic defect causing a reduced complex III assembly. In such patients' mitochondria the level of complex I was also reduced, interpreted by the authors as due to a failure to stabilize complex I in a supercomplex (24). In contrast, studies in the Scheffler laboratory have found that the absence of complex I in Chinese hamster mitochondria does not lead to any reduction in complex III levels (Potluri et al., unpublished).

What are other physiological consequences of supercomplex formation? A stringent test would require a comparison of normal mitochondria with mitochondria in which the individual complexes are present at normal amounts, but not aggregated into supercomplexes. Only a limited number of experimental systems exist for such a comparison. Zhang et al. (34) noted that the formation of $(CIII)_2(CIV)_2$ supercomplexes in yeast mitochondria was dependent on cardiolipin. Measurements of NADH oxidation in wild-type cells were "consistent with single functional unit kinetics of the respiratory chain," while cardiolipin-deficient mitochondria exhibited perturbed kinetics. Other authors (35) concluded that cardiolipin was not absolutely required for supercomplex formation in yeast, but greatly promoted their stability.

No studies are available on the influence of supercomplex formation by complexes I, III, and IV on the morphology of mitochondrial cristae, but the intriguing observation has been reported that dimer or higher aggregate formation by complex V is required for normal cristae morphology in yeast (36). From a different perspective, Bornhovd and colleagues reported that the mitochondrial membrane potential is altered when aggregate formation by complex V is prevented (37). It becomes apparent that the problem of supercomplex formation may have significance far beyond the kinetic aspects of the fluid versus the solid-state model of the ETC, or the stabilization of the individual complexes. A rigorous formulation of the relationship between cristae morphology and the performance of the electron transport chain will remain a challenge for the future.

5.2.3 Introduction to Bioenergetics

From a thermodynamic and electrochemical perspective, each step in the electron transport chain is associated with a free energy change, and each component undergoes a cycle of oxidation and reduction reactions; that is, it represents a redox couple. The properties of such a redox couple are conveniently described by the standard reduction potential, a quantity that can be easily related to an associated free energy change. For example, for the mobile cytochrome c the following describes the half-reaction and the corresponding standard reduction potential:

$$\text{cytochrome c }(\mathrm{Fe}^{3+}) + e^- \rightarrow \text{cytochrome c }(\mathrm{Fe}^{2+}) \quad E^{0'} = +0.235\,\mathrm{V}$$

A complete reaction consists of the sum of two half-reactions:

$$(1) \quad \text{succinate}^- \rightarrow 2e^- + 2\mathrm{H}^+ + \text{fumarate}^- \qquad E^{0'} = -0.031\,\mathrm{V}$$

$$(2) \quad \text{ubiquinone} + 2e^- + 2\mathrm{H}^+ \rightarrow \text{ubiquinol} \qquad E^{0'} = +0.045\,\mathrm{V}$$

$$(1) + (2) \quad \text{succinate} + \text{ubiquinone} \rightarrow \text{fumarate} + \text{ubiquinol} \qquad \Delta E^{0'} = +0.014\,\mathrm{V}$$

One can calculate the free energy change for this oxidation reduction reaction from the relationship

$$\Delta G = -nF\Delta E$$

where F is the faraday (1 F = 96,494 calories/mol). The standard reduction potentials are defined relative to the hydrogen half-reaction under standard conditions (pH = 0, 25°C, H_2(g) at 1 atm), and the prime superscript in the above symbols indicates that the values are for biological systems operating at pH 7.

It is informative and instructive to consider the electron transport chain from the point of view of an electron finding itself at a high potential (in NADH), dropping through a series of intermediate stages to its final destination, oxygen. This reaction, with reference to the various half-reactions and their associated standard reduction potential is shown in Figure 5.8. It is apparent from such a representation that there is not a very large drop between succinate and ubiquinone, and the free energy change may be too small for proton pumping (see below). The most substantial change in potential occurs in (a) the oxidation of NADH by ubiquinone and (b) the last step, reduction of oxygen to water. These steps are therefore virtually irreversible.

In the following, each complex will be discussed in turn, with an emphasis on composition, structure, and function. Some comparative data will also be

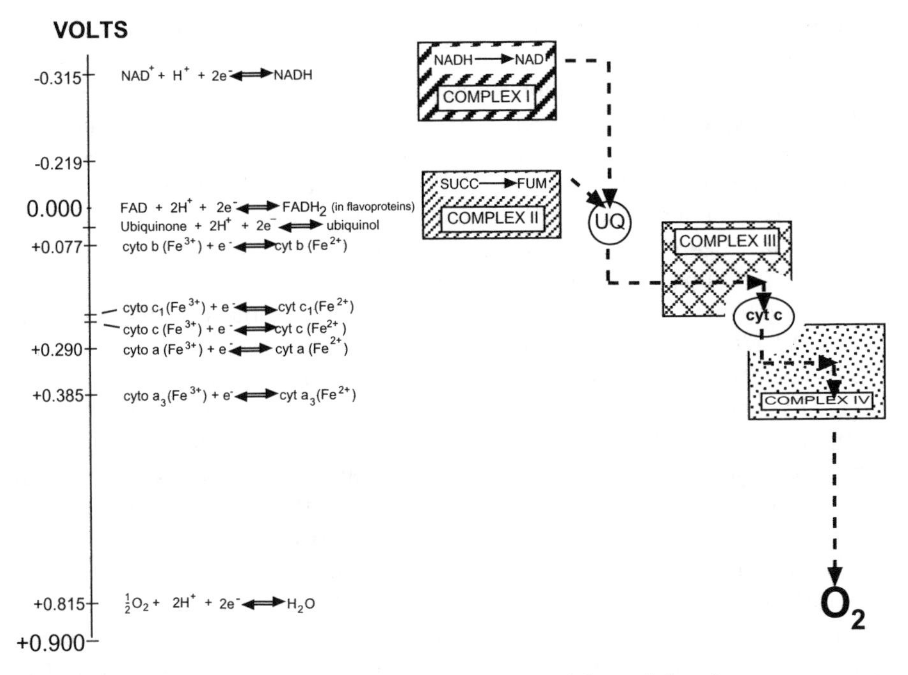

Figure 5.8 Half-reactions, standard reduction potentials, and the electron transport chain. The parameters were obtained from Table 15-4 in D. Voet and J. G. Voet, *Biochemistry*, 2nd ed., John Wiley & Sons, New York, 1995.

introduced to emphasize similarities and differences of these complexes in mitochondria of organisms from a broader phylogenetic range. A discussion of complex V will be deferred until after the section on the chemiosmotic hypothesis.

5.2.4 Complex I

Complex I is properly referred to as NADH-ubiquinone oxidoreductase with the inclusion of the substrates in the name. The overall reaction catalyzed by this complex can be described as follows:

$$\mathrm{NADH} + \mathrm{Q} + 5\mathrm{H}_i^+ \rightarrow \mathrm{NAD}^+ + \mathrm{QH}_2 + 4\mathrm{H}_o^+$$

Q and QH_2 refer to the oxidized and reduced form of ubiquinone, respectively. Of great significance is the number and subscript designation of the protons on either side of the equation, which will be made fully apparent in the section to follow. Here it suffices to state that the reaction is accompanied by the net transfer of four protons from the matrix side to the intermembrane space.

This apparently simple reaction requires a complex with 45 subunits (>900 kDa) in mammalian mitochondria. A similar complex in the membrane of prokaryotes has an estimated molecular mass of ~520 kDa, with only 14 subunits. They are homologous (orthologous?) to 14 subunits in the mammalian complex. Seven of these "core" subunits (designated as subunits ND1, ND2, ND3, ND4, ND4L, ND5, and ND6) are encoded by the mitochondrial genome in mammals and other organisms. The remaining 31 mammalian subunits have been referred to as "supernumerary," "accessory," or "ancillary." Nuclear mutations abolishing the activity of the mammalian complex I have been isolated in cells in tissue culture (38). Null mutations in two X-linked genes (*NDUFA1* and *NDUFB11*) encoding the accessory proteins MWFE and ESSS, respectively, have been characterized. The cDNAs encoding these small accessory peptides have been shown to complement the defect in these complex I-deficient Chinese hamster cells (28, 39–42). Thus, at least these two supernumerary subunits are absolutely essential for assembly of a functional complex I. All the nuclear encoded subunits of the bovine complex were originally cloned and sequenced as cDNAs, primarily in the laboratory of J. E. Walker (18, 19, 43, 44). Curiously, common yeasts such as budding and fission yeast use a completely different enzyme for electron transfer from NADH to ubiquinone, which does not pump protons out of the matrix. Therefore, molecular genetic studies in yeast and the customary homology cloning approaches were not possible. Another microorganism, *Neurospora crassa*, has been used as a model system for the study of this complex in preference to a mammalian source (beef heart), particularly when genetic studies and the use of mutants became powerful tools in the analysis (45–47). More recently, the genetic and biochemical analysis of complex I has made great strides in the yeast *Yarrowia lipolytica* (48–50). Complex I mutants have also been characterized from *Chlamydomonas* (51–53), and the subunit composition has been determined in several higher plants (54–56). The cloning and characterization of a large number of complex I genes and related genes from diverse organisms including prokaryotes has led to the construction of a detailed phylogenetic tree and very stimulating speculations about the evolution of this mitochondrial complex (56). The recent interest in human diseases resulting from complex I deficiency (57–61) has prompted the cloning of many of the corresponding human cDNAs or genes, and their mapping on various chromosomes (for a recent summary see (56, 61)). Both nuclear mutations and mitochondrial mutations have been found in human patients (see Chapter 7 for further discussion). A website dedicated to complex I will facilitate for the reader an entry into the voluminous and growing literature (http://www.scripps.edu/mem/biochem/CI/).

It is believed that all *eukaryotic* complexes have similar structure, and therefore there will be no distinction in the following discussion, unless specific differences are emphasized. The proton-pumping NADH:ubiquinone oxidoreductase in prokaryotes has fewer peptides (62), and it is thought that the 14 peptides found in *E. coli* represent the minimal number necessary for function.

Finel (63) and Friedrich (64–67) have published stimulating discussions of the organization and evolution of structural elements within complex I. These authors expand on hypotheses that complex I evolved by the combination of modules of preexisting domains of ancestral prokaryotic enzymes. However, these ideas addressed only the prokaryotic or "core" subunits. The additional peptides in higher eukaryotes may represent scaffolding and assembly factors not directly involved in electron transfer and proton transfer, or they may have a regulatory role. Several of them have by now been shown to be essential (see above), but on the whole their precise function remains a puzzle. Some of the accessory proteins found in mammals do not have counterparts in other organisms, and vice versa. It is noteworthy, however, that the total number of subunits jumps dramatically from ~14 in the prokaryotes to >40 in all eukaryotic systems examined so far. An exhaustive discussion of these and related issues can be found in the publication by Gabaldon et al. (56).

A few of the subunits have been found to have an independent function, or they have been implicated in apparently unrelated activities. For example, an acyl carrier protein was first identified in the bovine complex (68), and it has been proposed to play a role in lipoic acid metabolism (69). Another gene/protein, GRIM-19, was first found as a cell-death-regulatory gene induced by the interferon-beta and retinoic acid combination, and subsequently shown to be a subunit of complex I (70). Definite proof of its essential function in complex I assembly and activity was provided by Huang et al. (71), who demonstrated that homozygous deletion of GRIM-19 in mice was embryonic lethal due to the failure of assembly of complex I.

A comparison of peptide compositions in different organisms including *E. coli* is presented in Table 5.1. It is adapted from a recent comprehensive review of the complex I (56).

The overall, low-resolution structure of the complex has been derived from cryo-electron microscopy (72–74); it can be represented by a boot or L-shape, with the long arm integrated into the inner membrane and the short arm extending into the matrix (Figure 5.9). The short arm of the L functions as an NADH dehydrogenase, and it contains the FMN cofactor as well as seven to nine Fe–S clusters, not all of which can be resolved and characterized by EPR spectroscopy. The binding site for the substrate NADH is found on this subcomplex exposed to the mitochondrial matrix. This arm can also be referred to as the peripheral arm. The long, membrane-embedded arm plays a central role in proton translocation. The seven hydrophobic subunits encoded by mtDNA are part of the long arm. How electron transport in the peripheral complex and proton pumping are coupled remains a fundamental question. The suggestion has been made that two protons are pumped by a redox-driven, Q-cycle-related mechanism, while two others are transported by conformational transitions (65, 75).

Attempts to disrupt and solubilize portions of the complex I have made use of the chaotropic ion perchlorate or the detergent lauryl-dimethylamine oxide. With the use of perchlorate, three fractions have been obtained, referred to as

TABLE 5.1 Subunit Composition of Complex I

Name/Gene	Protein	*B. taurus*	*D. melanogaster*	*Yarrowia lipolytica*	*Neurospora crassa*	*Arabidopsis thaliana*	Bacteria
			mtDNA-Encoded Genes				
	ND1	++	+	++	++	++	NuoH
	ND2	++	+	++	++	+	NuoN
	ND3	++	+	++	++	+	NuoA
	ND4	++	+	++	++	+	NuoM
	ND4L	++	+	++	++	++	NuoK
	ND5	++	+	++	++	++	NuoL
	ND6	++	+	++	++	+	NuoJ
			Nuclear Genes				
NDUFS1/NUAM	75 kDa	++	+	++	++	++	NuoG
NDUFV1/NUBM	51 kDa	++	+	++	++	++	NuoF
NUCM/NDUFS2	49 kDa	++	+	++	++	++	NuoC
NDUFS3/NUGM	30 kDa	++	+	++	++	++	NuoD
NDUFV2/NUHM	24 kDa	++	+	++	++	++	NuoE
NDUFS7/NUKM	PSST	++	+	++	++	++	NuoB
NDUFS8/NUIM	TYKY	++	+	++	++	++	NuoI
NDUFA10/NUDM	42 kDa	++	+				
NDUFA9/NUEM	39 kDa	++	+	++		++	
NDUFS4/NUYM	18 kDa	++	+	++	++	++	
NDUFS5/NIPM	15 kDa	++	+	++	+	++	
NDUFS6/NUMM	13 kDa	++	+	+	+	+	
NDUFV3/NUOM	10 kDa	++	+				
NDUFB2/NIGM	AGGG	++	+				
NDUFB8/NIAM	ASHI	++	+	++	+		

NDUFS11/NESM	ESSS	++	+	++	+	+
NDUFC1/NIKM	KFYI	++	+			
NDUFA4/NUML	MLRQ	++	+			
NDUFB1/NINM	MNLL	++	+			
NDUFA1/NIMM	MWFE	++	+	+		++
NDUFB10/NIDM	PDSW	++	+	+	++	++
NDUFA8/NUPM	PGIV	++	+	++	++	+
NDUFAB1/ACPM	SDAP	++	+	++	++	+
NDUFB5/NISM	SGDH	++	+			
NDUFB9/NI2M	B22	++	+	+		+
NDUFB7/NB8M	B18	++	+	+		++
NDUFB6/NB7M	B17.2	++	+			
............/NB6M	B16.6	++	+	++		++
NDUFB4/NB5M	B15	++	+	++		
	B14.7	++	+			
NDUFA7/N4AM	B14.5a	++	+			
NDUFC2/N4BM	B14.5b	++	+			
NDUFA6/NB4M	B14	++	+			++
NDUFA5/NUFM	B13	++	+			
NDUFB3/NB2M	B12	++	+	++		+
NDUFA3/NI9M	B9	++		+	++	
NDUFA2/NI8M	B8	++	+			
NUXM				++	++	++
NUZM				++	++	
NURM					++	
						+ 7 other genes

Source: Reference 56.

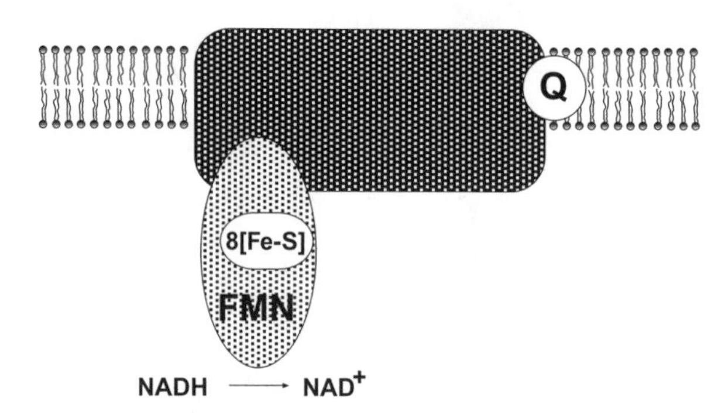

Figure 5.9 Highly simplified representation of complex I, the NADH–ubiquinone oxidoreductase. The entire mammalian complex has 45 polypeptides. The number of iron–sulfur centers is ~8 (depending on species).

the flavoprotein (Fp), the iron–sulfur protein (Ip), and a hydrophobic complex. The Fp complex consists of three peptides, (51, 24, and 9 kDa, respectively), and further analyses have confirmed the largest peptide to bind NADH and to contain the FMN and the [4Fe–4S] cluster. A second Fe–S cluster [2Fe–2S] is found in the 24-kDa peptide, while the 9-kDa peptide has no features or identifiable functions. Other detergent and salt combinations split the complex differently, yielding subcomplexes Iα and Iβ, and Iα can be subdivided further into Iγ and Iλ (19, 43). The Iα subcomplex appears to retain the capacity for transferring electrons from NADH to ubiquinone, while subcomplex Iβ is an inactive portion of the membrane complex. However, this oxidation of NADH is insensitive to rotenone, and the ubiquinone acts as an artificial electron acceptor from an abnormal site on the subcomplex. Most of the individual peptides, either nuclear encoded or mitochondrial, have been assigned to one or the other of these subcomplexes, but in view of the multitude of peptides a final picture of the arrangement of each of these peptides within the complex will require the crystal structure. The quest for crystals of the whole or portions of complex I from bacteria or eukaryotes has been a preoccupation of several of the prominent laboratories in this field for some time.

A major breakthrough has now been reported by Sazanov and Hinchcliffe (76), who succeeded in solving the structure of the hydrophilic (peripheral) subcomplex of complex I from *Thermus thermophilus* (Figure 5.10). It contains seven known hydrophilic subunits and a novel subunit that may be unique for this organism. The FMN moiety and nine Fe–S centers were localized precisely, defining a path (FMN–N3–N1b–N4–N5–N6a–N6b–N2) for electrons from FMN to the last Fe–S cluster N2, from which electrons are passed to ubiquinone. One cluster (N7) is too distant from the path to be directly involved. Another cluster (N1a) was proposed to be a temporary holding site for one of two electrons from reduced $FMNH_2$, thus shortening the half-life of the flavosemiquinone and preventing the formation of reactive oxygen species (superoxide) by electron transfer to oxygen. The reader is referred to this publication for many

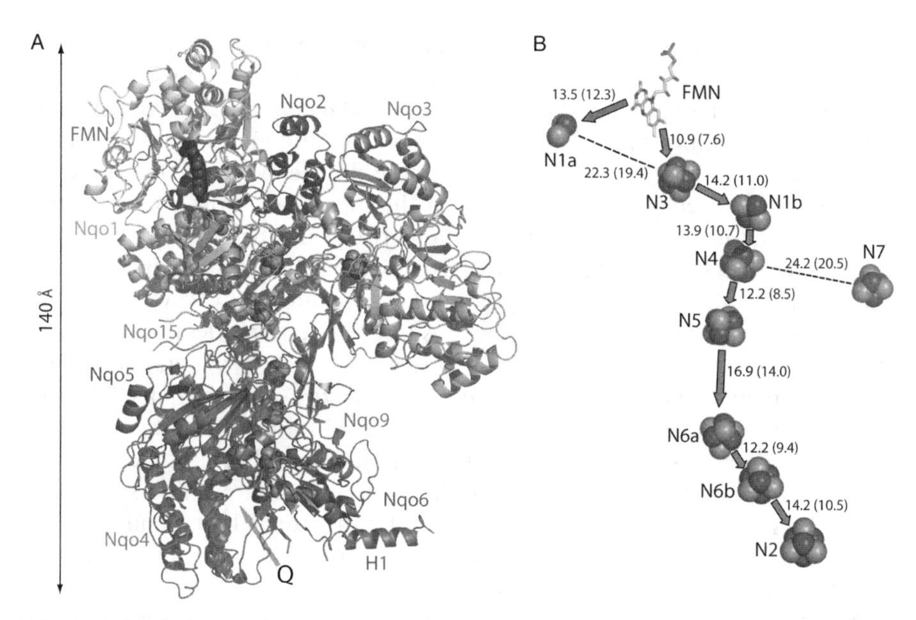

Figure 5.10 (A) Crystal structure of the peripheral subcomplex of respiratory complex I (NADH–ubiquinone oxidoreductase) from *Thermus thermophilus* at 3.3-Å resolution (74, 76). The subunits are labeled following the nomenclature for bacteria. (B) Location and distances between the 9 Fe–S clusters in this subdomain. (Reproduced from reference 76 with permission.) See color plates.

details on the structure. It provides answers to many questions, and it suggests that the evolutionary history may be an explanation for the apparently excessive number of Fe–S centers required for electrons transfer from FMN to ubiquinone. Alternatively, there may be an as yet undiscovered rationale linked to the mechanism of proton translocation by this enzyme.

Understanding the structure and function of this complex has been one of the challenges for which significant progress can be reported. Two major unsolved problems remain: (1) to understand the path of the protons pumped by the complex and (2) to elucidate the mechanism by which electron transport taking place exclusively in the peripheral subcomplex is coupled to proton translocation that must involve the integral membrane proteins. Another challenge is to understand the assembly of this complex from 38 subunits imported from the cytosol and seven subunits made in the mitochondrial matrix. What regulates the expression of the widely dispersed (unlinked) nuclear genes? What regulates the level of assembly of this complex? Is there a unique assembly pathway; that is, is it possible to identify assembly intermediates? Are the ND subunits inserted independently into the inner mitochondrial membrane, or is their synthesis and membrane integration coupled intimately to the assembly of subunits imported from the cytosol? Some answers to these questions have been obtained in recent years.

The first studies to address such questions were reported for *Neurospora crassa* (45). Effectively, gene knockouts can be achieved by site-specific integration of a hygromycin resistance marker, and the consequences of a missing subunit on assembly can be studied (47, 77). The mutant *nuo21* is of special interest because in the absence of this peptide the peripheral complex is assembled completely, while assembly of the membrane arm is only partial. With this mutant as starting material, the peripheral subcomplex can be prepared and purified conveniently. It can oxidize NADH when ferricyanide is used as an artificial electron acceptor, and it even has a low NADH: ubiquinone oxidoreductase activity. The latter is insensitive to piericidin A and thus believed to be due to ubiquinone binding to a site on the peripheral arm which is not normally exposed in an intact complex I. *Neurospora* mutants defective in a nuclear encoded complex I peptide have a reduced growth rate, but their mitochondria are not significantly respiration deficient. The presence of cytochromes and inhibition of respiration by antimycin and cyanide suggests a functioning electron transport chain downstream from ubiquinone, and the absence of a piericidin-sensitive complex I is overcome by an alternative NADH-ubiquinone oxidoreductase.

More recently the assembly of the mammalian complex has been studied in tissue culture systems. In Scheffler's laboratory (28, 39, 40) Chinese hamster mutants cells missing either the MWFE or the ESSS subunits in the membrane arm were shown to lack complex I by Blue Native gel electrophoresis. The mutants can be complemented with the respective cDNAs; moreover, the expression of these cDNAs can be regulated from an inducible promoter. When induced, the MWFE and ESSS subunits reach steady-state levels within 24 hours, but the appearance of the >900-kDa complex I on Blue Native gels is delayed by another ~24 hours, and full respiration measured by an oxygen electrode requires even more time (28; Potluri, unpublished). Some other subunits are missing when MWFE or ESSS are absent, presumably because they are unstable and rapidly turned over in the absence of assembly (their mRNAs are present at normal levels).

Other laboratories have examined complex I assembly and assembly intermediates in human cells from patients with isolated complex I deficiencies (see Chapter 7) (29, 30, 78–80). Such patients are alive because there is a residual activity; and in all cases examined so far, it appears that the reduced activity is due to a reduced steady-state level of intact complex I. The interpretation is that assembly is slowed down due to mutations in proteins encoded by nuclear genes (most of those are in the peripheral complex). Two precautionary statements may be in order. First, the assembly intermediates in such patient/mutant cell lines may not be exactly the same as those formed under normal conditions. Second, the analysis requires the use of Blue-Native gels and the solubilization of mitochondrial membranes by selected detergents. Mutations may introduce weakened associations in the complex that lead to dissociation into subcomplexes under the conditions of the experiment, but not *in vivo*.

Over the years the analysis of this complex and of its activity in mitochondria has been aided by the discovery of a variety of highly specific, naturally occurring inhibitors (Figure 5.11). The best known of these is rotenone, used as a fish poison; another is the antibiotic piericidin A. Some of these are being developed as potential insecticides and for various other pharmacological purposes (81). As discussed elsewhere in this volume, the "designer drug" MPTP (1-methyl-4-phenyl 1,2,3,6 tetrahydropyridine) can be converted to the active metabolite 1-methyl-4-phenyl prydinium (MPP^+) by monoamine oxidase and transported into dopaminergic neurons. At sufficiently high concentrations (mM) it is a specific inhibitor of complex I. This mechanism is thought to be responsible for the death of neurons in the substantia nigra and hence symptoms of Parkinsonism in humans and experimental primates (82, 83). A large and diverse series of specific inhibitors of complex I, some occurring naturally and some synthetics, have been discussed and investigated systematically by Friedrich et al. (84). On the basis of detailed kinetic analyses, these authors were able to divide the inhibitors into two classes: class I inhibitors (piericidin A, annonin VI, phenalamid A2, aurachins A and B, thiangazole, and fenpyroximate (synthetic)) inhibit complex I in a partially competitive manner with respect to ubiquinone; class II inhibitors (rotenone, phenoxan, aureothin, and benzimidazole (synthetic)) act in a noncompetitive manner. None of the inhibitors inhibit the NADH-ferricyanide reductase—that is, the peripheral membrane arm of the complex extending into the matrix. Instead, all inhibitors appear to arrest the transfer of electrons from the high-potential iron–sulfur cluster (N-2) to ubiquinone. Since the Fe–S clusters have not been completely characterized in this complex, it is not possible to be more specific. However, photoaffinity labeling studies with a rotenone derivative have suggested that this binding site is close to the peptide encoded by the mitochondrial ND1 gene. A more recent investigation of rotenone binding raising several still controversial issues related to conformational changes within the complex has been published by Grivennikova et al. (85). Specifically, these authors propose that one rotenone-specific site in complex I affects NADH oxidation by ubiquinone, while a second site is active in the ubiquinol–NAD^+ reductase reaction. This apparent violation of the principle of microscopic reversibility clearly deserves further clarification.

Detailed studies on the mechanism of inhibition by MPP+ analogues by Singer and colleagues (86) have led to the interpretation that there are two MPP+-binding sites, both of which must be occupied for complete inhibition. A "hydrophilic site" is defined by accessibility to relatively hydrophilic analogues, while a "hydrophobic site" may be less exposed to the aqueous environment. This interpretation is consistent with findings that the complex I can contain two ubisemiquinones. Miyoshi and colleagues have reexamined this question with the help of a new series of *N*-methyl-pyridinium and *N*-methyl-quinolinium analogues, with the goal of finding a more potent inhibitor effective at lower concentrations and, perhaps more significantly, the aim of finding an inhibitor which exhibits selectivity for only one of the two sites (87).

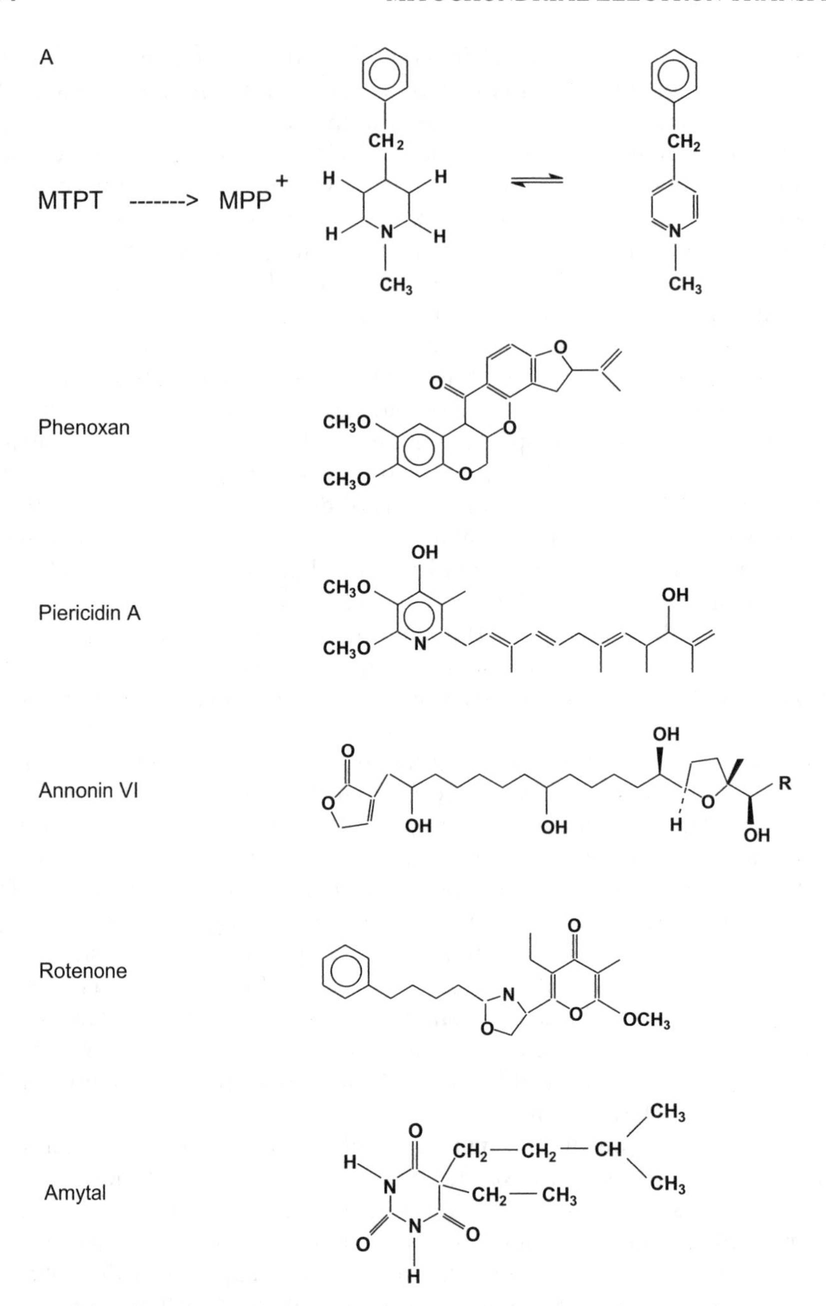

Figure 5.11 A and B: Structures of specific inhibitors of complex I frequently used in the study of this complex (modified from references 84 and 87). The compounds shown in part B are *N*-methyl-pyridinium analogues in which various side chains can be substituted at positions 2, 3, or 4.

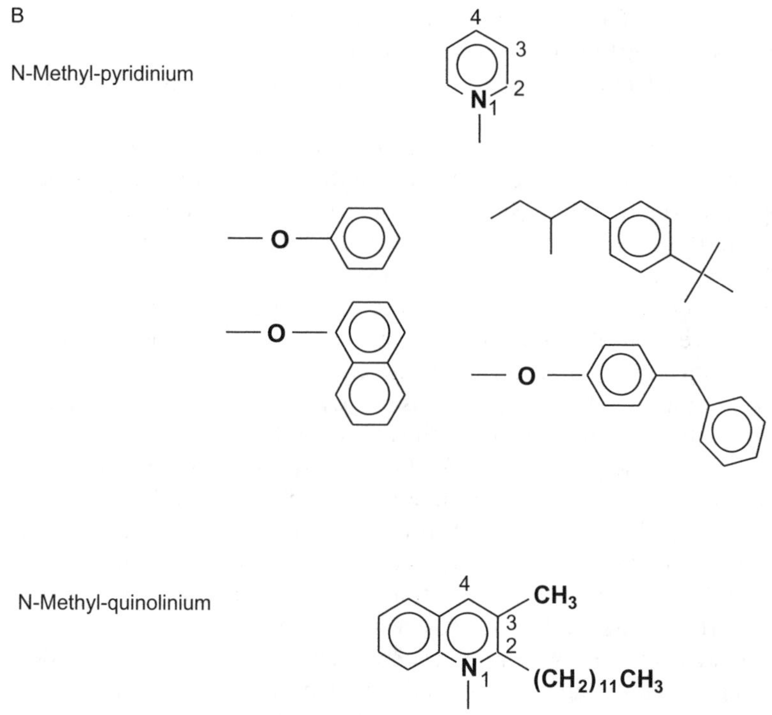

B

N-Methyl-pyridinium

N-Methyl-quinolinium

Figure 5.11 (*Continued*)

The results clearly support the two-site model, and significant progress toward achieving the second aim was also made. The presence of two ubiquinone binding sites is in good agreement with the acceptance of ubisemiquinone as an obligatory intermediate in the two-electron transfer from NADH to ubiquinone (88, 89).

5.2.5 Complex II

Complex II (succinate:ubiquinone oxidoreductase) consists of only four peptides and is thus the simplest of all the complexes of the ETC. The two largest peptides constitute the peripheral portion of the complex and function as the enzyme succinate dehydrogenase in the Krebs cycle. They are associated with the membrane through two integral membrane proteins, also referred to as the "anchor" proteins (see Figure 5.12). Electrons from the oxidation of succinate to fumarate are channeled through this complex to ubiquinone. Thus, complex II links the Krebs cycle directly to the ETC (90–92). A highly conserved and similar complex is also found as part of the electron transport system in bacterial membranes. The corresponding genes were first cloned in bacteria, and the bacterial systems (*Bacillus subtilis* and *Escherichia coli*) have been very useful model systems in the study of the structure, function, and assembly of this complex. Excellent authoritative reviews have been written by several authors (93–95).

Cofactors and metal ions make up the additional components of complex II. A flavin is linked covalently to the largest peptide (70 kDa), yielding the flavoprotein subunit (Fp). Singer and his colleagues established the chemical linkage in the early 1970s, and they also identified a histidine as the point of attachment (96–98). The timing and mechanism of this conjugation reaction was an open question until recently when the Fp gene (SDH1) was cloned in yeast. Conversion of His90 to Ser90 by site-directed mutagenesis prevented covalent attachment of FAD, but the assembly of the complex II in mitochondria was not affected (99–101). It was clearly shown that flavin attachment to the normal Fp peptide occurred in the mitochondrial matrix, after the removal of the targeting sequence (99). Additional experiments were interpreted to indicate that the enzyme responsible for this modification has to recognize a folded Fp peptide as one of its substrates, rather than a short peptide sequence surrounding the targeted histidine side chain. Nevertheless, this region of the peptide is highly conserved in evolution (see reference 102 for a review). When complex II was assembled in yeast with a noncovalently bound flavin, the complex was active only as a fumarate reductase, not as a succinate dehydrogenase (101). This result supports the idea that the covalent attachment of the flavin raises its midpoint potential to the level required for the oxidation of succinate (103).

The Fp subunit is intimately associated with the iron–protein subunit (Ip), made up of a peptide of 27 kDa containing three nonheme Fe–S centers: [2Fe–2S], [3Fe–4S], [4Fe–4S] (104). Each has a characteristic redox potential, and together these centers serve in the electron transport through the subunit (93, 98, 105). The Ip peptides from all species examined to date, and even from prokaryotes, have three highly conserved cysteine clusters (CI, CII, CIII) within the peptide. The first cluster is used to make the [2Fe–2S] center, and it resembles the [2Fe–2S] cluster also found in plant ferredoxins. The [4Fe–4S] center is made from the first, second, and third cysteines of cluster II and the third cysteine of cluster III, while the [3Fe–4S] center is made from the first and second cysteines of cluster III and the fourth cysteine of cluster II. These deductions were made from EPR studies made with the help of site-directed mutagenesis of bacterial sdh and frd iron–proteins and were subsequently confirmed in the crystal structures. Thus, the assembly of the clusters depends on the secondary and tertiary structure of the Ip subunit rather than the local peptide sequence. It should be noted here that the Fe–S clusters are assembled first by an independent mechanism (see Section 6.8) and then delivered to their final destination in Ip and various other proteins.

The Fp–Ip complex (SDH) can be dissociated from an isolated complex II by treatment with chaotropic ions as first shown by Hatefi and colleagues (92).

It functions as a succinate dehydrogenase, when an artificial electron acceptor such as ferricyanide or tetrazolium is included in the assay, but cannot interact directly with ubiquinones. To reconstitute ubiquinone reduction, the SDH complex must be combined with the two integral membrane proteins of complex II, $C_{II\text{-}3}$ and $C_{II\text{-}4}$, also referred to as QPs-1 and QPs-2 (or CybL and CybS). These "anchor" proteins are small peptides of molecular weight 15 and 12–13 kDa, respectively (in mammals), and sequence analysis suggested that each has three transmembrane segments, along with an N-terminal domain extending into the matrix for interaction with the SDH enzyme (106, 107).

Solubilization and reconstitution experiments, sequence analysis and hydropathy plots, and some earlier radiolabeling experiments by Merli et al. (108) suggested a domain structure for complex II that has been largely confirmed by X-ray crystallography. The first structure to be solved was the structure of the *E. coli* fumarate reductase (109). For good reasons (see below) the overall structure served as an excellent model for the structure of complex II in prokaryotes as well as eukaryotes. Since then the *E. coli* succinate dehydrogenase complex has also been solved (110–113), revealing in great detail the path of electrons from succinate via the flavin, three Fe–S cluster to ubiquinone (see Figure 5.13). Finally, a high-resolution crystal structure of a mammalian complex II (porcine heart) at 2.4-Å resolution was published by Sun

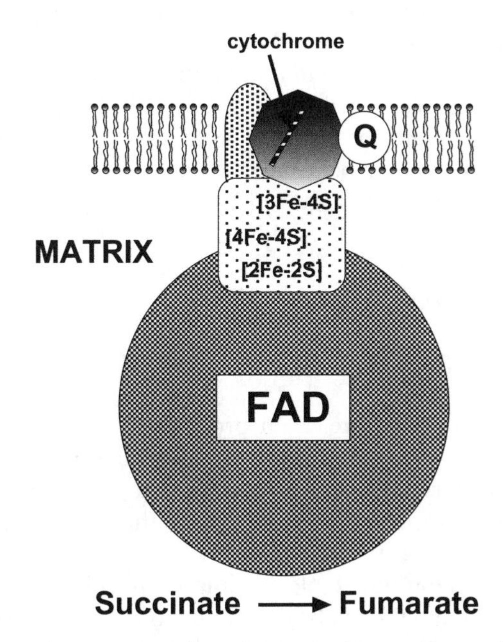

Figure 5.12 Schematic representation of complex II, succinate–ubiquinone oxidoreductase. Two integral membrane proteins serve to anchor the iron–protein (Ip) and the flavoprotein (Fp) of SDH.

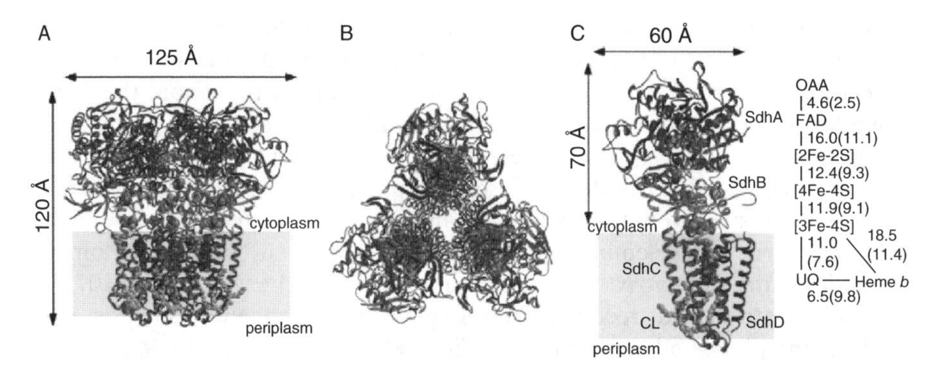

Figure 5.13 X-ray structure of succinate dehydrogenase (complex II) from *E. coli.* (A) The trimer viewed parallel to the membrane. (B) Perpendicular view of the trimer viewed from the cytoplasm. (C) View of a monomer showing the redox centers and the distances between them. (From reference 113 with permission.)

et al. (114). This structural determination has permitted a detailed analysis of binding sites for phospholipids, ubiquinone (two sites), and various inhibitors of the enzyme. A comparison with the *E. coli* enzyme has revealed many conserved structural components as well as informative differences.

There is some analogy with complex I, but on a much simpler scale. The peripheral subcomplex (Fp (SdhA), Ip (SdhB)) has the substrate binding site (succinate), the hydrogen acceptor (FAD), and 3 [Fe–S] centers for the conductance of electrons to the second and relatively simple membrane subcomplex consisting of the two anchor proteins. This membrane complex also contains a b-type cytochrome and the binding site(s) for ubiquinone. The involvement of the b-type cytochrome with electron transport to ubiquinone is still under investigation; it is attached to both anchor proteins. The C_{II-3} C_{II-4} peptides have highly conserved histidines linking the heme b to the simple membrane complex (102).

The reaction catalyzed by complex II is

$$\text{succinate} + \text{Q} \rightarrow \text{fumarate} + \text{QH}_2$$

In most eukaryotic cells the reaction proceeds from left to right, not only because of the relative concentrations of substrates and products, but also because the enzyme has diode-like properties that prevent electron flow in the opposite direction (115, 116). Understanding the biophysical basis for this behavior remains one of the challenges.

Specific inhibitors continue to play an important role in biochemical studies of electron transport and the analysis of mitochondrial functions. For complex II the most useful inhibitor has been the substrate analogue malonate, which binds competitively and specifically at the active site on the Fp subunit.

There are examples of organisms or conditions in which the reduction of fumarate becomes a physiologically relevant reaction. Facultative bacteria such as *E. coli* can proliferate under aerobic conditions and with a minimal supply of glucose when complete combustion to carbon dioxide and the most efficient exploitation of available energy resources is achieved by operating the Krebs cycle and oxidative phosphorylation. The interconversion of essential metabolites also requires the Krebs cycle. *E. coli* and other bacteria can also grow under anaerobic conditions when energy metabolism has to be altered drastically. Under such conditions the reduction of fumarate to succinate becomes an essential reaction. However, the reverse reaction is not achieved by the same SDH enzyme induced under aerobic conditions, but is instead catalyzed by a completely distinct complex named fumarate reductase, FRD. The two complexes are similar with respect to protein composition (Fp, Ip, and anchor proteins), but these peptides are encoded by two distinct operons. SdhA,[1] sdhB, sdhC, and sdhD are represented by four cistrons on a polycistronic transcript for succinate:ubiquinone oxidoreductase, and frdA, frdB, frdC, and frdD are encoded by four cistrons for fumarate:ubiquinone oxidoreductase). In *B. subtilis* there is only one anchor protein, but it is almost twice the size of the two *E. coli* peptides and may represent a gene fusion. The regulation of the expression of these operons by oxygen and glucose has been studied extensively in bacteria (94, 117, 118). It remains to understand in some detail how some changes in the primary sequence of these peptides can alter the microenvironment of the redox couples (Fe–S, FAD) to favor reactions in one direction or the other. Another noteworthy aspect of the SDH and FRD activities in bacteria is that ubiquinone serves as the electron acceptor from SDH under aerobic conditions, while reduced menaquinone serves as the electron donor via the FRD in the reduction of fumarate to succinate under anaerobic conditions. For a more detailed discussion of issues, the reader is referred to the authoritative review of Ackrell et al. (93), in which both prokaryotic and eukaryotic enzymes are discussed.

Another interesting situation arises for eukaryotic organisms, which spend part of their developmental cycle under relatively anaerobic conditions and another part of their life under aerobic conditions. For example, the sheep nematode *Haemonchus contortus* switches to a fermentative, predominantly anaerobic metabolism of the parasitic (adult) stages from the aerobic metabolism found in the free-living larvae (119). It appears that isozymes of the iron–protein of complex II are differentially expressed in larvae and adults, and it has been argued that a switch in isozymes is responsible for the change in complex II from functioning predominantly as a fumarate reductase to the conventional succinate:ubiquinone oxidoreductase under aerobic conditions. In a related species, *Ascaris suum*, a similar switch and the existence of isozymes for the flavoprotein have been demonstrated (120). The existence of multiple genes (isozymes) for complex II may be a general observation for

[1] The genes for complex II have been designated in different ways by different authors for various organisms. The designation SDH1, SDH2, SDH3, and SDH4 is proposed for the four genes encoding the subunits in the order of decreasing molecular mass of the peptides in yeast. sdhA, sdhB, sdhC, and sdhD have been used for the bacterial genes.

some protozoa, parasitic helminths, and marine organisms such as annelids and mussels, which spend their different life cycles under different conditions (121). It is not yet clear whether one or all of the subunits of complex II have to be changed. The problem is not only to change the direction of electron flow between succinate and fumarate, since the redox potential of all the participants must be considered. Hence, ubiquinone (UQ/UQH_2; $E^{0\prime} = +100\,\text{mV}$) is generally the acceptor in succinate oxidation, while menaquinone is the donor in bacteria. Eukaryotes engaged in mitochondrial fumarate reduction must synthesize a closely related rhodoquinone (RQ) (122). Since these quinones interact with the anchor proteins, a switch in energy metabolism during development may require not only a different set of anchor proteins, but also a new pathway for the synthesis of a different quinone. Since UQ and RQ differ only in one substituent on the quinone ring (the 5-methoxy group is replaced by an amino group in RQ), the branchpoint can be postulated to occur at the end of the pathway of UQ or RQ biosynthesis. In all eukaryotes examined to date, the finding of RQ has been indicative of a potential for fumarate reduction. For example, the detection of a significant quantity of RQ in the sporocysts of *Schistosoma mansoni* has been correlated with fumarate reduction as an essential pathway during the parasitic stage in the intermediate hosts (snails) where sporocysts behave as facultative anaerobes (121). On the other hand, in *Leishmania infantum* promastigotes it had also been suggested that fumarate reduction to succinate constitutes an electron sink during anoxia. However, the absence of RQ in these organisms suggested that this reaction was absent, and studies have confirmed that these organisms produce succinate exclusively via the Krebs cycle under aerobic conditions and go into a metabolic arrest during oxygen deprivation (123).

It has been speculated that fumarate reductase was the first activity in prokaryotes in evolution, with menaquinone as the cofactor. When oxygen concentrations were raised in the atmosphere, an oxidative metabolism evolved which included the Krebs cycle and specifically required the conversion of an FRD activity to an SDH activity. This was achieved by gene duplication, covalent attachment of the flavin, a raise in the standard redox potentials of the iron–sulfur clusters, and the utilization of ubiquinone instead of menquinone (124). The FRD activity was subsequently lost in eukaryotic mitochondria. However, it is conceivable that in a few facultative anaerobic eukaryotes a "reversed" evolution took place. A novel activity allowed the diversion of 5-hydroxy-6-methoxy-3-methyl-2-polyprenyl-1,4-benzoquinone to rhodoquinone instead of ubiquinone, while gene duplication and evolution yielded isopeptides that could be assembled to function as fumarate reductase again. A detailed examination of cloned genes and deduced peptide sequences will be required to resolve and support such a hypothesis.

5.2.5.1 Nuclear Versus Mitochondrial Location of Complex II Genes For a while it was correct to state that all complex II proteins were encoded by nuclear genes. This generalization has to be abandoned, because several different organisms have been found in which some of the complex II genes are found on the mitochondrial genome. In the red algae *Chondrus crispus* the sdhB and sdhC genes are found on the mt genome (125, 126); in a different red algae (*Porphyra purpurea*) and in the phylogenetically distant zooflagellate *Reclinomas americana,* the genes for the iron–protein and the two membrane anchor peptides are found in the mitochondria (127). These data are in agreement with the endosymbiont hypothesis for the origin of the mitochondrial genome, and they have been used to argue in favor of a monophyletic origin of mitochondria (see Chapter 2).

Eukaryotic genes for the Ip peptides (SDH2) in complex II were first cloned in 1989 (128), and their number has increased dramatically over the past few years. Similarly, the gene for the Fp peptide (SDH1) is available from numerous species. Cloning the SDH3 and SDH4 genes was initially more challenging, because the sequence conservation across species is poor, in contrast to the Ip and Fp peptides, and probes form one organism generally do not hybridize with genes/cDNA from another distantly related organism (102). An extensive review of the molecular genetics of complex II with an emphasis on gene cloning and regulation of gene expression has been published (102). Further updates with the progress from the Lemire group were published susequently (129, 130). The major conclusions can be found in the chapter on nuclear genes for mitochondrial proteins (Chapter 4).

Gene disruption experiments in yeast have provided yeast mutants with either the Fp, Ip, or anchor protein genes deleted, and such mutants can be used as recipients for novel gene constructs with specific alterations to investigate structure-function relationships (131–133). For example, a series of chimeric Ip proteins containing yeast and human sequences were investigated to delineate a sequence between the first and second cysteine cluster which appears to be species-specific and which, when replaced by a heterologous sequence, makes it impossible to assemble complex II (134).

An interesting mammalian cell mutant defective in SDH activity and assembly of complex II has been described for Chinese hamster fibroblasts in tissue culture (135). The mutant cells are respiration-deficient and unable to grow in medium in which glucose is limiting. The mutation has been established to result from a single nucleotide change converting a tryptophan codon to a stop codon, thus truncating the C_{II-3} peptide before the third transmembrane segment (136). The fate of the truncated peptide has not been determined, but the result is a complete failure of the assembly of complex II; not even an active SDH enzyme complex is formed, although the Fp and Ip peptides are made and imported normally into the mitochondria. This result can be contrasted with the findings with the *nuo21 Neurospora* mutant, where the peripheral complex is made in an active form but not attached to the membrane arm. The implications of this finding are that the Fp and Ip are not assembled separately before their attachment to the membrane. Instead, the matrix domains of the integral membrane proteins appear to participate in the folding and formation of the Fe–S centers in the Ip subunit and the simultaneous association with the Fp subunit. In this context it is worth noting that the Ip and Fp peptides cannot be separated without denaturation, and they have never been successfully reassembled *in vitro* into an active SDH enzyme.

In the past decade, human patients have been identified with partial complex II deficiencies due to mutations in the SDH genes (137). They will be discussed later in the context of mitochondrial diseases (Chapter 7). While some of these mutations lead to predictable pathologies from what we now understand about mitochondrial diseases, there are also mutations in the SDHB, SDHC, and SDHD genes that give rise to a very specific type of tumor: paragangliomas (138–140).

5.2.6 Complex III

Complex III is ubiquinone–cytochrome c oxidoreductase and is often named the bc_1 complex after the two cytochromes found within it; one also finds the simpler name cytochrome c reductase in the literature. The overall reaction catalyzed by this complex is

$$QH_2 + 2\,\text{cyt}\ c^{3+} + 2\,H_i^+ \Rightarrow Q + 2\,\text{cyt}\ c^{2+} + 4\,H_o^+$$

Similar to the reaction taking place with complex I, the oxidation of one of the substrates (QH_2) and the transfer of electrons to the mobile carrier (cyt c) is coupled to the transfer of protons across the inner mitochondrial membrane. The fate of the protons will be the subject of the discussion to follow, and a more detailed mechanism for proton translocation across the membrane is also deferred for the moment.

Complex III appears to be quite similar in different species including the yeast *Saccharomyces cerevisiae, Neurospora crassa*, animals and plants. A website devoted to complex III can be consulted for much recent information (http://www.life.uiuc.edu/crofts/bc-complex_site/). One peptide of the complex is encoded by the mitochondrial genome and incorporated as cytochrome b. The other peptides (8 in yeast, ~8 in *Neurospora*, 10 in mammals) are encoded by the nuclear genome, synthesized in the cytosol, and imported for assembly in the inner mitochondrial membrane. Both yeast and *Neurospora* have become model systems for the study of this complex because genetic manipulations allow detailed questions to be asked about (a) individual peptides and amino acids within each peptide and (b) their role in assembly, function, and activity. As described elsewhere in this volume, in yeast even the mitochondrial gene is now subject to deliberate mutational modification (Chapter 7).

Functionally, the most important subunits are the cytochromes b and c_1 and the Rieske iron–sulfur protein, since they are the only ones participating in electron transfer and in the accompanying proton translocation. This fact is emphasized by the finding that the corresponding complex in the bacterial electron transport chain has only those three proteins. The role of the other proteins is difficult to deduce, because they have no prosthetic groups. However, at least in *Neurospora* and potato the two large peripheral subunits appear to be processing proteases, and the largest subunit I has been shown to be involved in the import and processing of proteins from the cytosol (141, 142).

In the mammalian system, and to a large extent also in *Neurospora,* the elegant experimental approaches to studying complex III have primarily relied on solubilization, partial dissociation into subcomplexes, and reconstitution studies (141, 143). The original procedure by Hatefi for the isolation from mammalian mitochondria has been simplified and adapted to a smaller scale for many purposes. When too much lipid is removed during the purification, the activity of the complex declines. A most useful technique for isolating OX-PHOS complexes including complex III directly from homogenized human tissues is blue-native polyacrylamide gel electrophoresis (BN-PAGE), after solubilization of the membrane by neutral detergents (27, 144). It is suitable for the preparation of complexes in the microgram-to-milligram scale. As isolated from bovine and *Neurospora* mitochondria, the complex III is a dimer, a conclusion supported from electron microscopic studies of membrane crystals of the complex from *Neurospora*. It remains unclear whether dimerization is absolutely essential for the catalytic activity of the complex, but Covian and Trumpower have provided strong experimental evidence for the transfer of electrons between the two b_L hemes of the dimer (145).

Relatively mild detergent can dissociate two subunits from the bovine complex: the Rieske iron–sulfur protein (ISP) and a small ISP-associated protein. Further dissociation with 1.5 M guanidine yields a cytochrome c1 subcomplex (cytochrome c1, p9.2, p7.2, a core subcomplex, and several isolated subunits). Cytochrome b appears to remain with the core subcomplex, but a large fraction is lost in this procedure. When higher detergent concentrations are used to remove neutral lipids from the bovine complex, activity is lost, but can be restored by slow addition of phosphatidylcholine or phosphatidylethanolamine in Triton X-100. Quantitative measurements have suggested that a complex III dimer must be surrounded by a complete annulus of phospholipid. Less easily rationalized in terms of a structural model is the observation that bovine complex III has eight or nine tightly bound cardiolipin molecules per monomer. Their removal causes irreversible loss of activity.

All 11 mammalian complex III peptides have been purified on SDS-PAGE and used for partial sequencing and for the production of antibodies that can be used in the exploration of the topology of the complex in the inner membrane. Another experimental approach to topological studies has employed EPR techniques. An emerging model places the heme b_{562} near the middle of the lipid bilayer, and the other heme (b_{566}) and the Rieske iron sulfur cluster as well as cytochrome c_1 are on the "P-side" of the membrane (facing the intermembrane space).

Complex III, like complex I, has two reaction centers for ubiquinone, Q_N and Q_P. The significance of these multiple sites will become more apparent when the proton pumping mechanism will be discussed in greater detail.

Studies addressing problems related to the yeast complex III have been predominantly of a genetic nature. All eight nuclear genes for this complex have been cloned, and null mutants for each gene have been isolated. With the exception of the mutant affecting the acidic subunit VI, these mutants are respiration-deficient. Therefore, even the subunits without prosthetic groups are necessary for an active complex III, and one can speculate that they are required for assembly and stabilization of the complex. Among the many *pet* mutants analyzed (146), mutants with a defective ubiquinone–cytochrome c oxidoreductase activity can be found, although a direct selection or enrichment for such a specific defect is not possible. Thus, *pet* mutants having verified nuclear mutations must be further screened by biochemical assays to eliminate mutants affecting mitochondrial protein synthesis or the ubiquinone biosynthetic pathway (147). Among the mutants with a specific lesion in complex III activity, several classes have been distinguished:

1. Mutants with alterations in the eight nuclear genes for the protein subunits of the complex. As discussed above, null mutants are respiration deficient. In many of these the cytochrome b is also absent, presumably because it needs to be stabilized by association with other subunits.
2. A number of mutants are unable to make a cytochrome b detectable by spectroscopy, even though all structural genes in the nucleus and the COB gene on the mtDNA are normal. In another chapter on the expression of mitochondrial genes in yeast, it has been discussed that a variety of imported proteins are required for cytochrome b mRNA maturation (group I intron splicing and processing of the 5′ terminal) and for mRNA stabilization. The COB gene encodes a protein that appears to function both in 5′ end processing and mRNA stabilization (translational initiation?). Additional nuclear gene products are necessary for efficient translation of the COB mRNA, which in turn also affects the splicing of the intron, because the COB intron encodes a maturase (see Chapter 4). Another consequence may be a defect in cytochrome oxidase (complex IV) as a result of defective splicing of the COX1 mRNA.
3. Nuclear gene products are also required for the addition of the prosthetic groups to the various subunits. Hemes have to be covalently attached to apocytochromes, and another gene product, Bcs1p, has been implicated in the "maturation" of the Rieske protein. The precise function of the BCS1 gene product is not known.
4. Another set of proteins defined by nuclear mutations are involved in the "late stages of the assembly pathway" (147). This is a problem that so far has been defined exclusively by a genetic approach in yeast. Similar assembly factors have been identified for complexes IV and V. The mechanisms and pathways of assembly of the complexes in the ETC represent a major frontier in the quest for understanding mitochondrial biogenesis.

Cytochrome c reductase (complex III) from potato mitochondria has received scrutiny, because attention has been attracted more broadly to plant mitochondria. Similarities to the complex in mammals and yeast may not be surprising, but when sequencing information became available for comparative studies, the so-called core proteins from the potato complex showed the highest sequence identity with the mitochondrial processing peptidase (MPP) from mammals and yeast (142). These enzymes are involved in the cleavage of matrix targeting sequencing from precursors imported from the cytosol. It was subsequently verified that a bifunctional protein exists in potato mitochondria (all plants?) which plays a role in electron transport as well as in protein import. The complex is therefore referred to as the cytochrome c reductase/processing peptidase complex (see Section 4.6). It overall composition (at least 10 subunits) resembles the mammalian and fungal complexes.

A crowning achievement in the study of complex III was the completion of the complete crystal structure of the complex at 2.9-Å resolution by the groups of Yu and Deisenhofer (148). Following on the heels of the determination of the structure of complex IV (see below), it represents another major landmark in the understanding of the complexes of the mitochondrial electron transport chain. An atomic model has now become available to relate the proposed reaction mechanisms of QH_2 oxidation and the reduction of cytochrome c to the architecture and topology of the complex.

The model confirms, refines, and expands the stuctural features deduced from biochemical studies. In the crystal, two bc1 monomers interact to form a dimer with a twofold axis of symmetry perpendicular to the membrane. Each monomer has 12 transmembrane helices: Eight are from cytochrome b, one is from cytochrome c1, one is from the iron–sulfur protein, one from subunit 7, and one is still unassigned (Figure 5.14).

On one side of the membrane, more than half of the molecular mass of the complex extends 75 Å into the matrix. It consists mainly of the so-called core proteins that have a structural rather than a functional role. The peripheral core 2 proteins contribute significantly to the stabilization of the dimer, in addition to the interaction of the cytochrome b helices within the membrane. The two core proteins of the mammalian bc1 complex have homology to the α and β subunits of the mitochondrial matrix protein processing peptidase (MPP), and it is suggested that the soluble mammalian MPP has a similar structure. As discussed above, in plants the MPP is an integral part of complex III. A detailed side view of the complex is shown in Figure 5.15. On the outside of the membrane (P side), one finds the Rieske iron–sulfur cluster and the domain of cytochrome c_1 containing the heme. The electron density for this latter region is relatively poor, and the authors ascribe this problem to mobility in the crystal, which may in fact reflect a mobility required for interactions with, and electron transfer to, cytochrome c. Two additional heme irons could be located within the membrane domain, corresponding to heme b_L and heme b_H, respectively. The identification and distinction between the two hemes of cyt b was also aided and confirmed by the availability of difference density maps comparing the native crystals with crystals obtained with the inhibitors antimycin A or myxothiazol. The different targets of these inhibitors had been assigned from biochemical and spectroscopic studies (see below).

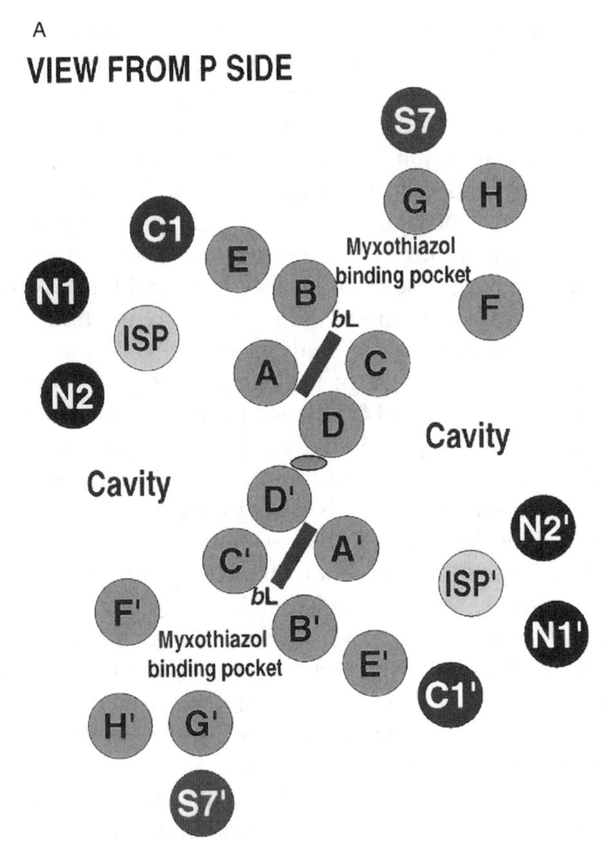

Figure 5.14 Complex III. Schematic representation of the transmembrane helices in the dimeric complex seen from the P and N (matrix) side (parts A and B, respectively) (82). See color plates.

Of special interest are the distances between the heme irons in b_H, b_L, and c_1 and the [2Fe–2S] cluster, in part to explain and confirm previous spectroscopic studies, and understand the rate of electron transfer. In the dimer the distances between two b_L's and two b_H's are comparable to the the b_H–b_L distance within a monomer, raising the possibility of electron transfer between the monomers.

The overall model with its "cavities" and "pockets" clearly invites further speculations about the accessibility of the complex to ubiquinone, antimycin, and myxothiazol and ultimately about the translocation of protons and the operation of the Q cycle.

Specific inhibitors of electron transport have been valuable as a means of arresting electron flux at known locations for the purpose of spectroscopic studies, for example, and to "isolate" a portion of the chain for detailed study. They have also been useful and revealing probes of structure, and in some

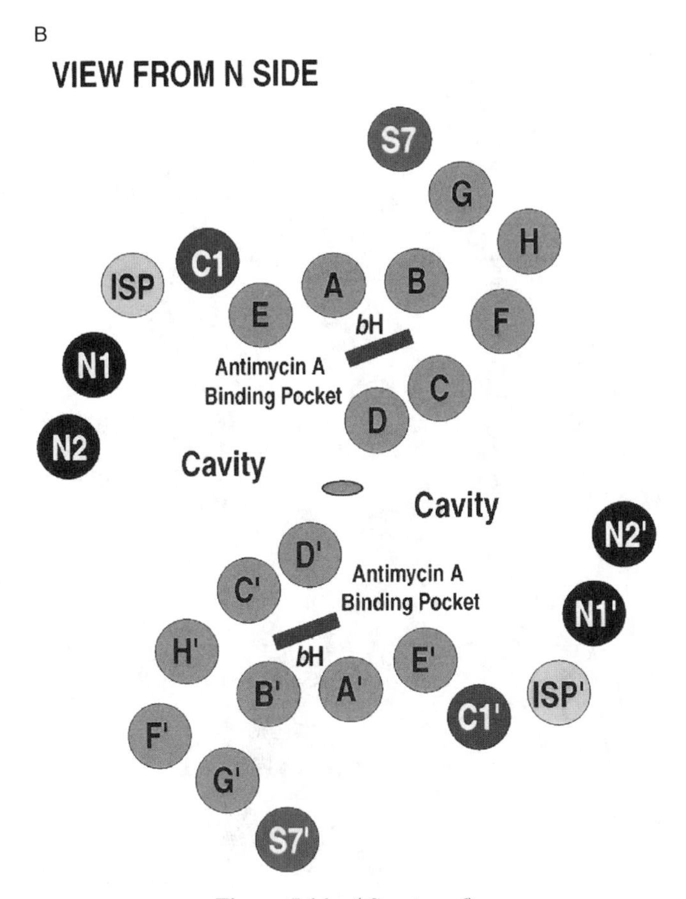

Figure 5.14 (*Continued*)

cases they have been useful to arrest electron flow within a complex to establish upstream and downstream portions of the complex. In the case of complex III, the antibiotic antimycin A is the best known of the inhibitors, but a variety of natural and synthetic inhibitors have been characterized. Initially their utility was in the definition of a specific block in the electron transport chain downstream from complexes I and II and upstream from cytochrome c, as well as in the localization of the "coupling sites" for oxidative phosphorylation. Later they were exploited in defining the distinction between center Q_N or center Q_P and in elucidating the protonmotive Q cycle (Figure 5.16).

The formulation of the Q cycle (149–152) required the postulate of two quinone reaction sites (see below). The Q_N or quinone reduction center is located on the matix side (N-side) of the inner membrane is associated with the recycling of half of the electrons back into the quinone pool and the uptake of protons from the matrix. It is inhibited by antimycin A, funiculosin, and

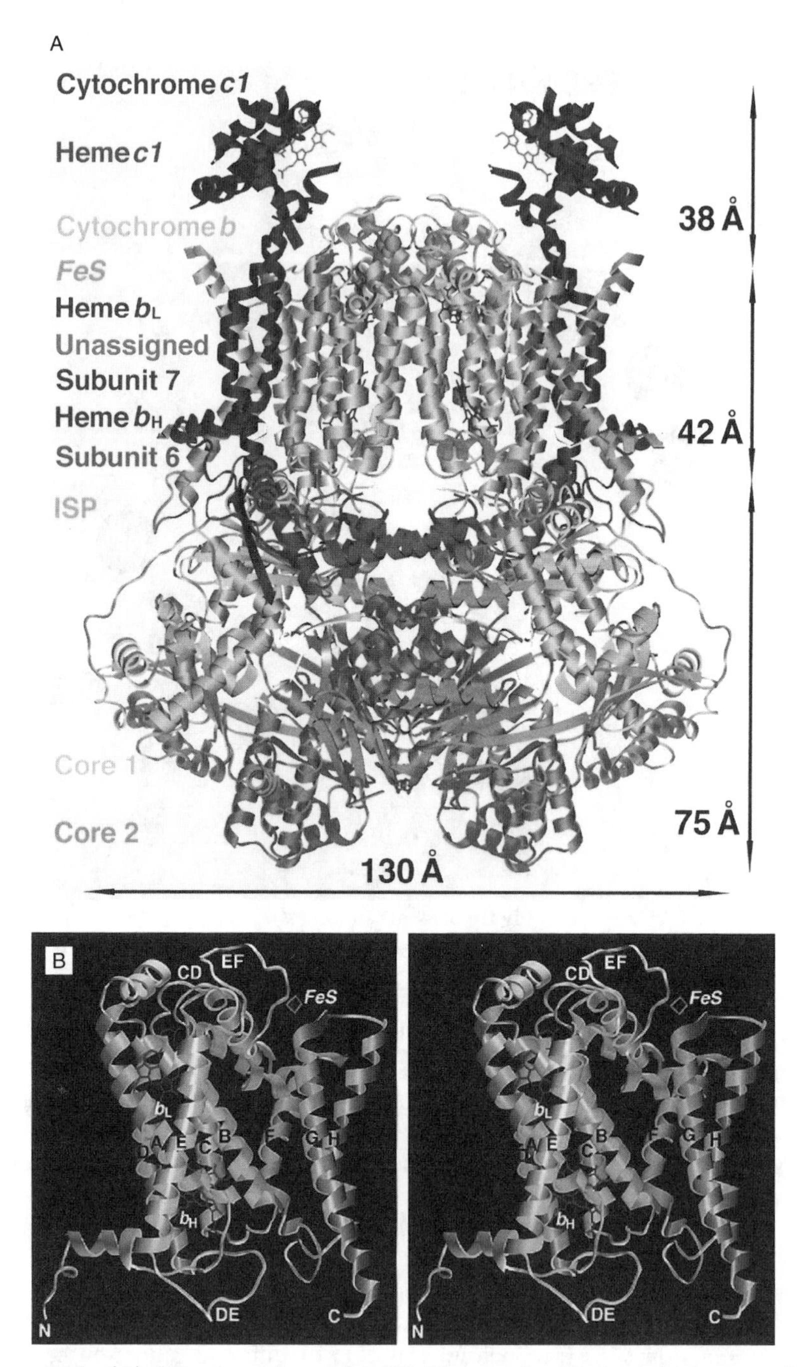

Figure 5.15 (A) Side view of complex III. The matrix side is at the bottom. Cytochrome c1 extends into the intermembrane space. The membrane-spanning helices of the various subunits are in the region indicated by the middle arrow (42 Å). (B) Stereoview of the transmembrane helices of a monomeric complex III. (C) Stereoview from the outside showing the myxothiazol binding pocket. (D) Stereoview from the matrix side showing the antimycin binding pocket. (From reference 147 with permission. The figure was generously provided by Dr. Yu.) See color plates.

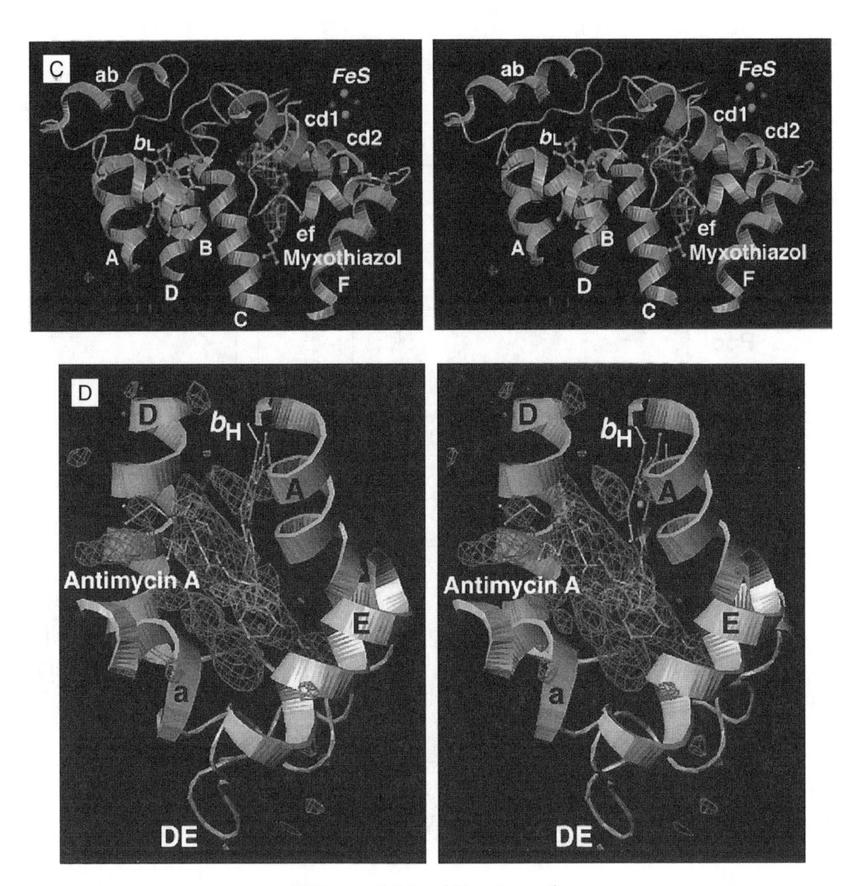

Figure 5.15 (*Continued*)

hydroquinoline-*N*-oxides, inhibitors that interfere with the electron transfer from heme b_H to Q or $QH^{\cdot-}$. At the Q_P center, electrons from reduced ubiquinone are accepted and divided into two pathways: half for recycling, and half for transfer via the iron–sulfur center and cyt c_1 to cytochrome c. It is located near the outer face of the inner membrane, and protons are released into the intermembrane space. A number of different compounds can inhibit at the Q_P center: 2-hydroxy-1,4-benzoquinone drivatives, stigmatellins, and MOA inhibitors containing the E-b-methoxyacrylate group (MOA-stilbene, myxothiazol). Some further discussion of the Q cycle can be found in Section 5.4.2.

A general review of the approaches and methodologies used with these inhibitors (Figure 5.17) has been written by Link et al. (153). Having established the specificity and efficacy of each inhibitor in arresting electron transport, additional studies have included light and EPR spectroscopy, the isolation of resistant mutants of a variety of species, the identification of mutated amino acid side chains in specific peptides from such mutants, characterization of

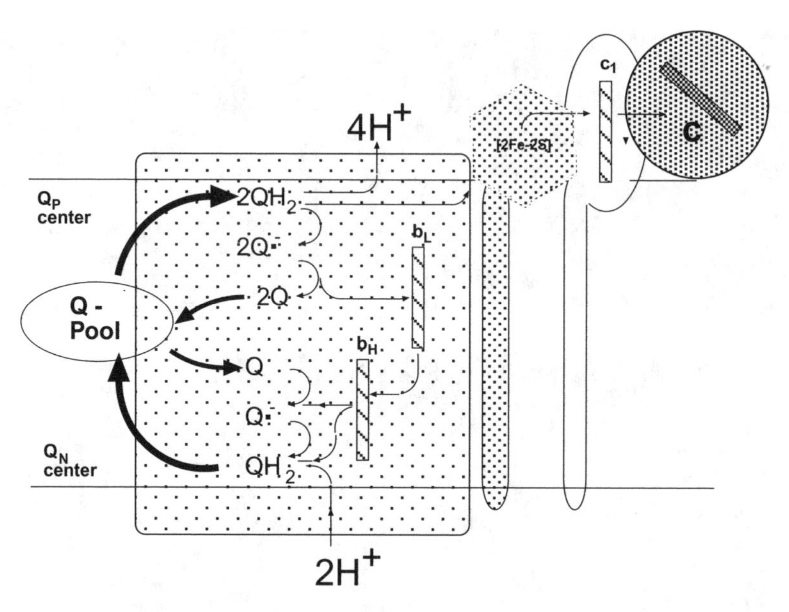

Figure 5.16 The Qcycle and electron flow through complex III. QH_2 is oxidized, and electrons are transferred to the mobile carrier cytochrome c (upper right in the figure). For a detailed explanation see the text. (After Trumpower (152).)

mutations induced by site-directed mutagenesis (154), and the synthesis of additional analogues—for example, antimycin A analogues (155).

5.2.7 Complex IV

Complex IV is generally referred to as cytochrome c oxidase. The overall reaction catalyzed is the following:

$$4\text{ cyt c}^{2+} + 4\text{ H}^{+}_{(s)\text{in}} + 4\text{ H}^{+}_{(p)\text{in}} + \text{O}_2 \rightarrow 4\text{ cyt c}^{3+} + 4\text{ H}^{+}_{(p)\text{out}} + 2\text{ H}_2\text{O}$$

Molecular oxygen is the terminal electron acceptor, the mobile carrier cyt c is reoxidized, and four protons are transferred to the intermembrane space. There are two processes involved. First, four electrons are donated on the IMS side, while four protons are taken up on the matrix side, resulting in a transfer of four positive charges across the membrane. Second, an average of one proton is pumped through the enzyme for each electron transferred to oxygen (156).

After three decades of study and analysis, this complex had the distinction of being the first complex of the ETC to have its high-resolution crystal structure determined at ~2.8-Å resolution. Two landmark papers on the bovine complex IV presented the largest and most complex membrane protein to be solved at this resolution in 1996 (157, 158). A monumental achievement, they

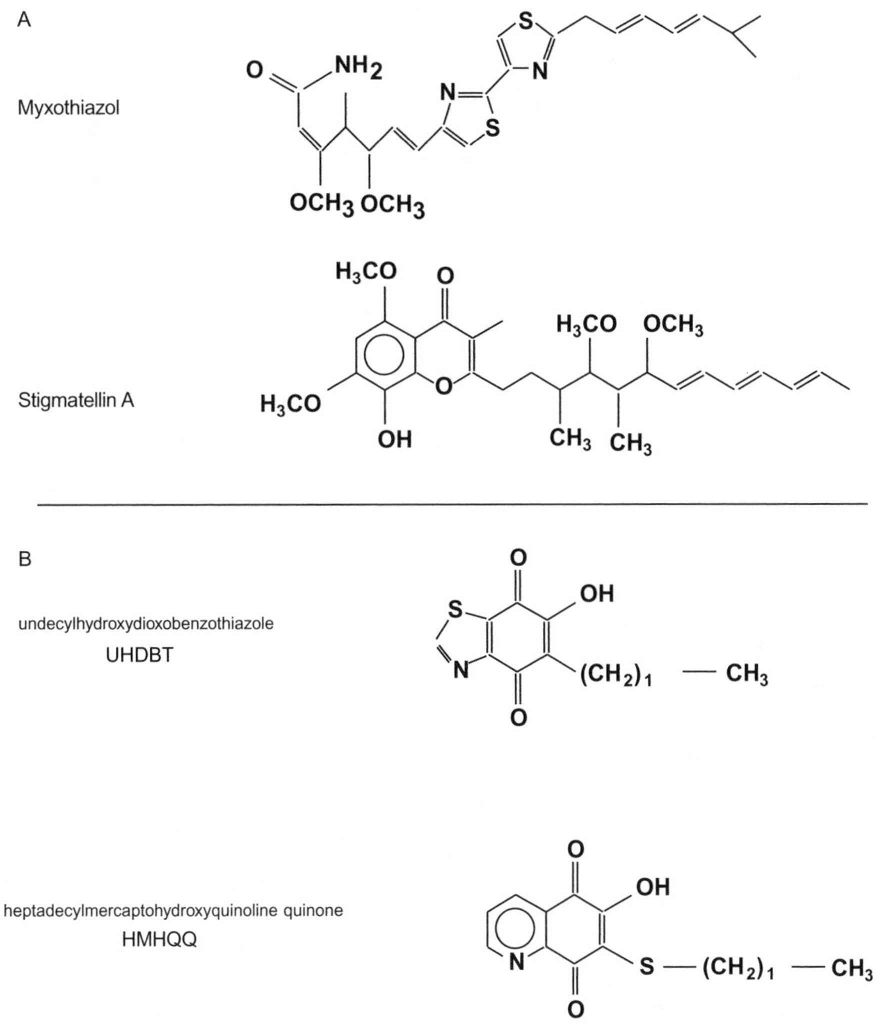
A
Myxothiazol
O
NH2
N
S
S
N
OCH3 OCH3
Stigmatellin A
H3CO
O
H3CO OCH3
H3CO
O
OH
CH3 CH3
B
undecylhydroxydioxobenzothiazole
UHDBT
O
OH
S
N
(CH2)1 — CH3
O
heptadecylmercaptohydroxyquinoline quinone
HMHQQ
O
OH
N
S — (CH2)1 — CH3
O

Figure 5.17 Different subclasses of inhibitors frequently used in the study of complex III. The fargets for the different drugs in groups A, B, and C are discussed in the text (see also Figure 5.14)

mark a major milestone in the history of electron transport and respiration. As always, these papers settled some older arguments and answered a number of questions quite definitively, but a wealth of new questions were raised to understand at the atomic level how electron transport is intimately coupled to proton pumping. Only high-resolution computer images with the added capability of rotating the image and zooming in and out of the structure can give a true sense of this achievement and its powerful influence on future thinking about these fundamental processes in bioenergetics. As one review

C

Antimycin A

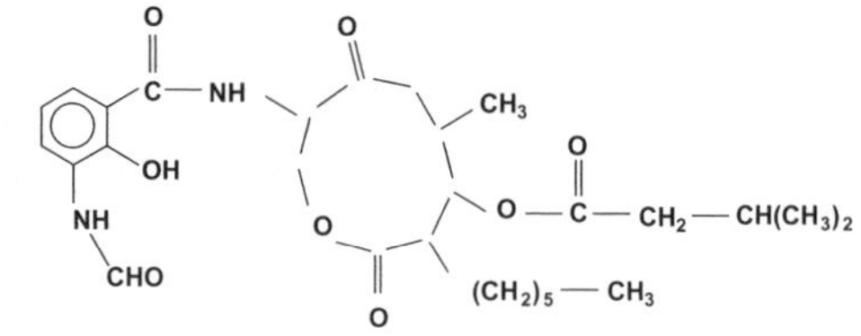

Funiculosin

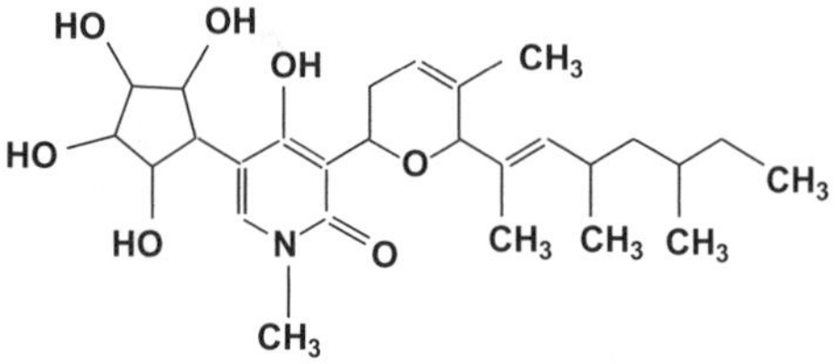

Heptylhydroxyquinoline-N-oxide

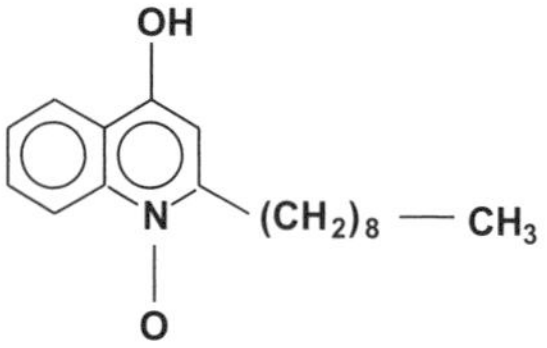

Figure 5.17 (*Continued*)

summarized the situation at the time: "The monster is subdued, but far from tamed" (159). In the meantime, more structures have been solved, and existing structures have been refined at higher resolution. The interested reader can find a wealth of information and links at a website dedicated to cytochrome oxidase (http://www-bioc.rice.edu/~graham/CcO.html).

The functionally homologous but simpler bacterial structure also became available in 1995 (160), by itself a major achievement. A remarkable conservation of structural features helps to emphasize those aspects most basic and essential for function. Certain differences will emphasize the requirement for regulatory mechanisms in higher organisms that may not exist or may be different in bacteria.

A discussion must start with the composition of the complex. The mammalian complex contains 13 subunits, the yeast complex is composed of 9 subunits, and one may expect to see further variations in the total number of subunits in different organisms. In all organisms the three largest subunits (I, II, III) are encoded by the mitochondrial genome and synthesized in the matrix. For the remaining subunits the nomenclature can be confusing to the uninitiated: In the mammalian complex we have subunits IV, Va, Vb, VIa, VIb, VIc, VIIa, VIIb, VIIc, and VIII; in yeast the subunits are IV, V, VI, VII, VIIa, and VIII.

And while a biochemist may refer to subunit VIIa in yeast, a geneticist might refer to it as the COX9 gene product.

A large number of biochemical and elegant spectroscopic analyses had established some time ago that the complex contained two hemes (aa_3) and two copper centers. In fact, the cytochromes a and a_3 were among the first described by D. Keilin in 1925. The X-ray structure has probably been most definitive in defining two additional metal centers containing Mg and Zn. Studies on the simpler bacterial cytochrome c oxidases and comparisons of homologous peptides have given clues about the association of these important metal centers with specific peptides which are now totally confirmed for yeast and mammals. Subunit I binds the heme a and heme a_3 prosthetic groups and also forms the Cu_B redox center. Subunit II binds the Cu_A center. Subunits I–III form the core of the enzyme. While subunits I and II can carry out the redox reactions and proton pumping, they become irreversibly inactivated during one cycle, and subunit III has been proposed to maintain the structural integrity of the complex (see reference 156 for a recent review and listing of many references). The other subunits may perform regulatory functions or play a role in insulation, stabilization, or assembly. Such a regulatory function is indicated by the existence of tissue-specific isoforms for the subunits VIa, VIIa, and VIII in mammals (161, 162).

In by now standard fashion, the yeast nuclear genes have been disrupted one at a time, and in the absence of either subunit IV, VI, VII, or VIIa there is no cytochrome c oxidase activity and cytochromes aa_3 are missing. A knockout of the COX8 gene (subunit VIII) causes only a 20% reduction of activity at normal aa_3 levels. There are two isoforms of subunit V, either one is required for activity, but complexes with either Va or Vb differ slightly in their kinetic and spectroscopic properties. The yeast complex may have additional subunits, making it more similar to the mammalian complex, but it has to be clarified whether they are *bona fide* subunits, scaffolding proteins required only for assembly, or simply contaminants.

For many years, laboratories had attempted to define the topology of these subunits within the inner membrane by reactions with membrane impermeable reagents, by cross-linking and co-precipitation, and by an exploration of the location of epitopes with a large variety of monoclonal antibodies (163, 164). In light of the solution of the crystal structure, these studies need not be reviewed here, but the development of the technology and reagents will continue to be useful tools in the examination of the enzyme, particularly in comparisons between the normal enzyme and enzymes derived from individuals with mitochondrial mutations in the genes for the three largest subunits (164).

There are 28 transmembrane helices contributed by the various subunits. Viewed from the top of the membrane, they form an irregular cluster surrounding the metal centers, in effect solubilizing the metal sites within the lipid bilayer and also "insulating" the electron path from the surrounding lipid layers. Many helices at the outside of the cluster probably have no function other than to serve in the assembly and stabilization of the complex. Notably, most of the transmembrane helices are not parallel to each other, and they are not vertical to the plane of the membrane. Three subunits (Va, Vb, and VIb) are not within the membrane, and they are associated as peripheral membrane proteins with extramembrane domains of the membrane subunits.

In considering the structure in detail, it is profitable to decompose the problem into separate structure–function relationships. The central issue is the path of electrons from the substrate ferrocytchrome c to oxygen. Another problem of fundamental importance is to understand the proton pumping activity associated with electron transport. Finally, regulatory functions, stabilization of the complex, and the assembly mechanism may be addressed from a consideration of the high-resolution structure. The emergence of the structure in two major stages reflects this order of priority.

The representation of the metal sites within the overall contours of the complex and in relationship to the membrane was the first major achievement (157), followed one year later by the entire structure (158) (Figures 5.18 and 5.19).

The two metal ions Mg^{2+} and Zn^{2+} are not redox centers. The Zn^{2+} ion is quite some distance away from the "action," and its major function appears to be in the stabilization of the three-dimensional structure of the complex. The Mg^{2+} ion, on the other hand, is between Cu_A and heme a_3, and, although not directly involved, it may have more than a simple stabilizing role, either in electron transport (and its direction) or in the coupling with proton transfer.

While the spatial arrangements and precise distances between the other metal centers were finally determined by the crystal structure, numerous ingenious spectroscopic studies had contributed to a path for the electrons which in a simplified form has the following appearance:

$$\text{ferrocyt c} \rightarrow Cu_A\text{–}Cu_A \rightarrow \text{heme a} \rightarrow \text{heme } a_3\text{–}Cu_B \rightarrow O_2$$

In this scheme the Cu_A center (with two copper ions) is located near the surface of the complex facing the intermembrane space and hence it is able to be brought in close contact with the reduced cytochrome c. The two heme groups are buried within the complex as electron carriers with a redox potential determined by the surrounding protein. The unique feature is the bimetallic heme iron–copper reaction center, which is also the site to which oxygen approaches for its activation and conversion to water. To understand more detail, one must understand the nature and redox behavior of the Cu_A–Cu_A center; and, most importantly, the mechanism of the transfer of electrons to molecular oxygen from the heme a_3–Cu_B combination.

The Cu_A–Cu_A center has the [2Cu–$2S_\gamma$] cyclic structure in which the two copper ions are coordinated with two cysteines, two histidines, a methionine, and a peptide carbonyl of a glutamate of subunit II. These ligands, except for the carbonyl group, had been proposed previously on the basis of mutagenesis experiments that were fully confirmed by the X-ray structures. The presence

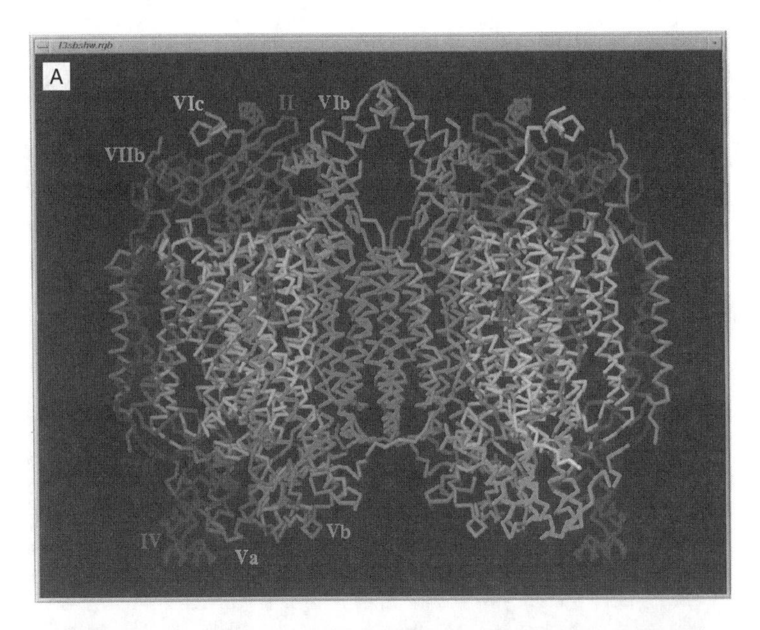

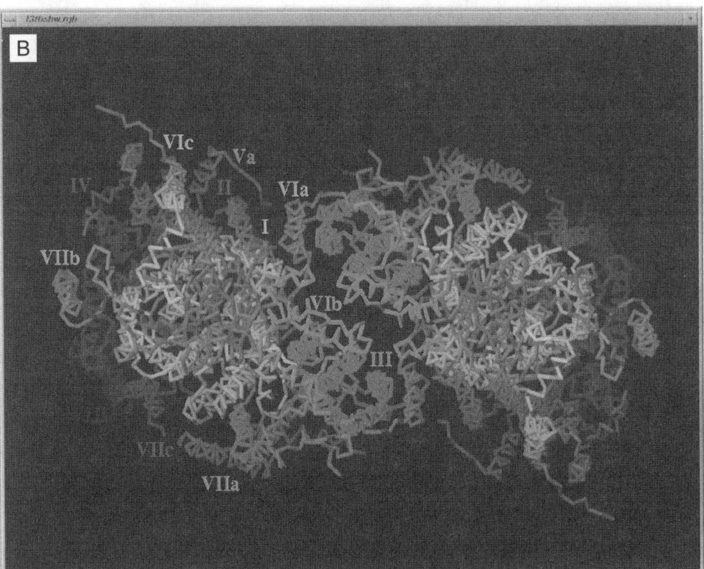

Figure 5.18 X-ray structure of complex IV. (A) Side view showing the transmembrane segments of the various subunits; matrix side at the bottom. (B) View from the cytosolic side. (The photographs were kindly provided by Dr. S. Yoshikawa of the Himeji Institute of Technology, Hyogo, Japan. From reference 157 with permission.) See color plates.

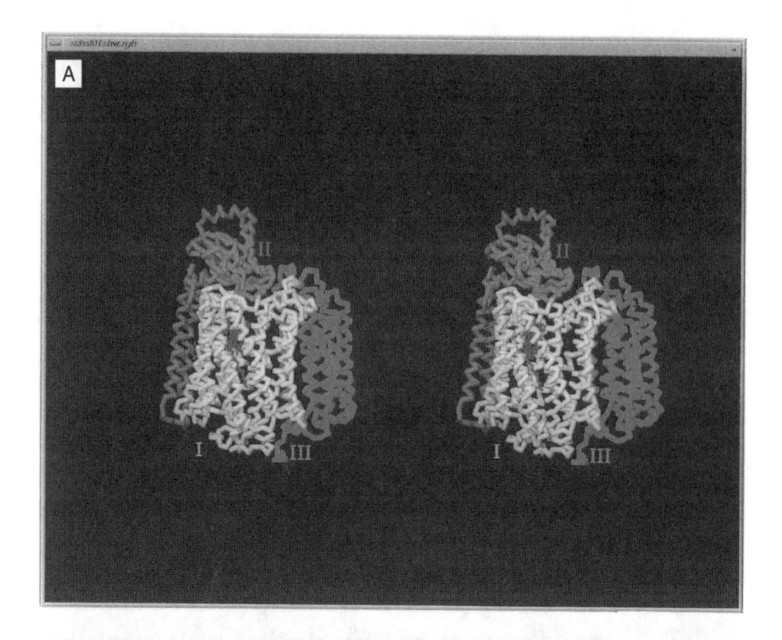

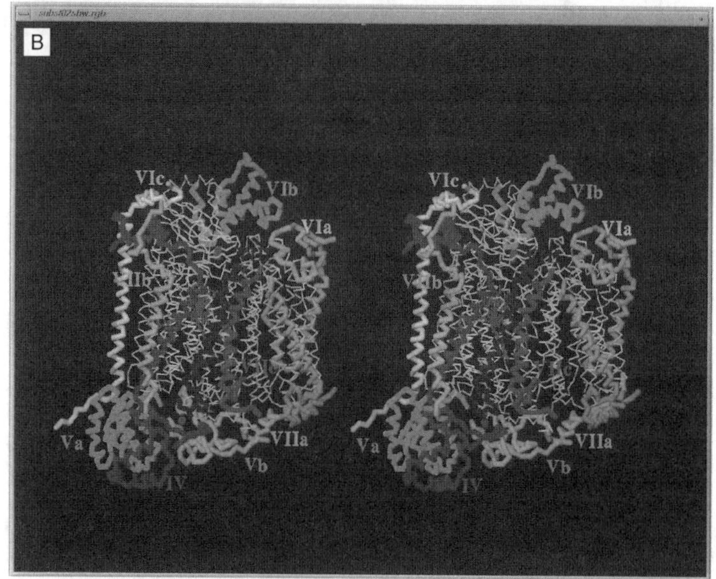

Figure 5.19 X-ray structure of complex IV. Stereo images of the C_α-backbone tracings. (A) Subunits I, II, and III. (B) Subunits IV, Va, Vb, VIa, Vib, Vic, VIIa, VIIb, and VIII. (The photographs were kindly provided by Dr. S. Yoshikawa of the Himeji Institute of Technology, Hyogo, Japan. From reference 157 with permission.) See color plates.

of two copper ions per site A had long been uncertain and controversial. An insightful account of the history of the Cu_A site in cytochrome oxidase has been written by H. Beinert (165). It is highly instructive in tracing the development of ideas about Cu in complex IV, in emphasizing the development of methodology and techniques (purification, spectroscopy, molecular genetics). Many general concepts in protein structure with special emphasis on metal–protein (ligand) interactions ultimately had to be applied to the understanding of a complex with seven metal ions, three of them Cu. It now appears that there are major similarities to the [2Fe–2S] centers, where one-electron transfers can be facilitated by delocalizing the charge between the two metal ions.

The most unusual and interesting structure is the binuclear heme a_3–Cu_B center, and major interest in this center arises from the nature of the reaction occurring there. It is usually presented as a simple addition of electrons and protons to oxygen:

$$4\,e^- + O_2 + 4\,H^+ \rightarrow 2\,H_2O$$

However, details of this process are more complicated (156, 166, 167). With 20% oxygen in the atmosphere, the organic material on the earth's surface would spontaneously burst into flame if the activation of molecular oxygen were simpler and energetically more favorable. The low normal reactivity of O_2 arises from the triplet electronic ground state of oxygen, with two unpaired electrons, which makes both single- and two-electron transfers from singlet-state donors kinetically slow. Thus, the thermodynamically highly favorable reduction of oxygen to water can be kept in check, except at high temperatures or at reaction centers such as the one under discussion. The latest scheme of oxygen activation proposed by Branden and colleagues is shown in Figure 5.20. The cycle starts with R (Fe(II)/Cu(I)). Oxygen binds to the heme iron (A), but the proximity of the copper ion allows two electrons from within the center to be transferred to oxygen, converting it to the peroxy ion (O_2^{2-}) and resulting in the oxidation of the metal ions to P_M (Fe(IV)/Cu(II), including a tyrosine radical at the catalytic site. The first of the four electrons from the upstream heme a enters together with a proton to form the intermediate F. In the F $\rightarrow$ OH transition another electron and a proton are added and the oxygen molecule is split. The proton redistribution leaves Fe(III) and Cu(II), each coordinated with a hydroxyl ion. The hydroxyl ions are released one at a time as water in two more steps ($O_H \rightarrow$ E and E $\rightarrow$ R) with the addition of one proton and one electron in each to return to the starting position Fe(II)/Cu(I). Many of these steps have been resolved by optical measurements at low temperatures, or by time-resolved Raman spectroscopy (168). For the more intricate biophysical, thermodynamic, and kinetic aspects of this problem the reader is referred to the specialized literature (e.g., references 156 and 167 and references therein). Not all aspects of the mechanism are resolved to everyone's satisfaction, especially with regard to the steps that drive the proton pumping. What makes the pump unidirectional? Recent expert reviewers have

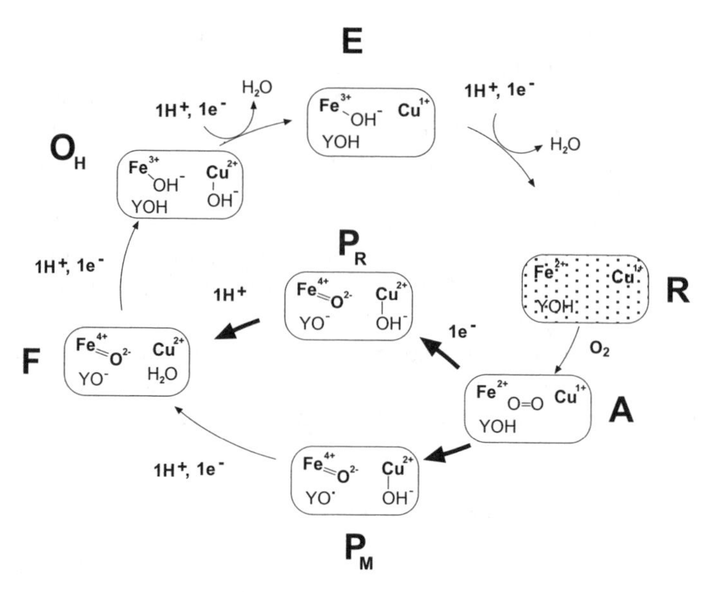

Figure 5.20 Proposed sequence of reactions for transferring four electrons from cytochrome c to an oxygen molecule to form two molecules of water. (Adapted from reference 155.)

referred to this enzyme as a "hysteric" enzyme; that is, its properties are dependent on its history, which depends (at least under experimental, *in vitro* conditions) on the rate of delivery of the electrons from cyt c.

A most important consequence arises when this overall process is perturbed by the premature departure of partially reduced oxygen. Partial reduction yields the superoxide radical ($O_2^{\cdot -}$), from which hydrogen peroxide (H_2O_2) or the hydroxyl radical ($OH^{\cdot}$) can be derived. These species are often referred to collectively as the "reactive oxygen species" (ROS), and their production is strongly implicated in oxygen toxicity and mutagenesis of mitochondrial DNA (see Chapter 7). A more elaborate discussion of the ROS and their reactions will be presented in Section 5.7. In the present context it is important to clarify whether reactive oxygen species produced in mitochondria arise from incomplete reduction of oxygen in complex IV. It has been estimated that a significant fraction (a few percent??) of oxygen entering mitochondria is reduced only partially, and this fraction may increase under abnormal conditions or as a result of electron transport complexes impaired by mutations. Normal individuals suffer from various afflictions when exposed to greater than 21% oxygen, and prolonged exposure can lead to lung damage. Premature babies exposed to higher-than-normal oxygen concentrations in incubators can develop retrolental fibroblasia (ocular damage) unless carefully monitored and given α-tocopherol as a free radical scavenger. It is generally accepted that mitochondrial ROS production does not occur at the site designed for reduction of oxygen in complex IV, but rather from the reaction of the highly soluble and diffusible oxygen with reduced upstream sites of the

ETC. In other words, a leakage of electrons can occur from high potential sites, especially in damaged or abnormal mitochondria.

The complete solution of the structure of complex IV in mammals and of the homologous simpler complex in bacteria has greatly stimulated the discussion of the mechanism of proton pumping coupled to electron transport (156, 158, 160, 167), but here even a structure at 1.8-Å resolution cannot give an absolutely definitive answer. Previous approaches employing mutagenesis had suggested the existence of two proton channels: one for protons that become associated with oxygen to form water, and another for the protons pumped across the membrane. From theoretical considerations, one would expect the proton channels to consist of a network of hydrogen-bonded side chains whose pK_a's would be controlled by conformational changes associated with redox reactions. Acidic groups on either side of the membrane must be alternately accessible, and their p*K* must vary depending on the oxidation state of the enzyme. Protons are shuttled through the protein complex along a string of residues that have been referred to as a "proton wire" (169). The relevant residues were identified primarily with the help of mutagenesis experiments on the bacterial enzyme(s). Several such networks were found, and included in these localized structures were cavities likely to contain water molecules that can participate in proton conduction. Since the original X-ray structures were reported in 1995, additional structures have now been examined of the enzyme in the fully reduced, fully oxidized, azide-bound, and carbon monoxide-bound states (170). These studies greatly refine the conformational changes associated with the redox reaction and contribute to the formulation of more specific proton pathways and models (see Figure 5.21). In fact, the power of modern computers is such that the behavior of the entire enzyme can nowadays be simulated by molecular dynamics, with various restraints imposed depending on the cost of the calculations (167). The proton pathways from the matrix to the a_3–Cu_B center appear to be resolved, but the subsequent path to the IMS remains to be further elucidated. The D pathway (Figure 5.21) appears to be involved in the pumping of all four protons. Although the pathways may be identified, there is still the challenge to understand how the rate and direction of movement are controlled by the electron flow. Finally, one must also identify a water channel for the escape of the product, H_2O, and possibly a pathway through which the substrate O_2 can approach the heme a_3–Cu_B reaction site. Potential pathways have been suggested from a consideration of the crystal structure, but none of them are wide enough in the static crystal structure to accommodate these molecules. Thus, rapid, reversible conformational changes have been postulated to operate in the opening and closing of such channels. Expert, comprehensive, and up-to-date reviews (156, 167, 171–173) should be consulted for further details.

Finally it should be noted that most of the more sophisticated biophysical studies have been performed with the prokaryotic complex from *Rhodobacter sphaeroidis*. In the eukaryotic enzymes the basic mechanisms of electron transport, oxygen reduction and proton pumping are likely to be similar, but superimposed on these will be regulatory mechanisms from ligand binding and

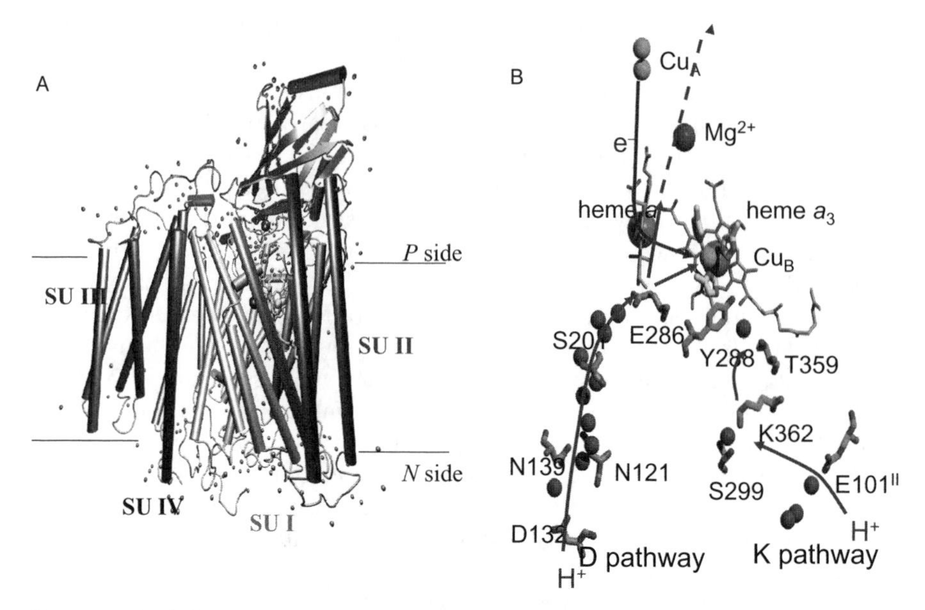

Figure 5.21 (A) Schematic view of complex IV with the four essential subunits (155). (B) The proposed D and K pathways for the entry of protons to form water and to be pumped across the membrane. The path from the region of the hemes to the outside is still poorly defined. (From reference 155 with permission.) See color plates.

protein phosphorylation (see Section 5.6). Cytochrome oxidase may have the capacity to regulate its own H^+/e^- ratio—that is, the efficiency with which proton pumping is coupled to electron flow; this may act as a protective mechanism to prevent the membrane potential to become too high (167). The specific role of the "supernumerary" subunits remains a challenge for the future.

In closing this section, one should not dwell on the outstanding issues and unsolved problems. Instead it should be recognized that since the discovery of the *Atmungsferment* by O. Warburg almost seven decades ago, this enzyme has been one of the most important objects of study in the area of bioenergetics. As such, it has attracted the attention of some of the greatest minds in the history of biochemistry. The whole structure at high resolution (1.8 Å) represents a crowning achievement. Words fail to describe it fully, but most institutions and many individual laboratories can now obtain the structure via the internet, or on computer disks for display on a workstation. When combined with three dimensional vision achieved with special goggles, the inspection and exploration of this structure becomes an awsome spectacle even for the nonspecialist.

5.2.8 The Assembly of the Electron Transport Chain Complexes

While the compositions, structures, and functions of the electron transport complexes are now understood in impressive detail, a remaining challenge is to understand their biogenesis. Gene expression has already been touched on in the chapter on the biogenesis of mitochondria. The following will make reference briefly to the mechanisms and pathways leading to the assembly of multisubunit complexes with integral and peripheral membrane subdomains. In addition, [Fe–S] centers, hemes, and other metal centers (Cu) have to be incorporated into these complexes. The synthesis of [Fe–S] centers and hemes will be covered in a separate section (6.8); the pathways for their synthesis are distinct up to the stage where the [Fe–S] cluster or heme is incorporated into the respective apo-proteins.

In general, one approach has been to characterize the intermediates that accumulate in the absence of a particular subunit for a complex of interest. It is a particularly feasible approach in yeast, *S. cerevisiae*, where specific knockouts can be made routinely and systematically. In other organisms, one is dependent on the often serendipitous finding of suitable mutants. Since partially assembled intermediates are generally not functional, their detection depends on a combination of electrophoretic fractionations and immunochemical identification of the constitutents of the assembly intermediate. Such fractionations require solubilizations in mild detergents and raise the possibility that existing (*in vivo*) unstable intermediates are destroyed in the process. It is also a common finding that when specific subunits are absent due to mutations, some or all of the remaining subunits fail to assemble and are rapidly degraded.

In favorable genetic model systems it becomes feasible to screen for mutants defective in oxidative phosphorylation, followed by a further screen for mutants missing a specific complex of the ETC or ATP synthase. Such screens and some serendipitous isolations have yielded mutants with isolated complex deficiencies where a subsequent search has failed to find mutations in the known structural genes for these complexes. From such observations it became clear that assembly factors or molecular chaperones participate in the biogenesis of these complexes. Furthermore, some of these factors are highly specific for a given complex; examples will be presented below.

Complex I assembly in mammalian mitochondria has been studied in several Chinese hamster cell mutants in tissue culture (28, 40, 41) and in human cell lines from patients with an isolated, partial complex I deficiency (29, 79, 174). In the future the yeast *Yarrowia lipolytica* (48, 175, 176) promises to offer the same powerful means of genetic manipulation for studying assembly that have proved to be so productive in the study of the other complexes in *S. cerevisae* (see below). Stimulating results have also been derived from the study of *Neurospora crassa* (47, 177, 178), but this system has proved to be less tractable for genetic manipulations. Kuffner et al. (178) made the suggestion that the matrix arm and the membrane arm are formed independently of each other and are joined in the course of assembly, but such an appealing and simple mechanism has not been confirmed for the mammalian complex. The subunit assembly must at some stage be coupled to the assembly of the 8–9 [Fe–S] clusters in the peripheral subcomplex and to the addition of he FMN cofactor in the 51-kDa subunit (*NDUFV1* gene in humans).

Two Chinese hamster cell mutants characterized in the Scheffler lab have shown some intriguing results. The absence of the "supernumerary" subunit MWFE (encoded by the *NDUFA1* gene) causes a total absence of complex I activity, even though the MWFE protein is only 70 amino acids long and represents less than 1% of the total mass of the complex. No prominent assembly intermediates have been detectable in these mutant cells by BN gel electrophoresis, although all of the subunits of the peripheral subcomplex appear to be present, as far as they could be detected by the available antisera (39, 41, 42). The existence of this null mutant has permitted the construction of a conditional complex I assembly system in which the expression of the MWFE protein is under the control of an inducible promoter, and thus complex I assembly can be initiated by the addition of the inducer doxycyclin to the cells (28). A surprising observation was that the appearance of a complex I (~900 kDa) detectable on BN gels was significantly delayed relative to the appearance of the MWFE protein, and there was a further delay of almost 24 hours of the complete restoration of respiration after the induction. In other words, the appearance of complex I in the mitochondria was not sufficient to restore respiration, and one can speculate that its assembly into the ETC and supercomplexes may require additional time (see reference 179). A similar picture emerges from the study of a mutant missing the ESSS subunit of complex I (40). An inducible system has also been constructed for this essential subunit and a considerable lag has been observed between the appearance of the protein in mitochondria, the detection of complex I on BN gels, and full restoration of respiration (Potluri and Scheffler, unpublished). Very recently, studies from the Nijtmans group (180) have identified assembly intermediates by the use of an inducible NDUFS3-GFP expression system in HEK 293 cells. Six subcomplexes in the assembly pathway were identified and partially characterized; two of these accumulated when mitochondrial protein synthesis was inhibited, demonstrating that mtDNA-encoded subunits are required for the formation of higher-molecular-weight complexes. The peripheral subcomplex is fully assembled only after membrane attachment with the help of ND1 and other ND subunits.

Earlier work in *Neurospora* had identified two genes/proteins required for complex I assembly: CIA84 and CIA30 (47, 178). The human homologue of CIA30 was characterized by Janssen et al. (181) and shown subsequently by Vogel et al. (78) to be a chaperone for complex I assembly. A mammalian homologue for CIA84 has not yet been found. It can be expected that many more assembly factors specific for complex I will be found in the future. Clinical laboratories in Europe and the United States have identified a substantial number of patients with an isolated complex I deficiency. All the known genes (mitochondrial and nuclear) for structural proteins of complex I have been examined in such patients (a majority of them boys), and no defects could be found in a large fraction. Thus, there must be additional genes/proteins responsible for the assembly of a functional complex I. Of all the known mapped and characterized genes for structural proteins, only two X-linked genes have been identified. The Scheffler laboratory has identified three X-linked genes in Chinese hamster cell lines required for complex I activity. Two of these have been characterized: *NDUFA1* (MWFE) and *NDUFB10* (ESSS) (see reference 40 for a summary). For the third complementation group the specific gene has yet to be found, although X-linkage was established by somatic cell hybridizations (182).

Complex II assembly would appear to be the smallest challenge since only four subunits are involved, and all are encoded by nuclear genes. Significant progress has been reported by the group of Lemire (129, 130). In the course of assembly, three [Fe–S] clusters must be incorporated into the IP subunit, a heme (b_L) group becomes coordinated with the two integral membrane subunits, and an FAD cofactor becomes covalently linked to the largest FP subunit. It appears that in the absence of the integral membrane subunit there is no assembly of a functional SDH activity (i.e. the peripheral subcomplex) (136).

Complex III has three catalytic subunits (cytochrome b, cytochrome c_1, and the Rieske protein) found in all organisms and seven or more supernumerary proteins whose function is still obscure. Zara and colleagues (183) have systematically investigated the effect of single or double deletions of these supernumerary proteins or cyt b on the formation of an active complex III. Some of these subunits were dispensable (QCR6p, QCR10p) as single deletions. Most double deletions had a significant effect on the assembly of a functional complex. A further analysis of what subunits are stable/present in these mutants led these authors to support an assembly scheme (184) that includes the formation of three subcomplexes characterized by the presence of cyt b, cyt c_1, or the two "core" proteins, respectively, their coalescence into a cyt bc_1 subcomplex, and finally the formation of the mature complex III with the addition of the Rieske protein and subunit 10. It is clear that the supernumerary subunits play an important role in the assembly and stabilization of subcomplexes. It is likely that at least some of the "supernumerary" subunits in complexes I and IV play similar roles.

The study of the assembly of complex IV has perhaps reached the most advanced stage from the efforts of several labs. At least 25 factors are required for complex IV assembly (in addition to the 12 (yeast) or 13 (mammals) subunits of the enzyme); these were identified largely from the analysis of COX-deficient yeast mutants and the identification of required genes other than the structural genes for known subunits. However, at least one interesting gene (*SURF-1/Shy1*) was also identified from a molecular–genetic analysis of human patients with Leigh syndrome, a mitochondrial disease associated with a complex IV deficiency (185–187). As described above, the three largest subunits (Cox1, Cox2, and Cox3) are highly conserved from prokaryotes to mammals; they are encoded by mtDNA and hence inserted into the membrane from the matrix side. The function of many of the other subunits is still a matter of speculation centered on their role in assembly, stabilization, and control of the enzyme activity. The assembly factors/chaperones have various names, depending on the history of their discovery, a phenotype, or some functional understanding. In yeast there are the Cox10, Cox11, Cox14, Cox15, Cox16, Cox17, Cox18, Cox19, Cox20, and Cox23 genes, the Pet100, Pet117, and Pet191 genes, the Shy1 gene (alias SURF-1 in mammals), Sco1 and Sco2 genes, Mba1 and MSS51 genes, the Oxa1 gene, and more. A thorough and up-to-date discussion of each of these can be found in the review by Herrmann and Funes (188), which references much of the pioneering work from various laboratories. Some of these are involved in the translation of Cox mRNAs in the matrix and the insertion of the peptides into the inner membrane. Others are required for heme a synthesis. The Sco gene products together with other chaperones are needed for the insertion of copper ions into the two centers. Herrmann and Funes describe "assembly lines" for Cox1 assembly (including heme a and copper) and Cox2 assembly (the second copper), the formation of several additional subcomplexes, and finally the completion of the mature complex with the addition of other subunits: zinc ions, magnesium ions, and cardiolipin. One could add dimerization and formation of supercomplexes in the ETC as further steps. Much remains to be learned. The SURF-1 gene product is absent in the Leigh syndrome patients, but such patients have some cytochrome oxidase activity, suggesting that it is not absolutely essential for complex assembly. This conclusion has been confirmed by experiments with *shy1* mutants in yeast. It is clearly not possible to do justice to all the elegant studies that have provided insights into the process of cytochrome oxidase biogenesis, but starting with some recent reviews the interested reader can gain rapid access to the literature (171, 188–190).

5.3 ELECTRON TRANSPORT IN OTHER ORGANISMS

The description provided above for the components of mitochondrial electron transport (there is also electron transport in chloroplasts) appears to be almost universally applicable to mitochondria in animals, plants, and most of the other species of eukaryotic organisms. However, this should not lull one into thinking that once one understands one mitochondrion, one understands them all. Yeast (*S. cerevisiae*) mitochondria lack a complex I of the kind described above. An alternate pathway for the oxidation of NADH in yeast will be presented below. Plant mitochondria have several additional capabilities which make them unique (191–193).

5.3.1 NAD(P)H Dehydrogenases

The first distinctive difference between plant and animal mitochondria is in the oxidation of NADH (194–197). Complex I in either system can accept NADH as substrate when it is produced in the matrix from reactions of the Krebs cycle, for example. NADH produced in the cytosol has to be transferred into the matrix by a shuttle system in animal mitochondria (see Section 6.9), but plant mitochondria have an external NADH dehydrogenase that can oxidize NADH and transfer reducing equivalents directly to ubiquinone or complex III. The reaction is insensitive to rotenone and generates ATP with a P/O ratio of 2 or less. A separate dehydrogenase dependent on Ca^{2+} on the outer surface of the inner membrane is specific for NADPH (194). No shuttles are required to deal with cytoplasmic sources of NADH (from glycolysis) or of NADPH (controlling the pentose shunt). These two activities are differentially sensitive to diphenyleneiodonium (DPI), with NADPH oxidation being irreversibly inhibited at submicromolar concentrations (198). The two enzymes appear to be single, relatively small polypeptides (32, 55 kDa) (194). A third rotenone-insensitive, nonphosphorylating (?) NADH dehydrogenase made up of two 43-kDa subunits has been localized on the matrix side of the inner membrane (199).

The full physiological implications of the existence of these external and internal nonphosphorylating NADH dehydrogenases in addition to complex I are not yet understood. Loosely speaking, these enzymes may participate in the efficient interconversion of a variety of intermediates required for rapid plant growth. The most puzzling aspect is the presence of two internal NADH dehydrogenases, one of which is very similar to complex I in animal mitochondria. Do both of these enzymes select substrates from a common pool, and what controls the relative affinity of NADH for one or the other enzyme? The rotenone-insensitive dehydrogenase appears to have a low affinity and thus may operate under certain conditions when NADH levels build up from a highly active Krebs cycle. It has been suggested that it is exclusively linked to the cyanide-resistant electron pathway (see below), providing a totally nonphosphorylating pathway for endogenous NADH oxidation (193).

5.3.2 A Cyanide-Insensitive Electron Pathway

Another, initially very puzzling finding was the observation of sometimes significant residual respiration by plant mitochondria in the presence of CO, azide, or cyanide (192, 200). Electrons from ubiquinol appear to have two choices, which may be regulated by physiological conditions or depend on the tissue type or developmental stage: They can either enter the conventional pathway via complex III and cytochrome oxidase to oxygen, or they can reach oxygen via a specific oxidase unaffected by the classical inhibitors. The characterization of the alternate oxidase was made more difficult because it had no characteristic absorption spectrum and did not exhibit electron paramagnetic resonance signals. The enzyme also appeared to be extremely labile, and not surprisingly it was not a soluble enzyme. The cloning of a cDNA encoding this protein, originally from the voodoo lily, *Sauromatum guttatum* (201), and subsequently from other species, has greatly facilitated the study of this enigmatic terminal oxidase. By expressing the voodoo oxidase cDNA in the yeast *S. pombe*, it could be demonstrated quite conclusively that a single polypeptide (~32 kDa) was sufficient for introducing alternate oxidase activity into the *S. pombe* mitochondria (202). It was cyanide- and antimycin-resistant. The protein is an integral membrane protein with two transmembrane segments. The N-terminal and C-terminal domains are exposed to the mitochondrial matrix. From sequence comparisons, highly conserved motifs have been found in the C-terminal domain which make it likely that the active site is composed of a coupled binuclear iron center similar to that found in ribonucleotide reductase or methane monooxygenase (see reference 202 for a collection of recent references). The expected ligands for the irons are two histidines, two glutamates, one aspartate, and two water molecules. The enzyme is active in the reduced form; changing the redox status of the sulhydryl–disulfide linkage affected the electron flux in the transgenic yeast mitochondria. The functional expression of this enzyme in yeast mitochondria in combination with site-directed mutagenesis will help to answer numerous questions about structure–function relationships in this enzyme.

Electron flow through the alternate pathway does not lead to ATP formation. In other words, there is no proton translocation across the inner membrane associated with this pathway. In some special plant tissues the pathway serves in thermogenesis (see Section 5.6.2), but like rotenone-insensitive NADH dehydrogenase in the matrix, the physiological significance of this pathway is still somewhat obscure. It has been considered to act as an overflow to drain off the energy of carbohydrates when they are in excess of demand (193). For example, the interconversion of intermediates via the Krebs cycle could continue without constraints by a membrane potential or ADP availability (see below). Since the contribution of this pathway depends on the tissue and is developmentally regulated—for example, in fruit ripening and storage organs (192)—much attention has been devoted to understanding the regulation of its activity. Transcriptional regulation of gene expression is an obvious focus of attention, now that the gene has been cloned. However, post-transcriptional, biochemical regulatory mechanisms are also under consideration. In the voodoo lily (*Sauromatum*) an inducer of heat production and cyanide-resistant respiration called "calorigen" had been known for over 50 years, and it was recently identified as salicylic acid (203). There is speculation that salicylic acid is a transcriptional inducer, but post-translational modifications of the protein may also be required (201).

5.3.3 NADH Oxidation in Yeasts

Budding yeast, *Saccharomyces cerevisiae*, the model system for almost everything a cell biologist wishes to study, does not have a complex I of the type described for animals. This became generally apparent from the absence of sensitivity to rotenone or piericidin and the lack of site I phosphorylation, although under some conditions the coupling site I may be induced (see below). Later, *S. cerevisiae* mtDNA proved to be the exception by having no genes encoding subunits for complex I which had been recognized in most other species. When NADH oxidation was measured with yeast mitochondria, both internal and external NADH dehydrogenase activities could be detected, that is, one enzyme had an active site facing the intermembrane space, while the other used NADH generated in the matrix. NADH itself cannot pass through the inner mitochondrial membrane.

The NADH dehydrogenases found in *S. cerevisiae* and also in *S. pombe* are likely to be related to the rotenone-resistant enzymes found in plants and bacteria. Some of these enzymes and their genes have been characterized (204–206). They contain FAD and reduce ubiquinone (Q_6) as well as a number of related and unrelated electron acceptors. A single polypeptide with apparent molecular mass 53 kDa believed to be the external enzyme was found to be induced about 5- to 10-fold in the absence of glucose. The internal enzyme is a single 57.2-kDa peptide (206). This yeast enzyme can be expressed as an active enzyme in *E. coli* (207); and most strikingly, it has been expressed from a mammalian expression vector in a complex I-deficient Chinese hamster cell mutant (208). It is localized in the hamster mitochondria and restores the capacity for respiration and oxidative phosphorylation in these cells. In other words, as a single peptide it functionally complements the entire complex I consisting of ~45 peptides.

Generalizations about "yeasts" should be taken with a grain of salt; there are many different species of yeasts, and some of them do appear to have a "conventional" complex I. The identification is based on several criteria. The detection of rotenone-sensitivity is suggestive; the isolation of large complexes (~700 kDa) is more convincing. Finally, sequencing the mitochondrial DNAs from several of these species has revealed the presence of ORFs with obvious homology to the mitochondrial genes for complex I (ND1, ND2, . . . , ND6, . . .) found in many organisms. A subdivision of yeast species into strictly aerobic yeasts and facultatively fermenting yeasts may relate to the capacity for complex I formation. The presence of a complex I-like activity has so far been deduced for the yeasts *Candida pinus, Cryptococcus albidud, Rhodotorula minuta, Trichosporon beigelii, Candida parapsilosis, Pichia guilliermondii, Clavispora lusitaniae, Hansenula polymorpha*, and others (209, 210). The yeast *Pichia pastoris* has attracted attention because of its useful attributes of inducible gene expression and thus its interest to the biotechnology field. Its bioenergetic properties have become a focus of attention only recently. It does not appear to have an alternative NADH oxidase and may have a complex I very similar to the typical vertebrate complex (211). A concerted effort by the group of Brandt and colleagues has made *Yarrowia lipolytica* a particularly influential model system in which most of the powerful recombinant DNA technology developed for yeasts can be applied to the molecular–genetic analysis of the structure and function of complex I (48, 49, 212, 213).

5.3.4 Energy Metabolism and NADH Oxidation in Trypanosomes

The unusual organization and expression of the mitochondrial genome in a group of protozoa with kinetoplasts has already been discussed in a previous chapter. There are two major taxonomic groups: the trypanosomatids (e.g.,

T. cruzi, T. brucei, L. tarentolae) and the cryotobiids (e.g., *T. borreli*). These parasites have a complex life cycle that includes stages in a vertebrate host and in an insect vector. Transitions from one stage to another are accompanied not only by morphological changes, but also by adaptations to the nutritional conditions encountered in the different hosts. Glucose is a major energy source, but amino acid catabolism can also serve to provide energy and other substrates. A unique feature is the compartmentalization of the reactions of the Embden–Meyerhof pathway in organelles referred to as glycosomes (214, 215). The fate of pyruvate then varies between members of the family and between different stages of the same species, and it depends also on the aerobic or anaerobic conditions encountered by the organism (216). Pyruvate may be completely oxidized to CO_2 by the classical pathway, but one also finds acetate as the end product and observes ATP generation from a combination of acetate:succinate-CoA transferase activity and the succinate:succinyl-CoA cycle. A detailed discussion of this subject is beyond the scope of this section.

Of special note here is the existence of alternate oxidases in trypanosmatid mitochondria which function in the maintenance of the redox balance in the glyoxisome. Reducing equivalents are moved from the glycosome to the mitochondria via a glycerol-3-phosphate/dihydroxyacetone phosphate shuttle, and an FAD-linked glycerol-3-phosphate dehydrogenase is used to donate electrons to the UQ/UQH_2 pool. UQH_2 may then be oxidized via the classical electron transport chain coupled to ATP production, or by a plant-like alternative oxidase. Thus, the long, slender bloodstream stage of *T. brucei* does not possess the usual complex III/complex IV respiratory chain, and it uses the cyanide-insensitive oxidase exclusively (216). In the procyclic (insect) stage of *T. brucei*, cytochromes and oxidative phosphorylation are observed readily, although there has been a debate over the apparent absence of a rotenone-sensitive NADH-dehydrogenase. More recent experiments suggest that procyclic *T. brucei* have a proton-translocating complex I, but with a subunit composition different from the mammalian complex and a decreased sensitivity to rotenone (217).

5.4 THE CHEMIOSMOTIC HYPOTHESIS

5.4.1 The Mitchell Hypothesis

We return now to the question raised earlier in this chapter about how the combustion of foods to CO_2 and water and the accompanying free energy change can be exploited in biological systems to do beneficial work in muscular contraction, biosynthetic reactions, and conductance of nerve impulses, for example. The experiments and ideas of Kalckar and Lipman had focused attention on ATP as the most useful and versatile intermediate in this interconversion of energy, and therefore the problem could be restated: How is respiration and the oxidation of carbon compounds coupled to the synthesis of ATP? (A large volume of research in bioenergetics and biochemistry is condensed in these simple summarizing statements.)

A theoretical foundation for the solution to this problem was proposed in 1961 by P. Mitchell (218), but the idea was so revolutionary at the time that it took another 10–15 years to convince most of the influential skeptics (as opposed to those who did not understand it at all) of its validity and, in turn, of its extremely broad applicability to a wide variety of basic biological mechanisms in oxidative phosphorylation, photosynthesis, and active transport. Other ingenious models had been proposed, and they coexisted for a while with the "Mitchell hypothesis." Models suggest experiments; and, with time, experimental observations led to models being discarded, modified, or even combined. Broadly speaking, three competing models existed during the early 1960s, and for convenience they can be named after their major proponents:

(1) Slater's chemical coupling hypothesis; (2) Boyer's conformational coupling hypothesis; and (3) Mitchell's chemiosmotic hypothesis. The first was essentially an extrapolation of insights gained from the understanding of substrate level phosphorylation (see below). The second was prompted by the failure to find experimental support for the first, and the third was a radically new synthesis of ideas.

When the glycolytic pathway had been elucidated, several energy conserving reactions had been established in which a free energy release resulting from an oxidation is coupled to the synthesis of ATP. The oxidation of glyceraldehyde-3-phosphate by NAD^+ converts an aldehyde to an acid group, but at the same time the enzyme glyceraldehyde-3-phosphate dehydrogenase "activates" inorganic phosphate by linking it to the acyl group. The reaction involves an acyl thioester intermediate that can be attacked by inorganic phosphate acting as a nucleophile. In a subsequent reaction, this phosphate was transferred from the high-energy mixed anhydride to ADP to form ATP. Thus,

$$\text{3-Phosphoglyceraldeyde} + NAD^+ + Pi \rightarrow NADH + H^+ + \text{1,3-bisphosphoglycerate} \quad [1]$$

$$\text{1,3-Bisphosphoglycerate} + ADP \rightarrow ATP + \text{3-phosphoglycerate} \quad [2]$$

Group transfer reactions and the idea of high-energy phosphate bonds (X ~ P) are illustrated here. The oxidation of X–H and activation of inorganic phosphate lead to the formation of X ~ P. The phosphate can then be transferred to an acceptor to form a lower-energy phosphate bond:

$$X\text{–}H + NAD^+ + P_i \Rightarrow X{\sim}P + NADH$$

$$X{\sim}P + Y\text{–}O\text{–}H \Rightarrow Y{\sim}P + X\text{–}O\text{–}H$$

where Y–O–H could be a sugar, an alcohol, or ADP.

For almost two decades, several prominent biochemistry laboratories sought to identify the mysterious compound X ~ P, which was postulated to be the intermediate trapping the free energy from mitochondrial electron transport (and oxidations). A mindset derived from classical biochemistry required the presence of a compound, X ~ P, even if very short-lived, as part of the mechanism of attaching inorganic phosphate to ADP. It should be pointed out that "X" could also represent a side chain of an amino acid of one or more subunits in the electron transport chain. For a while, phosphorylated histidine was a favorite, but in the end all efforts to identify such a chemical intermediate failed completely.

In the conformational coupling hypothesis the energy from oxidations and electron transport was thought to be "captured" in a high-energy conformational state of a protein (ATP synthase) which could then be used to drive the synthesis of ATP. It is noteworthy that a model for conformational coupling included two alternating sites for ADP and P_i binding, ATP synthesis and release (219). Conformational changes are still very much part of the mechanism of electron transport and proton pumping. On the other hand, it is true that "... the word conformational, applied to proteins, acquired a kind of magical significance, enabling proteins to accomplish anything (conveniently without the need to specify any biochemical mechanism) ..." (220). The fundamental question is concerned with the coupling of electron transport and ATP synthesis, and conformational changes in the complexes of the ETC and in the ATP synthase are not physically coupled.

P. Mitchell was able to approach the problem from a radically different direction. Restating a view expressed by van't Hoff, he has pointed out that "imagination and shrewed guess work are powerful instruments for acquiring scientific knowledge quickly and inexpensively ...". In his view of science, adopted from Popper, preconceived models are subject to constant experimental testing, ..." thus detecting and discarding the concepts that are false and retaining concepts that show by their survival that they are factually serviceable because they represent reality as far as it is known." Several concepts merged in an inspired hypothesis continuously supported by accumulating experimental evidence, which was almost nonexistent at the beginning. Numerous papers and reviews have appeared by him and his colleagues (149, 218, 220–227); a very readable and comprehensive review of the chemiosmotic hypothesis and the evolution of the ideas leading to it is the Ninth Sir Hans Krebs Lecture delivered in 1978 (220). It is probably fair to say that although Mitchell contributed to the experimental support of his hypothesis, his brilliant idea provoked many other talented individuals to either support or challenge it, and in the end it was verified as another major landmark in the intellectual landscape of bioenergetics. An excellent, exhaustive, and authoritative theoretical and experimental foundation for the chemiosmotic hypothesis can be found in the book by Nicholls and Ferguson (228). There have also been some voices who dispute the details of the story line presented above (229), and it will remain for the historians of science to sort out claims for priorities of ideas.

The first idea was that an enzyme embedded in a membrane in a unique orientation might catalyze a reaction by accepting substrates on one side of the membrane and release the products on the other side. This concept was referred to as "vectorial metabolism." It arose from two major biochemical mysteries: (a) oxidative phosphorylation and (b) the linkage between metabolism and transport. Even a "simple" group-transfer reaction occurring in a single aqueous phase was already considered a "chemiosmotic" process by Mitchell, because "the group-translocation pathway represents the field of action of a real through space force corresponding to the chemical potential gradient." Such microscopic osmotic processes could be converted to a macroscopic osmotic processe if the enzyme was "appropriately plugged through a membrane." Therefore, the free energy change associated with such a reaction might be used to drive the accumulation of a molecule or ion against a concentration gradient; that is under some conditions, an enzyme could work like a pump. Third, a concentration gradient of a solute across a membrane could be considered as a stored form of chemical free energy. If ions were the solutes in question, a concentration gradient might also be manifested as a membrane potential, and the membrane assumed properties similar to a charged capacitor. These ideas had been theoretically explored by Nernst. As Mitchell recalls (220), his ideas were further influenced by the proposal of electrochemical fuel cells made already in 1839 by Grove, as well as by Guggenheim's thermodynamic treatment of electrochemical cells and electric circuits, which split chemical reactions spatially into two half-reactions, connected internally by a specific conductor of one chemical species and connected by an outside circuit conducting another chemical species.

Progress with a number of simpler experimental systems added support to many of these individual concepts. Neurobiologists started to understand membrane potentials, sodium and potassium gradients across a membrane, and the mechanism of generating an action potential. Since the lipid bilayer is completely impermeable to ions, specific integral membrane proteins had to be identified to act as ion pumps and regulated gates whose combined action could explain such phenomena first for excitable membranes and then in more general situations. The Na^+–K^+ pump of the plasma membrane and the Ca^+ pump of the sarcoplasmic reticulum became classic examples of ion pumps that could use the hydrolysis of ATP to pump ions against their concentration gradients and against an electical gradient (active transport). These reactions were stoichiometric: A mole of ATP hydrolyzed was coupled to the transfer of a fixed number of ions across the membrane (3 Na^+ out of the cell, 2 K^+ into the cell in the case of the Na^+–K^+ pump). Later, simple proton pumps were discovered in the membranes of lysosomes, for example. Our understanding at an atomic level of resolution of how such pumps translocate ions across a membrane has advanced greatly. The pumps have been purified and have been reconstituted in pure synthetic membranes and vesicles. The corresponding genes have been cloned and sequenced. The corresponding amino acid sequences have led to predictions of structure and correlations of structure with function. Increasingly, high-resolution crystal structures have become available to aid in the interpretation of mechanism. It became apparent and demonstrable that such pumps could operate in reverse following the simple law of mass action. An artificially high experimentally established gradient could drive a reaction in the reverse direction and synthesize ATP from ADP and inorganic phosphate.

Another important concept contributed by Mitchell was the distinction between chemiosmotic and purely osmotic reactions. An ATPase pumping ions is an example of the former. Symport and antiport are examples of the latter, and they describe the coupled translocations of unrelated solutes across a membrane catalyzed by a porter: both solutes moving in one direction (sym), or two solutes moving in opposite directions (anti). The exchange of ADP and ATP across the mitochondrial membrane is a well-known, tightly coupled antiport system, while the uptake of glucose into intestinal mucosa mediated by Na^+ ions is one of the early examples of a symport system. These coupled reactions not only are vectorial, but also involve stoichiometric amounts of each participating solute. One can also "couple" a chemiosmotic reaction catalyzed by one enzyme to an osmotic reaction catalyzed by a spatially distant second enzyme: The Na^+ ion gradient established by the Na^+/K^+ ATPase can be used to transport glucose into a vesicle or cell by the Na^+/glucose symporter. From such insights it is not too far to realize that a proton gradient established by the electron transport chain could drive ATP synthesis by a biochemically and spatially distinct complex in mitochondria.

As more and more experiments contributed powerful support for the basic ideas about ion and electrical gradients across membranes and about active transport (pumps), or pumps running in reverse to make ATP, experimental systems applying these notions to mitochondria and chloroplasts were developed and perfected as well. Significantly, P. Mitchell himself stated that the chemiosmotic hypothesis was actually a byproduct of the chemiosmotic concepts of group translocation and vectorial metabolism (221). Two facts should be reemphasized at the outset: The ATP synthase of mitochondria resembles a simple ion pump (e.g., in the plasma membrane) only superficially. Its structure is significantly more complex, as will be described below. And, pumping protons out of the matrix by electron transport through metal ion centers in the ETC is a very complex system, with the only counterpart in chloroplasts and bacterial membranes.

Another instructive and influential model system developed in the 1970s was the purple membrane bacterium *Halobacter halobium.* Phosphorylation was driven by light energy absorbed by a photopigment soon identified as bacteriorhodopsin. Bacteriorhodopsin was shown to be a light-driven proton pump. A proton gradient in the dark could drive the synthesis of ATP in *Halobacter halobium.* In combination these two observations suggested that light energy was used to establish a proton gradient across the bacterial membrane (high on the inside, low on the outside), and a proton pump running in reverse was used to make ATP. The proteins involved were distinct and physically separable from each other. On the one hand, this system is clearly an illustration of the importance of a conformational change in a protein playing a role in energy conversion. The photon absorbed by the prosthetic group triggers an all-trans → 11-cis isomerization and hence a conformational change in the bacteriorhodopsin in the membrane. The conformational transitions accompanying excitation and return to the ground state are coupled to proton translocation. On the other hand, the conformational change is not linked directly to phosphorylation of ADP, as in the simple conformation-coupling hypothesis.

In the same vein, it is now clear that the passage of an electron through the various redox centers of complexes I, III, and IV may be accompanied by conformational changes throughout the complex (one or more subunits), and at least in the case of complex IV, these conformational transitions are believed to drive proton translocation. These changes need not be very dramatic and may involve only a limited number of side chains along a "channel" for proton translocation (see Figure 5.21). The situation is potentially more complicated and controversial for complexes I and III, because the participation of ubiquinone creates additional theoretical possibilities for proton translocation (see below).

A key aspect of Mitchell's chemiosmotic hypothesis is the requirement that respiratory (and photoredox) chains translocate protons across a membrane in one direction, while ATP synthesis is driven by proton flow in the opposite direction (Figure 5.22). A violation of this prediction would have been the end of the hypothesis. Testing the prediction was not trivial, but eventually much supportive evidence was accumulated. The Mitchell hypothesis is overwhelmingly accepted today, but some controversy may still exist about some of the quantitative aspects of the mechanism of oxidative phosphorylation. Expert reviews on this subject have been written by Hinkle (230, 231), and the most recent one appears to put most of these contentious issues to rest. Specific

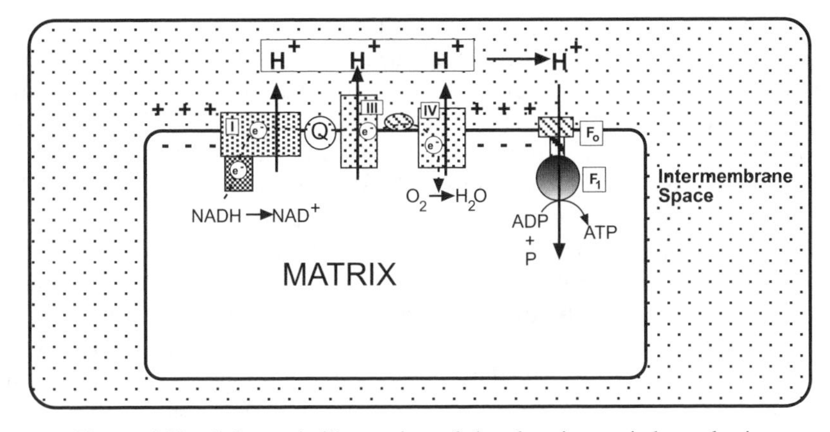

Figure 5.22 Schematic illustration of the chemiosmotic hypothesis.

questions can be formulated as follows: (1) How many protons are pumped out of the mitochondrial matrix for each pair of electrons passing from NADH to oxygen? (2) How many protons are pumped at each coupling site—that is, at complex I, complex III (Q-cycle), and complex IV? (3) How many protons are returned to the matrix through the ATP synthase to make one ATP? Theoretically, the oxidation of NADH by oxygen liberates enough free energy to make >6 ATPs from ADP and inorganic phosphate, but since even biological reactions are not 100% efficient, the expected number is ~3 for 50% efficiency.

Early in the investigations of respiration and oxidative phosphorylation, the phenomenon of "acceptor control" was recognized. Experimentally, it was observed that intact mitochondria supplied with abundant substrate such as succinate or β-hydroxybutyrate do not consume significant amounts of oxygen unless the "acceptor" ADP is present (inorganic phosphate is also needed, but routinely/typically present as part of the medium). Thus, it is possible to make precise measurements of oxygen consumption with limiting, known amounts of ADP. The number of molecules of ADP converted to ATP compared to the number of oxygen atoms converted to water yield a P/O ratio (or ADP/O ratio). Over the years there have been numerous reports reporting P/O ratios that were in the range of 2–3 for NADH oxidation and 1–2 for succinate oxidation. The recent review by Hinkle (231) summarizes these data and presents a critical discussion of the methodology, the experimental pitfalls, and the source of possible errors. In light of what we understand now, it is clear that the P/O ratio need not be an integer. A consensus value of ~2.5 is emerging for NADH-dependent oxidations, along with a value of ~1.5 for succinate-dependent oxidations. These experimental values may not only vary due to methodological differences and errors, but may even depend on the tissue of origin of the mitochondria (a P/O ratio of close to zero in mitochondria from brown adipose tissue is an extreme example; see below).

A significantly more challenging task is to measure the number of protons pumped out of the matrix. The method is the "respiratory pulse method," which measures a pH change in a mitochondrial suspension in the presence of a specific substrate following a brief pulse with a known quantity of oxygen. This measurement can be decomposed into a measurement of the total number of protons pumped for each pair of electrons passed from NADH to oxygen, or measurements of the number of protons pumped at each "coupling site," now generally accepted to be complex I, complex III, and complex IV. This number can be referred to as the H^+/O ratio or as the $H^+/2e^-$ ratio. Hinkle's review (231) can again be cited as the most critical and complete review of various reports over the past decades. According to Hinkle, "the arguments seem to be over, although not all groups have accepted the consensus explicitly." The accepted $H^+/2e^-$ ratios have values of 10 for NADH and 6 for succinate. From these, one can calculate that $H^+/2e^- = 4$ for site 1 (complex I), and this value has direct experimental support from measurements with submitochondrial particles (232). For complex IV the best experimental value for $H^+/$

$2e^-$ is 2.20 ± 0.2 (156, 233). Authoritative reviewers have raised the possibility that complex IV can control its own pumping efficiency (167); that is, it need not be a stoichiometric amount related to the redox reaction. It should be noted that complex IV also picks up four protons on the matrix side, referred to as substrate protons, to form water from two pairs of electrons and one oxygen molecule. As Hinkle points out, the situation at complex III is not controversial but not often clearly understood. The Q cycle (see below) predicts that for each $2e^-$ transferred from QH_2 to cytochrome c, four protons are released on the outside; however, two of those are transported electrogenically (pumped), while the other two are derived from the substrate (QH_2) and are therefore not transported. There is proton translocation as well as proton absorption or formation by the overall reactions at each coupling site, and therefore the count has to be made consistently. A summary and restatement of the above can be given with the help of the following equations:

$$\begin{aligned}
\text{Complex I:}\quad & 2\,\mathrm{NADH} + 2\,\mathrm{Q} + 10\,\mathrm{H}_i^+ \Rightarrow 2\,\mathrm{NAD}^+ + 2\,\mathrm{QH}_2 + 8\,\mathrm{H}_o^+ \\
\text{Complex III:}\quad & 2\,\mathrm{QH}_2 + 4\,\mathrm{cyt\ c}^{3+} + 4\,\mathrm{H}_i^+ \Rightarrow 2\,\mathrm{Q} + 4\,\mathrm{cyt\ c}^{2+} + 8\,\mathrm{H}_o^+ \\
\text{Complex IV:}\quad & 4\,\mathrm{cyt\ c}^{2+} + 8\,\mathrm{H}_i^+ + \mathrm{O}_2 \Rightarrow 4\,\mathrm{cyt\ c}^{3+} + 4\,\mathrm{H}_o^+ + 2\,\mathrm{H_2O} \\
\text{Sum:}\quad & 2\,\mathrm{NADH} + \mathrm{O}_2 + 22\,\mathrm{H}_i^+ \Rightarrow 2\mathrm{NAD}^+ + 2\,\mathrm{H_2O} + 20\,\mathrm{H}_o^+
\end{aligned}$$

It remains to reconcile the experimentally observed P/O ratios with the $H^+/2e^-$ ratios and the total number of protons pumped out of the matrix. Since one oxygen and two electrons are fixed values, one can also predict an H^+/P ratio at steady state, and therefore the number of protons returned to the matrix for each ATP molecule synthesized. Several comments are in order. It was recognized by Mitchell and others that mitochondrial membranes are not perfect insulators for protons, as viewed in the ideal model system, but instead have a variety of "proton leaks." These include specific symport or antiport systems and/or as yet poorly specified leakage channels. For example, the ADP/ATP antiporter has been reported to be electrogenic (234), while the phosphate/OH^- antiporter transports one proton into the matrix (235). Thus, the export of one ATP is coupled to a net import of one proton, and the H^+/P ratio has to include this reaction. Less than 10 protons pumped out by the electron transport chain are therefore available for ATP synthesis. Thermodynamic considerations and the estimated value of the electrochemical gradient (ΔpH and $\Delta\Psi$) predict that more than one proton must be returned to the matrix to drive the synthesis of one ATP. It is tempting to propose a value of 3 protons returned per ATP. This leads to a maximum value of 3 for the P/O ratio (from NADH oxidation), and in light of the above considerations a value of ~2.5 is rational. Proton pumping by the electron transport chain and ATP synthesis are physically distinct processes, and if the protons find other paths back into the mitochondrial matrix, ATP synthesis will be less than what would be maximally possible in a system with a perfectly insulating membrane, the

ETC and the ATPsynthase. A reevaluation of energy loss due to proton leakage has been made by Hinkle et al. (230). Both systematic experimental errors and theoretical considerations now make it possible to obtain fractional P/O ratios that are rational and do not violate the Mitchell hypothesis.

In summation, it can be stated that Mitchell's chemiosmotic hypothesis has stood the test of time, and therefore it "represents reality as far as it is known." Its major insight was that a proton gradient across the inner membrane can be set up independently by electron transport (like water moved uphill into a dam by evaporation and condensation), and this gradient can be used to drive ATP synthesis (water running through turbines to produce electricity).

Two points should be made more explicitly once again. In an initial presentation of the chemiosmotic hypothesis, the emphasis was on protons, and therefore on the "pH gradient" across the inner membrane. Since protons are charged, such a gradient also creates a membrane potential, $\Delta\Psi$. In considering the thermodynamic parameters for ATP synthesis, a driving force is therefore not only the concentration gradient represented by ΔpH, but there is also an electrochemical contribution from moving a positive charge from high to low potential. Other ions contribute to the overall electrochemical potential difference: Protons or hydroxyl ions participate in antiport or symport systems, exchanging protons for metal ions, hydroxyl ions for phosphate, and so on. Thus, the proton electrochemical gradient is conventionally denoted $\Delta\mu H$, and its components are $\Delta\Psi$ and ΔpH. A particularly useful experimental trick has been to use the ionophore nigericin as a K^+/H^+ antiporter. It can therefore decrease the ΔpH across the membrane, but not the $\Delta\Psi$. In contrast, the ionophore valinomycin is a potassium carrier, and it can collapse the portion of $\Delta\Psi$ resulting from excess K^+ on the outside, but not ΔpH. Such experiments will become relevant in the consideration of the control of oxidative phosphorylation and tightly coupled versus uncoupled mitochondria (Section 5). There is also a potential misunderstanding of the nature of the pH gradient. Excess protons are found only in the immediate vicinity of the surface of the inner membrane. In the bulk phase of the intermembrane space the pH is similar to that of the cytosol, since protons can presumably diffuse freely through the porin channels of the outer membrane.

There is one final point that could not have been made and appreciated until the structure of complex V had been more fully elucidated. A full discussion must be deferred until later when a detailed description of the composition, structure, and mechanism of action of complex V (ATP synthase) has been presented. There is an originally unexpected "symmetry mismatch" between the threefold rotational symmetry of the $\alpha_3\beta_3$ subunits of the F1 subcomplex and the number of c-type subunits in the F_0 subcomplex in the membrane (the number is variable in different organisms, but it is not a multiple of 3) (236–238). Such structural data suggest that the average number of protons passing through complex V per 120° turn is not an integer, and the H^+/P ratio may therefore be non integer as well.

5.4.2 The Q Cycle

The role of ubiquinone as a mobile carrier between complexes I and II and complex III has been mentioned before, but its role in proton translocation has not been emphasized so far. Two properties of ubiquinone are significant for the understanding of the Q cycle, proposed originally by Mitchell (149) and critically reviewed and modified by Trumpower and Brandt (152, 239) and others (240). First, the carrier is lipid-soluble in the oxidized and the reduced form. The quinone moiety can therefore diffuse from one side of the membrane to the other and carry hydrogen atoms with it. Second, the unprotonated semiquinone is a relatively stable intermediate, which itself cannot diffuse through the membrane, because of its negative charge. The complete Q cycle actually is the sum of two cycles, shown in Figure 5.23. The various oxidation states of ubiquinone are shown (Figure 5.6).

There are 10 isoprenoid units in the naturally occurring ubiquinone, making an aliphatic carbon chain of 40 carbons with regularly spaced double bonds and methyl side chains. Such a chain is much longer than what is required to span a lipid bilayer, and the usual representation of a Q diffusing through a schematic membrane many diameters wider than the Q is likely to give a totally wrong impression of the "diffusion" of Q "across" the membrane. Ignoring this problem for the moment, a consideration of the Q cycle is conveniently started with reduced QH_2 supplied by complex I or complex II on the matrix side of the membrane (step 1). Diffusion across the membrane brings it to the other side and in a position to give up two protons to the intermembrane space, passing one electron to the iron–sulfur protein (ISP, Rieske protein) of complex III (step 2). This electron is passed on to the cyt c_1 of the complex III and finally to cytochrome c. The semiquinone in the unprotonated form gives up a second electron to the first of two hemes, b_L, and becomes oxidized to Q (step 3); the electron is passed to the second heme b_H. Q diffuses back to the matrix side, and there it is reduced to the semiquinone by the heme b_H (step 4). During this first half of the Q cycle the net reaction is the oxidation of QH_2 to the seminquinone, the reduction of one cytochrome c, and the transfer of two protons to the intermembrane space. A repeat of steps 1–3 starting with QH_2 generates Q, another reduced cyt c, and a reduced heme b_H, and it releases two more protons in the intermembrane space. At this time, however, the heme b_H is transferring its electron to the semiquinone, accompanied by the pick-up of two protons from the matrix side to form ubiquinol. The summation of the two series of reactions is

$$QH_2 + 2\ \text{cyt}\ c_{ox} + 2\ H^+_{in} \Rightarrow Q + 2\ \text{cyt}\ c_{red} + 4\ H^+_{out}$$

The net reaction is that four protons are translocated to the outside (intermembrane space) for every pair of electrons passing through complex III from QH_2 to cytochrome c (239). A recent up-to-date and highly technical review of the Q cycle discusses a detailed mechanism for the behavior of complex III

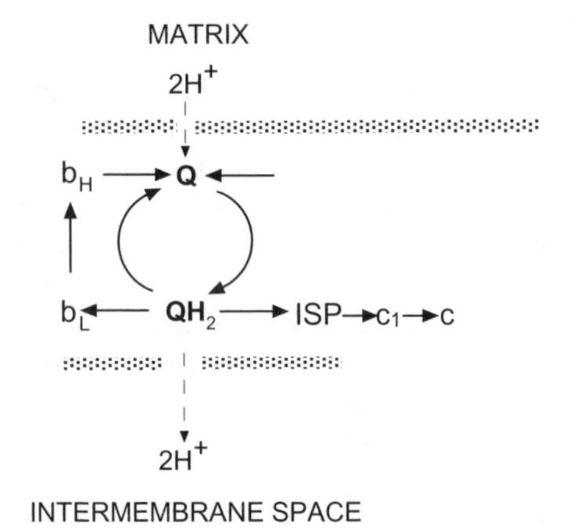

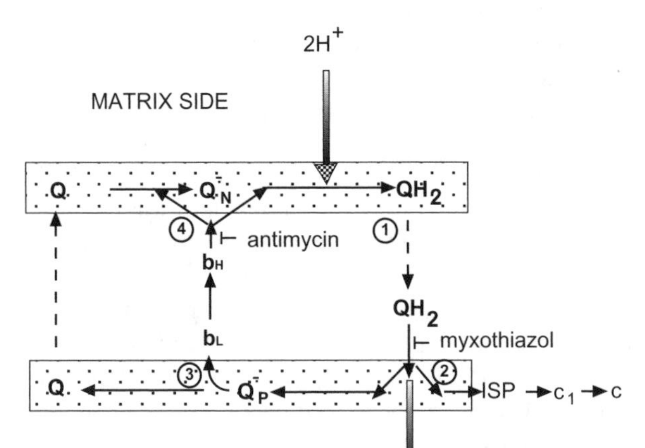

Figure 5.23 Another view of the proposed Q cycle by Trumpower and colleagues.

in light of very detailed structural information from X-ray crystallography (241). Furthermore, the reviewer draws attention to the structural similarity of the ubiquinone binding sites in the bc_1 and the prokaryotic photosynthetic reaction center (RC). The latter can be studied kinetically by using short flashes of light, allowing the identification of intermediate states. The interested reader is referred to this review, which illuminates many of the remaining controversial aspects of the mechanism, for example, the sequential or concerted oxidation of ubiquinol, and the prevention of a "short circuit" of the second electron into cytochrome c_1 instead of to the heme b_l. This problem has been addressed by other authors as well; a very readable and lucid discussion of several different possibilities for short circuiting in complex III is presented by Osyczka et al. (242). The solution requires either gating mechanisms

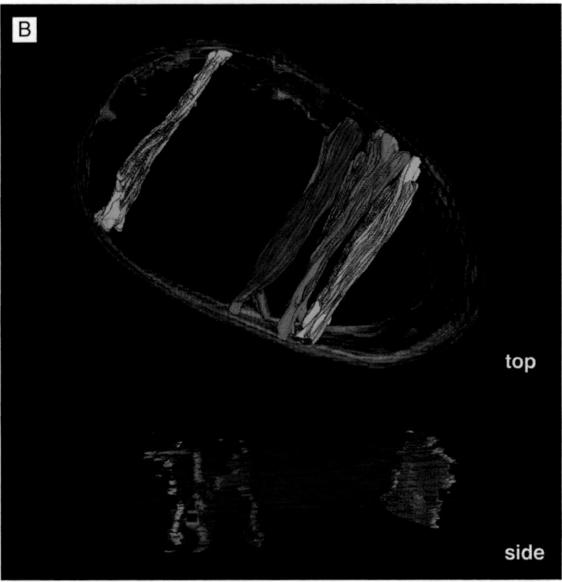

Figure 3.5 (A) Mitochondria seen by electron microscope tomography. Cells were derived from chick CNS. [The original figure was provided by G. Perkins, University of California, San Diego; with permission from Academic Press (Perkins et al. (5)).] (B) The 3-D reconstruction was achieved after the volume was segmented manually contouring into the regions bounded by the outer, inner, and cristal membranes. [For details the reader is referred to the original publication (Perkins et al. (5).] In the bottom of the figure, individual lamellar sections of crista are shown, together with the tubular portions which connect the disks to the inner boundary membrane. (The photographs were generously provided by Dr. G. Perkins, University of California, San Diego; with permission by Academic Press.)

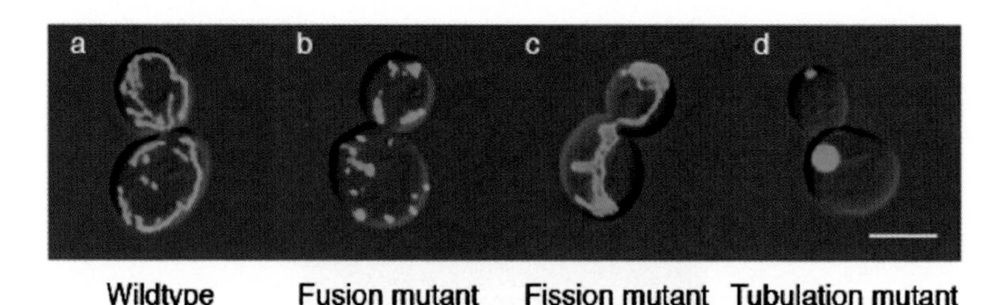

Figure 3.11 GFP targeted to the matrix in wild-type yeast cells and in various mutants: fusion mutant (*fzo1*Δ), fission mutant (*dnm1*Δ), tabulation mutant (*mmm1*Δ); GFP fluorescence images were superimposed on images from a differential interference contrast microscope. (From the review of Okamoto and Shaw (59).)

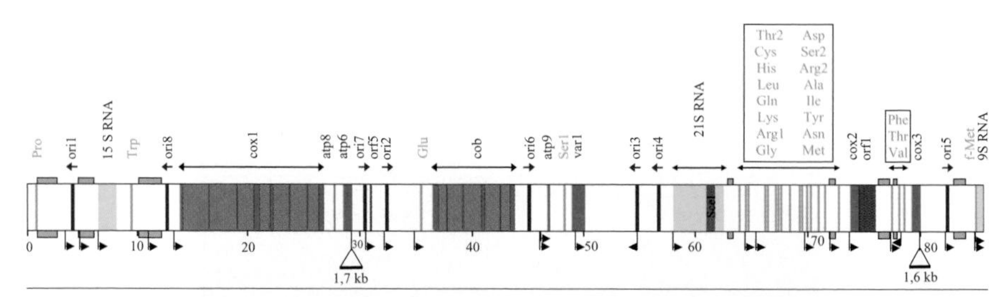

Figure 4.3 Schematic representation of the yeast mitochondrial genome map, strain FY1679, presented as a linearized map (35). Exons of protein coding genes are indicated in red (cox1, cox2, cox3, cob, atp9, var1), introns are gray; tRNA genes (green) are labeled with the corresponding amino acid, 9S, 15S, and 21S RNA (yellow) are labeled. Two significant deletions are found in this strain, and flags indicate transcription initiation sites. For details the original paper (35) should be consulted. (Reproduced from reference 35, with permission.)

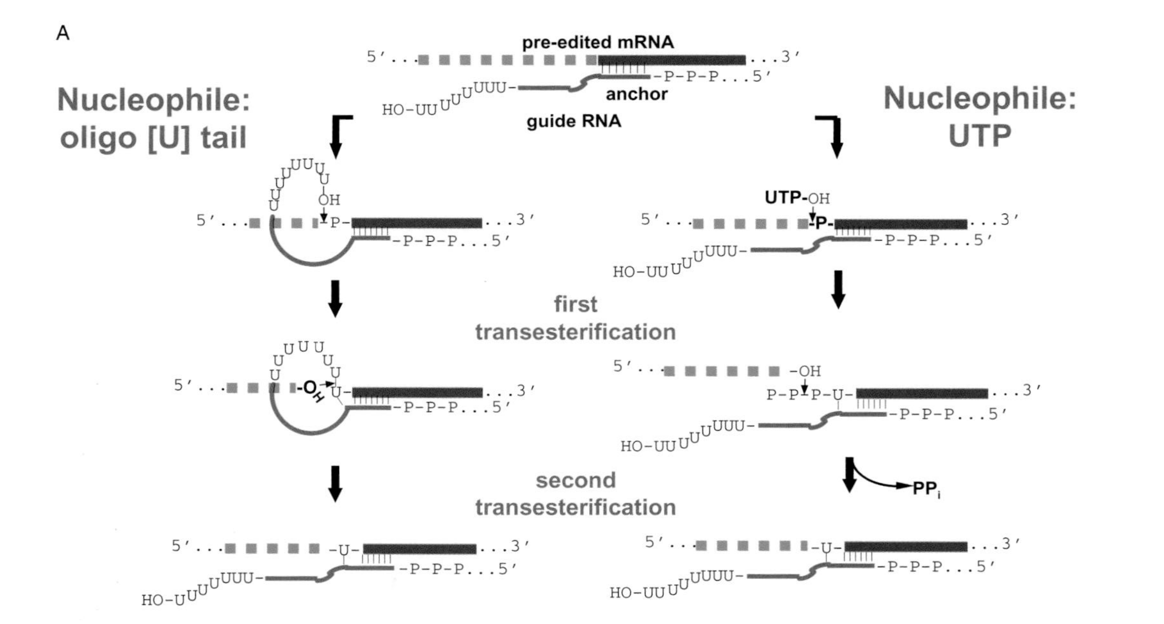

Figure 4.9 RNA editing in Trypanosomes (see reference 201 for details). (A) Double transesterification models, showing only the U-insertions. (B) The modified enzyme cascade model showing U-deletion, or U-addition or misedited U-addition as alternatives. In the simplest version the U-tail of the guide RNA is shown as a free overhang, but it is also possible that it may pair with a purine-rich sequence of the pre-edited RNA. The number of Us added to the 5′ fragment in the U-addition pathway (middle) is show as 13, but this number is actually variable. If the trimming is precise, correct guided products are obtained. If trimming is incomplete or excessive, a misedited product may result. (The figures were provided by Dr. L. Simpson. With permission from Oxford University Press.)

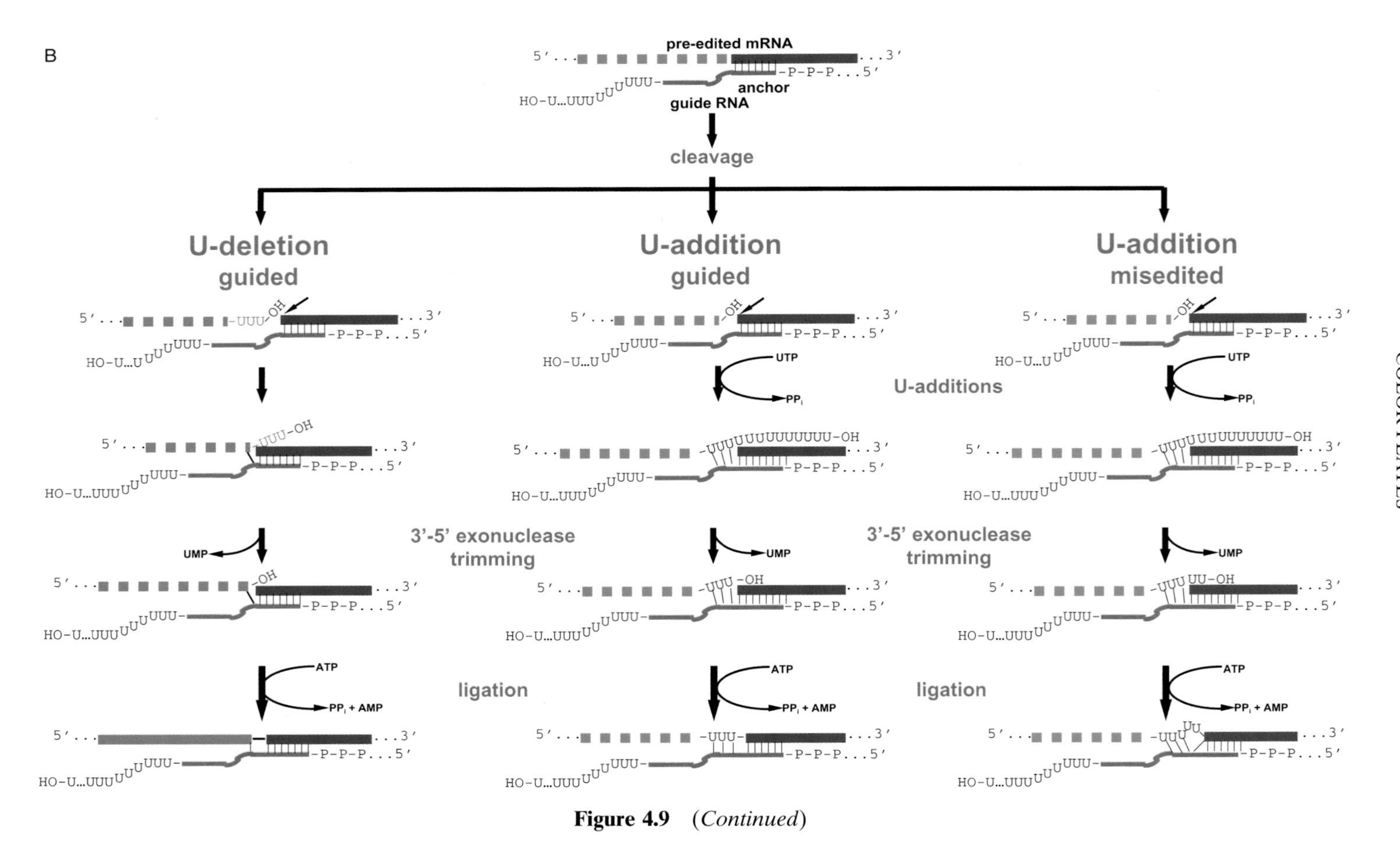

Figure 4.9 *(Continued)*

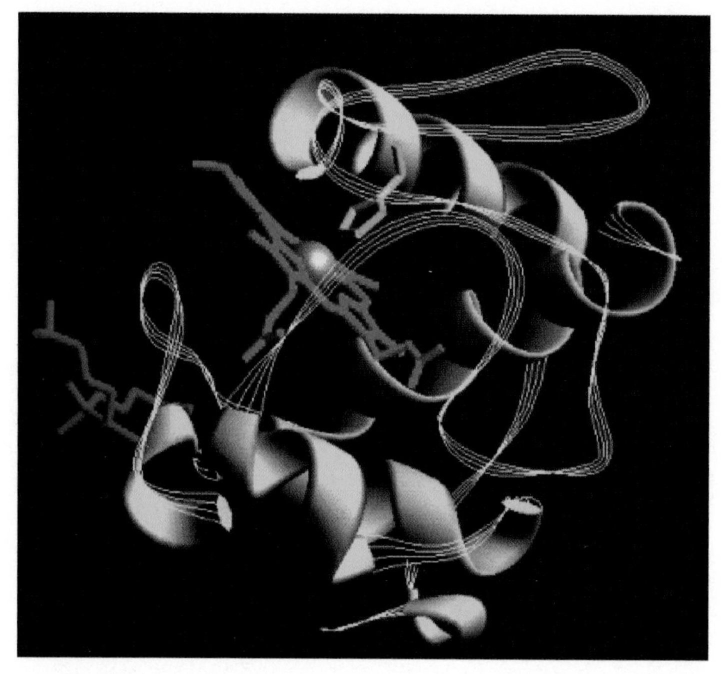

Figure 5.2 Structure of cytochrome c showing the association of the heme with the polypeptide chain.

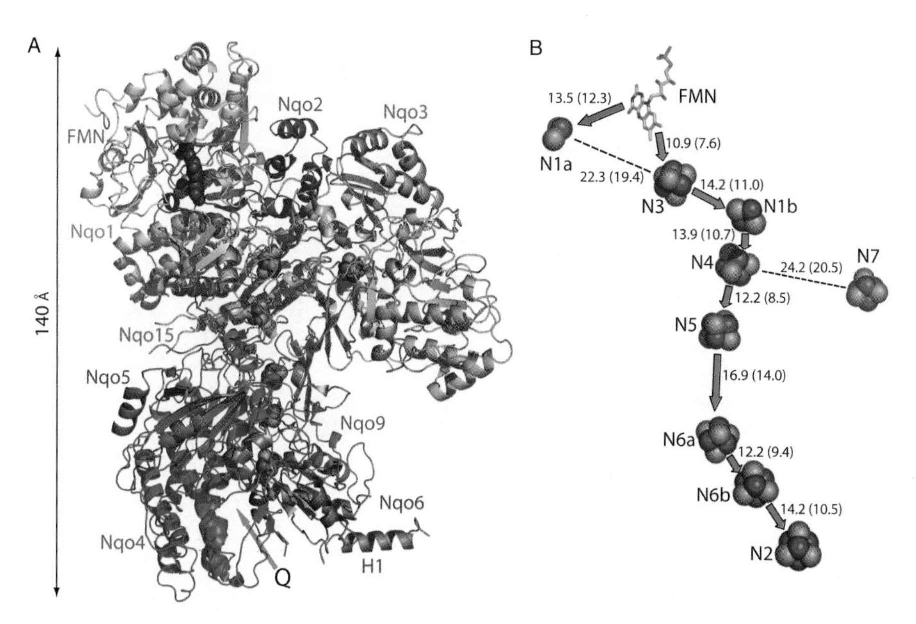

Figure 5.10 (A) Crystal structure of the peripheral subcomplex of respiratory complex I (NADH–ubiquinone oxidoreductase) from *Thermus thermophilus* at 3.3-Å resolution (74, 76). The subunits are labeled following the nomenclature for bacteria. (B) Location and distances between the 9 Fe–S clusters in this subdomain. (Reproduced from reference 76 with permission.)

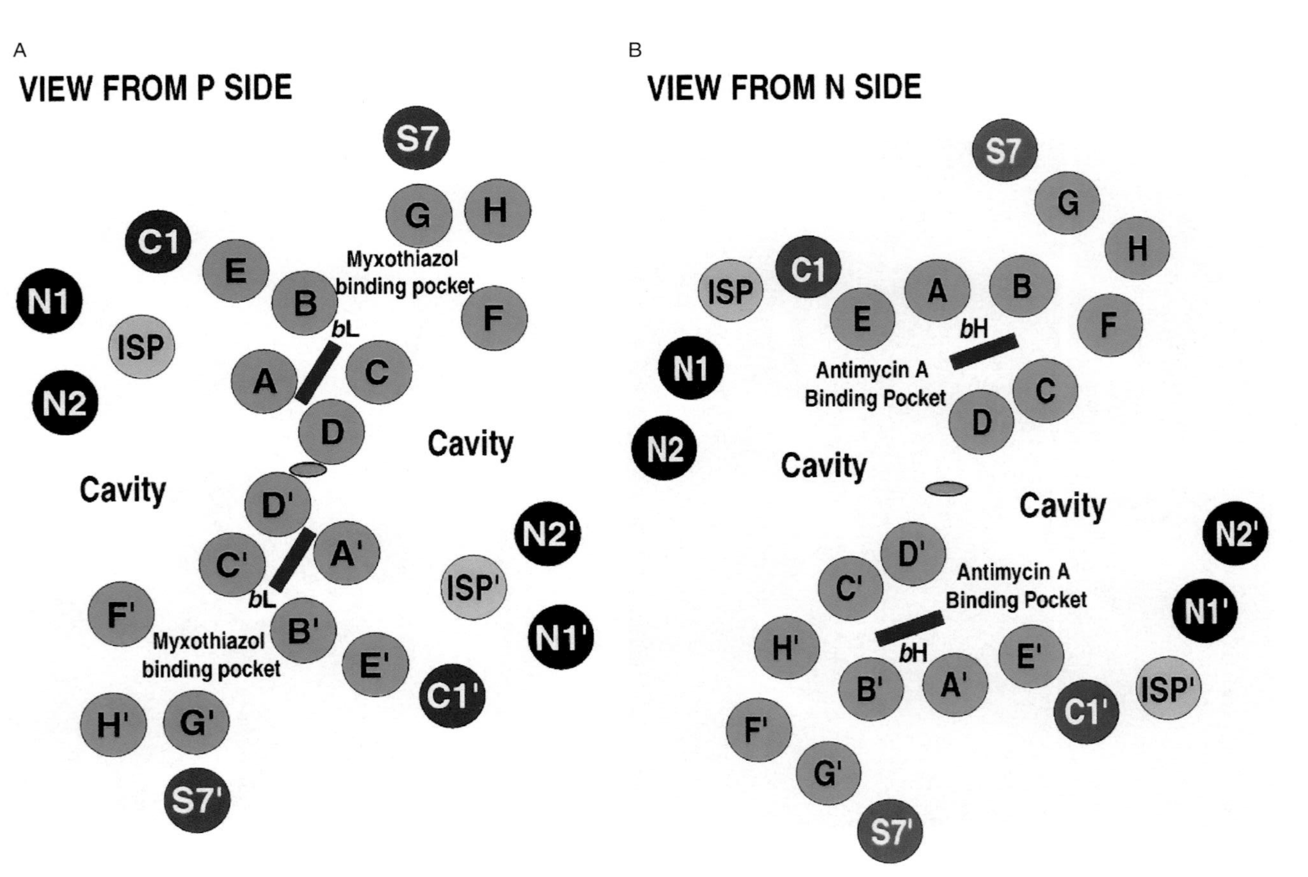

Figure 5.14 Complex III. Schematic representation of the transmembrane helices in the dimeric complex seen from the P and N (matrix) side (parts A and B, respectively) (82).

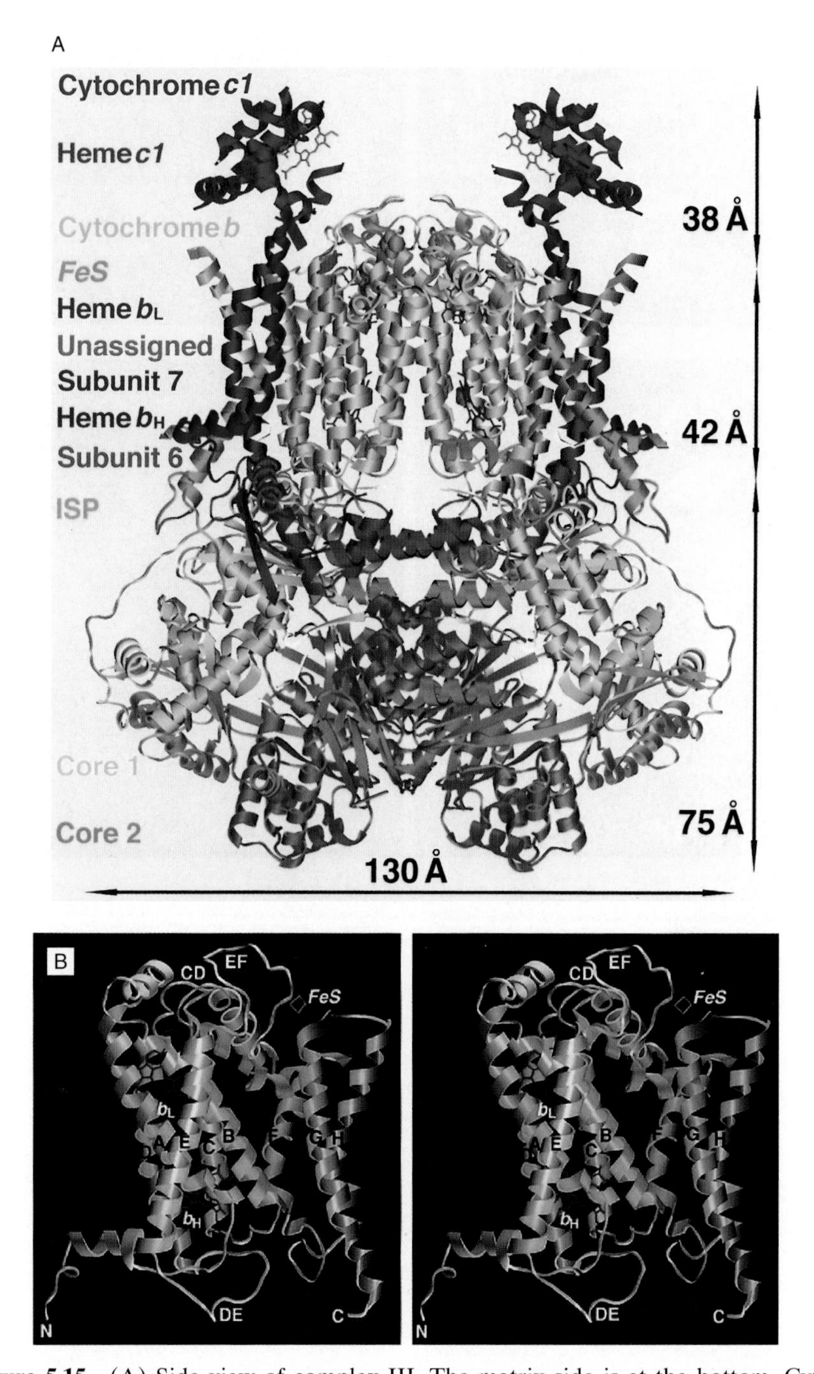

Figure 5.15 (A) Side view of complex III. The matrix side is at the bottom. Cytochrome c1 extends into the intermembrane space. The membrane-spanning helices of the various subunits are in the region indicated by the middle arrow (42 Å). (B) Stereoview of the transmembrane helices of a monomeric complex III. (C) Stereoview from the outside showing the myxothiazol binding pocket. (D) Stereoview from the matrix side showing the antimycin binding pocket. (From reference 147 with permission. The figure was generously provided by Dr. Yu.)

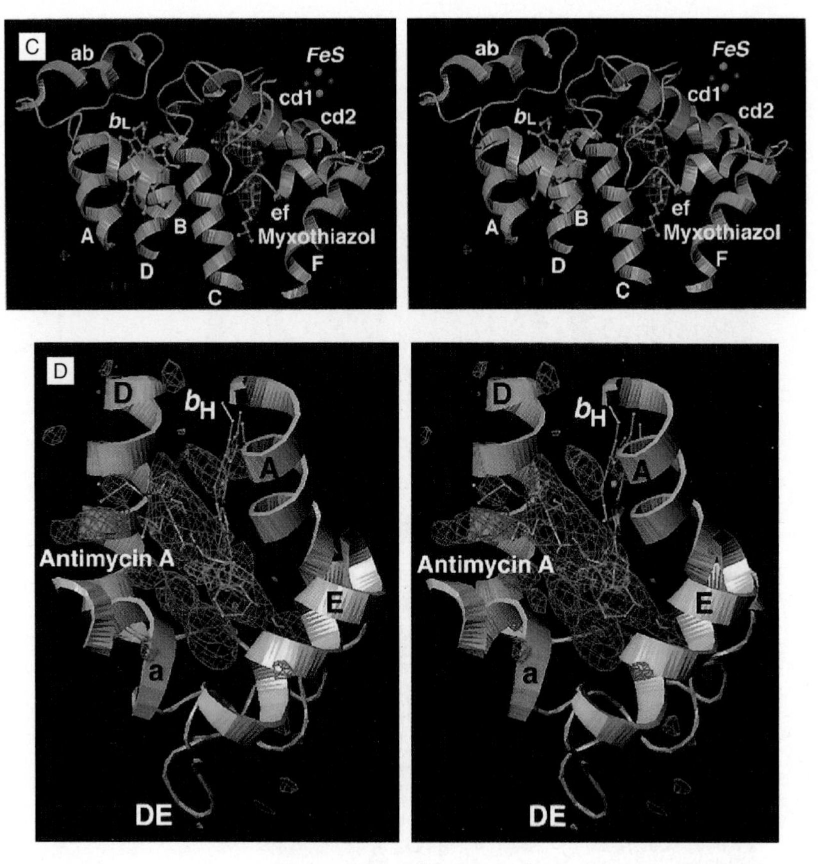

Figure 5.15 (*Continued*)

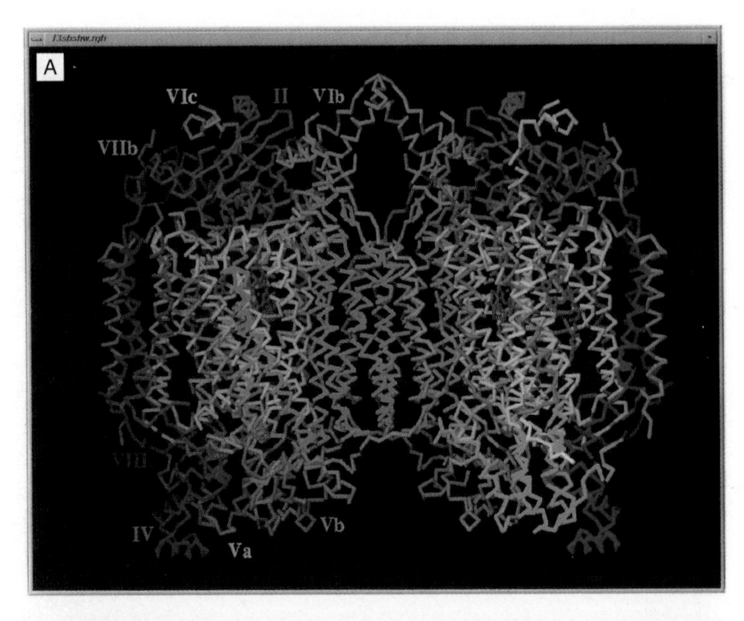

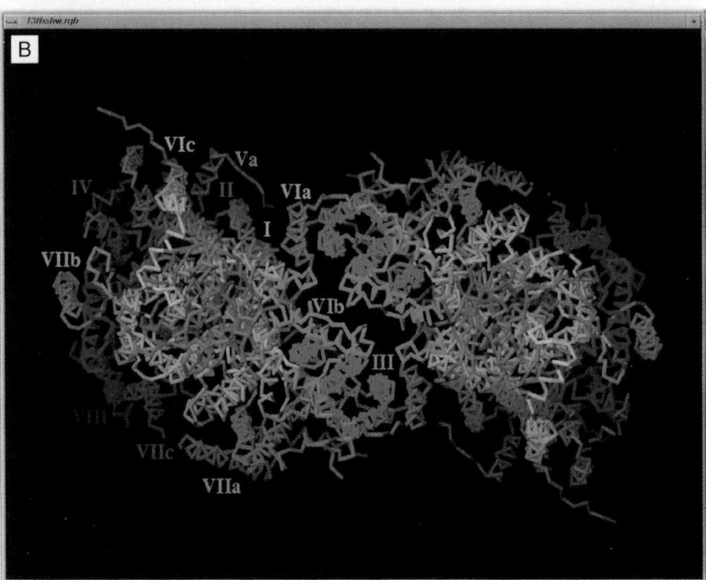

Figure 5.18 X-ray structure of complex IV. (A) Side view showing the transmembrane segments of the various subunits; matrix side at the bottom. (B) View from the cytosolic side. (The photographs were kindly provided by Dr. S. Yoshikawa of the Himeji Institute of Technology, Hyogo, Japan. From reference 157 with permission.)

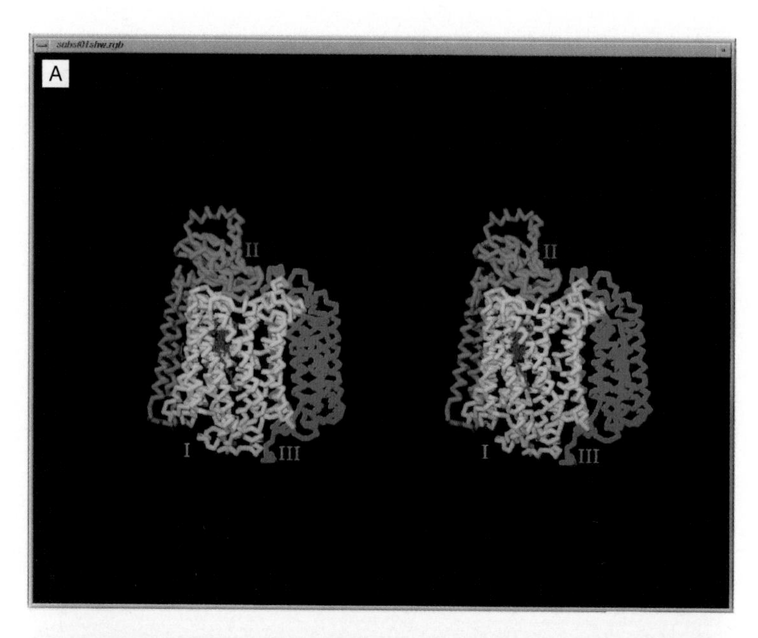

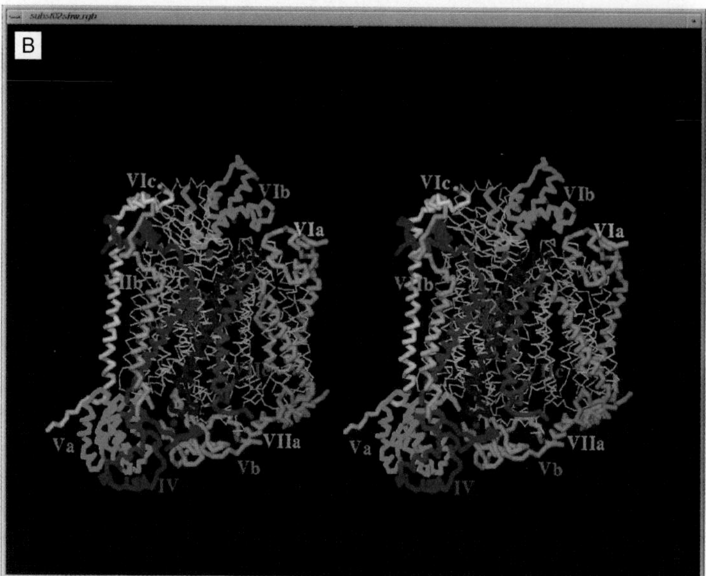

Figure 5.19 X-ray structure of complex IV. Stereo images of the C_α-backbone tracings. (A) Subunits I, II, and III. (B) Subunits IV, Va, Vb, VIa, Vib, Vic, VIIa, VIIb, and VIII. (The photographs were kindly provided by Dr. S. Yoshikawa of the Himeji Institute of Technology, Hyogo, Japan. From reference 157 with permission.)

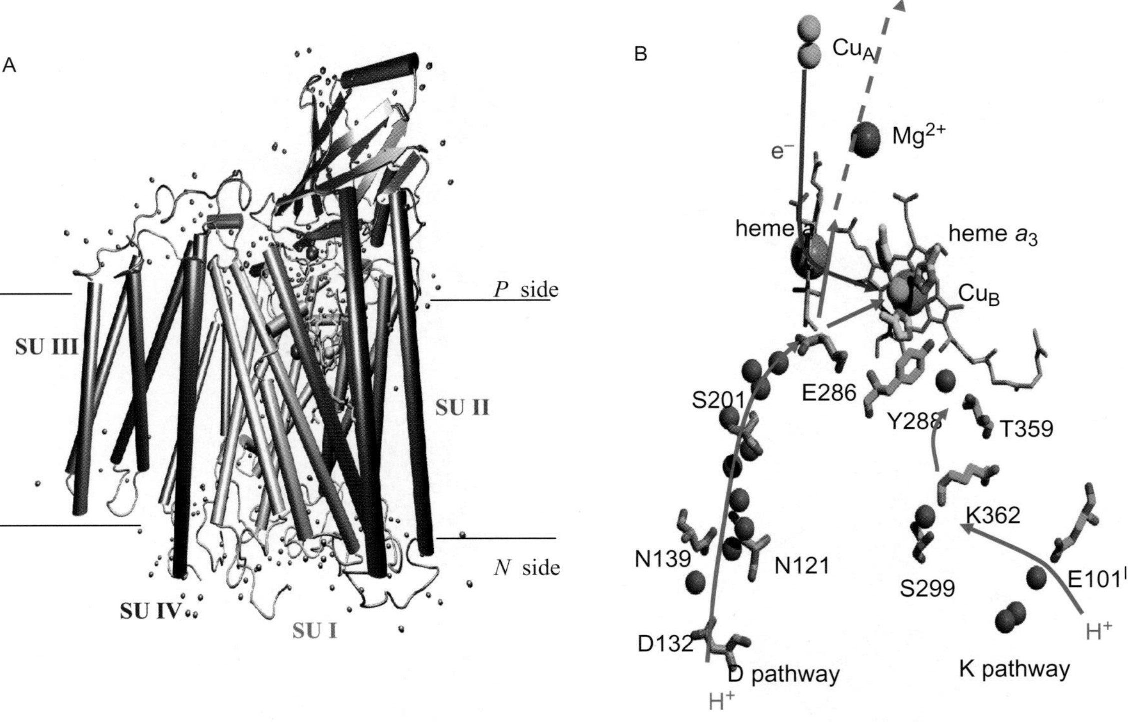

Figure 5.21 (A) Schematic view of complex IV with the four essential subunits (155). (B) The proposed D and K pathways for the entry of protons to form water and to be pumped across the membrane. The path from the region of the hemes to the outside is still poorly defined. (From reference 155 with permission.)

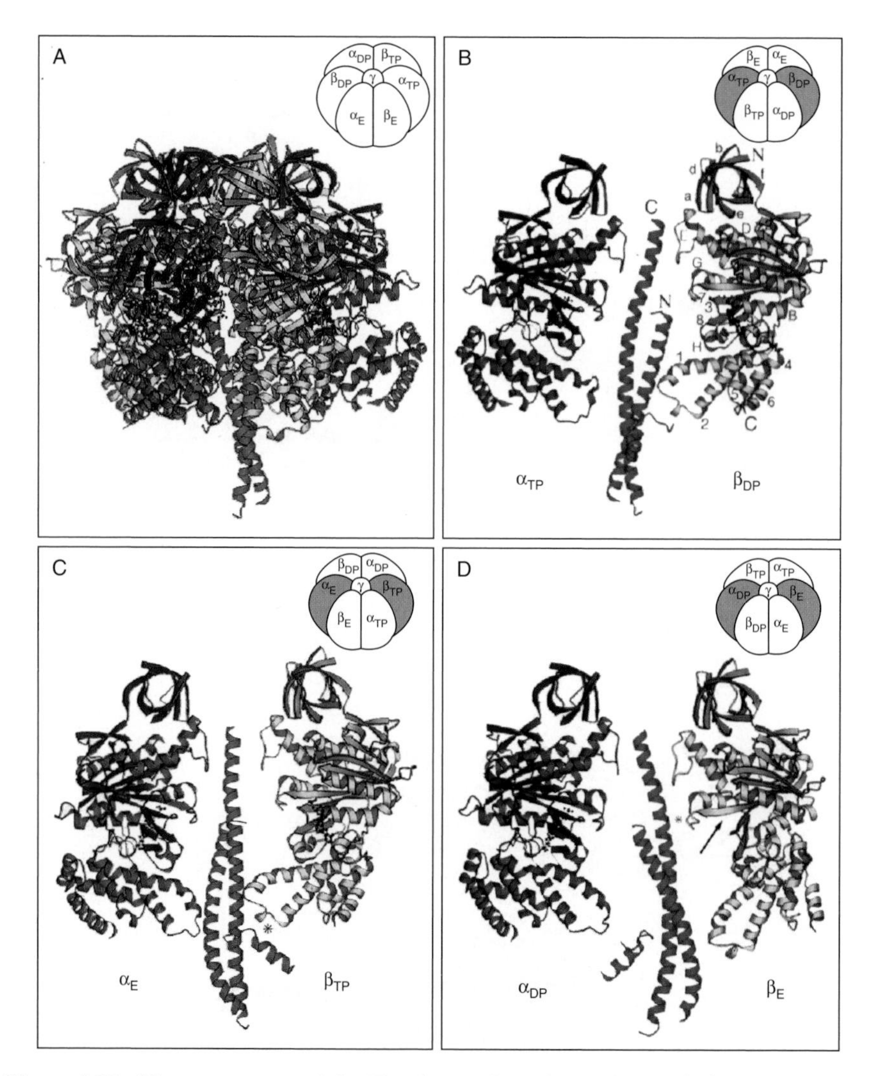

Figure 5.27 X-ray structure of the F_1 subcomplex of complex V. (A) Ribbon model showing the $\alpha_3\beta_3$ hexamer and the γ subunit constituting the rotating shaft. (B–D) Three different views of the shaft (γ subunit) and side views of two opposite α and β subunits. The image emphasizes the asymmetry and the different contacts made by each of the six subunits with the central shaft. (From reference 263 with permission.)

and even double gating mechanisms, or concerted electron transfer with no semiquinone intermediate. The proposal of a "Fe–S lock"—that is, a shuttling of the Fe–S domain of the Rieske protein between two conformations (see reference 242 for many references related to this topic)—represents a popular model, but Crofts and colleagues also consider a model in which a quinone serves as gate by moving about the Q_0 site (240, 243).

The cytochrome bc1 complex is found as a dimer both in the crystal structure and in solution. This has led several authors to consider an electron exchange between cytochrome b's from adjacent monomers (145, 244).

The protonmotive Q cycle originally proposed by Mitchell and later modified by Trumpower has made it into the standard textbooks, and it appears to have gained general acceptance. Arguments in support of this formulation, especially with reference to the two inhibitors antimycin and myxothiazol binding at the "center N" and the "center P," respectively, have been critically reviewed by Brandt and Trumpower (239). Nevertheless, some doubts about its validity have been raised by experiments by Matsuno-Yagi and Hatefi (245, 246), again with special attention focused on observations made with the two inhibitors. Their proposed model has the attractive feature that the ubiquinol does not have to execute as much motion within the membrane and the complex. It also places the path for proton translocation within the cyt b with its eight transmembrane helices, and it removes this function from the mobile carrier ubiquinol. Further support for this model has not been forthcoming in the recent literature.

5.4.3 Probing the Mitochondrial Membrane Potential with Fluorescent Dyes

Opposition to Mitchell's ideas arose in part because of misunderstandings about ion transport, membrane potential, and the principle of electroneutrality in solutions formulated by classical physical chemistry. The latter principle is never violated in the bulk of the solutions on either side of the membrane. Ion transport separates charges across an extremely thin membrane. These charges separated by a thin insulating layer create a system analogous to a charged capacitor. Because of the extreme thinness of biological membranes, an electrical difference of 250 mV across 5–10 nm corresponds to >250,000 V/cm, a very strong electric field by any standard.

The membrane potential is not the same for all mitochondria or constant at all times for a given mitochondrion. A direct measurement would not only provide strong experimental support for the chemiosmotic hypothesis, but would allow continuous monitoring of the $\Delta\Psi$ under varying conditions and manipulations of mitochondria. Direct electrophysiological experiments with isolated mitochondria have been reported, but they tend to be challenging (247, 248). More convenient methods have been developed over the years. These methods initially gave only qualitative results in most hands, but increasingly quantitative data can be obtained today with appropriate instrumentation. The basis for these optical methods is the existence of dyes that accumulate in mitochondria in a membrane potential-dependent manner. Another advantage of the use of such dyes is that they can be used in living cells to explore mitochondria *in vivo*. The use of dyes to investigate mitochondria goes back to the last century, when mitochondria were first seen without being recognized for their function (see Chapter 1).

A fundamental property of the useful dyes is that they are lipophilic and carry a positive charge that can be delocalized throughout the molecule by resonance structures. Therefore they not only can pass through membranes, but also will be actively transported across a membrane with a $\Delta\Psi$ (249, 250). In the case of mitochondria, this leads to their accumulation in the mitochondria. Thus, if the plasma membrane potential can be eliminated, uptake and fluorescence of such a dye can become a measure of the mitochondrial membrane potential. At the simplest level of analysis, one can stain mitochondria and determine their intracellular distribution, their behavior in response to disruptions of the cytoskeleton, or their behavior during cell division (251). Two types of quantitative measurements should be distinguished: (1) One can measure total, integrated fluorescence from a fixed number of cells and obtain information about the overall mitochondrial activity in the population of cells; and (2) one can monitor individual cells, and even individual mitochondria within a cell to address questions about intercellular and intracellular heterogeneity. The choice of the appropriate dye is an important consideration, and even then very precise quantitative measurements and their interpretation are subject to theoretical limitations which should not be ignored. Experimentally, such measurements are not as simple as measuring concentrations from absorbances and the use of Beer's law (see reference 228 for an expert discussion of the theory and methodology).

The first dye to be used extensively was rhodamine 123 (Rh123) (252) (Figure 5.24). A serendipitous observation was soon exploited by Chen's laboratory to demonstrate the specificity for mitochondria in living cells. Its uptake into mitochondria was dependent on $\Delta\Psi$, since various ionophores and uncouplers known to dissipate the membrane potential could be shown to prevent Rh123 uptake. Similarly, inhibitors of the electron transport chain (rotenone, antimycin, cyanide) can also reduce or eliminate the uptake of Rh123. It was observed that when glycolysis produces sufficient ATP, a membrane potential could still be generated from the reverse reaction of the F_0F_1 ATPase, a reaction that can be inhibited by oligomycin. In the presence of the H^+/K^+ antiporter nigericin, Rh123 could be used to show hyperpolarization of the membrane, since the proton gradient can be converted into a K^+ gradient with less inhibitory effect on electron transport. Thus, Rh123 fluorescence was a faithful indicator of membrane potential in mitochondria. As described in more detail in the original literature (summarized in reference 253), the effects of a plasma membrane potential can be eliminated in 137 mM K^+, and for some experiments the inhibitor oubain has to be included to shut off the Na^+–K^+ pump in the plasma membrane.

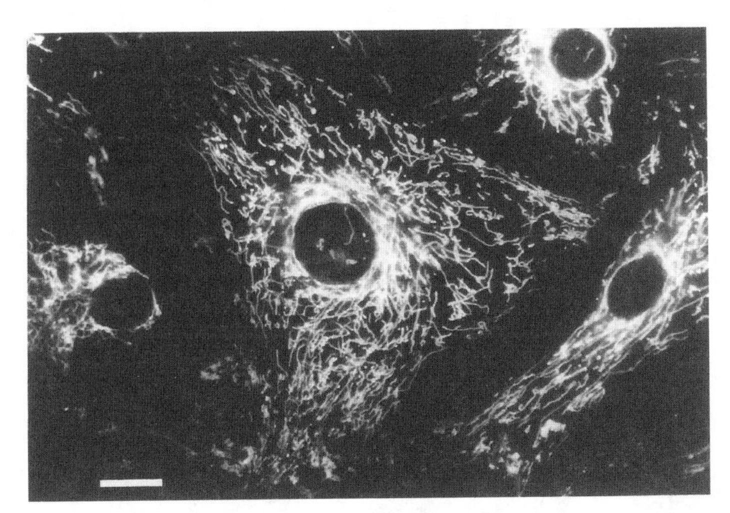

Figure 5.24 Mitochondria stained with rhodamine 123. (Photograph provided by Dr. Lan Bo Chen.)

Literally thousands of different dyes from the color photography industry were tested. Several others will be discussed below. However, Rh123 (Figure 5.25A) has remained a favorite for two main reasons. It appears to be the least toxic to cells, measured by a variety of criteria: survival after weeks of exposurel; DNA, RNA, and protein synthesis; operation of the secretory pathway, maintenance of the cytoskeleton; and cell motility. It may be of special interest that it appears more toxic to cancer cells. It also appears to be unique in being unable to stain mitochondria in the absence of $\Delta\Psi$ (i.e., when azide and oligomycin are used to shut down all mechanisms to produce it), while most of the other dyes tested still give some staining of mitochondria. Differences in lipophilicity and the presence of special lipids within the inner membrane (cardiolipin) have been offered as explanations (253), but the result remains partially empirical at this time. A word of caution is in order. Studies have clearly shown that these dyes may interact with individual complexes of the electrontransport chain—for example, complex V (W. Allison, personal communiation). Thus, inhibitory effects of these dyes in long-term studies should not be ignored.

The concentrations of dyes, the length of exposure of the cells to the dye, the excitation wavelength, and the emission spectrum are all parameters that have to be considered. For Rh123 the typical conditions are: 0.1 mg/ml for 3 hours, excitation by blue light (485 nm), and a greenish fluorescence is measured with the filter normally used for fluorescein dyes (251). For details the original papers should be consulted.

Experiments with Rh123 have shown that all mitochondria in a given cell are homogeneous with respect to fluorescence intensities—that is, have identical membrane potentials, regardless of shape, size, and intracellular location.

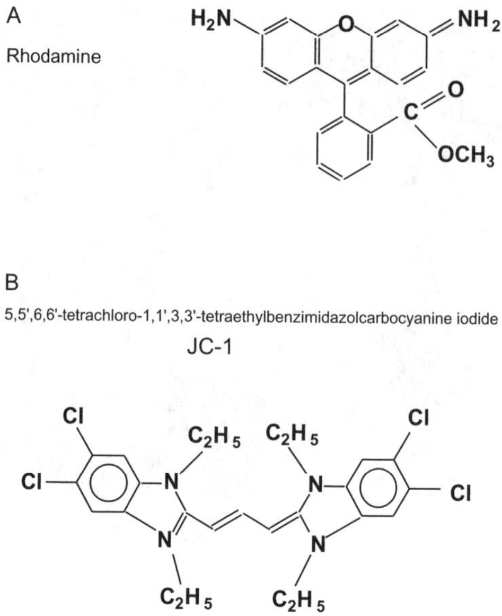

A

Rhodamine

B

5,5',6,6'-tetrachloro-1,1',3,3'-tetraethylbenzimidazolcarbocyanine iodide

JC-1

Figure 5.25 Structures of (A) rhodamine 123 and (B) the dye JC-1.

On the other hand, in a population of cells, significant differences in $\Delta\Psi$ between individual cells have been observed (251) (see below). Not surprisingly, mitochondria in cardiac and skeletal muscle cells have the highest membrane potentials, followed by smooth muscle, macrophages, hepatocytes, fibroblasts, resting neuronal cells, glial cells, and keratinocytes. The lowest $\Delta\Psi$s were found in resting T and B lymphocytes. Differences have been observed between resting cells and rapidly proliferating cells, and also in the course of cell differentiation. The most dramatic change occurs when myoblasts differentiate into myotubes, suggesting an activation of mitochondria which is still under active investigation. Another significant change in Rh123 fluorescence and hence $\Delta\Psi$ is observed at some stage in the progression of a cell toward apoptosis (see Section 7.4), when the permeability transition creates large pores for the dissipation of all ion gradients. A large number of tumor cells have been investigated and compared with their normal counterparts (251). In general, carcinoma cells were found to have higher membrane potentials, but the significance is still not completely clear. It may be related to their proliferative potential.

In recent years, several other dyes have found applications for greater sensitivity and discriminations between mitochondria within a given cell and even between regions of the inner membrane in a single mitochondrion. The dyes can be divided into two classes: fast dyes and slow dyes. An example of the first is Rh123, a molecule that can partition into the lipid bilayer in the presence of an electric field and respond almost instantly to a change in membrane potential by a change in spectroscopic characteristics (quantum yield, absorption or emission spectrum). The slow dyes form aggregates in specific environments, again accompanied by dramatic changes in optical properties compared to the monomer in solution. One of these is the cyanine dye 5,5′,6,6′-tetrachloro-1,1′,3,3′-tetraethylbenzimidazoleocarbocyanine iodide, mercifully referred to as JC-1 (254) (Figure 5.25B). It was initially used as a sensitizer for silver-halide-based photography. Theoretical treatments attempting to explain the aggregation and spectral changes are available (see reference 255 for references). Several aspects of the uptake and behavior of JC-1 by isolated mitochondria and intact living cells were examined (254). When coupled, energized mitochondria were titrated with JC-1, an absorption peak corresponding to the monomer (527 nm—green) was observed first, which was not affected by manipulations of the membrane potential. It can be taken as a measure of mitochondrial volume (255). At higher concentrations, a second peak (590 nm—red) appeared which corresponds to J-aggregate formation, and this J-aggregate fluorescence was proportional to the membrane potential over the range of 30–180 mV. Less quantitative estimates with a fluorescence microscope are based on mitochondria appearing orange in the energized state and green in the uncoupled state. Most surprisingly, within a certain range of average $\Delta\Psi$, mitochondria were detected with apparently heterogeneous staining, indicative of local fluctuations in $\Delta\Psi$ within a single mitochondrion. Similarly, J-aggregate formation and red fluorescence in mitochondria of living cells were sensitive to the electrochemical gradient; that is, the aggregates failed to appear in the presence of either CCCP, dinitrophenol, or azide plus oligomycin, and so on, or they could be dissociated when these uncouplers or inhibitors were added later. Combinations of ionophores were used to show that aggregate formation was sensitive to $\Delta\Psi$ but not to ΔpH alone.

Significantly, a number of different cell types examined exhibited both intercellular and intracellular heterogeneity in staining with JC-1, which had not been detected previously with Rh123 or other cyanine dyes (e.g., reference 256). The explanation offered was that detection systems in the past (photography, the human eye) may have been used outside of the linear response range—that is, were saturated and hence incapable of discriminating local differences. Therefore, even intracellular heterogeneity must now be taken seriously, and an explanation must be sought in terms of different microenvironments for mitochondria within a cell, of the type already described by sensitive determinations of Ca^{2+} gradients. Dynamic, multichannel continuous recordings may in the future be able to address these questions.

An even more challenging problem arises from the observation of low and high membrane potentials (or at least an uneven distribution of J-aggregates) within a single mitochondrion (254). Proton diffusion is extremely rapid, and localized proton circuits or "respiration hot spots" within a mitochondrion are almost certainly transient phenomena. How can a slow dye respond to these, and how are such states maintained at least during the interval required for photography or other detection methods?

5.5 ATP SYNTHASE (F_1F_0-ATPASE)

5.5.1 Introduction

Reference has been made repeatedly to a distinct complex (V) playing a crucial role in oxidative phosphorylation. An enzyme had to be responsible for the synthesis of ATP from ADP and inorganic phosphate. Somehow, this endergonic reaction had to be coupled to redox reactions and electron transport. The acceptance of Mitchell's chemiosmotic hypothesis and the elucidation of the biochemical and structural properties of this enzyme complex evolved somewhat in parallel, and today we understand that a physically separable complex can catalyze this reaction driven simply by a proton gradient and membrane potential, which can be set up artificially across a vesicle membrane by the choice of buffers or by the incorporation of a light-driven proton pump from the purple membrane bacterium *Halobacter halobium*. ATP synthase was first "seen" by electron microscopists around 1962. Lollipop-like structures were identified by their spherical heads on the matrix side of the inner membrane, and these were attached to the membrane by a slender stalk. Pioneering work in Racker's laboratory had succeeded in dissociating the heads from the stalk. The heads were soluble in water, and they could be reassociated with the membrane in the presence of an essential protein known as the "oligomycin sensitivity conferral protein" (OSCP). OSCP had first been isolated and characterized in the laboratory of Tzagoloff (257–259). Inhibition of the ATP synthesis by oligomycin had become a diagnostic test for this mitochondrial enzyme, and therefore the physiological significance of the required reassembly factor was apparent. Contrary to a long-held belief, oligomycin binds to the c-ring and not to OSCP.

The complete complex was separated in the fractionation scheme developed by Hatefi (13, 14), and the combined efforts of these early investigators led to the now well-established separation of the complete complex V into a soluble F_1-ATPase, and an insoluble membrane complex referred to as F_0. The F_0 complex has the ability to translocate protons across the membrane from their high potential on the outside; and when coupled to the F_1 subcomplex, ATP synthesis can be achieved.

It was quickly apparent that this structure was significantly more complex than the first ion pumps being characterized: (a) the Na^+–K^+ pump in the plasma membrane consisting of two subunits ($\alpha\beta$) and (b) the Ca^{2+} pump in the sarcoplasmic reticulum. The F_1-ATPase has at least eight different peptides (Figure 5.26). The relative positions of $\alpha, \beta, \gamma, \delta, \varepsilon$ (present in the ratio 3:3:1:1:1) were established earlier, while the positions of F6, d, and OSCP were determined only recently. The mitochondrial F_0 subcomplex (Figure 5.26) has a variable number of different subunits depending on the species (11 in mammals), of which three essential proteins (a, b, c) are homologous and functionally related to the three proteins in the bacterial complex, and the remainder are accessory proteins necessary for assembly, stabilization, and

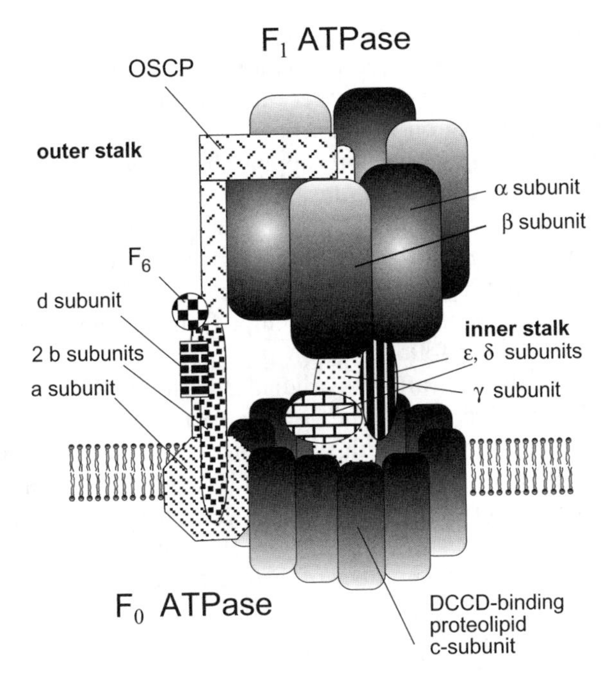

Figure 5.26 Schematic representation of complex V. The F_1-ATPase extends into the mitochondrial matrix and is connected to the F_0 complex in the inner membrane by a stalk. The major peptides associated with each of the subcomplexes are discussed in the text. The interaction of the OSCP subunit with the b_2 complex and the outside domain of an α subunit is also emphasized.

control. The combination of biochemical analyses and the identification of relevant structural genes have contributed to the determination of the composition of the complex in various organisms (238). All three core proteins in F_0 are encoded by the mitochondrial genome in *Saccharomyces cerevisiae* and plants, while one of these genes has been transferred to the nucleus in animals and fungi. The establishment of the stoichiometric ratios for the F_0 subcomplex peptides proved to be quite challenging (260). Three distinct subunit types (a, b, c) are present in the ratio ab_2c_n, where different values for n, ranging from 10 to 14, have been reported for different organisms.

In the literature on this complex, and especially in reviews, reference is frequently made to information from mammalian mitochondria, yeast mitochondria, chloroplasts, and the *E. coli* complex in the inner bacterial membrane. The most recent and authoritative review on the composition, gene expression and assembly with an emphasis on the yeast and *E. coli* systems is by Ackerman and Tzagolof (238). The nomenclature is not yet completely uniform, and there is potential for some confusion. For example, the F_1-ATPase in all examples has five subunits (αβγδε), but the mitochondrial ε subunit has no counterpart (sequence, function) in either chloroplast or *E. coli* F_1 or F_0, and the δ subunits of the bacterial and chloroplast F_1-ATPase are related to the mitochondrial OSCP, which until recently was counted among the F_0 stalk peptides. Thus, depending on which schematic structure is shown in various texts, one can find the OSCP subunit capping the mammalian complex, or the δ subunit capping the bacterial complex (261). At the same time, it is connected to the highly elongated b subunits of the F_0 complex extending upward from the base. The positioning of the ε subunit as part of the stalk was originally supported by cross-linking studies. The mitochondrial δ and the bacterial ε subunits are related. Table 5.2 is an attempt to make the cross-referencing easier in the discussion to follow.

More than three decades after its original isolation by Racker's group (see reference 262 for an account of the early work) the studies on the bovine F_1-ATPase culminated in 1994 with the publication of the structure at 2.8-Å resolution by Walker's group at Cambridge (263, 264). This achievement was recognized by a Nobel Prize in 1997. Over the years the structure evolved from being the globular head of a lollipop roughly 100 Å in diameter, to an increasingly refined description of a number of distinct features based on interpretation of data from enzyme kinetics, biochemistry, biophysics, immunochemistry, cryo-electron microscopy, X-ray crystallography, and nanotechnology. For

TABLE 5.2 Subunit Composition of ATP Synthase

Comparison and Nomenclature of Complex V Subunits in Various Organisms			
Bacteria (*E. coli*)	Animal Mitochondria	Yeast Mitochondria	Plant Chloroplasts
α(55.2)[a]	α(55.1)	α(55.3)	α(56.8)
β(50.1)	β(51.6)	β(52.5)	β(53.9)
γ(32.4)	γ(30.2)	γ(30.6)	γ(38.1)
δ(19.3)	OSCP	OSCP(20.9)	δ
ε(14.9)	δ(15.1)	δ(14.5)	ε(14.9)
—	ε(5.7)	ε(6.6)	—
—	Inhibitor protein		—
a	a (ATPase-6)	Subunit 8 (5.87)	a
b	b	Subunit 6 (27.9)	b, b′
c	c (ATPase-9)	Subunit 9 (7.79)	c
—	d	d (19.66)	—
—	e		—
—	f	b (P25)	—
—	g		—
—	A6L		—
—	F6		—

[a]The numbers in parentheses indicate the molecular mass of the peptides in kilodaltons.

example, the subunit stoichiometry and the determination of nucleotide binding sites suggested the presence of a threefold axis of rotation, and it was reasonable to make this axis parallel to the stalk. Now that an almost complete crystal structure of this complex is available, the description could go into considerable detail, but reference to schematic diagrams is still invaluable and instructive.

5.5.2 X-Ray Structure

The single subunits γ and ε form a structure that interacts with components of the F_0 complex in the membrane and with the α and β subunits of the F_1 on the matrix side, thus being part of the stalk seen by electronmicroscopy. The N- and C-terminal segments of the γ subunit form extended α-helices forming a left-handed, antiparallel, very loose coiled coil (with an extending C-terminal), and this rod-shaped core has the α and β subunits arranged alternately around it like the segments of an orange (Figure 5.27).

The δ and ε subunits appear not well-ordered and may be substoichiometric in the original crystals (263). In other words, the purification of the F_1-ATPase requires a rupture of the stem structure, with the δ and ε subunits becoming randomly distributed and disordered. Thus, their precise orientation and interactions are still under investigation. It is significant that the coiled-coil core of γ with its extended C-termial helix is somewhat bent and by itself has no threefold symmetry. Therefore the symmetrical $\alpha_3\beta_3$ hexamer is slipped onto an asymmetrical camshaft, and when the camshaft rotates (see below), the individual subunits do not make equivalent contacts at any given instant. Thus, the nucleotide binding sites have different affinities representing phases in the turnover of the active site.

The α and β subunits are similarly folded: A six-stranded β barrel at the N-terminal is connected through a nucleotide binding domain to a bundle of 7 (α) or 6 (β) helices at the C-terminal. The nucleotide binding domains are made up of a nine-stranded β-sheet with nine associated α-helices, and they are recognizable from similiarities with other known nucleotide binding proteins such as RecA, as well as from the presence of typical loops and motifs (263). They are at the interfaces between the α and β subunits (i.e., between the slices of the orange). There are three catalytic sites at which nucleotide turnover occurs, as well as three ATP binding sites where nucleotide binding appears to be required for the maintenance of symmetry and structure and to prevent abortive complex formation. A proposed physiological regulatory role for these noncatalytic sites has been questioned (265). Some of the amino acid side chains involved in nucleotide binding had been identified from previous studies such as affinity labeling with 2-azido-ATP, or site-directed mutagenesis (see reference 266 for earlier references to the primary literature), and the complete structure not only confirmed previous deductions, but allowed a detailed reconstruction of all the contacts between the phosphates, ribose, and the adenine base.

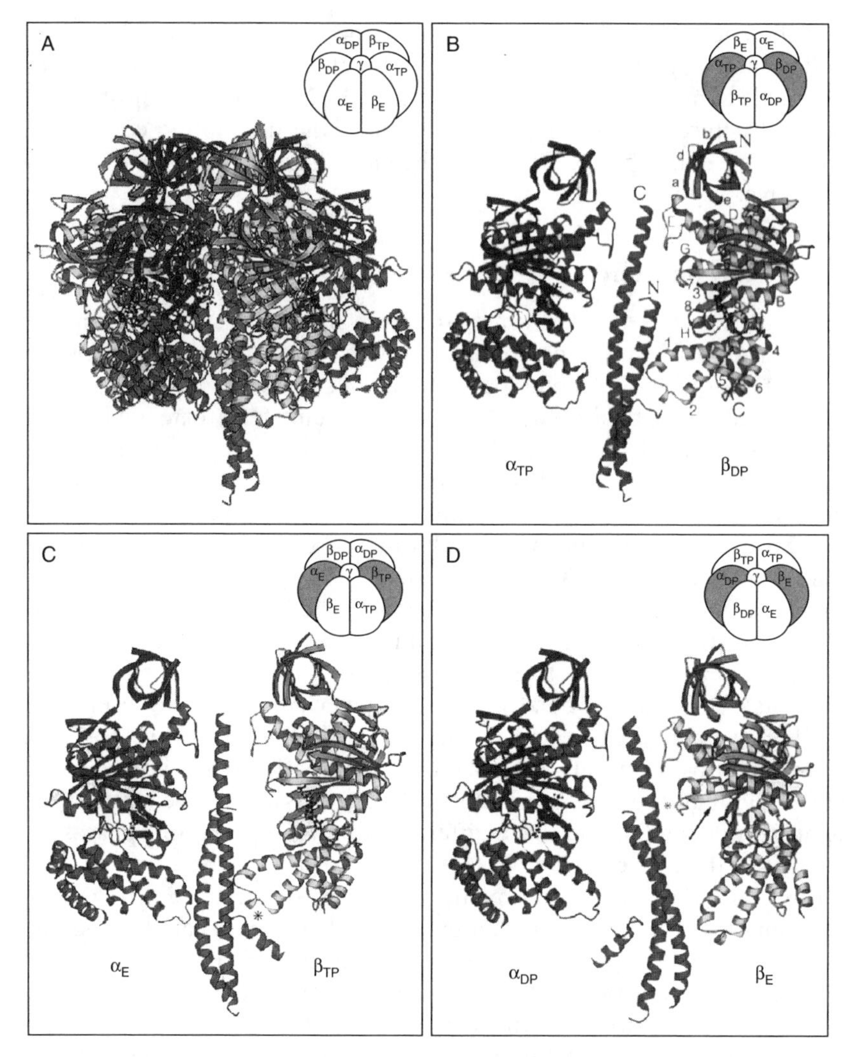

Figure 5.27 X-ray structure of the F_1 subcomplex of complex V. (A) Ribbon model showing the $\alpha_3\beta_3$ hexamer and the γ subunit constituting the rotating shaft. (B–D) Three different views of the shaft (γ subunit) and side views of two opposite α and β subunits. The image emphasizes the asymmetry and the different contacts made by each of the six subunits with the central shaft. (From reference 263 with permission.) See color plates.

Since the original structure of the F1 subcomplex was published, numerous refinements have been made. The characterization of structures binding various inhibitors has allowed us to go from a single snapshot of the enzyme to a more detailed view of the behavior of the enzyme during a catalytic cycle (267, 268). The structure of the specific inhibitory protein IF1 was first solved as an isolated protein (269), and subsequently in a complex with the F1-ATPase, where

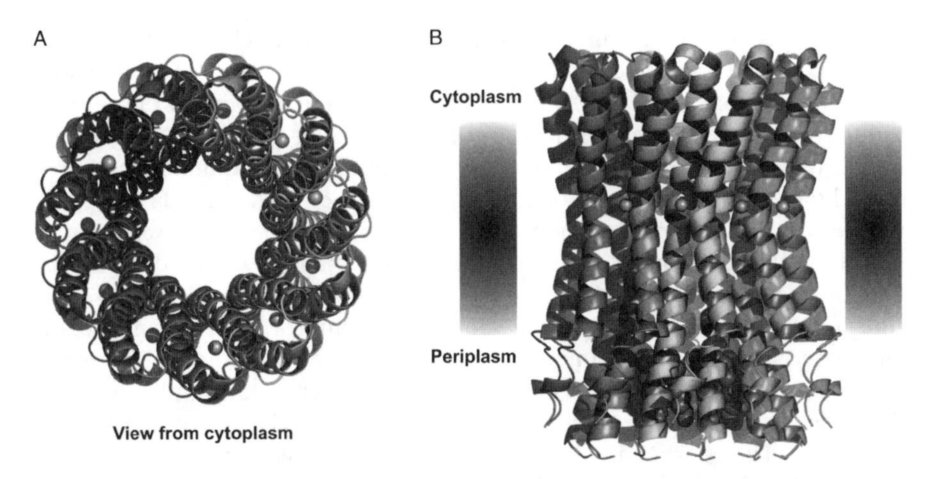

Figure 5.28 Structure of the rotor ring of F-type Na^+-ATPase from *Ilyobacter tartaricus*. (From reference 275 with permission.) (A) View perpendicular to the membrane from the cytoplasmic side. Two of the 11 subunits are labeled. (B) Side view. Bound Na^+ ions are shown as blue spheres, and the blue molecules inside the ring on the periplasmic side represent detergent molecules.

it binds in the α–β interphase and also contacts the γ subunit (270). On another front, the structure of the peripheral stalk or stator has come into a much better focus in recent years (271–273). A partial structure at 2.8-Å resolution includes portions of the subunits b, d, and F6. New insights are derived from these studies about the location of the oligomycin-sensitivity conferring protein (OSCP). It is located near the top of the F_1 domain, interacting with the N-terminal region of one α-chain and with the uppermost region of the b subunit in the stator (see Figure 5.25).

The structure of the intact complex V has been refined first by cryo-electron microscopy (274); and as discussed below, much progress has been made toward a high-resolution structure of the F_0 subcomplex including the c subunits. For practical/technical reasons the high-resolution structure of the c-ring has been obtained so far for a related F-type Na^+-ATPase of *Ilyobacter tartaricus* (275) (see Figure 5.28) and for the V-type Na^+-ATPase of *Enterococcus hirae* (276). The structure of the c-monomer had been deduced for the *E. coli* protein from MNR studies in chloroform:methanol:water (277).

5.5.3 ATP Synthesis and Catalytic Mechanisms

Before discussing more detailed aspects of the structure, it is necessary to become familiar with the current thinking about the mechanism by which the ATP synthase carries out the simple reaction

$$ADP + P_i \Rightarrow ATP + H_2O$$

on the matrix side of the inner mitochondrial membrane. The problem is that this reaction is endergonic and thus requires an input of energy. The chemiosmotic hypothesis identified a source of this energy and introduced the general notion that this enzyme was a proton pump running in reverse. When this notion was verified experimentally, the challenge became to understand precisely how the coupling of proton flow through the complex and ATP synthesis were achieved. The fractionation of the F_1 ATPase subcomplex was a direct indication that this complex contained the active site for ATP binding, and in the absence of any other driving force it could catalyze the hydrolysis of ATP (the reverse reaction). With the F_0 portion of complex V being an integral membrane complex, it became the obvious choice for a pathway of protons through the inner membrane. The connecting stalk/camshaft in turn must be the structure that couples these two subcomplexes mechanically and conformationally. The question is, How? The history can be found in several detailed reviews (278, 279) that appeared prior to the publication of the crystal structure. With the information from the crystal structure the focus of the discussions could obviously be sharpened considerably (263–265, 280–283).

At this point the subunit stoichiometry of the F_1 complex and the threefold symmetry of the arrangement of the α and β subunits come into consideration. A "rotating" three-site model was first proposed by Boyer in 1982 (284), as an extension of an "alternating" site model proposed earlier (285, 286). Nucleotide binding studies, chemical probes, and ligand/inhibitor binding studies had suggested that the α and β subunits were conformationally asymmetric, and this concept was fully substantiated by the confirmation of the stoichiometry $\alpha_3\beta_3\gamma\delta\varepsilon$ in the F_1 complex plus stalk. Ignoring the δ and ε peptides for the moment, the coiled-coil of the ends of the γ peptide forms an asymmetrical axis on which the $\alpha\beta$ hexamer is impaled (see above); that is, each α and/or β subunit is exposed to a different surface of this shaft. The original formulation of the model speculation about rotations of the inner portion of the enzyme relative to the outer portions had already been advanced (278), and it has since been spectacularly confirmed that a "central rotor of radius ~1 nm, formed by the γ subunit, turns in a stator barrel of radius ~5 nm formed by the three α and three β subunits." Rotary motion of the F1-ATPase was observed directly (Figure 5.29). In the presence of ATP a fluorescent actin filament attached to the γ subunit was shown to rotate for more than 100 revolutions in an anticlockwise direction when viewed from the membrane side (287).

This elegant and definitive experiment was performed with an $\alpha_3\beta_3\gamma$ complex from a thermophilic bacterium expressed in *E. coli*. The complex was attached to a microscope slide like an upside-down mushroom, and the stalk (γ subunit) was coupled to fluorescently labeled actin via a multivalent streptavidin molecule binding biotin on the actin and on the γ subunit. The model system has demonstrated rotation of the γ subunit relative to the alternating α/β hexamer. The system reconstituted the smallest molecular rotary motor described to date. It can be noted here that there are now three types of rotary motors found in nature. The flagellar motor of bacteria is very different and will not

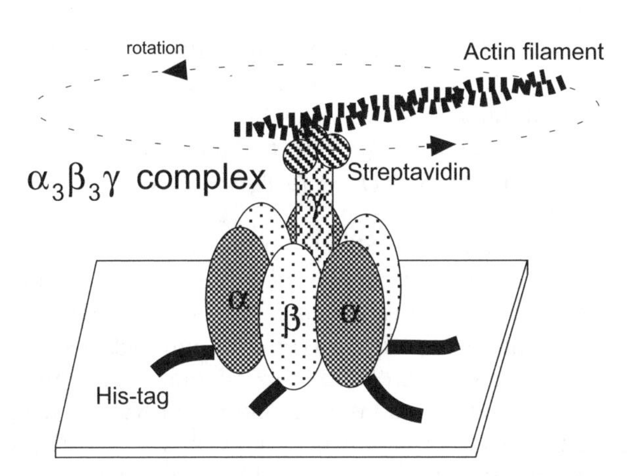

Figure 5.29 Demonstration of the rotation of the shaft (γ subunit) relative to the F1 complex. (Adapted from reference 287.)

be further discussed. The F-ATPases typically synthesize ATP driven by a proton motive force, while the V-type ATPases hydrolyze ATP to generate an electrochemical gradient across a membrane. The latter two are operating on similar principles. The discussion will focus on the mitochondrial complex V, an F-ATPase. In the above model system, ATP hydrolysis by the a/β subunits caused the γ subunit to rotate, and one can therefore imagine this rotation to be the reverse of the rotation caused by a proton flux driving ATP synthesis. Following the original publication describing rotary motion, several more elegant demonstrations of rotary motion have appeared (288–293). Some of these model systems have included the c subunits of the rotor, demonstrating their rotation when attached to the γ subunit with the help of ε. These studies are beautiful illustrations of the power of nanotechnology applied to the study of single molecules, but they have not only provided a picture, but have yielded quantitative data fully reconciling mechanistic and kinetic studies on ATP synthesis by this enzyme (283, 290, 291). The reaction is completely reversible. Hydrolysis of one ATP causes a 120° rotation of γ, and the rotation of γ in the opposite direction can be driven by proton flux coupled to ATP synthesis from ADP and P_i. The 120° rotation can be decomposed into two sub-steps of 80° and 40°, where the 80° rotation is brought about by the binding of ATP to a single empty site, and the 40° rotation is associated with the release of ADP from the site. Under physiological conditions the bacterial enzyme is estimated to rotate at ~100 Hz, and averaged over time all three catalytic sites are occupied, two by ADP and one by ATP. All three sites are involved in driving the rotation of γ. Pure β subunits do not catalyze hydrolysis at a significant rate, and even the $\alpha_3\beta_3$ complex shows suboptimal activity and altered affinities and kinetics. In the complete structure, "multisite catalysis" is now accepted as the dominant mode of catalysis *in vivo*, based on the estimated intracellular concentrations of substrates and the measured values for the K_m of the enzyme.

Multisite catalysis is characterized by an impressive (~5 orders of magnitude) increase in the rate of hydrolysis attributed to the highly cooperative interaction between the subunits and their catalytic sites.

There appears to be a consensus that the binding of ADP and P_i at one of the three active sites of the F_1-ATPase provides the energy for combining the two substrates to make ATP. In other words, ATP formation is spontaneous after binding and does not require the input of additional energy. The reaction is a simple reversal of hydrolysis and has no additional covalent intermediates. A nearby glutamate participates in acid catalysis. The electrostatic repulsion between the phosphate groups is overcome by interactions with amino acid side chains in the active site, and the perfect alignment presumably makes the formation of a covalent bond and the elimination of water entropically favorable. The strong binding energy, however, now makes the release of the ATP unfavorable and slow. It is the release of ATP which is believed to be the step driven by the proton translocation and rotation. The release must be triggered by a change in affinity for ATP, which can be brought about by a change in the conformation of the protein subunits. The three orientations of a given β subunit relative to γ can be shown to correspond to the progression from ATP-occupied ⇒ ADP-occupied ⇒ empty, the reaction sequence expected for the hydrolysis reaction.

An authoritative and exhaustive review in 1993 by Boyer (278) has provided a summary of the biochemical observations related to the ATP synthase mechanism and provided critical comments on each. An update by the same author incorporating information from the crystal structure has appeared subsequently (294), and highlights in the development of this penetrating and detailed analysis can be found in the acceptance speeches and publications of the 1997 Nobel Prize winners in Chemistry. Very significant contributions and refinements have also been contributed by Senior and colleagues (see reference 282 for an expert review). They are testimony to the ingenuity and insights of a large number of investigators keeping pace with or even staying ahead of the structural studies, until this amazing molecular engine and its inner workings are now revealed in exquisite detail. Once more it should be emphasized that biochemical studies were constantly reviewed and tested for consistency with the structural studies. The clarification of the subunit stoichiometry was followed by cross-linking studies under various conditions, demonstrating conformational changes associated with the presence of, for example, Mg^{2+}-ADP in the catalytic site, contrasted to the presence of the nonhydrolyzable ATP analog AMP-PNP. When the β and γ subunits were cross-linked, enzyme activity was severely inhibited, a result now fully explainable in terms of the rotary motion associated with one cycle of activity. Finally, although most of the biophysical studies were performed with bacterial enzymes, it is a consensus that the other F- and V-type ATPases operate on the same principles, with perhaps notable differences in kinetics and mechanisms of regulation.

The noncatalytic nucleotide binding sites have a structural role and are likely to play a part in regulating or modulating the stability and/or the structure of this enzyme (266, 278). Nucleotide exchange at the noncatalytic sites is slow. One is reminded of the role of one nonhydrolyzable GTP in the $\alpha\beta$ tubulin dimer. Nevertheless, while a physiologically relevant regulatory role has not yet been demonstrated, abolition of the noncatalytic nucleotide binding sites by site-directed mutagenesis has shown that they are not dispensable. Such a modified enzyme ($\alpha_3\beta_3\gamma$ complex) exhibits an initial ATPase activity, but this activity decays rapidly to zero. This observation has been interpreted to indicate an entrapment of MgADP in a catalytic site (see reference 295 for additional references to investigations of the specific role of the noncatalytic sites).

A schematic representation of the three-site model is shown in Figure 5.30 for the synthesis of ATP (Figure 5.30A) and for the hydrolysis of ATP (Figure 5.30B). (see 281, 282, for details).

5.5.4 The F_0 Subcomplex and Proton Flow

Attention now must focus on the stator (b_2OSCP) in the periphery and on the F_0 subcomplex (ab_2c_n) for which complete high-resolution crystal structures are not yet available. The F_0 subcomplex provides a path (channel) for the protons. Proton flow through the F_0 channel must be converted into a torque on the rotor ($\gamma\varepsilon c_n$) in the center. Thus, two questions can be formulated: (1) What is the path of the protons through the F_0 subcomplex? (2) How does the translocation of protons from one side of the membrane to the other produce a torque? To avoid confusion, the following discussion will refer to the essential subunits of the bacterial complex, and the reader is referred to Table 5.2 for the corresponding subunits in the mitochondrial complex in various organisms.

While δ and ε subunits were dispensable in the demonstration of F_1-ATPase ($\alpha_3\beta_3\gamma$) as a rotary motor, the ε subunit is likely to be necessary for the coupling between the F_1 and F_0 subcomplexes. It can be cross-linked via its C-terminal domain to the $\alpha_3\beta_3$ barrel and via its N-terminal domain to the γ subunit. Sequence conservation in bacterial and chloroplast ε subunits has focused attention on an N-terminal domain, and studies employing site-directed mutagenesis have shown that mutations (e.g., εH38C) affect coupling between proton flux and ATPase (266). Cryo-electron microscopy has shown a major shift of the ε subunit relative to the α and β subunits when viewed down the threefold symmetry axis upon binding of different nucleotides (266).

To function as a highly specific inhibitor of the ATP synthase in mitochondria, oligomycin (Figure 5.31) requires the presence of the OSCP subunit (oligomycin-sensitivity conferring protein); it was shown to be equivalent to the bacterial δ subunit. Since it blocks proton flow, it had been associated with the proton translocating mechanism, but it seems clear now that the inhibition is indirect. The δ subunit is located "on top" of the a_3b_3 barrel, facilitating the interaction of the tip of the b_2 complex with the $\alpha_3\beta_3$ barrel (see Figure 5.26). A role in the assembly of the ATP synthase is discussed briefly below.

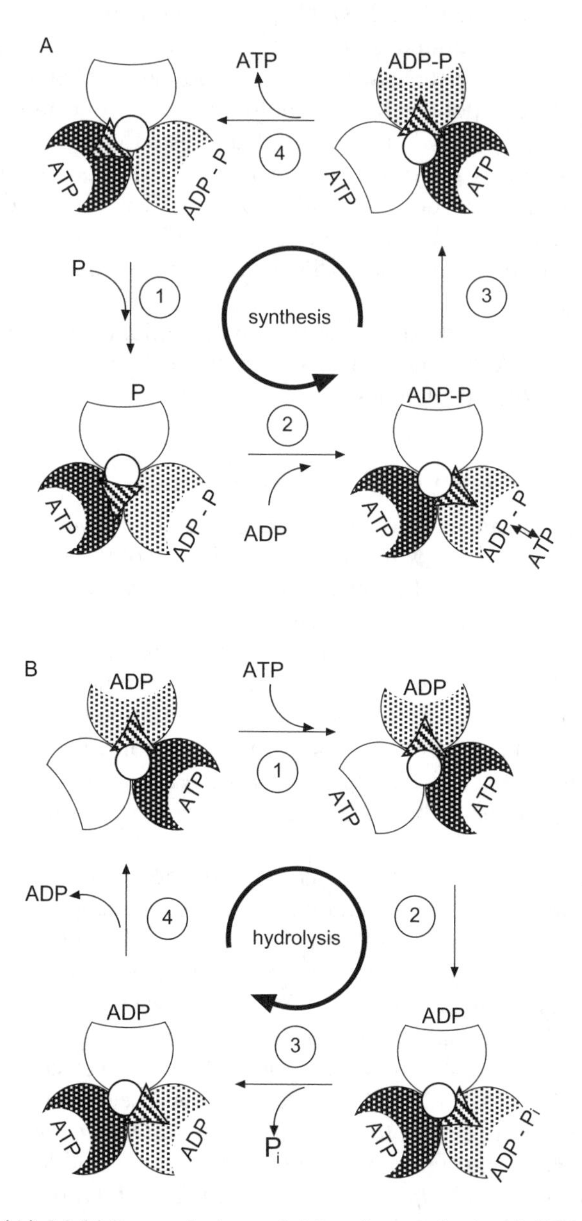

Figure 5.30 (A) Multisite catalysis model for the synthesis of ATP. (B) Multisite catalysis model for the hydrolysis of ATP. In this model all three catalytic sites are occupied, except for a very transient period when the product has left and before a new substrate enters the site. (Adapted from reference 281.)

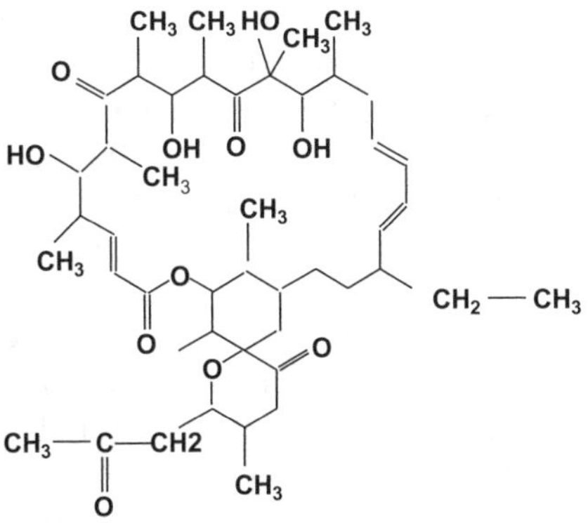

Figure 5.31 Structure of oligomycin.

The bovine F_0 subcomplex had first been prepared in Racker's laboratory by a combination of sonication and urea treatment. A more homogeneous preparation suitable for structural analysis has been devised by the Cambridge group of J. Walker. The final complex from bovine heart contains nine different proteins (a–g, F6, A6L) (296). The stoichiometry for the mammalian mitochondrial F_0 complex had also been established independently (e.g., reference 260). All of the subunits have been expressed in bacteria from cloned cDNAs, and successful re-constitution experiments making various subcomplexes as well as the entire F_1F_0-ATPase have been reported (296). Complete information for the yeast mitochondrial F_1F_0-ATPase peptides (238) is based primarily on genetic analyses of respiration-deficient yeast mutants, followed by biochemical experiments. Additional gene products were found to be necessary for the assembly of the complex, but were not present in the purified complex (238, 297).

From comparisons with the simpler complex in bacteria, one can conclude that three essential peptides make up the proton-translocating hydrophobic F_0 core (298). Subunit a (271 residues in bacteria) has multiple transmembrane regions, most likely an even number (4–8). This places both the N-terminal and the C-terminal on the same side of the membrane. With the help of fusion proteins, the ends have been placed on the outside (periplasmic side) of the bacterial membrane (298). It is not required for the binding of the F_1. The b subunit may be present as a dimer. In bacteria, it has been deduced from the amino acid sequence that subunit b has a single transmembrane segment near the N-terminal, with the remaining domain forming a predominantly α-helical, highly charged (hydrophilic) secondary structure. Early ideas about its interactions with the γ and δ subunits of F_1 (298) have been abandoned in favor of a model placing the two b subunits on the outside of the α/β complex where it acts as the stator (see Figure 5.26). A high-resolution structure of this stator has been published recently (271). The a and b subunits can be cross-linked. They are attached to the outside of the c ring (see below). A high-resolution structure for this interaction is not yet available in spite of the intense interest focused on this interaction. Protons translocated into the matrix are very likely taking a path that includes the interface between the a subunit and the c_n ring. The b_2 dimer is not believed to be involved in forming the proton channel. Thus, understanding this specific protein–protein interaction promises to shed much light on the mechanism of torque production by proton translocation.

The c subunit is present in multiple copies (10–14, depending on the species) compared to the a subunit; 10–14 c-subunits are associated to form the c ring. Within its short sequence (79 residues in bacteria) two transmembrane α-helices are predicted, flanking a polar loop region. Such a structure has been verified by NMR studies of the monomer in chloroform:methanol:water (277). Using this structure for molecular modeling structures for the c ring were proposed. A high-resolution structure for the c ring of a related F-type Na^+-ATPase of *Ilyobacter tartaricus* was finally reported in 2005 (275) (Figure 5.28), while Murata et al. (276) published the corresponding structure for the V-type Na^+-ATPase of *Enterococcus hirae*. Some significant differences between these structures and the mitochondrial c_n ring can still be expected. First, the number of monomers per ring may be different; and second, the mammalian ATP synthase is driven by protons.

Even before detailed crystallographic structures were available, several revealing structure–function studies were carried out with the bacterial c subunit modified by site-directed mutagenesis. For example, a Q42E mutant c-peptide allowed normal assembly of the F_1F_0-ATPase, but ATP hydrolysis was uncoupled from proton translocation (the bacterial enzyme is fully reversible depending on physiological conditions) (298). The well-known inhibition of the F_1F_0 ATPase by DCCD is due to a unique reaction of Asp61 (D61) with dicyclohexylcarbodiimide (DCCD). Even if only a single c-subunit is derivatized by DCCD, the ATPase activity of the whole complex is abolished. Studies of the isolated subunit c in a chloroform–methanol–water mixture were encouraged by experiments suggesting that the isolated protein retains features of the native protein, including the reactivity of D61 with DCCD, and the inhibition of this reaction by the I28T mutation. Experiments of this type and mutational analysis at this Asp residue have been used to argue that the carboxyl group of D61 is a participant in proton translocation. At this time it appears to be the best candidate (277, 298). Similarly, a mutational analysis of the bacterial a subunit has identified the R210 residue as an absolutely essential residue for proton translocation. Thus, the single a subunit and 9–12 c subunits form a path for proton flux across the membrane.

Proton flow in such a molecular motor must produce a torque. Proton transfer must be associated with a rotation or displacement of an a-subunit relative to a c subunit. The model must also include the experimental fact that several protons and several c subunits are involved in a 120° displacement (more than one proton has to be moved across the membrane for one ATP to be made). Junge and his colleagues as well as Oster and colleagues have proposed ingenious solutions in which biochemistry meets engineering (280, 299, 300); they await further experimental verification. The reader is referred to the original literature for details on two major models under consideration that can be labeled as the "power stroke model" and the "Brownian ratchet model." Evidence from available crystal structures and theoretical considerations favor the ratchet model (299). In this model the c ring executes stochastic rotational motions. There are two access channels for ions on opposite sides of the stator (a subunit), but they are not collinear. After an ion has entered from one side, the c ring has to turn through some angle to allow this ion to escape on the other side. The unequal concentration and potential of the ions on either side of the membrane drives a flux of ions in one direction, and it also biases the Brownian rotational fluctuations of the c ring to produce a net rotation in one direction. The emerging structural details are clearly of great importance for defining the proton channels, the relevant side chains on the a and c subunits (already partially determined from mutagenesis experiments), and the possible conformational changes associated with this mechanism.

Finally, two challenges remain that are closely related in some aspects. Thermodynamic considerations strongly suggest that on the order of three protons must be translocated across the membrane for every ATP produced, or, for every 120° turn of the rotor. This had led to predictions at an earlier stage of exploration that the c ring should have a number of subunits that are a multiple of three (i.e., 9 or 12). The experimental finding of 10 or 14 subunits in the c_n ring raises the issue of a "symmetry mismatch": The symmetry of the c-ring does not match the symmetry of the $\alpha_3\beta_3$ barrel of the F_1 subcomplex. How many c subunits pass the single a subunit of the stator in a 120° turn? The number must be non-integer, and hence the number of protons passing from one side to the other on average must also be non-integer. There are considerations that this symmetry mismatch is an essential characteristic of the torque-producing mechanism, but the problem will keep biophysicists occupied in the future.

Upon reaching the end of this chapter, one should expect the reader to be left with a sense of awe, without, however, being pushed into the camp of those who believe in "intelligent design." Francois Jacob compared evolution to the work of a "tinkerer," and it is indeed amazing what several billion years of tinkering can achieve. It is also truly "awesome" to think of the billions of cells in our body, each containing tens of thousands of little motors going continuously at 50–100 rps, thus sustaining a turnover of an amount of ATP/ADP per day that is equivalent to or greater than our body weight.

5.5.5 Assembly of Complex V and Dimerization

The biogenesis of ATP synthase raises several questions that have already come up in the discussion of the other complexes of the ETC. One has to consider (a) the expression of multiple genes and (b) the assembly of multiple subunits making up the subcomplexes of the mature enzyme. In eukaryotes, mitochondrial genes as well as nuclear genes contribute to the biogenesis; and furthermore, the nuclear genes are typically dispersed on different chromosomes (in contrast to *E. coli*, where they form an operon). The control and coordinate expression of such genes is briefly discussed elsewhere (Chapter 4). In yeast the nuclear genes of ATP synthase appear to be constitutively expressed, regardless of the carbon source, in contrast to the nuclear genes for the electron transport complexes. At the same time, there appears to be no specific mechanism to coordinate the expression of nuclear genes and mitochondrial genes in yeast. A more detailed discussion can be found in the review by Ackerman and Tzagoloff (238).

It seems intuitive that the assembly of a complex of 14 or more subunits would occur in an orderly fashion, but the challenge has been to elucidate the assembly pathway. Work in *S. cerevisiae* has been pioneering by taking advantage of the powerful methodologies of molecular genetics. Thus, it is possible to investigate gene knockouts and determine how far assembly can proceed in the absence of a particular subunit. As a first approximation, one can state that the core subunits of the F_1 subcomplex (α, β, γ) and the core subunits of the F_0 subcomplex (c, a, b) assemble independently, and at some stage they are joined and combined with other subunits to form the functional complex. An F_1 subcomplex with ATPase activity can be made in yeast mutants lacking mtDNA (ρ^o mutants). This activity is essential in combination with the adenine nucleotide transporter to generate a membrane potential for protein import and for the maintenance of mtDNA in ρ^- mutants. Alternatively, the assembly (and hence activity) of ATP synthase has been found to be defective in mutants even though all known structural genes were normal. Thus, it became apparent that assembly factors or molecular chaperones are involved (a conclusion equally applicable to the assembly of the other complexes of the ETC). These assembly factors are not found in the final complex. Two other properties of such factors are noteworthy: (1) For example, the factors Atp11p and Atp12p are specifically required only for the assembly of the F_1 subcomplex, in contrast to chaperones such as mtHsp70 and Hsp60; (2) because of their restricted function, their deletion is not necessarily lethal under conditions where glycolysis can satisfy the energy requirements of the cells. The most explicit and detailed description of the ATP synthase assembly in yeast can be found in a review by the authors who have also contributed much of the experimental support for the model (238).

There are two ATP synthase subunits whose significance has been appreciated only recently. The subunits e and g are required for dimerization of complex V, detected by blue-native gel electrophoresis. In this context, the most intriguing observation was made that ATP synthase dimers are playing a essential role in the establishment of the morphology of mitochondrial cristae (37, 38, 301, 302). Yeast mutants lacking either e or g are devoid of lamellar cristae, with the inner membrane folded into onion-like structures or highly abnormal folds. How ATP synthase dimerization can lead to the "zippering up" and alignment of cristae membranes remains a challenge for the future.

5.6 CONTROL OF RESPIRATION AND OXIDATIVE PHOSPHORYLATION

5.6.1 General Considerations

It is quite apparent that the energy demands of a given cell or tissue will vary over time, and the need for regulatory mechanisms becomes obvious. A distinction to be made at the beginning of this section is to separate mechanisms involved in long-term control, in the course of development and tissue differentiation, from short-term control mechanisms operating on a time scale of seconds to hours. A prominent example is muscle, but other tissues and organs in mammals and other organisms may also exhibit fluctuations in respiration and oxidative phosphorylation in response to regulatory signals from hormones, growth factors, or the nervous system. A subtle but very important aspect of this problem is the balancing between (a) the capture of the free energy from combustion in the form of ATP and (b) the release of some of that energy in the form of heat. This is of course the basic mechanism for maintaining the body temperature in warm-blooded animals, and specific tissues and mechanisms have evolved for this task.

The long-term control in differentiating tissues, or in the adaptation of a microorganism such as yeast to changing environmental conditions, is likely to be based on the control of gene expression and the rate of biogenesis of mitochondria or the components of the inner membrane. In Chapter 3 a discussion of the morphology of mitochondria and of the density and shape of cristae has made reference to the likely explanation that the density of cristae reflects the capacity for respiration of the particular cells examined. One could add to this the control of the copy number of mtDNA per cell.

The present discussion will focus on short-term control. A highly readable and thoughtful review of this subject with many primary references has been written by Brown (303). It concludes with the statement that "there is no simple answer to the question 'what controls respiration?'." The question becomes increasingly more complex as one considers an isolated mitochondrion, a uniform cell population in the controlled environment of the culture medium, or an organ. The answer is likely to vary also depending on the origin of the mitochondria. Even more variations will have to be considered when one compares mitochondria from different organisms. One may also have to make a distinction between (a) studies with isolated mitochondria where extremes of conditions can be explored and (b) *in vivo* situations where the parameters cannot be so well-controlled or are not even precisely known. In the following discussion, therefore, the emphasis will be on the basic principles that will have to be taken into consideration when the question of the regulation of mitochondrial activity is raised.

A formal analysis of the problem that has led to important insights and conclusions is based on the theoretical framework now called Metabolic Control Analysis (304, 305). The "Control of Flux" was a landmark paper by

Kacser and Burns first published in 1973 and then released as a recent updated and annotated reprint (304). It begins with the emphatic statement that "flux is a systemic property and questions of its control cannot be answered by looking at one step in isolation—or even each step in isolation", and it ultimately challenges our familiar and simplifying concept of a "rate-controlling enzyme" or of a "bottleneck in the pathway." In its explicit formulation it constitutes a highly mathematical treatment seeking to express how sensitive a flux in a pathway is to an enzyme's concentration or activity. A flux control coefficient is introduced as a quantitative measure of the system's sensitivity to modulations of one of its components (enzyme activity/concentration), and it is a measure of the importance of this enzyme for control of flux through the pathway, regardless of whether and how this enzyme is controllable. A coefficient of 1.0 would indicate that an enzyme is controlling the pathway and that in practice a 1% increase in enzyme concentration/activity would lead to a 1% increase in flux. Such an enzyme would constitute the typical "bottleneck" in a pathway. Conversely, a coefficient of 0 would indicate that the enzyme is in excess and that changing its concentration fractionally would not lead to a significant change in the flux. It appears that the control coefficient for many enzymes is small, hence heterozygous individuals are normal and mutations inactivating one allele are classified as recessive. Post-transcriptional mechanisms may influence such simple considerations based on gene dose. The formal treatment also introduces elasticity coefficients for each enzyme, formerly known as the controllability coefficients, which roughly define the sensitivity of an enzyme to all the metabolites and effectors that interact with it. The reader is referred to the original literature for a formal discussion of the terminology and mathematical treatment.

While the treatment is highly theoretical and lends itself to explicit computer simulations, it is also highly suitable for the interpretation of experimental data from well-known pathways such as glycolysis and oxidative phosphorylation. Input substrate concentrations can be varied, specific inhibitors can be used to modulate the activity of individual enzymes in the pathway, and more recently it has become possible to manipulate individual enzyme levels by genetic methodologies (e.g., in yeast). On the other hand, the best computer model is useless if it is too simple to reflect the reality inside the cell.

Applications of metabolic control analysis to oxidative phosphorylation have been numerous over the years, and experimental systems have included isolated mitochondria, intact cells, perfused tissues, and whole living organisms. Recent examples by Brand and his colleagues describe (a) the metabolic control analysis of the effects of cadmium on oxidative phosphorylation in mitochondria isolated from potato tubers (306) and (b) the control of oxidative phosphorylation in liver mitochondria, hepatocytes, and other tissues (307–312). Before discussing the formalism used by these groups and the conclusions derived from this "top-down approach" (i.e., "starting from the top and analysing the whole system"), a more conventional set of considerations will be presented.

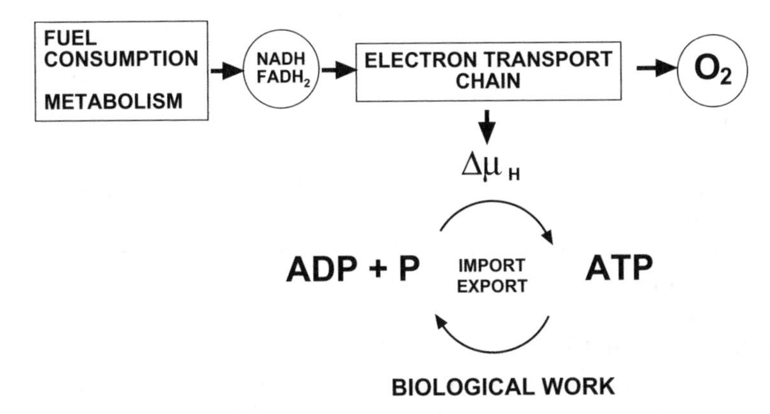

Figure 5.32 Schematic view of the major reactions in the consideration of the control of respiration and oxidative phosphorylation.

5.6.1.1 The Role of Substrates It is useful to start with a broad view of the overall problem in which the major reactions are grouped as shown in Figure 5.32. With this perspective, one can ask a few simple questions that may or may not be easy to answer, depending on the experimental system chosen. For example, some definite answers may be obtainable with isolated mitochondria, but the *in vivo* conditions may be less well known.

Based on some early work by Lardy and Chance with isolated mitochondria, the idea evolved that the major controlling reaction is the rate of consumption of ATP and hence the availability of ADP (and P_i). In other words, when oxidizable substrates (fuels) were supplied, the rate of oxygen consumption depended mainly on the supply of ADP + P_i. Failure to provide a substrate to complex V was assumed to back up the electron transport chain by mass action and because electron transport and ATP synthesis were tightly coupled. If the $\Delta\mu_H$ could not be dissipated by ATP synthases, further proton translocation and electron transport was inhibited upstream from the cytochrome oxidase. For example, the antibiotic oligomycin has been known for a long time to inhibit respiration by binding specifically to a peptide of the F_0F_1 ATP synthase and inhibiting ATP synthesis (see Section IV above). Mitochondria supplied with a fuel (e.g., β-hydroxybutyrate) and an excess of ADP + P_i are said to be in “state 3” and maximum respiration (oxygen consumption) is observed. In the absence of ADP the mitochondria are said to be in “state 4,” and oxygen consumption can be an order of magnitude lower. Under such conditions the rate of respiration is controlled by the proton leakage through other porters, namely, channels—often ill-defined. It is clear that meaningful experiments require intact mitochondria, and mitochondria that have been stored for a while or have been frozen exhibit increased proton leakage and hence less and less “tight coupling.” One way to increase respiration in the absence of ADP is to abolish the proton gradient. This can be accomplished in several ways. The addition of nigericin (a K^+/H^+ antiporter) can collapse the ΔpH without destroying the membrane potential $\Delta\Psi$. Alternatively, the classical uncouplers dinitrophenol or FCCP (fluoro carbonyl cyanide phenyl hydrazone) can collapse both ΔpH and most of the $\Delta\Psi$ by providing a path for protons across the lipid bilayer. The action of naturally occurring proteins in brown adipose tissue that cause proton leaks and thus are uncouplers will be elaborated below in the discussion of thermoregulation.

ATP consumption occurs mostly in the cytosol. The adenine nucleotide carrier/transporter (ANC/T), also known as the ATP/ADP antiport system, therefore can become a potential regulatory site. It controls the flux of ATP out of the mitochondria, controls the flux of ADP into mitochondria, and is electrogenic, hence sensitive to the membrane potential. Influx of ADP also transports a proton into the matrix and decreases ΔpH, thus stimulating electron transport. The phosphate carrier is electroneutral, and at least in liver its capacity far exceeds the requirements for ATP synthesis. It is not likely to be a regulator of OXPHOS. The structure and mechanism of these carriers will be discussed in further detail in the following chapter.

Typically the ratio of [ATP/ADP] *in vivo* is very high (~100:1 in the cytosol; is it different in the matrix?). As a result, changes in relative levels of ATP are small compared to relative changes in ADP. In addition, the ATP concentration may be buffered by the pool of creatine phosphate (depending on the tissue). Controversy has centered around the question of whether the relevant parameter is the change in [ATP], the change in [ADP], a change in the [ATP]/[ADP] ratio, or even a combination including [P_i]. A general consensus appears to favor the view that the control of respiration by ATP consumption is mainly due to changes in [ADP] inside the mitochondria. The enzyme adenylate kinase catalyzes the reaction

$$2\text{ADP} \leftrightarrow \text{ATP} + \text{AMP} \qquad (\Delta G'^0 \sim 0)$$

and under physiological conditions a small change in [ATP] is accompanied by a large (relative change in [AMP]; thus, AMP is a sensitive allosteric regulator for a number of cytosolic enzymes in intermediary metabolism (glycolysis). Another parameter that has been defined is the energy charge:

$$\text{Energy charge} = \frac{1}{2}\frac{(2[\text{ATP}]+[\text{ADP}])}{([\text{ATP}]+[\text{ADP}]+[\text{AMP}])}$$

It is a measure of what fraction of the adenylate pool is in the form of ATP. It can vary from 0 to 1, but in "healthy" cells it is typically in the range of 0.87–0.94 (313). The question is, What mechanisms maintain this parameter in such a narrow range? Other formal thermodynamic treatments have led to a "near equilibrium hypothesis" (314, 315) for the overall scheme presented above. In the old-fashioned way of thinking, the question becomes, What is the rate-limiting step? Nowadays one would want to know the flux control coefficients for each step/enzyme.

The K_m of isolated cytochrome oxidase (absence of $\Delta\mu_H$) for oxygen has been estimated to be in the micromolar range, and an apparent K_m for oxygen in isolated mitochondria has been measured to be in the same range, or even significantly lower (differing in tightly coupled mitochondria from uncoupled mitochondria). It is believed that most cells have oxygen concentrations far above these values, although steep gradients within tissues may exist or be created temporarily. Most likely, however, oxygen is not a rate-limiting substrate in mammalian tissues. Different considerations may apply when anoxia is encountered as a result of injuries. In several tissues (brain, liver, kidney), respiration has been measured under conditions when the oxygen supply was far below the physiological levels. Initial rates of oxygen consumption were normal, but a subsequent decline was observed, probably due to secondary effects.

Looking at the beginning of the scheme, one may ask whether the supply of NADH is ever a rate-limiting factor. With isolated mitochondria *in vitro*, such conditions can be created easily, but under physiological conditions *in vivo* it is less certain whether NADH is in short supply. (This statement may not apply to microorganisms such as yeasts, molds, fungi, and so on, where environmental conditions are likely to include periods of starvation.) The electron transport chain is limited by the intrinsic rate of electron transfer between the various centers, and it cannot be "pushed" by supplying excess NADH. At any rate, this substrate cannot by supplied directly. Instead, substrates such as β-hydroxybutyrate or succinate are added to isolated mitochondria; cells or tissues can be given glucose. As will be discussed elsewhere in this volume, there are elaborate feedback mechanisms that can shut down the Krebs cycle when NADH levels rise above a threshold.

The parameters discussed up till now have been the immediate substrates of oxidative phosphorylation: NADH, ADP, P_i, protons, and oxygen. So far, their influence on respiration and oxidative phosphorylation has been viewed as a mass-action effect in a system of constant cycling. NADH is re-supplied by many diverse reactions that are individually regulated at various levels, and the ADP supply appears to depend on the biological work performed by the cell. Protons exert an effect because they are substrates of vectorial reactions operating near equilibrium. One can now ask whether the specific activity of the individual complexes can be modulated by various potential effectors—Ca^{2+} ions, free fatty acids, hormones, adenine nucleotides—or even by protein phosphorylation and associated allosteric mechanisms.

Cytoplasmic calcium levels are controlled by extracellular signals such as hormones and growth factors, and most prominently by electrical signals in the case of muscle. Mitochondria have an electrogenic uniporter for import and a Ca^{2+}/Na^+ exchanger for efflux (316, 317) which keep the matrix concentration of Ca^{2+} in the range of 100 nM to 1 μM. Increases in free Ca^{2+} are thought to stimulate various dehydrogenases in the Krebs cycle; pyruvate dehydrogenase may be stimulated indirectly by Ca^{2+} by a conversion to the unphosphorylated, more active form through an activation of a phosphatase. Therefore, the NADH supply is increased, but not necessarily the rate of respiration. The effect of cytosolic Ca^{2+} on muscle activity is well known, and muscle contraction utilizes ATP. There are some observations with liver mitochondria suggesting that Ca^{2+} may be involved in changes in mitochondrial volumes following a hormonal signal (317), and this effect in turn is believed to be responsible for increased electron flow. However, alternative explanations have been offered (303).

The [ATP]/[ADP] ratio may have an influence on the relative activities of kinase/phosphatase couples controlling specific enzymes such as pyruvate dehydrogenase (see above), or either of these nucleotides may be an effector in the control of Krebs cycle enzymes. For example, ADP has been shown to be an allosteric activator of isocitrate dehydrogenase by decreasing the K_m for isocitrate. Again, the NADH supply is controlled and elevated at higher ADP concentrations. Both ATP and ADP have been shown to be allosteric effectors of cytochrome oxidase (318–321).

A special mechanism influencing the [ATP]/[ADP] ratio is the activity of so-called futile cycles. Such cycles require a substrate (e.g., fructose-6-phosphate), a specific kinase, and a specific phosphatase that may be independently regulated. By combining the two activities, a net ATPase mechanism is generated, and ATP wastage can be regulated. Increasing the rate of such a futile cycle can lead to increases in the rate of respiration.

Because cytochrome oxidase does not operate on a reaction near equilibrium, it has been considered as a bottleneck and rate-limiting enzyme in electron transport. An intriguing observation is that complex IV is the only complex with tissue-specific subunits (isozymes) (164, 322). Subunits VIa, VIIa, and VIII have been characterized to have a liver (L) form and a heart (H) form. With reconstituted cytochrome oxidase from either bovine heart or liver, the laboratory of Kadenbach has demonstrated that ADP interacts with and stimulates the heart enzyme but not the liver enzyme (323, 324). The effectiveness of intra-liposomal ADP on reconstituted COX was interpreted to indicate an interaction with the matrix domain of the enzyme, and specifically with the 17 N-terminal amino acids of subunit VIa-H. This interpretation was strengthened by the demonstration that a monoclonal antibody directed against the VIa-H but not VIa-L form can also enhance the activity of the enzyme, presumably by inducing a conformational change in the heart enzyme (318).

An even more general speculation has been proposed by Kadenbach's group: all of the nuclear-encoded subunits of complex IV may serve as targets for allosteric effectors, with no role in the catalytic function of the enzyme (324). Such a proposal is based on the comparison of prokaryotic cytochrome oxidases with eukaryotic cytochrome oxidases. In *Paracoccus denitrificans* two subunits are sufficient for activity, while in eukaryotes a variable number of additional subunits (5 in *Dictyostelium*, 6 in yeast, 10 in mammals) are found (325).

There have been claims over the years that the H^+/e^- stoichiometry of cytochrome oxidase from bovine heart is regulated by the intramitochondrial

[ATP]/[ADP] ratios (e.g., reference 326). In this case, an effector not only would regulate the activity of an enzyme, but would also control the stoichiometry of the reaction (a "slipping pump"). How many protons are pumped out of the matrix for each pair of electrons reaching oxygen? To reemphasize: A P/O ratio that is not an integer can be rationalized by the chemiosmotic hypothesis, but a variable H^+/e^- ratio for a given complex, and the reaction carried out by this complex, would be quite unorthodox. A recent statement by expert reviewers revives this idea: "cytochrome oxidase has the ability to control its own pumping efficiency . . . by 'intrinsic uncoupling' via proton backflow through the protein in response to a build-up in membrane potential . . ." (167). An alternative mechanism for inducing slippage is protein phosphorylation (see below).

Returning to the formalism of metabolic control analysis by Brand and colleagues, the simplest version considers three blocks of reactions: substrate oxidation (succinate oxidation), ATP turnover (using exogenous hexokinase as ATP consumer), and proton leakage, all connected by the common intermediate Δp, the protonmotive force, measured as $\Delta\Psi$ by the distribution (and fluorescence) of methyltriphenylphosphonium. Titrations with an inhibitor of substrate oxidation (malonate), with an inhibitor of the F_0F_1 ATPase (oligomycin), and in the presence of various uncoupler concentrations yield experimental parameters that can be used to calculate flux control coefficients over respiration rate. The analysis confirms earlier conclusions that control over respiration rate (O_2 consumption) is shared between the three blocks of reactions: There is no single, clearly outstanding rate-limiting step. When respiration is maximum, ATP turnover and substrate oxidation are about equally important (control coefficients ~0.5), while proton leakage is almost insignificant. When the respiration rate is 60% of maximum, ATP turnover dominates (control coefficient ~0.7), while substrate oxidation and proton leakage are equally but less influential. Conditions must be specified to talk about control of OXPHOS.

The authors also discuss their results in terms of coupling efficiency and P/O ratios using the same dataset. Not surprisingly, it is strongly negatively controlled by proton leakage and strongly positively controlled by ATP turnover, while it is insensitive to substrate oxidation. In other words, effectors acting only on substrate oxidation will not affect the coupling efficiency, but the efficiency can be increased by increased ATP turnover.

While the simple model introduced above is a powerful model to investigate the actions of hormones and of other effectors on OXPHOS in a given tissue ("Once you get the hang of it, both the theory and experimental applications are simple" (307)), the simplest model treats the entire electron transport chain as "one enzyme" and hence is not concerned with control coefficients of individual complexes (and mobile carriers such as cytochrome c). However, the latter problem becomes of interest in the understanding of mitochondrial diseases, when the activity of individual complexes may be affected by specific mitochondrial mutations (see Chapter 8). Korzenievski and his colleagues have elaborated the model considerably by decomposing the substrate oxidation into three reactions involving complexes I, III, and IV (327). Theoretical calculations for muscle mitochondria respiring on pyruvate were made, and control coefficients were calculated for each of the complexes (I, III, and IV) and were calculated separately for substrate dehydrogenation (pyruvate to generate NADH), complex V, the ATP/ADP carrier, the phosphate carrier, proton leakage, and ATP turnover (hexokinase in the presence of glucose). The highest flux control coefficient was calculated for complex III in state 3 mitochondria: 0.26. In other words, control is again distributed over many reactions. Under physiological conditions (state 3.5 mitochondria), control coefficients are generally less than 0.1 for the complexes I, III, IV, and V, while the coefficient for ATP turnover was 0.8. Availability of ADP thus controls the respiration rate. More significantly, the low control coefficient for complex IV (0.07 in state 3) can explain why the respiration rate is almost unaffected when the activity of complex IV is titrated with an inhibitor to the level of ~80% inhibition; at >80% inhibition, respiration declined sharply. Such a threshold effect may be highly relevant for the understanding of the physiological consequences of heteroplasmy and mitochondrial mutations (see, for example, references 311 and 328). Lastly, a reader interested in the application of the theory of non-equilibrium thermodynamics to the problem of the mitochondrial electron transport chain is referred to the treatment by Jin and Bethke (310). The authors claim that the resulting expression describes "the nonlinear dependence of flux on electrical potential gradient, its hyperbolic dependence on substrate concentration, and the inhibiting effects of reaction products."

5.6.2 The Uncoupling Proteins in Warm-Blooded Animals

The proton gradient established by electron transport can be dissipated in two ways: ATP synthesis through complex V, or proton leakage through the inner membrane without the production of ATP to be used for other chemical and biological work. By mechanisms described above, respiration may be controlled by ATP/ADP levels and turnover in tightly coupled mitochondria. As a first approximation, one can describe tightly coupled mitochondria as mitochondria in which proton leakage outside of complex V is minimized. The combustion of carbon compounds to CO_2 is also accompanied by the release of a fixed amount of heat, the inevitable byproduct of chemical reactions occurring at finite rates. The free energy change associated with the conversion of glucose to CO_2 and water is distributed between the heat released and the useful free energy temporarily stored in the ATP pool. Additional heat will be released from the hydrolysis of ATP. Thus, a mechanism is, in principle, provided to increase our body temperature above ambient temperatures.

It has been recognized for some time that a constant body temperature in warm-blooded animals is maintained not only by conducting excess heat away (perspiration, etc.), but that at low temperatures additional heat must be generated. Two mechanisms have been identified: (1) thermogenesis by shivering in effect increases ATP turnover and respiration in muscles, hence the production of heat; and (2) nonshivering thermogenesis was discovered in newborns and hibernating animals, and the primary role of brown adipose tissue was established. Brown adipocytes were subsequently characterized to possess a unique mechanism for increasing respiration without ATP synthesis (uncoupling). A specific protein was found to be induced in adipose cells, which could transfer protons across the inner mitochondrial membrane—that is, effectively short circuit the proton circuit, resulting in increased respiration and heat production. Under these conditions, UCP1 constitutes up to 10% of the membrane protein. The role of this uncoupling protein (UCP-1, or thermogenin) in brown fact in relation to development, cold adaptation, diet-induced thermogenesis, and genetic obesity has been studied extensively, and a number of authoritative reviews have been written (e.g., references 329–332). On the other hand, it also became apparent that adult animals and humans do not have significant amounts of brown adipose tissue, and hence the problem of thermoregulation in adult mammals could not be easily explained with the help of UCP-1. It was argued that in adults thermoregulation may occur in other tissues and cell types, perhaps by means of a similar mechanism involving a related protein. Two orthologous proteins, UCP-2 and UCP3, have now been discovered in mammals (333–335). In contrast to UCP-1, they are expressed in a wide variety of tissues, but at relatively low concentrations. Their physiological functions will be addressed below.

The history, arguments, and experimental evidence leading to the identification of UCP-1 in brown adipose tissue have been expertly reviewed by Nicholls and Locke (329). The presence of an uncoupler was suspected by the observation that brown adipose cells had insufficient ATPase activity (total) to account for the observed oxygen consumption coupled to ATP turnover. Free fatty acids were first thought to be involved in the uncoupling mechanism, but later a "nucleotide-induced recoupling" was added and distinguished from the "fatty acid-induced uncoupling." The binding and activity of nucleotides was a defining characteristic that led to the first isolation of UCP-1 from mitochondria in 1978. A high-affinity, saturable binding site for [3H]-GDP on the outer surface of the inner membrane could be demonstrated to be identical to a component in the inner membrane responsible for short circuiting the membrane for protons—that is, with activity as a proton uniporter.

A small protein of molecular mass 33 kDa was found to be present exclusively in brown adipose tissue from a variety of mammals, accounting for 6–14% of the inner membrane protein. Following the discovery by Nicholls and his group, significant contributions to the characterization of the protein and its activity have been made by Klingenberg and co-workers (see references 336, 337, and 338) for recent publications with references to past work). It was shown that H^+ transport requires free fatty acids as obligatory cofactors and is inhibited by adenosine and guanine di- or triphosphates. This "cofactor" model has since been challenged by some investigators. A recent study with highly purified UCP1 in reconstituted liposomes yielded results supporting a model in which long-chain fatty acid anions are "flipped" across the membrane. This flippase model proposes that fatty acid anions are transported by the UCP1, and after protonation they can "diffuse" back through the lipid bilayer as the uncharged form (339). Nucleotide binding affinity decreases with pH. The pH sensor for nucleotide binding has been identified and has been incorporated into a model for the orientation of the UCP transmembrane helices in the membrane (337). The UCPs share homology with the ADP/ATP transporter (antiporter) and many other members of the mitochondrial carrier family (336, 340). Both proteins bind nucleotides, but in one case the nucleotide is a substrate capable of binding to either side, while in the other it is a regulator binding only the cytosolic side. Originally thought to form functional dimers in the inner membrane, with a single binding site per dimer, more recently the monomer has been favored as the functional form of the carrier. The entire carrier family is characterized by being relatively short peptides with three repeat structures of ~100 residues, each repeat containing two transmembrane helices. A monomer therefore has six transmembrane helices; the N-terminal and C-terminal domains extend into the intermembrane space. The connecting loops on the matrix side are made up of ~40 highly polar residues. A further discussion of the ATP/ADP transporter can be found in Chapter 6.

Uncoupling by UCP-1 can be regulated on at least two levels. The carrier itself is sensitive to the proton gradient and the pH in the intermembrane space as well as to nucleotide concentrations and fatty acids, and such mechanisms lend themselves to rapid short-term control. A plausible signaling pathway has norepinephrine bind to the β-receptor, causing a rise in cAMP, activation of a protein kinase which controls lipolysis (by a lipase), and eventually accumulation of free fatty acids. Nucleotide inhibition is released, and oxidation of acyl carnitine and acyl-CoA can procede at an increased rate (329). There is, however, also a long-term control of thermogenic capacity related to the dietary status of an animal such as a rat, or its adaptation to prolonged exposure to cold. The correlation of thermogenic capacity with the levels of the UCP-1, along with β-adrenergic stimulation of brown adipocyte metabolism and respiration, has been the subject that has received much attention. Nicholls and Locke (329) present the topic from a more physiological perspective. In the meantime, the cloning of the UCP-1 gene has permitted an analysis of its promoter and the control of its expression. An up-to-date review (330) lists thyroid hormone and retinoic acid response elements as having a direct influence, while catecholamines act via a cAMP pathway. Insulin, glucocorticoids, and IGF-I may modulate gene expression via more indirect pathways, and CCAAT enhancer binding proteins and peroxisome proliferation activator receptor γ2 have also been implicated in control. UCP-1 mRNA appears to have a short half-life, permitting rapid fluctuations in UCP-1 mRNA steady-state levels depending on the ambient temperature, food intake, and hormone release. By regulating the level of these proteins by any one of a number of mechanisms, it therefore becomes possible to fine-tune the proton leakage in mitochondria and, as a result, uncouple respiration and heat production from ATP synthesis.

After the cloning of the mammalian gene(s) for the UCPs, it became possible to express these proteins in yeast mitochondria to study their function, and even in bacteria to produce larger quantities for biochemical studies in liposomes. Various inconsistencies in the results obtained appear to be due to differences in codon usage in these organisms, leading to proteins that cannot be converted to the functionally native state (338). Thus, the recent studies of Breen et al. (341) employed a strain of *E. coli* (Rosetta), which uses human codons to produce an authentic rat/mammalian protein in bacteria for purification and reconstitution into liposomes.

The mammalian UCP1 in yeast was demonstrated to retain its properties: proton and chloride transport, high-affinity binding of nucleotides, and the dependence of conductance on nucleotides and fatty acids (342). Site-directed mutagensis was employed to investigate systematically the functional role of various amino acid side chains. Chemical modifications had already implicated several cysteine residues in activity, and models based on dithiol–disulfide interconversions had been proposed. The replacement of all seven cysteine residues by serine was subsequently found to have no effect on activity or regulatory characteristis of the UCP-1 in yeast (342). Studies in yeast have also addressed the problem of how these proteins are imported into mitochondria and oriented in the inner membrane. They lack a removable signal sequence. Instead, the first loop between transmembrane helices 1 and 2 acts a signal sequence for import, and the second matrix loop is responsible for insertion into the inner membrane (343).

The physiological function of the other mitochondrial uncoupling proteins, UCP2, UCP3, . . . is still subject to debate, although some consensus is emerging. A comprehensive and expert recent review (335) refers the reader to more than a dozen other reviews that have appeared in the past five years. Although these orthologues have been found in several mammalian tissues (as well as in other animlas and plants), their abundance is low (~0.01–0.1% of the membrane proteins). Their tertiary structure in the inner membrane is similar to that of UCP1. Investigators agree that these carriers do not transport protons except in the presence of specific activators. *In vitro* studies revealed that purine nucleotides are inhibitory and fatty acids are probably required for activation. Another prominent group of activators includes reactive oxygen species (ROS) and free-radical-derived alkenals (335). The same controversy exists about the mechanism of proton transport/conductance. On the one hand, there is the view that they act as gated proton channels, and an alternative mechanism proposes fatty acid cycling back and forth across the membrane, as suggested for UCP1 (339). Mice in which UCP2 or UCP3 are knocked out are not severely affected in the laboratory. It may be amusing ("uncoupling the agony from the ecstacy") to learn that mice lacking UCP3 become more tolerant to MDMA (3,4-methylenedioxymethamphetamine, nicknamed "ecstasy") (344). The recreational use of amphetamine-type stimulants by humans can produce a marked and sometimes lethal increase in body temperature.

A succinct formulation and critical discussion of the various functions for UCP2 and UCP3 can be found in the reviews by Brand and Esteves (335, 345). These authors consider four major roles for UCP2/UCP3: (1) a role in thermogenesis and protection against obesity; (2) control of mitochondrial ROS production and protection against oxidative damage; (3) mediation of ROS signaling and insulin secretion; and (4) transport or export of fatty acids and fatty acid peroxides. These mechanisms are clearly not exclusive of each other (although the issue of proton channel vs fatty acid "flipping" remains to be resolved). Their relationship to UCP1, along with the role of UCP1 in brown adipose tissue, invited initial speculation about their role in thermogenesis, but there is agreement that gross thermoregulation does not depend on UCP2/UCP3. The most convincing support for this conclusion is derived from UCP2 and UCP3 knockout mice that have normal responses to exposure to low temperatures (and are not obese). The reaction of UCP3 knockout mice to MDMA (ecstacy) described above nevertheless does not allow a complete disregard for UCP3 in thermoregulation. Another argument has been made from observations in birds: They lack brown adipose tissue and UCP1, but have a UCP (avUCP) that is 70% identical in sequence to mammamlian UCP2/UCP3. The regulation and properties of this avUCP are highly suggestive of a role in thermogenesis (e.g., their level and proton conductance are up-regulated when penguins are immersed in cold water). The connection of UCP activity to obesity is obvious in a superficial sense. Normal food intake, lack of exercise, and tightly coupled mitochondria would be expected to lead to enhanced fatty acid biosynthesis.

The second postulated role of UCPs invokes their capacity for uncoupling for the control of the mitochondrial membrane potential ($\Delta\Psi$) and the generation of reactive oxygen species (ROS). They act as safety valves preventing the buildup of an excessive potential that is believed to cause excessive ROS production. Thus, they protect us against ROS-induced damage that may lead to neurodegenerative diseases, diabetes, and aging (346). What is the pathway of UCP activation? Brand and colleagues have proposed a detailed pathway in which superoxide or the protonated hydroperoxyl radical initiates a series of reactions involving poly-unsaturated fatty acyl groups and the ultimate production of hydroxynonenal. GDP-sensitive UCP activation has been shown to result from exposure to superoxide or hydroxynonenal. It may also be considered that a rise in membrane potential above a threshold activates the UCPs directly, without the direct participation of any ROS.

UCP2 knockout mice have been shown to have elevated ATP levels in pancreatic islet cells. These cells also exhibit increased insulin secretion stimulated by glucose. Such a finding can be incorporated into a widely accepted model for glucose-stimulated insulin secretion. For a detailed discussion of this model the reader must consult the specialized literature (see references 335 for references).

Fatty acid transport is invoked to provide a mechanism for exporting free fatty acid anions from the matrix, if the supply exceeds the capacity for oxidation. Alternatively, excess fatty acids "in the cytosol" can be protonated and "flipped" into mitochondria. These simple explanations still lack convincing experimental support, and they rely on the postulated capacity of UCPs to transport fatty acids as opposed to being activated by fatty acids. It is also not clear what is meant by excess and free (?) fatty acids. Normal import of fatty acids into mitochondria occurs via the well-known carnitine shuttle.

In summary, there are plausible suggestions for the functions of UCP2/UCP3 in human physiology and pathology, but in view of the relatively minor phenotypes shown by UCP2 or UCP3 knockout mice, the effects in humans may also be subtle, and it remains to be seen whether the treatment of obesity or a significant protection from ROS damage by activation of UCPs can be achieved.

5.6.3 Uncoupling in Other Organisms

***5.6.3.1 In* Saccharomyces cerevisiae** The growth of yeast on a variety of carbon sources (glucose, glycerol, acetate, ethanol) requires a large number of adjustments to optimize growth rates under conditions that may be constitute a carbon limitation or an energy limitation. Theoretical considerations of the need to maintain redox balance, along with the observed discrepancy between ATP requirements for biomass formation with expected yields of ATP production from substrate oxidation, has suggested that oxidative processes uncoupled from ATP production must be operative (347). Among the mechanisms considered have been futile cycles (see above) and increased proton leakage. In a series of papers, Prieto et al. (347–349) have provided details on the activation of a proton-conductance pathway by ATP. At low [ATP]/[Pi] ratios, binding of ATP to cytochrome oxidase alters its kinetic properties; but at higher ratios, ATP binding to the outer side of the inner membrane can increase membrane permeability and cause the collapse of ΔpH or $\Delta\Psi$ (348).

Understanding the properties of the endogenous ATP-induced proton channel in yeast mitochondria was an important prerequisite for the successful expression and study of the mammalian UCP-1 in yeast (342). In the presence of phosphate, the two uncouplers could be clearly distinguished. One can also note that the identity of the newly isolated UCP-2 gene product was confirmed by its expression in yeast, where its activity as an uncoupler was even more pronounced than that of UCP-1 (333).

5.6.3.2 In Plants A previous section has emphasized that plant mitochondria have all of the components found in animal mitochondria, making oxidative phosphorylation in plant mitochondria a very similar process. At the same time, plant mitochondria have several unique features, most notably a cyanide-insensitive electron pathway not coupled to phosphorylation. Therefore, there appears to be no need for a mechanism requiring a specific uncoupling protein, when oxidative reactions have to be speeded up at certain developmental stages, in specific tissues, or under defined physiological conditions. However, a first report on the cloning of a potential uncoupling protein from a potato flower has recently appeared (350). It is 44% and 47% identical to the human UCP1 and UCP2, respectively, and is induced after exposure to cold temperatures. Although there is still uncertainty and controversy about the physiological significance of cyanide-resistant respiration (see above), there is one class of examples where the operation of this uncoupled process for the apparently sole purpose of heat production is understood in physiological and functional terms. In the male reproductive structure of cycads and in the flowers or inflorescences of species belonging to the families *Annonaceae, Araceae, Aristolochiaceae, Cyclanthaceae,* and *Nympheaceae*, an accumulated supply of starch is metabolized (glycolysis) and oxidized by the alternate pathway in at times a matter of hours. The temperature in the structures involved can reach 15°C above the ambient temperature. This thermogenesis is useful in at least two ways. In some instances it simply permits the plant to flower under climactically adverse conditions—that is, early in a season; in other examples it has been clearly established that the elevated temperatures help to volatilize insect attractants (e.g, the putrescent odor of skunk cabbage) to gain the attention of pollinating insects (191, 193, 351).

5.7 REACTIVE OXYGEN SPECIES

Entering the keyword ROS (reactive oxygen species) in a PubMed search leads to the recovery of ~13,600 citations (October 2006). Of these, ~2200 papers were published in 2006. ROS can be blamed for a lot of problems, and they may have some positive functions (352). Oxygen sensing itself has been proposed to depend on mitochondria-generated reactive oxygen species (ROS) but independent of oxidative phosphorylation (353, 354). We know what they are. We have a good understanding of how and where they are produced: as an inevitable byproduct of respiration in mitochondria, but in lesser amounts also at other sites in a cell. However, knowing that some of them are very reactive and potentially destructive, we do not yet understand very well in detail how they exert their pathological effects. An expert and comprehensive treatment on free radicals in biology and medicine can be found in the book by Halliwell and Gutteridge (355). It is clear that no attempt can be made here to summarize the content of approximately 900 pages of this book, let alone the additional work from the past 7–8 years. There is progress and optimism (356).

The major reactive oxygen species (ROS) are the superoxide radical, $O_2^{\cdot-}$, hydrogen peroxide, H_2O_2, and the hydroxyl radical, $OH^{\cdot}$. When protonated, the superoxide radical becomes a hydroperoxyl radical ($pK_a = 4.8$). In fact, the only very reactive species among them is the hydroxyl radical, which is believed to be responsible for most of the actual damage done to biological macromolecules. It is so reactive that damage is restricted to a small radius limited by diffusion of the radical. The risk posed by the other two species comes primarily from their spread and subsequent reactions, which ultimately lead to the formation of the hydroxyl radical. Attempts to react the superoxide in aqueous solution directly with biologically relevant molecules have been frustrated. Therefore, the proposal has been made that a series of reactions ultimately result in hydroxyl radical formation (357, 358):

$$2\,O_2^{\cdot-} + 2\,H^+ \Rightarrow O_2 + H_2O_2 \qquad \text{(reaction 1)}$$

$$O_2^{\cdot-} + Fe(III) \Rightarrow O_2 + Fe(II) \qquad \text{(reaction 2)}$$

$$Fe(II) + H_2O_2 \Rightarrow Fe(III) + OH^- + OH^{\cdot} \qquad \text{(reaction 3)}$$

Summation of reactions 2 and 3 yields

$$O_2^{\cdot-} + H_2O_2 + H^+ \Rightarrow O_2 + H_2O + OH^{\cdot} \qquad \text{(reaction 4)}$$

Reactions 2 and 3 have been postulated as intermediate reactions because reaction 4 has been shown to depend on the presence of redox metals. Reaction 3 is the Fenton reaction, and it emphasizes again that hydrogen peroxide is by itself less reactive; but upon encountering the ubiquitous metal ions in a cell, it generates the highly reactive hydroxyl radical. Instead of Fe(II), reaction 3 can also take place with Cu(I) as the reducing agent. It should be noted that ligands of these metal ions can affect the reaction rate significantly (355).

Oxygen is a true substrate of complex IV of the mitochondrial electron transport chain, and naively it may be assumed that it is the site in a cell where oxygen can be activated to the superoxide radical. Instead, it is widely accepted that electrons "leaking" from the electron transport chain to oxygen at high potential upstream sites are responsible for the majority of ROS produced in mitochondria (358–361). Thus, complexes I ((362, 363) and III (364, 365), and recently complex II (366, 367), have been implicated in the generation of ROS. The proposed mechanism has mutated over the years. In consideration of the small size and zero charge of the oxygen molecule, it can potentially penetrate deeply into protein domains of these complexes. But where does the errant electron originate from? Among the suspects are the free radical semiquinone intermediate of ubiquinone (coenzyme Q) and the free radical intermediates formed by the flavin moieties in complex I (and II). Alternatively, it has been proposed that electrons may escape from the reduced [Fe–S] center N2 (368). It is quite likely that a semiquinone intermediate of the Q cycle is the source of electrons in ROS production by complex III (364). Recent experiments and arguments in favor of complex I as a major source of ROS tend to favor the flavin intermediate as the donor (366, 369, 370). McLennan and Degli also implicate complex II, which had long been ignored as a source of ROS.

The superoxide radical is also generated "accidentally" by reactions of oxygen at microsomal cytochromes P450, as well as by less well defined or understood reactions (auto-oxidation reactions) involving catecholamines, ascorbic acid, and reduced flavins, often with the participation of a metal ion. The toxicity of such compounds has been exploited in some physiologically useful reactions. The best-known example is the generation of superoxide, radicals, hydrogen peroxide, and HOCl by activated macrophages, presumably to kill their target cells (bacteria, fungi, etc.) before phagocytosis. Several peroxisomal oxidases generate hydrogen peroxide—for example, D-amino acid oxidase, xanthine oxidase, and the dehydrogenase in the first step of very long chain fatty acid oxidation. In some systems the superoxide radical has been proposed as an intercellular signal. In a more indirect way, the $O_2^{\cdot-}$ ion can remove nitric oxide, another free radical, and an intercellular messenger in neurotransmission, in the immune response, and in the regulation of smooth muscle contraction (see below). Finally, reactive oxygen species have been assigned a central role in the mechanisms of apoptosis (see Chapter 7), with arguments still unresolved about whether they are responsible for signaling and cell killing or whether their increased generation is the consequence of another damaging event affecting mitochondria and perhaps other cellular constituents.

A variety of protective enzymes have evolved to guard against damage by ROS, or even to repair damage (e.g., in DNA) resulting from their reactivity. Present-day aerobic organisms presumably arose from anaerobes by evolving antioxidant defense mechanisms among others, as the oxygen content of the atmosphere on earth increased to the present 21%. The existence of antioxidant defense mechanisms in aerobic organisms is therefore crucial for survival or longevity. Activity in the field concerned with ROS and antioxidants continues to be high, because many fundamental questions remain unanswered (359).

There are various controls to maintain the mitochondrial membrane potential below dangerous levels by uncoupling proteins and other mechanisms to prevent leakage of electrons from high potential sites to oxygen. Another protective mechanism in cells is to sequester iron and copper ions in such a way that they cannot react readily with superoxide or hydrogen peroxide. Thus, availability and location of "free" metal ions in a cell thus can determine the site of damage, because the hydroxyl radical formed is too short-lived to diffuse far from its site of origin. For example, metal ions bound to DNA can initiate free radical damage in the nucleus. The discovery of superoxide dismutase (SOD) by McCord and Fridovich (371) represented a major beakthrough in the field. These enzymes greatly accelerate the dismutation reaction:

$$2\,O_2^{\cdot-} + 2\,H^+ \rightarrow H_2O_2 + O_2$$

Catalases (see below) then convert the hydrogen peroxide to water and oxygen. Two SODs have since been characterized in some detail. A cytosolic form (SOD1) is distinguished by having copper and zinc at the active site

(Cu,Zn-SOD), while a mitochondrial form (SOD2) contains manganese (Mn-SOD). SOD1 may also be localized in the IMS. The cytosolic SOD is encoded by a gene on human chromosome 21, and tissue levels of Cu,Zn-SOD are elevated ~50% in Down's syndrome patients. Since transgenic mice overexpressing human SOD also have some neuromuscular abnormalities reminiscent of Down's syndrome, it has been concluded that too much of this protective enzyme can also be deleterious. At the same time, certain mutated forms of SOD1 are associated with amyotrophic lateral sclerosis (ALS) (372). It remains puzzling why such mutant SOD1 proteins affect primarily the mitochondrial capacity of motoneurons (373–375). A further discussion of the role of SOD and other scavenging enzymes in neurodegenerative diseases and in aging will be deferred to Chapter 7.

Catalases are enzymes defining peroxisomes, where they remove hydrogen peroxide generated by peroxisomal oxidations. However, the bulk of the H_2O_2 in mammalian cells is removed by glutathione peroxidases. A selenocysteine residue is required at their active site, where H_2O_2 is used to oxidize glutathione (GSH) to oxidized glutathione (GSSG). GSH can be regenerated from GSSG by the oxidation of NADPH, catalyzed by glutathione reductase. Thus, H_2O_2 levels can influence the mitochondrial ratio of GSH/GSSG (a measure of the redox potential), which in turn may control sulfhydryl and disulfide bond formation and hence activity of certain key enzymes.

It is generally considered that the generation of ROS and the activity of antioxidant defenses may have to be carefully balanced *in vivo*. A very slight imbalance in favor of ROS may be responsible for normal aging. A more severe imbalance is said to result in oxidative stress. Oxidative stress may result in the induction of antioxidant defense enzymes and mechanisms, but beyond a certain level it can increase intracellular calcium levels and free iron levels, as well as mtDNA mutations, escalating the problem to catastrophic proportions. A major unsolved and still controversial issue is whether increased ROS production is a primary consequence of mitochondrial mutations and dysfunction, or whether a primary defect in ROS scavenging activity is responsible for an abnormal respiratory activity. Clearly, specific mutations in structural genes for subunits of the electron transport complexes could increase the resistance to electron flow and hence cause the escape of electrons from flavin or semiquinone radicals. However, the electron leakage is likely to occur upstream from the mutated complex; in other words, the mutation is not directly responsible for electron leakage from that complex. For example, isolated bc1 complexes from bovine heart and from wild-type and mutant yeast strains (in which the midpoint potentials of the cytochrome b hemes and the Rieske iron–sulfur cluster were altered) were examined by Sun and Trumpower (376). In the absence of inhibitors, about 3–5% of the electrons escaped to oxygen to form superoxide anion instead of reducing cytochrome c. The authors concluded that changes in the midpoint potentials of the redox components that accept electrons during ubiquinol oxidation may slow down the activity of the enzyme but have only small effects on the rate of superoxide formation. It should also be noted that stigmatellin, binding at the ubiquinol oxidation site of the bc1 complex, prevented superoxide formation by the isolated complex. When the bc1 complex is inhibited by antimycin in intact mitochondria, increased ROS production is probably due to a back-up and leakage from complex I. Specific inhibitors of the complexes of the electron transport chain have been shown to cause increased ROS production. When rotenone occupies the quinone binding site of complex I, the flavin semiquinone is likely to become an electron donor. Similarly, the complex II inhibitor alpha-tocopheryl succinate has been shown to stimulate ROS production followed by apoptosis (367). The rate of ROS production is increased when oligomycin is used to inhibit the ATP synthase. This inhibition hyperpolarizes the membrane. An increased $\Delta\Psi$ will effectively increase the resistance to electron flow in the electron transport chain, stalling electrons at upstream sites from which they can escape to oxygen.

Reports on the relationship of ROS production to apoptosis are too numerous to summarize here (see also Section 7.4). Are they a signal to, or a side effect of, the apoptosis pathway? To cite just one example (377): Depriving neurons of nerve growth factor triggers apoptosis by Bax-induced cytochrome c release and activation of a caspase cascade. The authors claim that "a Bax-dependent increase in mitochondrial-derived ROS is an important component of the apoptotic cascade." How does Bax induce ROS production? Does the loss of cytochrome c lead to membrane hyperpolarization? Are the ROS needed to promote further steps in the progression of apoptosis, or is their appearance incidental?

A very clever and stimulating experimental model system has recently been reported by two groups (378–380). An error-prone DNA polymerase γ in a mouse model was shown to increase mutations in mtDNA quite dramatically, with the result that the life span of such mice was shortened significantly. At the same time, there was no apparent increase in ROS production, and the typical symptoms of oxidative stress were not detectable (379, 380). These results argue against an "error catastrophe" hypothesis that postulates that a steady increase in mtDNA mutations would in turn lead to increased oxidative stress (ROS production) causing the accelerated aging process.

The whole field of mitochondria, oxidative stress, and disease deserves a book of its own. It continues to be a challenge in many pathological cases to establish clear cause-and-effect relationships: (1) What is the primary cause or mechanism resulting in oxidative stress? (2) If oxidative stress means increased levels of reactive oxygen species, what are the targets: DNA, proteins, or lipids, or all of the above? (3) What kind of damage (faulty protein) can then act "catalytically" and increase the rate of accumulation of further damage by ROS-independent mechanisms? From a practical point of view, there are several challenges. The hydroxyl radical is too short-lived to be measured. Hydrogen peroxide can be measured in a cellular extract, but such an assay obscures the intracellular localization of this ROS. Superoxide anions are also too short-lived to be determined in extracts, but their production can be assessed *in situ* in cultured cells by introducing a reagent (e.g., hydroethidine) that will be oxidized by the superoxide to produce a fluorescent product. In such an assay, one measures a rate of accumulation of fluorescent product (i.e., the rate of superoxide production) and not its steady-state level. A precise subcellular localization of the source of $O_2^{-\cdot}$ production represents an additional challenge.

From a medical point of view, there have been numerous efforts to reduce mitochondrial oxidative damage by targeting antioxidants specifically to mitochondria. The general strategy is to use lipophilic, membrane-permeable cations (e.g., hexyltriphenylphosphonium cation) similar to those used for staining mitochondria to promote uptake and accumulation in the organelle. These cations are linked covalently to organic molecules that can act as free radical scavengers (MitoQ, MitoE, MitoBPN, etc.). Pioneering work has been done by Murphy (381–383), and a recent review by Reddy (384) can also be consulted for entry into the relevant literature.

5.8 NITRIC OXIDE (NO)

In 1987 the first physiological function for NO (nitric oxide) was identified as a vasodilator of blood vessels, and in 1992 it gained the distinction of being declared the "molecule of the year" in recognition of its many diverse and important roles. The original discovery earned a Nobel Prize for three out of the four principal early investigators. In the meantime, NO has been shown to act also on mitochondria in very significant direct and indirect ways that will be summarized briefly in the following paragraphs.

Nitric oxide is formed by the enzyme nitric oxide synthase (NOS) using arginine as a substrate and producing NO and citrulline. Required cofactors are: tetrahydrobiopterin (BH4), FAD, FMN, NADPH, and oxygen. NOS constitute a family of enzymes comprised of neuronal NOS (*n*NOS or NOS-1), inducible NOS (iNOS or NOS-2), and endothelial NOS (eNOS or NOS-3). There has been a long-lasting controversy about the existence of a distinct mitochondrial NOS (mNOS). One authoritative and exhaustive review discusses the contentious issues regarding mNOS and comes to a negative conclusion (385), although publications affirming the existence of a distinct mNOS also continue to appear (e.g., 386–388). Support for a mitochondrial NOS has been provided in plants (Arabidopsis) by Guo and Crawford (389). These authors made a NOS-GFP fusion protein that could be localized convincingly in mitochondria. Sequence analyses of eukaryotic genomes have identified unique genes for the three isoforms of NOS (NOS-1, NOS-2, NOS-3) and failed to find another that could encode a mitochondrial isoform. NOS copurified with mitochondria (contaminant?) "looks like" nNOS (386), but this problem has not yet been definitely settled. Colocalization by immunocytology cannot distinguish whether the detected NOS is in the inside (matrix) or bound on the outside. Post-translational modifications such as acylation have been reported (386) which may explain the attachment to mitochondrial membranes, but do not resolve the problem of localization. Various stimuli are known to affect the expression of the NOS isoforms, but no unique and specific physiological signal has been discovered targeting exclusively mNOS expression. As Giulivi et al. (388) put it, "NO production in mitochondria is spatially regulated by mechanisms that determine subcellular localization of mtNOS, likely acylation and protein-protein interactions, in addition to transcriptional regulation as nNOS." They also claim that NO rapidly decays in mitochondria. On the other hand, it was recently reported that mNOS activity is regulated by the membrane potential; that is, mNOS is a voltage-dependent enzyme (387). Brookes (385) also addresses various other relevant issues including substrate availability, with a special focus on the intersection of the reactions catalyzed by NOS with reactions of the urea cycle, and the conditions in the matrix (pH and pO_2).

Why the high level of interest? What are the actions of NO on mitochondria? At a physiological level, NO regulates blood flow and hence the supply of substrates to mitochondria. It has also been proposed that NO directly regulates the affinity of oxygen to hemoglobin and thus the availability of oxygen. These mechanisms will not be further discussed here.

NO is an inhibitor of cytochrome oxidase (COX), and a regulated synthesis of NO in the matrix could be a potential mechanism for controlling respiration. The most direct action is the competition of NO for the oxygen binding site on COX (complex IV) (390, 391). However, as pointed out by Brookes, there is a paradox: At low $[O_2]$, NO can compete with oxygen at the binuclear site of COX, but this $[O_2]$ is too low for NOS to produce NO (if the reported K_m of mNOS for O_2 is correct). When $[O_2]$ is high enough for NOS activity, it will also outcompete NO at the COX binding site. Thus, it is likely that if NO is to play a role in the control of COX activity, it will have to be produced by enzymes outside of the mitochondria (with lower K_m values for O_2).

The inhibition is freely reversible and dependent on the relative concentrations of these two gases in the matrix. Competitive inhibition has been demonstrated with the isolated enzyme, with mitochondria, nerve terminals, cells in culture, and tissues. NO effectively increases the apparent K_m of the enzyme for oxygen, and the NO/COX system has been proposed to serve as the acute oxygen sensing system in cells (392). Difficulties in measuring NO concentrations *in vivo* have made it problematic to state whether NO concentrations reach the level required for effective inhibition (see reference 391 for further discussion). A mathematical modeling approach by Antunes et al. (393) leads to the suggestion that there is a "moderate, but not excessive, overlap between the roles of NO in COX inhibition and in cellular signaling" under normal conditions, but under conditions of COX deficiency a significant inhibition of COX activity/respiration is predicted. A recent reevaluation of the interaction of NO with COX has concluded that NO interacts either with ferrous heme (competitive) or with the oxidized copper (noncompetitive) where it is reduced to nitrite (394). In this context an interesting recent study reports that COX (from yeast and rat liver) can produce NO from nitrite under hypoxic conditions (395). These authors therefore suggest that mitochondrially produced NO is responsible for hypoxic signaling.

It should be noted that NO is a hydrophobic gaseous free radical that is relatively stable and capable of diffusing across large distances within a cell and even between cells. It binds to ferrous ions within heme proteins (COX, hemoglobin, and guanylate cyclase) and reacts rapidly with superoxide to produce peroxynitrite. The latter reacts readily with thiol groups on proteins (or glutathione) to produce nitrosothiols (391). The known cytotoxicity of NO may in large part result from its conversion to peroxynitrite. It is assumed that such covalent modifications are generally irreversible and hence may lead to irreversible inhibition of respiratory complexes (396). A discussion of the consequences of the modification of thiol groups in mitochondrial proteins is a subject by itself and the focus of considerable attention (383, 397–399).

The interactions of NO with mitochondria and COX considered so far describe short-term and direct effects. A long-term effect has now become apparent that may surpass the short-term effects in significance. In a diverse group of cells (brown adipocytes, 3T3, HeLa), exposure to the NO donor *S*-nitrosoacetyl penicillamine (SNAP) caused a three- to sixfold increase in mtDNA/cell and an increase in the biogenesis of functionally active mitochondria (390, 400). NO was found to activate guanylate cyclase and hence elevate cGMP levels. This signal induced the peroxisome proliferator-activated receptor γ coactivator 1α (PGC-1α) and PGC-1β, followed by expression of the nuclear respiration factors NRF-1 and NRF-2. These transcription factors play crucial roles in the transcription of nuclear genes encoding subunits of the electron transfer chain complexes and the mitochondrial transcription factor A (Tfam alias mtTFA) (401–403). Details on the mechanisms remain to be elucidated, but a very plausible signaling pathway now includes eNOS as a master regulator of mitochondrial biogenesis, with important implications for understanding physiological phenomena such as thermogenesis, obesity, diabetes, response to aerobic exercise, calorie restriction, senescence, and more. In this context the emphasis so far has been on eNOS, since in experiments with eNOS-negative (–/–) mice the anticipated responses to calorie restriction were not observed (404).

5.9 THE ROLE OF SPECIFIC LIPIDS

Over the years the purification of various complexes from the electron transport chain or complex V, and also of individual carriers such as the nucleotide transporter, has yielded preparations that not unexpectedly contained various lipids, since efforts were made to retain biological activity and native conformations. As higher-resolution structures were obtained by X-ray crystallography, specific lipids were consistently localized in association with these complexes, suggesting more than a simple, random role in the integration of the protein complex in a lipid bilayer. Again and again, such preparations contained cardiolipin (e.g., 405–408). Cardiolipin has become the "signature lipid of mitochondria" (see Section 6.6), and it became "conventional wisdom" to postulate that cardiolipin may be essential for either the assembly of these complexes in the membrane or their maintenance in their functional conformation (e.g., 409, 410).

An experimental test for this idea became possible when genetic approaches were used to eliminate cardiolipin by mutating the specific enzymes required for its biosynthesis (see Section 6.6). In mammalian cells in tissue culture, this has been achieved by Ohtsuka et al. (411, 412), who isolated Chinese hamster mutants with a defect in phosphotidylglycerophosphate synthase. Such cells had depleted cardiolipin levels and exhibited mitochondrial dysfunction. Complex I appeared to be the most seriously affected. In contrast, the initial studies with a yeast mutant containing a defective cardiolipin synthase gene, *cls1*, gave the surprising result that such mutants were viable and capable of respiration; that is, they grew in the presence of either nonfermentable or fermentable carbon sources. These molecular–genetic studies in yeast have now been greatly expanded. The relevant enzymes and genes required for the biosynthesis have been identified to be used for genetic manipulations (413). It emerges that the biosynthesis of cardiolipin is highly regulated by several mechanisms. Some of these could be expected to be linked to mitochondrial biogenesis in general, but there are also specific controls mediated by inositol and by the mitochondrial pH (414, 415). Thus, mutant mitochondria with a defective assembly of the electron transport chain exhibit decreased cardiolipin synthesis. While cardiolipin may not be absolutely essential for the assembly and activity of the electron transport chain, experiments suggest that it plays a role in stabilizing respiratory chain supercomplexes (36, 416). Needless to say, its implication in the process of apoptosis has also been investigated (417–419). One hypothesis is that the oxidation of cardiolipin by ROS reduces cytochrome c binding to the inner membrane and thus enhances cytochrome c release.

A combination of HPLC, mass spectroscopy, and selective enzymatic cleavage has shown that cardiolipin from different organisms and different tissues contains only one or two types of fatty acids (out of four per molecule of CL), suggesting a "high degree of structural uniformity and molecular symmetry" (420). Is this critical? One affirmative answer comes from the elucidation of the defect in the Barth syndrome in humans, an X-linked cardio- and skeletal myopathy (415, 416, 421–423). Progress in our understanding of the molecular basis for this syndrome came from the identification/cloning of the human gene (TAZ, encoding tafazzin), followed by the finding, characterization, and manipulation of homologous genes in yeast and *Drosophila*. As discussed further in Section 6.6, cardiolipin is synthesized *de novo*, but is then "remodeled" to contain an increased proportion of unsaturated fatty acids (e.g., linoleic acid). Tafazzin is required for this remodeling of cardiolipin. It has been characterized as a 1-palmitoyl-2-[14C]linoleoyl-phosphatidylcholine: monolysocardiolipin linoleoyltransferase (424). *Drosophila melanogaster* mutant flies defective in the tafazzin gene exhibited reduced locomotor activity, and their mitochondrial cristae were morphologically abnormal in their indirect flight muscles (425).

A more general and authoritative review of the role of lipids in determining the structure and function of integral membrane proteins and protein complexes has been written by Palsdottir and Hunte (426). It classifies lipids as annular, nonannular, and integral protein lipids and discusses in detail how lipids influence the vertical positioning of protein complexes, the stabilization of oligomeric and supercomplexes, and possible functional roles. The reader will find a wealth of information on basic principles governing the structure of biological membranes.

REFERENCES

1. Keilin, D. (1966) *The History of Cell Respiration and Cytochrome*, 1st ed., Cambridge University Press, New York.

1a. Oertmann (1887)

2. Fruton, J. S. (1972) *Molecules and Life: Historical Essays on the Interplay of Chemistry and Biology*, 1st ed., Wiley-Interscience, New York.
3. Fruton, J. S. (1999) *Proteins, Enzymes and Genes: The Interplay of Chemistry and Biology*, 1st ed., Yale University Press, New Haven, CT.
4. Crane, F. L., Hatefi, Y., Lester, R. L., and Widmer, C. (1957) *Biochim. Biophys. Acta* **25**, 220–221.
5. Hatefi, Y. (1963) *Adv. Enzymol.* **25**, 275–328.
6. Wainio, W. W. (1970) *The Mammalian Mitochondrial Respiratory Chain*, Academic Press, New York.
7. Beinert, H. (1977) Iron–sulfur centers of the mitochondrial electron transfer system—Recent developments. In Lovenberg, W., editor. *Iron–Sulfur Proteins*, Academic Press, New York, pp. 61–100.
8. Beinert, H., Holm, R. H., Munck, E., and Münck, E. (1997) *Science* **277**, 653–659.
9. Lill, R. and Muhlenhoff, U. (2005) *TIBS* **30**, 133–141.
10. Gerber, J. and Lill, R. (2002) *Mitochondrion* **2**, 71–86.
11. Muhlenhoff, U. and Lill, R. (2000) *Biochim. Biophys. Acta* **1459**, 370–382.
12. Rieske, J. S., Hansen, R. E., and Zaugg, W. S. (1964) *J. Biol. Chem.* **239**, 3017–3022.
13. Lill, R. and Muhlenhoff, U. (2006) *Annu. Rev. Cell Dev. Biol* **22**, 457–486.
14. Hatefi, Y. (1976) The enzymes and the enzyme complexes of the mitochondrial oxidative phosphorylation system. In Martonosi, A., editor. *The Enzymes of Biological Membranes*, Plenum Press, New York.
15. Hatefi, Y., Galante, Y. M., Stiggal, D. L., and Ragan, C. I. (1979) *Methods Enzymol.* **56**, 577–602.
16. Hatefi, Y. (1985) *Annu. Rev. Biochem.* **54**, 1015–1069.
17. Ragan, C. I. (1976) *Biochem. J.* **154**, 295–305.
18. Walker, J. E., Skehel, J. M., and Buchanan, S. K. (1995) *Methods Enzymol.* **260**, 14–34.

19. Carroll, J., Fearnley, I. M., Shannon, R. J., Hirst, J., and Walker, J. E. (2003) *Mol. Cell Proteomics* **2**, 117–126.
20. Chomyn, A., Mariottini, P., Cleeter, M. W. J., Ragan, C. I., Matsuno-Yagi, A., Hatefi, Y., Doolittle, R. F., and Attardi, G. (1985) *Nature* **314**, 592–597.
21. Chomyn, A., Cleeter, M. W., Ragan, C. I., Riley, M., Doolittle, R. F., and Attardi, G. (1986) *Science* **234**, 614–618.
22. Hackenbrock, C. R., Chazotte, B., and Gupte, S. S. (1986) *J. Bioenerg. Biomembr.* **18**, 331–368.
23. Chazotte, B. and Hackenbrock, C. R. (1991) *J. Biol. Chem.* **266**, 5973–5979.
24. Schagger, H., De Coo, R., Bauer, M. F., Hofmann, S., Godinot, C., and Brandt, U. (2004) *J. Biol. Chem.* **279**, 36349–36353.
25. Schagger, H. (2002) *Biochim. Biophys. Acta* **1555**, 154–159.
26. Schagger, H. (2001) Blue-native gels to isolate protein complexes from mitochondria. In Pon, L. A. and Schon, E. A., editors. *Mitochondria*, Academic Press, San Diego, pp. 231–242.
27. Schagger, H. (1995) *Methods Enzymol.* **260**, 190–202.
28. Yadava, N., Houchens, T., Potluri, P., and Scheffler, I. E. (2004) *J. Biol. Chem.* **279**, 12406–12413.
29. Antonicka, H., Ogilvie, I., Taivassalo, T., Anitori, R. P., Haller, R. G., Vissing, J., Kennaway, N. G., and Shoubridge, E. A. (2003) *J. Biol. Chem.* **278**, 43081–43088.
30. Ugalde, C., Janssen, R., Van den Heuvel, L. P., Smeitink, J. A., and Nijtmans, L. G. (2004) *Hum. Mol. Genet.* **13**, 659–667.
31. Cardol, P., Matagne, R. F., and Remacle, C. (2002) *J. Mol. Biol.* **319**, 1211–1221.
32. Cruciat, C. M., Brunner, S., Baumann, F., Neupert, W., and Stuart, R. A. (2000) *J. Biol. Chem.* **275**, 18093–18098.
33. Lenaz, G. and Genova, M. L. (2006) *Am. J. Physiol. Cell. Physiol.* **292**, C1221–1239.
34. Zhang, M., Mileykovskaya, E., and Dowhan, W. (2005) *J. Biol. Chem.* **280**, 29403–29408.
35. Pfeiffer, K., Gohil, V., Stuart, R. A., Hunte, C., Brandt, U., Greenberg, M. L., and Schagger, H. (2003) *J. Biol. Chem.* **278**, 52873–52880.
36. Everard-Gigot, V., Dunn, C. D., Dolan, B. M., Brunner, S., Jensen, R. E., and Stuart, R. A. (2005) *Eukaryot. Cell* **4**, 346–355.
37. Bornhovd, C., Vogel, F., Neupert, W., and Reichert, A. S. (2006) *J. Biol. Chem.* **281**, 13990–13998.
38. Scheffler, I. E. (1986) Biochemical genetics of respiration -deficient mutants of animal cells. In Morgan, M. J., editor. *Carbohydrate Metabolism in Cultured Cells*, Plenum Press, London, pp. 77–109.
39. Scheffler, I. E., Yadava, N., and Potluri, P. (2004) *Biochim. Biophys. Acta* **1659**, 160–171.
40. Potluri, P., Yadava, N., and Scheffler, I. E. (2004) *Eur. J. Biochem.* **271**, 3265–3273.
41. Yadava, N., Potluri, P., Smith, E., Bisevac, A., and Scheffler, I. E. (2002) *J. Biol. Chem.* **277**, 21221–21230.

42. Au, H. C., Seo, B. B., Matsuno-Yagi, A., Yagi, T., and Scheffler, I. E. (1999) *Proc. Natl. Acad. Sci. USA* **96**, 4354–4359.
43. Hirst, J., Carroll, J., Fearnley, I. M., Shannon, R. J., and Walker, J. E. (2003) *Biochim. Biophys. Acta* **1604**, 135–150.
44. Carroll, J., Shannon, R. J., Fearnley, I. M., Walker, J. E., and Hirst, J. (2002) *J. Biol. Chem.* **277**, 50311–50317.
45. Schulte, U. and Weiss, H. (1995) *Methods Enzymol.* **260**, 3–14.
46. Marques, I., Duarte, M., and Videira, A. (2003) *J. Mol. Biol.* **329**, 283–290.
47. Videira, A. and Duarte, M. (2002) *Biochim. Biophys. Acta* **1555**, 187–191.
48. Brandt, U., Abdrakhmanova, A., Zickermann, V., Galkin, A., Drose, S., Zwicker, K., and Kerscher, S. (2005) *Biochem. Soc. Trans.* **33**, 840–844.
49. Abdrakhmanova, A., Zickermann, V., Bostina, M., Radermacher, M., Schagger, H., Kerscher, S., and Brandt, U. (2004) *Biochim. Biophys. Acta* **1658**, 148–156.
50. Kerscher, S., Drose, S., Zwicker, K., Zickermann, V., and Brandt, U. (2002) *Biochim. Biophys. Acta* **1555**, 83–91.
51. Cardol, P., Vanrobaeys, F., Devreese, B., Van Beeumen, J., Matagne, R. F., and Remacle, C. (2004) *Biochim. Biophys. Acta* **1658**, 212–224.
52. Remacle, C., Baurain, D., Cardol, P., and Matagne, R. F. (2001) *Genetics* **158**, 1051–1060.
53. Remacle, C., Baurain, D., Cardol, P., and Matagne, R. F. (2001) *Genetics* **158**, 1051–1060.
54. Rasmusson, A. G., Heiser, V., Zabaleta, E., Brennicke, A., and Grohmann, L. (1998) *Biochim. Biophys. Acta Bio-Energ.* **1364**, 101–111.
55. Eubel, H., Jansch, L., and Braun, H. P. (2003) *Plant Physiol.* **133**, 274–286.
56. Gabaldon, T., Rainey, D., and Huynen, M. A. (2005) *J. Mol. Biol.* **348**, 857–870.
57. Smeitink, J. A., Zeviani, M., Turnbull, D. M., and Jacobs, H. T. (2006) *Cell Metab.* **3**, 9–13.
58. (2004) *Oxidative Phosphorylation in Health and Disease*, Landes Bioscience, Georgetown, TX.
59. Smeitink, J. A., van den Heuvel, L. W., Koopman, W. J., Nijtmans, L. G., Ugalde, C., and Willems, P. H. (2004) *Curr. Neurovasc. Res.* **1**, 29–40.
60. Smeitink, J., Sengers, R., Trijbels, F., and Van den Heuvel, L. (2001) *J. Bioenerg. and Biomembr.* **33**, 259–266.
61. Robinson, B. H. (1998) *Biochim. Biophys. Acta* **1364**, 271–286.
62. Yagi, T., Yano, T., Di Bernardo, S., and Matsuno-Yagi, A. (1998) *Biochim. Biophys. Acta Bio-Energ.* **1364**, 125–133.
63. Finel, M. (1998) *Biochim. Biophys. Acta* **1364**, 112–121.
64. Friedrich, T. and Bottcher, B. (2004) *Biochim. Biophys. Acta* **1608**, 1–9.
65. Friedrich, T. (2001) *J. Bioenerg. Biomembr.* **33**, 169–177.
66. Friedrich, T. and Scheide, D. (2000) *FEBS Lett.* **479**, 1–5.
67. Friedrich, T., Abelmann, A., Brors, B., Guenebaut, V., Kintscher, L., Leonard, K., Rasmussen, T., Scheide, D., Schlitt, A., Schulte, U., and Weiss, H. (1998) *Biochim. Biophys. Acta* **1365**, 215–219.

68. Runswick, M. J., Fearnley, I. M., Skehel, J. M., and Walker, J. E. (1991) *FEBS Lett.* **286**, 121–124.

69. Wada, H., Shintani, D., and Ohlrogge, J. (1997) *Proc. Natl. Acad. Sci. USA* **94**, 1591–1596.

70. Fearnley, I. M., Carroll, J., Shannon, R. J., Runswick, M. J., Walker, J. E., and Hirst, J. (2001) *J. Biol. Chem.* **276**, 38345–38348.

71. Huang, G., Lu, H., Hao, A., Ng, D. C., Ponniah, S., Guo, K., Lufei, C., Zeng, Q., and Cao, X. (2004) *Mol. Cell Biol.* **24**, 8447–8456.

72. Grigorieff, N. (1998) *J. Mol. Biol.* **277**, 1033–1046.

73. Guénebaut, V., Schlitt, A., Weiss, H., Leonard, K., and Friedrich, T. (1998) *J. Mol. Biol.* **276**, 105–112.

74. Sazanov, L. A., Carroll, J., Holt, P., Toime, L., and Fearnley, I. M. (2003) *J. Biol. Chem.* **278**, 19483–19491.

75. Yagi, T. and Matsuno-Yagi, A. (2003) *Biochemistry* **42**, 2266–2274.

76. Sazanov, L. A. and Hinchliffe, P. (2006) *Science* **311**, 1430–1436.

77. Marques, I., Duarte, M., Assuncao, J., Ushakova, A. V., and Videira, A. (2005) *Biochim. Biophys. Acta* **1707**, 211–220.

78. Vogel, R. O., Janssen, R. J., Ugalde, C., Grovenstein, M., Huijbens, R. J., Visch, H. J., Van den Heuvel, L. P., Willems, P. H., Zeviani, M., Smeitink, J. A., and Nijtmans, L. G. (2005) *FEBS J.* **272**, 5317–5326.

79. Vogel, R., Nijtmans, L., Ugalde, C., van den, H. L., and Smeitink, J. (2004) *Curr. Opin. Neurol.* **17**, 179–186.

80. Ugalde, C., Vogel, R., Huijbens, R., van den Heuvel, B., Smeitink, J., and Nijtmans, L. (2004) *Hum. Mol. Genet.* **13**, 1–12.

81. Degli Esposti, M. (1998) *Biochim. Biophys. Acta* **1364**, 222–235.

82. Davis, G. C., Williams, A. C., Markey, S. P., Ebert, M. H., Caine, E. D., Reichert, C. M., and Kopin, I. J. (1979) *Psychiatry Res.* **1**, 649–654.

83. Langston, J. W., Ballard, P., Tetrud, J. W., and Irwin, I. (1983) *Science* **219**, 979–980.

84. Friedrich, T., Van Heek, P., Leif, H., Ohnishi, T., Forche, E., Kunze, B., Jansen, R., Trowitzsch-Kienast, W., Hofle, G., Reichenbach, H., and Weiss, H. (1994) *Eur. J. Biochem.* **219**, 691–698.

85. Grivennikova, V. G., Maklashina, E. O., Gavrikova, E. V., and Vinogradov, A. D. (1997) *Biochim. Biophys. Acta Bio-Energ.* **1319**, 223–232.

86. Gluck, M. R., Krueger, M. J., Ramsay, R. R., Sablin, S. O., Singer, T. P., and Nicklas, W. J. (1994) *J. Biol. Chem.* **269**, 3167–3174.

87. Miyoshi, H., Inoue, M., Okamoto, S., Ohshima, M., Sakamoto, K., and Iwamura, H. (1997) *J. Biol. Chem.* **272**, 16176–16183.

88. Vinogradov, A. D., Sled, V. D., Burbaev, D. S., Grivennikova, V. G., Moroz, I. A., and Ohnishi, T. (1995) *FEBS Lett.* **370**, 83–87.

89. De Jong, A. M. and Albracht, S. P. (1994) *Eur. J. Biochem.* **222**, 975–982.

90. Hanstein, W. G., Davis, K. A., Ghalambor, M. A., and Hatefi, Y. (1971) *Biochemistry* **10**, 2517–2524.

91. Davis, K. A. and Hatefi, Y. (1971) *Biochemistry* **10**, 2509–2516.

92. Davis, K. A. and Hatefi, Y. (1972) *Arch. Biochem. Biophys.* **149**, 505–512.
93. Ackrell, B. A. C., Johnson, M. K., Gunsalus, R. P., and Cecchini, G. (1992) Structure and function of succinate dehydrogenase and fumarate reductase. In Muller, F., editor. *Chemistry and Biochemistry of Flavoenzymes*, CRC Press, Boca Raton, FL, pp. 229–297.
94. Hederstedt, L. and Ohnishi, T. (1992) Progress in succinate:oxidoreductase research. In Ernster, L., editor. *Molecular Mechanisms in Bioenergetics*, Elsevier Science Publishers B.V., Amsterdam.
95. Ohnishi, T. (1987) *Curr. Topics Bioenerg.* **15**, 37–65.
96. Walker, W. H. and Singer, T. P. (1970) *J. Biol. Chem.* **245**, 4224–4225.
97. Kenney, W. C., Walker, W. H., and Singer, T. P. (1972) *J. Biol. Chem.* **247**, 4510–4513.
98. Singer, T. P. and Johnson, M. K. (1985) *FEBS. Lett.* **190**, 189–198.
99. Robinson, K. M. and Lemire, B. D. (1996) *J. Biol. Chem.* **271**, 4055–4060.
100. Robinson, K. M. and Lemire, B. D. (1996) *J. Biol. Chem.* **271**, 4061–4067.
101. Robinson, K. M., Rothery, R. A., Weiner, J. H., and Lemire, B. D. (1994) *Eur. J. Biochem.* **222**, 983–990.
102. Scheffler, I. E. (1998) *Progr. Nucl. Acid Res. Mol. Biol.* **60**, 267–315.
103. Blaut, M., Whittaker, K., Valdovinos, A., Ackrell, B. A. C., Gunsalus, R. P., and Cecchini, G. (1989) *J. Biol. Chem.* **264**, 13599–13604.
104. Johnson, M. K., Morningstar, J. E., Bennett, D. E., Ackrell, B. A., and Kearney, E. B. (1985) *J. Biol. Chem.* **260**, 7368–7378.
105. Beinert, H. (1990) *FASEB J.* **4**, 2483–2491.
106. Yu, L., Wei, Y.-Y., Usui, S., and Yu, C.-A. (1992) *J. Biol. Chem.* **267**, 24508–24515.
107. Cochran, B., Capaldi, R. A., Ackrell, B. A. C., and Ackrell, B. A. (1994) *Biochim. Biophys. Acta* **1188**, 162–166.
108. Merli, A., Capaldi, R. A., Acckrell, B. A. C., and Kearney, E. B. (1979) *Biochemistry* **18**, 1393–1400.
109. Iverson, T. M., Luna-Chavez, C., Cecchini, G., and Rees, C. (1999) *Science* **284**, 1961–1966.
110. Cecchini, G. (2003) *Annu. Rev. Biochem.* **72**, 77–109.
111. Cecchini, G., Schroder, I., Gunsalus, R. P., and Maklashina, E. (2002) *Biochim. Biophys. Acta (BBA)—Bioenergetics* **1553**, 140–157.
112. Tornroth, S., Yankovskaya, V., Cecchini, G., and Iwata, S. (2002) *Biochim. Biophys. Acta* **1553**, 171–176.
113. Yankovskaya, V., Horsefield, R., Tornroth, S., Luna-Chavez, C., Miyoshi, H., Leger, C., Byrne, B., Cecchini, G., and Iwata, S. (2003) *Science* **299**, 700–704.
114. Sun, F., Huo, X., Zhai, Y., Wang, A., Xu, J., Su, D., Bartlam, M., and Rao, Z. (2005) *Cell* **121**, 1043–1057.
115. Sucheta, A., Ackrell, B. A. C., Cochran, B., and Armstrong, F. A. (1992) *Nature* **356**, 361–362.
116. Ackrell, B. A. C., Armstrong, F. A., Cochran, B., Sucheta, A., and Yu, T. (1993) *FEBS Lett.* **326**, 92–94.
117. Spiro, S. and Guest, J. R. (1990) *FEMS Microbiol. Rev.* **75**, 399–428.

118. Green, J., Trageser, M., Six, S., Unden, G., and Guest, J. R. (1991) *Proc. R. Soc. London Ser. B* **244**, 137–144.
119. Roos, M. H. and Tielens, A. G. M. (1994) *Mol. Biochem. Parasitol.* **66**, 273–281.
120. Saruta, F., Kuramochi, T., Nakamura, K., Takamiya, S., Yu, Y., Aoki, T., Sekimizu, K., Kojima, S., and Kita, K. (1995) *J. Biol. Chem.* **270**, 928–932.
121. Van Hellemond, J. J. and Tielens, A. G. M. (1994) *Biochem. J.* **304**, 321–331.
122. Van Hellemond, J. J., Klockiewicz, M., Gaasenbeek, C. P. H., Roos, M. H., and Tielens, A. G. M. (1995) *J. Biol. Chem.* **270**, 31065–31070.
123. Van Hellemond, J. J., Van der Meer, P., and Tielens, A. G. M. (1997) *Parasitol.* **114**, 351–360.
124. Gest, H. (1980) *FEMS Microbiol. Lett.* **7**, 7–77.
125. Viehmann, S., Richard, O., Boyen, C., and Zetsche, K. (1996) *Curr. Genet.* **29**, 199–201.
126. Leblanc, C., Boyen, C., Richard, O., Bonnard, G., Grienenberger, J.-M., and Kloareg, B. (1995) *J. Mol. Biol.* **250**, 484–495.
127. Burger, G., Lang, B. F., Reith, M., and Gray, M. W. (1996) *Proc. Natl. Acad. Sci. USA* **93**, 2328–2332.
128. Gould, S. J., Subramani, S., and Scheffler, I. E. (1989) *Proc. Natl. Acad. Sci. USA* **86**, 1934–1938.
129. Lemire, B. D. and Oyedotun, K. S. (2002) *Biochim. Biophys. Acta (BBA)—Bioenergetics* **1553**, 102–116.
130. Lemire, B. D. and Oyedotun, K. S. (2001) *Biochem. Biophys. Acta* **1553**, 102–116.
131. Lombardo, A. and Scheffler, I. E. (1989) *J. Biol. Chem.* **264**, 18874–18877.
132. Daignan-Fornier, B., Valens, M., Lemire, B. D., and Bolotin-Fukuhara, M. (1994) *J. Biol. Chem.* **269**, 15469–15472.
133. Bullis, B. L. and Lemire, B. D. (1994) *J.Biol.Chem.* **269**, 6543–6549.
134. Saghbini, M., Broomfield, P. L. E., and Scheffler, I. E. (1994) *Biochemistry* **33**, 159–165.
135. Soderberg, K., Ditta, G. S., and Scheffler, I. E. (1977) *Cell* **10**, 697–702.
136. Oostveen, F. G., Au, H. C., Meijer, P.-J., and Scheffler, I. E. (1995) *J. Biol. Chem.* **270**, 26104–26108.
137. Ackrell, B. A. (2002) *Mol. Aspects Med.* **23**, 369–384.
138. Baysal, B. E. and Myers, E. N. (2002) *Microsc. Res. Tech.* **59**, 256–261.
139. Baysal, B. E., Willett-Brozick, J. E., Lawrence, E. C., Drovdlic, C. M., Savul, S. A., McLeod, D. R., Yee, H. A., Brackmann, D. E., Slattery, W. H. 3., Myers, E. N., Ferrell, R. E., and Rubinstein, W. S. (2002) *J. Med. Genet.* **39**, 178–183.
140. Baysal, B. E. (2002) *J. Med. Genet.* **39**, 617–622.
141. Schulte, U. and Weiss, H. (1995) *Methods Enzymol.* **260**, 63–70.
142. Braun, H. P., Emmermann, M., Krust, V., and Schmitz, U. K. (1992) *EMBO J.* **11**, 3219.
143. Schagger, H., Brandt, U., Gencic, S., and Von Jagow, G. (1995) *Methods Enzymol.* **260**, 82–96.
144. Schagger, H., Cramer, W. A., and Von Jagow, G. (1994) *Anal. Biochem.* **217**, 375.

145. Covian, R. and Trumpower, B. L. (2005) *J. Biol. Chem.* **280**, 22732–22740.
146. Tzagoloff, A. and Dieckmann, C. L. (1990) *Microbiol. Rev.* **54**, 211–225.
147. Tzagoloff, A. (1995) *Methods Enzymol.* **260**, 51–63.
148. Xia, D., Yu, C.-A., Kim, H., Xia, J.-Z., Kachurin, A. M., Zhang, L., Yu, L., Deisenhofer, J., Yu, C. A., and Xian, J. Z. (1997) *Science* **277**, 60–66.
149. Mitchell, P. (1976) *J. Theoret. Biol.* **62**, 327.
150. Mitchell, P. (1975) *FEBS Lett.* **59**, 137–139.
151. Von Jagow, G., Link, T. A., and Ohnishi, T. (1986) *J. Bioenerg. Biomembr.* **18**, 157–179.
152. Trumpower, B. L. (1990) *J. Biol. Chem.* **265**, 11409–11412.
153. Link, T. A., Haase, U., Brandt, U., and Von Jagow, G. (1993) *J. Bioenerg. Biomembr.* **25**, 221–232.
154. Brasseur, G. and Brivet-Chevillote, P. (1994) *FEBS Lett.* **354**, 23–29.
155. Miyoshi, H., Tokutake, N., Imaeda, Y., Akagi, T., and Iwamura, H. (1995) *Biochim. Biophys. Acta* **1229**, 149–154.
156. Branden, G., Gennis, R. B., and Brzezinski, P. (2006) *Biochim. Biophys. Acta* **1757**, 1052–1063.
157. Tsukihara, T., Aoyama, H., Yamashita, E., Tomizaki, T., Yamaguchi, H., Shinzawa-Itoh, K., Nakashima, R., Yaono, R., and Yoshikawa, S. (1995) *Science* **269**, 1069–1074.
158. Tsukihara, T., Aoyama, H., Yamashita, K., Tomizaki, T., Yamaguchi, H., Shinzawa-Iyoh, K., Nakashima, R., Yaono, R., Yoshikawa, S., Yamashita, E., and Shinzawa-Itoh, K. (1996) *Science* **272**, 1136–1144.
159. Ferguson-Miller, S. (1996) *Science* **272**, 1125.
160. Iwata, S., Ostermaier, C., Ludwig, B., and Michel, H. (1995) *Nature* **376**, 660–669.
161. Schlerf, A., Droste, M., Winter, M., and Kadenbach, B. (1988) *EMBO J.* **7**, 2387–2391.
162. Schillace, R., Preiss, T., Lightowlers, R. N., and Capaldi, R. A. (1994) *Biochim. Biophys. Acta* **1188**, 391–397.
163. Capaldi, R. A. (1990) *Annu. Rev. Biochem.* **59**, 569–596.
164. Capaldi, R. A., Marusich, M. F., and Taanman, J. W. (1995) *Methods Enzymol.* **260**, 117–132.
165. Beinert, H. (1997) *Eur. J. Biochem.* **245**, 521–532.
166. Babcock, G. T. and Wikstrom, M. (1992) *Nature* **356**, 301–309.
167. Hosler, J. P., Ferguson-Miller, S., and Mills, D. A. (2006) *Annu. Rev. Biochem.* **75**, 165–187.
168. Varotsis, C., Zhang, Y., Appelman, E. H., and Babcock, G. T. (1993) *Proc. Natl. Acad. Sci. USA* **90**, 237–241.
169. Gennis, R. B. (1998) *Science* **280**, 1712–1713.
170. Yoshikawa, S., Shinzawa-Itoh, K., Nakashima, R., Yaono, R., Yamashita, E., Inoue, N., Yao, M., Fei, M. J., Peters Libeu, C., Mizushima, T., Yamaguchi, H., Tomizaki, T., and Tsukihara, T. (1998) *Science* **280**, 1723–1729.
171. Richter, O. M. and Ludwig, B. (2003) *Rev. Physiol. Biochem. Pharmacol.* **147**, 47–74.

172. Mills, D. A. and Ferguson-Miller, S. (2003) *FEBS Lett.* **545**, 47–51.

173. Wikstrom, M., Ribacka, C., Molin, M., Laakkonen, L., Verkhovsky, M., and Puustinen, A. (2005) *Proc. Nat. Acad. Sci.* **102**, 10478–10481.

174. Ugalde, C., Vogel, R., Huijbens, R., Van Den, H. B., Smeitink, J., and Nijtmans, L. (2004) *Hum. Mol. Genet.* **13**, 2461–2472.

175. Brandt, U. (2006) *Annu. Rev. Biochem.* **75**, 69–92.

176. Kerscher, S., Grgic, L., Garofano, A., and Brandt, U. (2004) *Biochim. Biophys. Acta* **1659**, 197–205.

177. Videira, A. (1998) *Biochim. Biophys. Acta* **1364**, 89–100.

178. Kuffner, R., Rohr, A., Schmiede, A., Krull, C., Schulte, U., Scacco, S., Vergari, R., Scarpulla, R. C., Technikova-Dobrova, Z., Sardanelli, A., Lambo, R., Lorusso, V., and Papa, S. (1998) *J. Biol. Chem.* **283**, 409–417.

179. Vogel, F., Bornhovd, C., Neupert, W., and Reichert, A. S. (2006) *J. Cell Biol.* **175**, 237–247.

180. Vogel, R. O., Dieteren, C. E. J., van den Heuvel, L. P. W. J., Willems, H. G. M., Smeitink, J. A. M., Koopman, W. J. H., and Nijtmans, L. G. J. (2007) *J. Biol. Chem.* **282**, 7582–7590.

181. Janssen, R., Smeitink, J., Smeets, R., and van den Heuvel, J. (2002) *Hum. Genet.* **110**, 264–270.

182. Day, C. and Scheffler, I. E. (1982) *Somat. Cell Genet.* **8**, 691–707.

183. Zara (2004) *Eur. J. Biochem.* **271**, 1209–1218.

184. Berden, J. A., Schoppink, P. J., and Grivell, L. A. (2006) A model for the assembly of ubiquinol–cytochrome c oxidoreductase in yeast. In Palmieri, F. and Quagliariello, C., editors. *Molecular Basis of Biomembrane Transport*, pp. 195–208.

185. Tiranti, V., Hoertnagel, K., Carrozzo, R., Galimberti, C., Munaro, M., Granatiero, M., Zelante, L., Gasparini, P., Marzella, R., Rocchi, M., Bayona-Bafaluy, M. P., Enriquez, J. A., Uziel, G., Bertini, E., Dionisi-Vici, C., Franco, B., Meitinger, T., and Zeviani, M. (2001) *Am. J. Hum.Genet.* **63**, 1621.

186. Tiranti, V., Galimberti, C., Nijtmans, L. G., Bovolenta, S., Perin, M. P., and Zeviani, M. (1999) *Hum. Mol. Genet.* **8**, 2533–2540.

187. Zhu, Z., Yao, J., Johns, T., Fu, K., De Bie, I., Macmillan, C., Cuthbert, A. P., Newbold, R. F., Wang, J., Chevrette, M., Brown, G. K., Brown, R. M., and Shoubridge, E. A. (1998) *Nat. Genet.* **20**, 337–343.

188. Herrmann, J. M. and Funes, S. (2005) *Gene* **354**, 43–52.

189. Barrientos, A. (2003) *IUBMB Life* **55**, 83–95.

190. Robinson, B. H. (2000) *Pediatr. Res.* **48**, 581–585.

191. Moore, A. L. and Rich, P. R. (1980) *TIBS* **5**, 284–288.

192. Laties, G. G. (1982) *Annu. Rev. Plant Physiol.* **33**, 519–555.

193. Douce, R. and Neuburger, M. (1989) *Annu. Rev. Plant Physiol. Mol. Biol.* **40**, 371–414.

194. Soole, K. L. and Menz, R. I. (1995) *J. Bioenerg. Biomembr.* **27**, 397–406.

195. Leach, G. R., Krab, K., and Moore, A. L. (1994) *Biochem. Soc. Trans.* **22**, 406S.

196. Moller, I. M., Rasmusson, A. G., and Fredlund, K. M. (1993) *J. Bioenerg. Biomembr.* **25**, 377–384.

197. Siedow, J. N. (1995) Bioenergetics: The mitochondrial electron transfer chain. In Levings, C. S. I. and Vasil, I. K., editors. *The Molecular Biology of Plant Mitochondria*, Kluwer Academic Publishers, Dordrecht, pp. 281–312.
198. Roberts, T. H., Fredlund, K. M., and Moller, I. M. (1995) *FEBS Lett.* **373**, 307–309.
199. Menz, R. I. and Day, D. A. (1996) *J. Biol. Chem.* **271**, 23117–23120.
200. Henry, M. F. and Nyns, E. J. (1975) *Sub-Cell. Biochem.* **4**, 1–65.
201. Rhoads, D. M. and McIntosh, L. (1991) *Proc. Natl. Acad. Sci. USA* **88**, 2122–2126.
202. Albury, M. S., Dudley, P., Watts, F. Z., and Moore, A. L. (1996) *J. Biol. Chem.* **271**, 17062–17066.
203. Raskin, I., Ehmann, E., Melander, W., and Meeuse, B. J. D. (1987) *Science* **237**, 1601–1602.
204. De Vries, S. and Grivell, L. A. (1988) *Eur. J. Biochem.* **176**, 377–384.
205. Marres, C. A. M., De Vries, S., and Grivell, L. A. (1991) *Eur. J. Biochem.* **195**, 857–862.
206. De Vries, S., Van Witzenburg, R., Grivell, L. A., and Marres, C. A. M. (1992) *Eur. J. Biochem.* **203**, 587–592.
207. Kitajima-Ihara, T. and Yagi, T. (1998) *FEBS Lett.* **421**, 37–40.
208. Seo, B. B., Kitajima-Ihara, T., Chan, E. K. L., Scheffler, I. E., Matsuno-Yagi, A., and Yagi, T. (1998) *Proc. Natl. Acad. Sci. USA* **95**, 9167–9171.
209. Nosek, J. and Fukuhara, H. (1994) *J. Bact.* **176**, 5622–5630.
210. Buschges, R., Bahrenberg, G., Zimmermann, M., and Wolf, K. (1994) *Yeast* **10**, 475–479.
211. Mar Gonzalez-Barroso, M., Ledesma, A., Lepper, S., Perez-Magan, E., Zaragoza, P., and Rial, E. (2006) *Yeast* **23**, 307–313.
212. Zwicker, K., Galkin, A., Drose, S., Grgic, L., Kerscher, S., and Brandt, U. (2006) *J. Biol. Chem.* **281**, 23013–23017.
213. Brandt, U. (2006) *Annu. Rev. Biochem.* **75**, 69–92.
214. Clayton, C. E. and Michels, P. (1996) *Parasitol. Today* **12**, 465–471.
215. Opperdoes, F. R. (1995) Carbohydrate and energy metabolism in aerobic protozoa. In Marr, J. J. and Muller, M., editors. *Biochemistry and Molecular Biology of Parasites*, Academic Press, San Diego, pp. 19–32.
216. Tielens, A. G. M. and Van Hellemond, J. J. (1998) *Parasitol. Today* **14**, 265–271.
217. Beattie, D. S. and Howton, M. M. (1996) *Eur. J. Biochem.* **241**, 888–894.
218. Mitchell, P. (1961) *Nature* **191**, 144–148.
219. Boyer, P. D. (1977) *Annu. Rev. Biochem.* **46**, 957–966.
220. Mitchell, P. (1979) *Eur. J. Biochem.* **95**, 1–20.
221. Mitchell, P. (1977) *Annu. Rev. Biochem.* **46**, 996–1005.
222. Mitchell, P. (1980) *Ann. NY Acad. Sci.* **341**, 564–584.
223. Mitchell, P. (1976) *Biochem. Soc. Trans.* **4**, 399–430.
224. Mitchell, P. (1972) *J. Bioenerg.* **3**, 5–24.
225. Mitchell, P. and Moyle, J. (1967) *Nature* **213**, 137–139.
226. Mitchell, P. (1967) *Fed. Proc.* **26**, 1370–1379.

227. Mitchell, P. (1966) *Biol. Rev. Camb. Philos. Soc.* **41**, 445–502.

228. Nicholls, D. G. and Ferguson, S. J. (2002) *Bioenergetics*, 3rd ed., Academic Press, San Diego.

229. Prebble, J. (2002) *Trends Biochem. Sci.* **27**, 209–212.

230. Hinkle, P. C., Kumar, M. A., Resetar, A., and Harris, D. L. (1991) *Biochemistry* **30**, 3576–3582.

231. Hinkle, P. C. (2005) *Biochim. Biophys. Acta* **1706**, 1–11.

232. Galkin, A. S., Grivennikova, V. G., and Vinogradov, A. D. (1999) *FEBS Lett.* **451**, 157–161.

233. Antonini, G., Malatesta, F., Sarti, P., and Brunori, M. (1993) *Proc. Natl. Acad. Sci. USA* **90**, 5949–5953.

234. Klingenberg, M. and Rottenberg, H. (1977) *Eur. J. Biochem.* **73**, 125–130.

235. Coty, W. A. and Pedersen, P. L. (1974) *J. Biol. Chem.* **249**, 2593–2598.

236. Stock, D., Leslie, A. G. W., and Walker, J. E. (1999) *Science* **286**, 1700–1705.

237. Junge, W., Panke, O., Cherepanov, D. A., Gumbiowski, K., Muller, M., and Engelbrecht, S. (2001) *FEBS Lett.* **504**, 152–160.

238. Ackerman, S. H. and Tzagoloff, A. (2005) *Prog. Nucleic. Acid. Res. Mol. Biol.* **80**, 95–133.

239. Brandt, U. and Trumpower, B. (1994) *Crit. Rev. Biochem. Mol. Biol.* **29**, 165–197.

240. Crofts, A. R. (2004) *Photosynth. Res.* **80**, 223–243.

241. Mulkidjanian, A. Y. (2005) *Biochim. Biophys. Acta* **1709**, 5–34.

242. Osyczka, A., Moser, C. C., and Dutton, P. L. (2005) *TIBS* **30**, 176–182.

243. Berry, E. A., Guergova-Kuras, M., Huang, L., and Crofts, A. R. (2000) *Annu. Rev. Biochem.* **69**, 1005–1075.

244. Gong, X., Yu, L., Xia, D., and Yu, C. A. (2005) *J. Biol. Chem.* **280**, 9251–9257.

245. Matsuno-Yagi, A. and Hatefi, Y. (1997) *J. Biol. Chem.* **272**, 16928–16933.

246. Matsuno-Yagi, A. and Hatefi, Y. (1997) *J. Biol. Chem.* **271**, 6164–6171.

247. Szabo, I., Bathori, G., Wolff, D., Starc, T., Cola, C., and Zoratti, M. (1995) *Biochim. Biophys. Acta* **1235**, 115–125.

248. Ballarin, C. and Sorgato, M. C. (1995) *J. Biol. Chem.* **270**, 19262–19268.

249. Haydon, D. A. and Hladky, S. B. (1972) *Q. Rev. Biophys.* **5**, 237–282.

250. Skulachev, V. P. (1971) *Curr. Topics Bioenerg.* **4**, 127–190.

251. Berger, K. H. and Yaffe, M. P. (1996) *Experientia* **52**, 1111–1116.

252. Chen, L. B., Summerhayes, I. C., Johnson, L. V., Walsh, M. L., Bernal, S. D., and Lampidis, T. J. (1982) *Cold Spring Harbor Symp. Quant. Biol.* **XLVI**, 141–155.

253. Chen, L. B. (1988) *Annu. Rev. Cell Biol.* **4**, 155–181.

254. Reers, M., Smiley, S. T., Mottola-Hartshorn, C., Chen, A., Lin, M., and Chen, L. B. (1995) *Methods Enzymol.* **260**, 406–417.

255. Albracht, S. P., Mariette, A., and De Jong, P. (1997) *Biochim. Biophys. Acta* **1318**, 92–106.

256. Bereiter-Hahn, J. and Voth, M. (1994) *Microsc. Res. Tech.* **27**, 198–219.

257. MacLennan, D. H. and Tzagoloff, A. (1968) *Biochemistry* **7**, 1603–1610.

258. Tzagoloff, A., Byington, K. H., and MacLennan, D. H. (1968) *J. Biol. Chem.* **243**, 2405–2412.

259. Tzagoloff, A., MacLennan, D. H., and Byington, K. H. (1968) *Biochemistry* **7**, 1596–1602.

260. Hekman, C., Tomich, J. M., and Hatefi, Y. (1991) *J. Biol. Chem.* **266**, 13564–13571.

261. Ogilvie, I., Aggeler, R., and Capaldi, R. A. (1997) *J. Biol. Chem.* **272**, 16652–16656.

262. Racker, E. (1976) *A New Look at Mechanisms in Bioenergetics*, Academic Press, New York.

263. Abrahams, J. P., Leslie, A. G. W., Lutter, R., and Walker, J. E. (1994) *Nature* **370**, 621–628.

264. Leslie, A. G. W. and Walker, J. E. (2000) *Philos. Trans. R. Soc. Lond. B Biol. Sci.* **355**, 465–471.

265. Weber, J. and Senior, A. E. (1997) *Biochim. Biophys. Acta* **1319**, 19–58.

266. Capaldi, R. A., Aggeler, R., Turina, P., and Wilkens, S. (1994) *Trends in Biochem. Sci.* **19**, 284–289.

267. Gledhill, J. R. and Walker, J. E. (2006) *Biochem. Soc. Trans.* **34**, 989–992.

268. Gledhill, J. R. and Walker, J. E. (2005) *Biochem. J.* **386**, 591–598.

269. Cabezon, E., Runswick, M. J., Leslie, A. G., and Walker, J. E. (2001) *EMBO J.* **20**, 6990–6996.

270. Cabezon, E., Montgomery, M. G., Leslie, A. G., and Walker, J. E. (2003) *Nat. Struct. Biol.* **10**, 744–750.

271. Dickson, V. K., Silvester, J. A., Fearnley, I. M., Leslie, A. G., and Walker, J. E. (2006) *EMBO J.* **25**, 2911–2918.

272. Silvester, J. A., Dickson, V. K., Runswick, M. J., Leslie, A. G., and Walker, J. E. (2006) *Acta Crystallograph. Sect. F Struct. Biol. Cryst. Commun.* **62**, 530–533.

273. Walker, J. E. and Dickson, V. K. (2006) *Biochim. Biophys. Acta* **1757**, 286–296.

274. Rubinstein, J. L., Walker, J. E., and Henderson, R. (2003) *EMBO J.* **22**, 6182–6192.

275. Meier, T., Polzer, P., Diederichs, K., Welte, W., and Dimroth, P. (2005) *Science* **308**, 659–662.

276. Murata, T., Yamato, I., Kakinuma, Y., Leslie, A. G. W., and Walker, J. E. (2005) *Science* **308**, 654–659.

277. Fillingame, R. H., Jones, P. C., Jiang, W., Valiyaveetil, F. I., and Dmitriev, O. Y. (1998) *Biochim. Biophys. Acta* **1365**(1–2), 135–142.

278. Boyer, P. D. (1993) *Biochim. Biophys. Acta* **1140**, 215–250.

279. Pedersen, P. L. and Amzel, L. M. (1993) *J. Biol. Chem.* **268**, 9937–9940.

280. Junge, W., Lill, H., and Engelbrecht, S. (1997) *TIBS* **22**, 420–423.

281. Senior, A. E., Nadanaciva, S., and Weber, J. (2002) *Biochem. Biophys. Acta* **1553**, 188–211.

282. Weber, J. and Senior, A. E. (2003) *FEBS Lett.* **545**, 61–70.

283. Senior, A. E. and Weber, J. (2004) *Nature Struct. Mol. Biol.* **11**, 110–112.

284. Gresser, M. J., Meyers, J. A., and Boyer, P. D. (1982) *J. Biol. Chem.* **257**, 12030–12038.

285. Repke, K. R. H. and Schon, R. (1974) *Acta Biol. Med. Ger.* **33**, K27–K38.
286. Adolfsen, R. and Moudrianakis, E. N. (1976) *Arch. Biochem. Biophys.* **172**, 425–433.
287. Yasuda, R., Yoshida, M., and Kinosita, K., Jr., Noji, H. (1997) *Nature* **386**, 299–302.
288. Yasuda, R., Noji, H., Yoshida, M., Kinosita, K., Jr., and Itoh, H. (2001) *Nature* **410**, 898–904.
289. Ueno, S., Suzuki, T., Kinosita, K., Jr., and Yoshida, M. (2005) *Proc. Natl. Acad. Sci. USA* **102**, 1333–1338.
290. Diez, M., Zimmermann, B., Borsch, M., Konig, M., Schweinberger, E., Steigmiller, S., Reuter, R., Felekyan, S., Kudryatsev, V., Seidel, C. A. M., and Graber, P. (2004) *Nature Struct. Mol. Biol.* **11**, 135–141.
291. Nishizaka, T., Oiwa, K., Noji, H., Kimura, S., Muneyuki, E., Yoshida, M., and Kinosita, K., Jr. (2004) *Nature Struct. Mol. Biol.* **11**, 142–148.
292. Sambongi, Y., Iko, Y., Tanabe, M., Omote, H., Iwamoto-Kihara, A., Ueda, I., Yanagida, T., Wada, Y., and Futai, M. (1999) *Science* **286**, 1722–1724.
293. Nishio, K., Iwamoto-Kihara, A., Yamamoto, A., Wada, Y., and Futai, M. (2002) *Proc. Natl. Acad. Sci. USA* **99**, 13448–13452.
294. Boyer, P. D. (1997) *Annu. Rev. Biochem.* **66**, 717–749.
295. Matsui, T., Muneyuki, E., Honda, M., Allison, W. S., Dou, C., and Yoshida, M. (1997) *J. Biol. Chem.* **272**, 8215–8221.
296. Walker, J. E., Collinson, I. R., Van Raaij, M. J., and Runswick, M. J. (1995) *Methods Enzymol.* **260**, 163–190.
297. Law, R. H. P., Manon, S., Devenish, R. J., and Nagley, P. (1995) *Methods Enzymol.* **260**, 133–163.
298. Fillingame, R. H. (1992) *J. Bioenerg. Biomembr.* **24**, 485–491.
299. Junge, W. and Nelson, N. (2005) *Science* **308**, 642–644.
300. Oster, G. and Wang, H. (2003) *Trends Cell Biol.* **13**, 114–121.
301. Paumard, P., Vaillier, J., Coulary, B., Schaeffer, J., Soubannier, V., Mueller, D. M., Brethes, D., Di Rago, J.-P., and Velours, J. (2002) *EMBO J.* **21**, 221–230.
302. Arselin, G., Giraud, M. F., Dautant, A., Vaillier, J., Brethes, D., Coulary-Salin, B., Schaeffer, J., and Velours, J. (2003) *Eur. J. Biochem.* **270**, 1875–1884.
303. Brown, G. C. (1992) *Biochem. J.* **284**, 1–13.
304. Kacser, H. and Burns, J. A. (1997) *Biochem. Soc. Trans.* **23**, 341–365.
305. Heinrich, R. and Rapoport, T. A. (1974) *Eur. J. Biochem.* **42**, 89–95.
306. Brand, M. D. and Kesseler, A. (1995) *Biochem. Soc. Trans.* **23**, 371–376.
307. Brand, M. D., Chien, L.-F., and Rolfe, D. F. S. (1993) *Biochem. Soc. Trans.* **21**, 757–762.
308. Papin, J. A., Price, N. D., Wiback, S. J., Fell, D. A., and Palsson, B. O. (2003) *Trends Biochem. Sci.* **28**, 250–258.
309. Gnaiger E. and Kuznetsov, A. V. (2002) *Biochem. Soc. Trans.* **30**, 252–258.
310. Jin, Q. and Bethke, C. M. (2002) *Biophys. J.* **83**, 1797–808.
311. Korzeniewski, B. (2002) *Mol. Biol. Rep.* **29**, 197–202.

312. Kudin, A., Vielhaber, S., Elger, C. E., and Kunz, W. S. (2002) *Mol. Biol. Rep.* **29**, 89–92.

313. Atkinson, D. E. (1977) *Cellular Energy Metabolism and Its Regulation*, Academic Press, New York.

314. Klingenberg, M. (1961) *Biochem. Z.* **335**, 231–272.

315. Erecinska, M. and Wilson, D. F. (1982) *J. Membr. Biol.* **70**, 1–14.

316. Crompton, M. (1985) *Curr. Top. Membr. Transp.* **25**, 231–276.

317. McCormack, J. G., Halestrap, A. P., and Denton, R. M. (1990) *Physiol. Rev.* **70**, 391–425.

318. Anthony, G., Reimann, A., and Kadenbach, B. (1993) *Proc. Natl. Acad. Sci. USA* **90**, 1652–1656.

319. Arnold, S. and Kadenbach, B. (1999) *FEBS Lett.* **443**, 105–108.

320. Ludwig, B., Bender, E., Arnold, S., Huttemann, M., Lee, I., and Kadenbach, B. (2001) *Eur. J. Chem. Biol* **2**, 392–403.

321. Kadenbach, B. (2003) *Biochim. Biophys. Acta* **1604**, 77–94.

322. Linder, D., Freund, R., and Kadenbach, B. (1995) *Comp. Biochem. Physiol. B Biochem. Mol. Biol.* **112**, 461–469.

323. Rohdich, F. and Kadenbach, B. (1993) *Biochemistry* **32**, 8499–8503.

324. Kadenbach, B., Huettemann, M., Arnold, S., Lee, I., and Bender, E. (2000) *Free Radic. Biol. Med.* **29**, 211–221.

325. Saccone, C., Pesole, G., and Kadenbach, B. (1991) *Eur. J. Biochem.* **195**, 151–156.

326. Frank, V. and Kadenbach, B. (1996) *FEBS Lett.* **382**, 121–124.

327. Korzeniewski, B. and Mazat, J. P. (1996) *Biochem. J.* **319**, 143–148.

328. Wiedemann, F. R. and Kunz, W. S. (1998) *FEBS Lett.* **422**, 33–35.

329. Nicholls, D. G. and Locke, R. M. (1984) *Physiol. Rev.* **64**, 164.

330. Silva, J. E. and Rabelo, R. (1997) *Eur. J. Endocrinol.* **136**, 251–264.

331. Cannon, B. and Nedergaard, J. (2004) *Physiol. Rev.* **84**, 277–359.

332. Nedergaard, J. and Cannon, B. (1992) The uncoupling protein thermogenin and mitochondrial thermogenesis. In Ernster, L., editor. *Molecular Mechanisms in Bioenergetics*, Elsevier Science, London, pp. 385–420.

333. Fleury, C., Neverova, M., Collins, S., Raimbault, S., Champigny, O., Levi-Meyrueis, C., Bouillaud, F., Seldin, M. F., Surwit, R. S., Ricquier, D., and Warden, C. H. (1997) *Nat. Genet.* **15**, 269–272.

334. Esteves, T. C. and Brand, M. D. (2005) *Biochim. Biophys. Acta (BBA)—Bioenergetics* **1709**, 35–44.

335. Brand, M. D. and Esteves, T. C. (2005) *Cell Metab.* **2**, 85–93.

336. Klingenberg, M., Winkler, E., and Huang, S. (1995) *Methods Enzymol.* **260, 89**, 369–389.

337. Winkler, E., Wachter, E., and Klingenberg, M. (1997) *Biochemistry* **36**, 148–155.

338. Klingenberg, M. and Echtay, K. S. (2001) *Biochim. Biophys. Acta* **1504**, 128–143.

339. Garlid, K. D., Jaburek, M., Jezek, P., and Varecha, M. (2000) *Biochim. Biophys. Acta* **1459**, 383–389.

340. Wohlrab, H. (2005) *Biochim. Biophys. Acta* **1709**, 157–168.

341. Breen, E. P., Gouin, S. G., Murphy, A. F., Haines, L. R., Jackson, A. M., Pearson, T. W., Murphy, P. V., and Porter, R. K. (2006) *J. Biol. Chem.* **281**, 2114–2119.

342. Arechaga, I., Raimbault, S., Prieto, S., Levi-Meyrueis, C., Zaragoza, P., Miroux, B., Ricquier, D., Bouillaud, F., and Rial, E. (1993) *Biochem J.* **296**, 693–700.

343. Ledesma, A., De Lacoba, M. G., and Rial, E. (2002) *Genome Biol.* **3**, reviews 3015.1–3015.9.

344. Mills, E. M., Banks, M. I., Sprague, J. E., and Finkel, T. (2003) *Nature* **426**, 403–404.

345. Esteves, T. C. and Brand, M. D. (2005) *Biochim. Biophys. Acta* **1709**, 35–44.

346. Brookes, P. S. (2005) *Free Radic. Biol. Med.* **38**, 12–23.

347. Prieto, S., Bouillaud, F., Ricquier, D., and Rial, E. (1992) *Eur. J. Biochem.* **208**, 487–491.

348. Prieto, S., Bouillaud, F., and Rial, E. (1995) *Biochem. J.* **307**, 657–661.

349. Prieto, S., Bouillaud, F., and Rial, E. (1996) *Arch. Biochem. Biophys.* **334**, 43–49.

350. Laloi, M., Klein, M., Riesmeier, J. W., and Muler-Rober, B. (1997) *Nature* **389**, 135–136.

351. Meeuse, B. J. D. (1975) *Annu. Rev. Plant. Physiol.* **26**, 117–126.

352. Rhee, S. G. (2006) *Science* **312**, 1882–1883.

353. Brunelle, J. K., Bell, E. L., Quesada, N. M., Vercauteren, K., Tiranti, V., Zeviani, M., Scarpulla, R. C., and Chandel, N. S. (2005) *Cell Metab.* **1**, 409–414.

354. Guzy, R. D., Hoyos, B., Robin, E., Chen, H., Liu, L., Mansfield, K. D., Simon, M. C., Hammerling, U., and Schumacker, P. T. (2005) *Cell Metab.* **1**, 401–408.

355. Halliwell, B. and Gutteridge, J. M. (1999) *Free Radicals in Biology and Medicine*, 3rd ed., Oxford University Press, Oxford.

356. Finkel, T. (2003) *Curr. Opin. Cell Biol.* **15**, 247–254.

357. Saltman, P. (1989) *Semi. Hematol.* **26**, 249–255.

358. Halliwell, B. and Cross, C. E. (1994) *Environ. Health Perspect.* **102**, 5–12.

359. Rodriguez, R. and Redman, R. (2005) *Proc. Natl. Acad. Sci USA* **102**, 3175–3176.

360. Scandalios, J. G. (2002) *Trends Biochem. Sci.* **27**, 483–486.

361. Jezek, P. and Hlavata, L. (2005) *Int. J. Biochem. Cell Biol.* **37**, 2478–2503.

362. Grivennikova, V. G. and Vinogradov, A. D. (2006) *Biochim. Biophys. Acta* **1757**, 553–561.

363. Kussmaul, L. and Hirst, J. (2006) *Proc. Natl. Acad. Sci USA* **103**, 7607–7612.

364. Chen, Q., Vazquez, E. J., Moghaddas, S., Hoppel, C. L., and Lesnefsky, E. J. (2003) *J. Biol. Chem.* **278**, 36027–36031.

365. Muller, F. L., Liu, Y., and Van Remmen, H. (2004) *J. Biol. Chem.* **279**, 49064–49073.

366. McLennan, H. R. and Degli, E. M. (2000) *J. Bioenerg. Biomembr.* **32**, 153–162.

367. Neuzil, J., Wang, X. F., Dong, L. F., Low, P., and Ralph, S. J. (2006) *FEBS Lett.* **580**, 5125–5129.

368. Genova, M. L., Pich, M. M., Bernacchia, A., Bianchi, C., Biondi, A., Bovina, C., Falasca, A. I., Formiggini, G., Castelli, G. P., and Lenaz, G. (2004) *Ann. NY Acad. Sci.* **1011**, 86–100.

369. Adam-Vizi, V. and Chinopoulos, C. (2006) *Trends Pharmacol. Sci.* **27**, 639–645.

370. Galkin, A. and Brandt, U. (2005) *J. Biol. Chem.* **280**, 30129–30135.

371. McCord, J. M. and Fridovich, I. (1969) *J. Biol. Chem.* **244**, 6059–6055.

372. Bacman, S. R., Bradley, W. G., and Moraes, C. T. (2006) *Mol. Neurobiol.* **33**, 113–131.

373. Ferri, A., Cozzolino, M., Crosio, C., Nencini, M., Casciati, A., Butler Gralla, E., Rotilio, G., Selverstone Valentine, J., and Carri, M. T. (2006) *Proc. Natl. Acad. Sci. USA* **103**, 13860–13865.

374. Manfredi, G. and Xu, Z. (2005) *Mitochondrion* **5**, 77–87.

375. Dawson, V. L. (2004) *Nat. Med.* **10**, 905–906.

376. Sun, J. and Trumpower, B. L. (2003) *Arch. Biochem. Biophys.* **419**, 198–206.

377. Kirkland, R. A. and Franklin, J. L. (2006) *Exp. Neurol.* **204**, 458–461.

378. Trifunovic, A., Wredenberg, A., Falkenberg, M., Spelbrink, J. N., Rovio, A. T., Bruder, C. E., Bohlooly, Y., Gidlof, S., Oldfors, A., Wibom, R., Tornell, J., Jacobs, H. T., and Larsson, N. G. (2004) *Nature* **429**, 417–423.

379. Trifunovic, A., Hansson, A., Wredenberg, A., Rovio, A. T., Dufour, E., Khvorostov, I., Spelbrink, J. N., Wibom, R., Jacobs, H. T., and Larsson, N. G. (2005) *Proc. Natl. Acad. Sci. USA* **102**, 17993–17998.

380. Kujoth, G. C., Hiona, A., Pugh, T. D., Someya, S., Panzer, K., Wohlgemuth, S. E., Hofer, T., Seo, A. Y., Sullivan, R., Jobling, W. A., Morrow, J. D., Van Remmen, H., Sedivy, J. M., Yamasoba, T., Tanokura, M., Weindruch, R., Leeuwenburgh, C., and Prolla, T. A. (2005) *Science* **309**, 481–484.

381. Murphy, M. P. (2006) *Am. J. Physiol. Heart Circ. Physiol.* **290**, H1754–H1755.

382. Ross, M. F., Da Ros, T., Blaikie, F. H., Prime, T. A., Porteous, C. M., Severina, I. I., Skulachev, V. P., Kjaergaard, H. G., Smith, R. A., and Murphy, M. P. (2006) *Biochem. J.* **400**, 199–208.

383. James, A. M., Cocheme, H. M., and Murphy, M. P. (2005) *Mech. Ageing Dev.* **126**, 982–986.

384. Reddy, P. H. (2006) *J. Biomed. Biotechnol.* **2006**, 31372.

385. Brookes, P. S. (2004) *Mitochondrion* **3**, 187–204.

386. Haynes, V., Elfering, S., Traaseth, N., and Giulivi, C. (2004) J *Bioenerg. Biomembr.* **36**, 341–346.

387. Valdez, L. B., Zaobornyj, T., and Boveris, A. (2006) *Biochim. Biophys. Acta* **1757**, 166–172.

388. Giulivi, C., Kato, K., and Cooper, C. E. (2006) *Am. J Physiol Cell Physiol* **291**, C1225–C1231.

389. Guo, F. Q. and Crawford, N. M. (2005) *Plant Cell* **17**, 3436–3450.

390. Clementi, E. and Nisoli, E. (2005) *Comp. Biochem. Physiol. A Mol. Integr. Physiol.* **142**, 102–110.

391. Brown, G. C. (2001) *Biochim. Biophys. Acta* **1504**, 46–57.

392. Lahiri, S., Roy, A., Baby, S. M., Hoshi, T., Semenza, G. L., and Prabhakar, N. R. (2005) *Prog. Biophys. Mol. Biol.* **91**, 249–286.
393. Antunes, F., Boveris, A., and Cadenas, E. (2004) *Proc. Natl. Acad. Sci. USA* **101**, 16774–16779.
394. Mason, M. G., Nicholls, P., Wilson, M. T., and Cooper, C. E. (2006) *Proc. Natl. Acad. Sci. USA* **103**, 708–713.
395. Castello, P. R., David, P. S., McClure, T., Crook, Z., and Poyton, R. O. (2006) *Cell Metab* **3**, 277–287.
396. Burwell, L. S., Nadtochiy, S. M., Tompkins, A. J., Young, S. M., and Brookes, P. S. (2005) *Biochem. J.* **394**, 627–634.
397. Hurd, T. R., Costa, N. J., Dahm, C. C., Beer, S. M., Brown, S. E., Filipovska, A., and Murphy, M. P. (2005) *Antioxid. Redox. Signal.* **7**, 999–1010.
398. Hurd, T. R., Filipovska, A., Costa, N. J., Dahm, C. C., and Murphy, M. P. (2005) *Biochem. Soc. Trans.* **33**, 1390–1393.
399. Beer, S. M., Taylor, E. R., Brown, S. E., Dahm, C. C., Costa, N. J., Runswick, M. J., and Murphy, M. P. (2004) *J. Biol. Chem.* **279**, 47939–47951.
400. Nisoli, E., Clementi, E., Paolucci, C., Cozzi, V., Tonello, C., Sciorati, C., Bracale, R., Valerio, A., Francolini, M., Moncada, S., and Carruba, M. O. (2003) *Science* **299**, 896–899.
401. Scarpulla, R. C. (2002) *Biochim. Biophys. Acta* **1576**, 1–14.
402. Scarpulla, R. C. (2002) *Gene* **286**, 81–89.
403. Scarpulla, R. C. (2006) *J. Cell. Biochem.* **97**, 673–683.
404. Nisoli, E., Tonello, C., Cardile, A., Cozzi, V., Bracale, R., Tedesco, L., Falcone, S., Valerio, A., Cantoni, O., Clementi, E., Moncada, S., and Carruba, M. O. (2005) *Science* **310**, 314–317.
405. Robinson, N. C., Zborowski, J., and Talbert, L. H. (1990) *Biochemistry* **29**, 8962–8969.
406. Yue, W. H., Zou, Y. P., Yu, L., and Yu, C. A. (1991) *Biochemistry* **30**, 2303–2306.
407. Eble, K. S., Coleman, W. B., Hantgan, R. R., and Cunningham, C. C. (1990) *J. Biol. Chem.* **265**, 19434–19440.
408. Beyer, K. and Klingenberg, M. (1985) *Biochemistry* **24**, 3821–3826.
409. Schlame, M. and Haldar, D. (1993) *J. Biol. Chem.* **268**, 74–79.
410. Schlame, M., Brody, S., and Hostetler, K. Y. (1993) *Eur. J. Biochem.* **212**, 727–735.
411. Ohtsuka, T., Nishijima, M., and Akamatsu, Y. (1993) *J. Biol. Chem.* **268**, 22908–22913.
412. Ohtsuka, T., Nishijima, M., Suzuki, K., and Akamatsu, Y. (1993) *J. Biol. Chem.* **268**, 22914–22919.
413. Schlame, M., Rua, D., and Greenberg, M. L. (2000) *Prog. Lipid Res.* **39**, 257–288.
414. Gohil, V. M., Hayes, P., Matsuyama, S., Schagger, H., Schlame, M., and Greenberg, M. L. (2004) *J. Biol. Chem.* **279**, 42612–42618.
415. Li, G., Chen, S., Thompson, M. N., and Greenberg, M. L. (2006) *Biochim. Biophys. Acta* **1771**, 432–441.
416. McKenzie, M., Lazarou, M., Thorburn, D. R., and Ryan, M. T. (2006) *J. Mol. Biol.* **361**, 462–469.

417. Iverson, S. L. and Orrenius, S. (2004) *Arch Biochem Biophys.* **423**, 37–46.
418. Domenech, O., Sanz, F., Montero, M. T., and Hernandez-Borrell, J. (2006) *Biochim. Biophys. Acta* **1758**, 213–221.
419. Choi, S. Y., Gonzalvez, F., Jenkins, G. M., Slomianny, C., Chretien, D., Arnoult, D., Petit, P. X., and Frohman, M. A. (2006) *Cell Death Differ.* **14**, 597–606.
420. Schlame, M., Ren, M., Xu, Y., Greenberg, M. L., and Haller, I. (2005) *Chem. Phys. Lipids* **138**, 38–49.
421. Barth, P. G., Valianpour, F., Bowen, V. M., Lam, J., Duran, M., Vaz, F. M., and Wanders, R. J. (2004) *Am. J. Med. Genet. A* **126**, 349–354.
422. Brandner, K., Mick, D. U., Frazier, A. E., Taylor, R. D., Meisinger, C., and Rehling, P. (2005) *Mol. Biol. Cell* **16**, 5202–5214.
423. Hauff, K. D. and Hatch, G. M. (2006) *Prog. Lipid Res.* **45**, 91–101.
424. Xu, Y., Malhotra, A., Ren, M., and Schlame, M. (2006) *J. Biol. Chem.* **281**, 39217–39224.
425. Xu, Y., Condell, M., Plesken, H., Edelman-Novemsky, I., Ma, J., Ren, M., and Schlame, M. (2006) *Proc. Natl. Acad. Sci. USA* **103**, 11584–11588.
426. Palsdottir, H. and Hunte, C. (2004) *Biochim. Biophys. Acta (BBA)—Biomembr.* **1666**, 2–18.

6

METABOLIC PATHWAYS INSIDE MITOCHONDRIA

Summary

Many reactions taking place inside of mitochondria occupy a central position on any metabolic chart. The subject is well covered in most standard biochemistry textbooks and is undoubtedly a very familiar subject to many readers. Biochemical pathways are taken for granted, and little time is devoted nowadays to the painstaking analytical studies that were necessary to establish the reaction sequences and the various intermediates and branch points. It is not the intention to make up for this deficiency here. However, the present treatment of mitochondria would be incomplete without reference to the major biochemical pathways, and to the variety of enzymes localized in the matrix or the inner mitochondrial membrane. Fatty acid oxidation and the Krebs cycle (tricarboxylic acid cycle) were in many ways the defining reactions for mitochondria. Other major pathways or cycles include the urea cycle, heme biosynthesis, part of the reactions leading to steroid, dolichol and ubiquinone biosynthesis, and the synthesis of a specific lipid, cardiolipin, which is found almost exclusively in mitochondrial membranes. A relatively recent insight is that all iron–sulfur centers (Fe–S clusters) originate in mitochondria.

In almost all cases (Fe–S clusters being the exception) the pathways were established some time ago. Attention is now on regulatory phenomena, on detailed structural analyses of the many different enzymes involved, and on the cloning of the corresponding cDNAs and genes to be used in diagnostic tests in the diagnosis of inborn errors of metabolism.

Mitochondria, Second Edition, by Immo E. Scheffler

6.1 INTRODUCTION

Until about the middle of the twentieth century, investigators preparing crude extracts from muscle and liver homogenates for the purpose of biochemical and enzymatic studies did not pay any attention to subcellular compartmentalization and its potential significance. Thus, when Claude and Deduve at the Rockefeller Institute perfected techniques for isolating relatively pure mitochondria, a new era in biochemistry began. In a short time span the enzymes for several important metabolic pathways were localized within the mitochondria, starting with the pioneering work of Lehninger and Kennedy on fatty acid oxidation. The citric acid cycle (TCA cycle) had already been postulated by Krebs in 1937, building on earlier work by Szent Gyorgyi and others, but again, Lehninger and Kennedy, taking advantage of having pure, isolated mitochondria, demonstrated in 1948 that the entire citric acid cycle takes place inside the mitochondria. Together with the spectroscopic identification of cytochromes in mitochondria, these discoveries led to the recognition of mitochondria as the "powerhouse of the cell." Mitochondria became the first subcellular organelles outside of the nucleus to be functionally characterized. It took not much longer to identify lysosomes as the organelles containing hydrolytic enzymes, although the role of lysosomes in the overall economy of cells did not become apparent until later. Peroxisomes were also characterized by the discovery of a few key enzymes such as catalase, but the description of their full complement of enzymes took several decades. Finally, the endoplasmic reticulum (ER) and Golgi apparatus were placed firmly in the secretory pathway by the classical pulse-chase experiments of Jamieson and Palade and the pioneering biochemical studies of Sabatini and Blobel. The detailed biochemical and functional descriptions of the ER and Golgi constitute highlights of the past two decades.

6.2 THE KREBS CYCLE

A description of the citric acid cycle is found in all elementary and advanced textbooks, and its formulation ranks among the major achievements in the study of metabolism. It was the second cycle to be postulated by Krebs, since he had also discovered the urea cycle a few years earlier. When it was first proposed by Krebs in 1937, there were still some major areas of uncertainty, especially in the production of citrate, which Krebs formulated as the following reaction:

$$\text{pyruvate} + \text{oxaloacetate} \rightarrow \text{citrate} + CO_2$$

It was not until coenzyme A had been discovered by Kaplan and Lipmann in 1945 that Ochoa and Lynen were able to show in 1951 that acetyl-CoA was the intermediate in the reaction with oxaloacetate to form citrate.

Today the reactions of the citric acid cycle are firmly established, and the cycle itself occupies a central position on any metabolic chart. When it is first presented to students, it is frequently emphasized that the constituents of the cycle act catalytically—that is, are themselves not being increased or consumed. Subsequently, it becomes necessary to add to the catabolic functions of the cycle anabolic functions as well; that is, cycle intermediates are being drained off to form amino acids and numerous precursors for other biosynthetic reactions. For example, succinyl-CoA is a precursor for heme synthesis. At the same time, the α-keto dicarboxylic acids of the cycle can be formed from the transamination of amino acids, and oxaloacetate can also be produced from the carboxylation of phosphoenolpyruvate in certain tissues such as liver.

Reactions of the TCA cycle are also essential for complementing the reactions of the glyoxylate cycle in plants. This cycle has been speculated to be an evolutionary precursor for the TCA cycle, still sharing with it some reactions. However, the significant aspect of the glyoxylate cycle is that a novel reaction can take place with isocitrate as the substrate. Instead of being followed by two decarboxylation reactions, isocitrate is split by isocitrate lyase to form succinate and glyoxylate. Glyoxylate condenses with a second acetyl-CoA to enter the cycle to form malate. The net result is the operation of a cycle that converts two acetyl groups to succinate. The glyoxylate cycle takes place in glyoxysomes, which are distinct organelles of plants related to peroxisomes and found prominently in germinating seedlings. It permits the utilization of fats and oils stored in the seeds to make carbohydrates before plant growth and development has formed green leaves for photosynthesis. The succinate produced has to leave the glyoxisomes and enter mitochondria, where it is converted to oxaloacetate that can be withdrawn to form phosphoenolpyruvate for gluconeogenesis.

Is there anything left to discover about the citric acid cycle? All the enzymes have been purified from various organisms, and many of the corresponding genes have been or are being cloned. Catalytic mechanisms have been defined in considerable detail, and crystal structures of many enzymes are available. All but one enzyme of the citric acid cycle are soluble enzymes in the mitochondrial matrix. The exception is succinate dehydrogenase (see Chapter 5), which is part of complex II and forms the peripheral membrane component of this complex. Its characteristics as an electron transfer complex have been discussed elsewhere in this volume. Structure–function analyses of this complex are now greatly enhanced by the completion of the crystal structure. Understanding the assembly of the complex from precursors made in the cytosol and imported into mitochondria, the addition of Fe–S clusters to one subunit, FAD to another, and a b-type heme to the integral membrane subunits continue to pose questions.

For the theoretically inclined, the regulation of the fluxes of metabolites through the cycle presents a second challenge, especially in light of (a) new insights about the "robustness" of metabolic pathways and (b) the emphasis on control coefficients for the various enzymes as opposed to the simpler view of rate limiting enzymes. How is the activity of the cycle adjusted to various conditions in a given tissue such as muscle, and how is the cycle's capacity regulated in different tissues?

Before some answers to these questions can be proposed, a presentation of the cycle and a limited discussion of the individual steps are in order. It is logical to start with one reaction outside of the cycle that produces the important intermediate acetyl-CoA. The overall reaction is

$$\text{pyruvate} + \text{CoA} + \text{NAD}^+ \rightarrow \text{NADH} + \text{acetyl-CoA} + \text{CO}_2$$

The enzyme involved is a multienzyme complex named pyruvate dehydrogenase, consisting of 132 subunits (30 E_1 dimers, 60 E_2 dimers, and 6 E_3 dimers in mammals) with a total molecular mass of several hundred thousand kilodaltons. Five coenzymes are necessary: thiamine pyrophosphate, lipoamide, CoA, FAD, and NAD. Substrates and intermediates are shuttled between the subunits, facilitated by the highly organized, three-dimensional structure of this complex. For a detailed discussion of the structure and mechanism the reader is referred to standard biochemistry textbooks, or to the original literature.

Figure 6.1 presents an overview of the intermediates and their sequence in the citric acid cycle. In this scheme the individual reactions have been numbered, and some brief comments about salient features of each reaction and the corresponding enzyme will have to suffice for the present treatment.

1. Citrate synthase catalyzes the condensation of acetyl-CoA with oxaloacetate to form citrate, a tricarboxylic acid from which a common name of the cycle has been derived. The X-ray structure of the enzyme has been determined, and considerable detail about the mechanism is known. The reaction can be described as a mixed aldol–Claisen condensation involving the stabilization of the enol form of acetyl-CoA as an intermediate. Our current understanding of enzymes and their active sites also makes it clear why a specific chiral carbon is formed with the carboxymethyl group from acetyl-CoA constituting the pro-S arm. A consequence of this stereochemistry is that the two carbons entering the cycle from acetyl-CoA will not be liberated as CO_2 during the first turn of the cycle.

2. Aconitase catalyzes the reversible interconversion between two structural isomers, citrate and isocitrate. Since citrate is prochiral, the two carboxymethyl groups of the symmetrical cityrate are distinguishable, and the hydroxyl group is moved to the methylene carbon from oxaloacetate rather than to the methylene group of the original acetyl-CoA. The enzyme is of special interest because a nonheme iron–sulfur cluster ([4Fe–4S]) participates in the active site of the enzyme, although no electron transfer takes place. A specific Fe(II) in the cluster serves to bind and stabilize the substrates and intermediates. More recently it was discovered that a cytosolic aconitase also serves as a sensor of iron in the cytosol. Before its identity was established, this

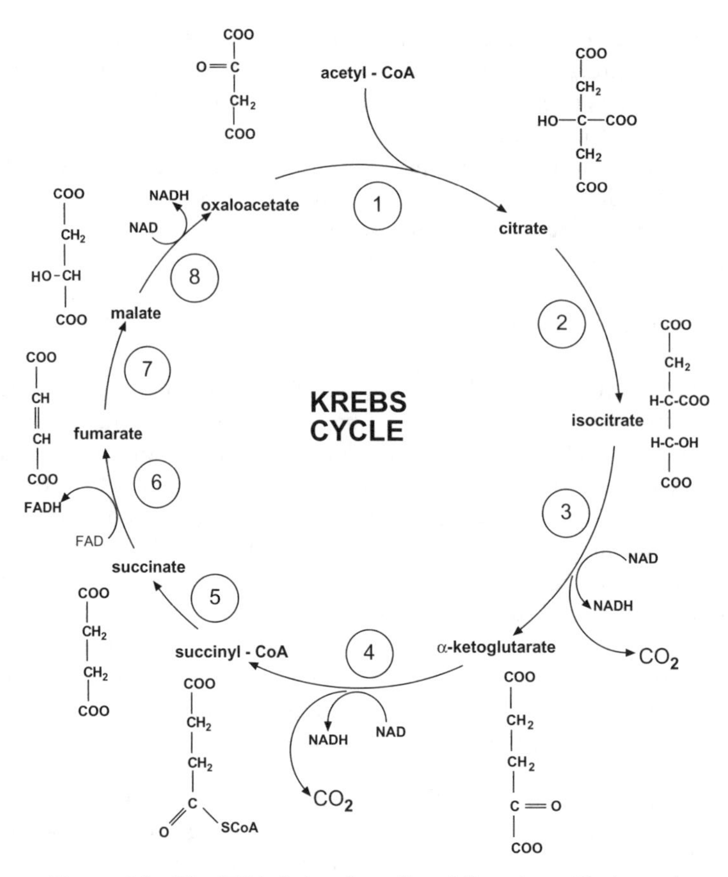

Figure 6.1 The TCA (tri carboxylic acid) cycle or Krebs cycle.

iron-responsive protein (IRP) had been shown to interact with iron-responsive elements (IREs) on certain mRNAs. These IREs can be recognized as specific hairpin structures. When bound to IREs in the 5′ untranslated region of the mRNAs for ferritin or 5-aminolevulinate synthase, it prevents translation of these mRNAs. Cytosolic aconitase binds to IREs when the iron concentration is low; at high iron concentrations, aconitase is more tightly folded around the [4Fe–4S] cluster and incapable of binding the mRNA. An IRE is also found in the 3′ untranslated region of the transferrin receptor mRNA. Binding of aconitase (low iron) prevents turnover of the mRNA. At high iron concentrations, degradation of the transferrin receptor mRNA is promoted. Thus, aconitase as the iron sensor controls the transport of iron into the cell, the storage of iron, a step in heme biosynthesis, and probably other pathways related to iron (see references 1–3 for recent reviews).

3. Isocitrate dehydrogenase occurs in two forms in mammalian tissues. An NAD-dependent form is found exclusively in mitochondria; a second form

utilizes NADP and is localized both in mitochondria and the cytosol. The reaction produces the first of two CO_2 molecules produced per turn of the cycle, and it links a decarboxylation to the oxidation of an alcohol to a keto group. The first of three NADH molecules is also produced here.

4. The enzyme α-ketoglutarate dehydrogenase is a multifunctional enzyme complex like the pyruvate dehydrogenase; and the third enzyme, dihydrolipoyl dehydrogenase, is identical to that in PDH. The first two subunits are similar, with specificity for the succinyl instead of acetyl moiety, and the final product is another high-energy thioester, succinyl-CoA. NADH and CO_2 are the other products of the reaction. At this point, two CO_2 molecules have been produced, accounting for the total input of two carbons from acetyl-CoA, but the stereochemical considerations related to the formation of citrate and the subsequent reactions can account for the observations that no radioactive CO_2 is produced from [^{14}C] acetyl-CoA during the first turn of the cycle.

5. Succinyl-CoA synthetase carries out a substrate-level phosphorylation with the slight twist that GDP is the phosphate acceptor:

$$\text{succinyl-CoA} + P_i + \text{GDP} \rightarrow \text{succinate} + \text{GTP} + \text{CoA}$$

There is persuasive evidence from isotope tracer experiments that the inorganic phosphate displaces the CoA to form succinyl-phosphate. The phosphate is transferred to GDP via a 3-phospho-histidine intermediate on the enzyme.

6. Succinate is oxidized to fumarate by the membrane-linked SDH complex. A large (70-kDa) flavoprotein forms the active site, and a covalently linked flavin becomes the hydrogen acceptor to form $FADH_2$. The flavin is linked to a histidine side chain in a highly conserved portion of the protein. It is immediately reoxidized with the release of protons and the transfer of two electrons to the iron–sulfur centers of the iron–protein subunit of SDH. The Ip subunit has three nonheme iron–sulfur centers ([2Fe–2S], [3Fe–4S], and [4Fe–4S]) and constitutes a prime example of a protein conducting electrons in the absence of a heme group. The two integral membrane proteins of complex II are associated with a cytochrome, but it is still controversial whether this cytochrome participates in the conduit of the electrons to the ubiquinone (CoQ). Succinate dehydrogenase is the only Krebs cycle enzyme directly linked to the electron transfer chain. A more detailed discussion of this complex can be found in Chapter 5.

7. Fumarate is hydrated by the addition of H_2O to the double bond, catalyzed by fumarase. Controversy about the mechanism has centered on the question of whether the addition of OH^- is the first step, leading to the formation of a carbanion that appears now to have been confirmed by ^{18}O exchange experiments.

8. The final reaction catalyzed by malate dehydrogenase constitutes the last oxidation step and hence production of another NADH.

A final accounting of all the reactions with the inclusion of the pyruvate dehydrogenase reactions leads to the following summation:

$$\text{pyruvate} + P_i + \text{GDP} + 4\,\text{NAD}^+ + \text{FAD} \Rightarrow 3\,CO_2 + 4\,\text{NADH} + \text{FADH}_2 + \text{GTP}$$

It should be reemphasized that pyruvate (C3) has become converted to three CO_2 molecules without any consumption of oxygen. If a steady supply of NAD^+ was available, glucose could thus be broken down to carbon dioxide in the absence of oxygen. The cost of NAD^+ makes this impossible, and in reality it has to be obtained by recycling. Hence, oxidation of NADH by complex I of the electron transfer chain is essential to keep the citric acid cycle going. Accumulation of NADH causes a strong inhibition of pyruvate dehydrogenase and α-ketoglutarate dehydrogenase, and the author's laboratory has isolated mammalian cell mutants with defects in the electron transfer chain based on the almost total inhibition of the Krebs cycle (4). Substrate and product concentrations are the immediate controlling factors determining the flux through the cycle, and ATP, ADP, and Ca^{+2} are allosteric effectors of citric acid cycle enzymes. Overall, oxygen consumption, NADH oxidation, and ATP synthesis are tightly coupled not only to each other, but also to the citric acid cycle by the above mechanisms.

In the end, Lavoisier was correct in viewing respiration as combustion, but little did he suspect that many steps are necessary to make this "burning" of glucose in a living cell such a controlled process. The many steps involved not only ensure control over a "slow burn," but also permit the efficient capture of the free energy released in the process, along with its interconversion and utilization in many biological processes. It all started with attempts to understand respiration, followed by a curiosity about the biochemistry and energetics of muscular contraction. As a final thought, it may be worthwhile to be made aware that an average person turns over more than his/her own bodyweight of ATP per day, depending on the level of activity.

6.3 FATTY ACID METABOLISM

Fatty acid oxidation was one of the first metabolic pathways firmly localized in mitochondria by the pioneering studies of Lehninger and Kennedy in the late 1940s. Since fats and fatty acids derived from them are obvious energy sources, their degradation in mitochondria must have been suggestive. Soon thereafter the localization of the Krebs cycle enzymes in the same organelle allowed the integration of fatty acid oxidation with the TCA cycle via the important intermediate acetyl-CoA. Peroxisomes had not been recognized at that time, and therefore the discovery of β-oxidation of fatty acids in peroxisomes did not confuse the issue. It is now known that peroxisomes in animal cells "specialize" in the degradation of very long fatty acids (>22 carbon atoms), and they may even just shorten them to be accepted by the mitochondrial system. In plant cells, however, fatty acid degradation is restricted to peroxisomes and glyoxysomes. The latter are specialized peroxisomes with special significance in metabolism during seed germination.

Before considering the details, it will be instructive to take a more global view. The fatty acids are first made available in the cytosol—for example, from the hydrolysis of cholesterol esters of low-density lipoprotein (LDL) in lysosomes or from the degradation of phospholipids by lipases. Their transport into mitochondria is not a trivial issue (see below). The final product of their degradation is acetyl-CoA that can feed into the Krebs cycle. Under different conditions, fatty acids can also be synthesized, also starting from acetyl-CoA. The biosynthesis of fatty acids does not occur simply by the reversal of the reactions of the β-oxidation, an observation that generally applies to other pathways such as glycolysis and gluconeogenesis. Different enzymes are utilized; and most significantly, the biosynthesis of fatty acids takes place in the cytosol. During elongation in the cytosol the acyl group is attached to a large multifunctional protein (in animal cells) taking the place of CoA.

Fatty acid degradation generates acetyl-CoA as the final product, but intermediates are also attached to CoA during the entire process. Thus, the first step in fatty acid degradation requires an activation:

$$\text{Fatty acid} + \text{CoA} + \text{ATP} \rightarrow \text{acyl-CoA} + \text{AMP} + \text{PP}_i$$

Several acyl-CoA-sythetases (thiokinases) serve in this activation in the cytosol. To reach the mitochondrial matrix, a shuttle system has evolved, which is shown in Figure 6.2. The acyl group is transferred to carnitine, liberating CoA. Acyl-carnitine is transported into the mitochondria by a specific carrier protein; and once it is inside, the transesterification is reversed to form carnitine and acyl-CoA. The carrier protein is operating like an antiporter, transporting free carnitine out of the mitochondria.

The β-oxidation reactions (Figure 6.3) are well known and found in all standard biochemistry textbooks. A brief listing of all reactions and some select comments will suffice here.

The first reaction is an an elimination of hydrogen to form a double bond between the α and β carbons of acyl-CoA.

$$\text{acyl-CoA} + \text{FAD} \Rightarrow \textit{trans}\text{-}\Delta 2\text{-enoyl-CoA} + \text{FADH}_2 \qquad (1)$$

The FAD is a cofactor of acyl-CoA dehydrogenase, and the $FADH_2$ is reoxidized via two proteins, the electron transfer flavoprotein (ETF) and the ETF: ubiquinone oxidoreductase transferring electrons to ubiquinone (to yield QH_2), which is reoxidized via the electron transfer chain. Superficially, the acyl-CoA dehydrogenase therefore resembles succinate dehydrogenase in the use of flavin as the immediate acceptor of hydrogen, followed by electron transfers via iron–sulfur centers to ubiquinone.

$$\textit{trans}\text{-}\Delta 2\text{-enoyl-CoA} + \text{H}_2\text{O} \Rightarrow \text{3-L-hydroxyacyl-CoA} \qquad (2)$$

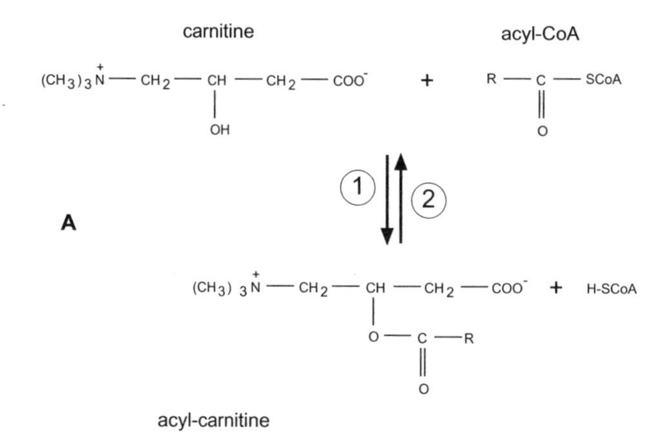

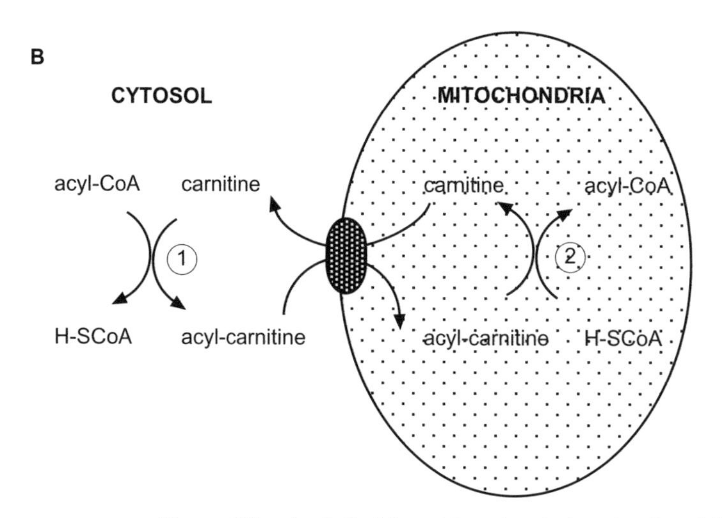

Figure 6.2 Acyl-CoA/carnitine shuttle in mitochondria.

This hydration reaction resembles to formation of malate from fumarate and is catalyzed by enoyl-CoA hydratase. It is followed by an oxidation to β-ketoacyl-CoA:

$$\text{3-L-hydroxyacyl-CoA} + \text{NAD}^{+} \Rightarrow \beta\text{-ketoacyl-CoA} + \text{NADH} \qquad (3)$$

In the final step, another CoA is a second substrate, and the reaction splits off a two-carbon unit to form acetyl-CoA, leaving acyl-CoA, but now two carbon atoms shorter than at the start:

$$(C_{n})\ \text{acyl-CoA} + \text{CoA} \Rightarrow (C_{n-2})\ \text{acyl-CoA} + \text{acetyl-CoA} \qquad (4)$$

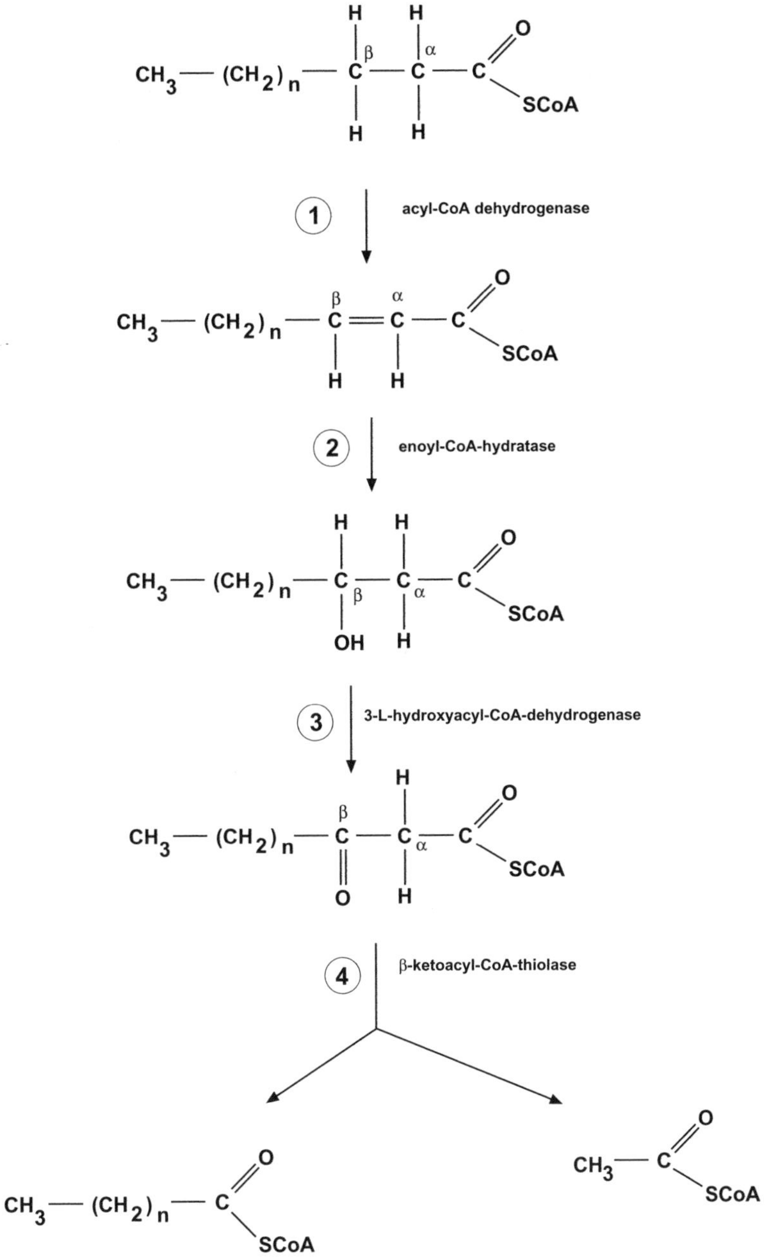

$CH_3-(CH_2)_n-CH_2(\beta)-CH_2(\alpha)-C(=O)SCoA$

① acyl-CoA dehydrogenase

$CH_3-(CH_2)_n-CH(\beta)=CH(\alpha)-C(=O)SCoA$

② enoyl-CoA-hydratase

$CH_3-(CH_2)_n-CH(OH)(\beta)-CH_2(\alpha)-C(=O)SCoA$

③ 3-L-hydroxyacyl-CoA-dehydrogenase

$CH_3-(CH_2)_n-C(=O)(\beta)-CH_2(\alpha)-C(=O)SCoA$

④ β-ketoacyl-CoA-thiolase

$CH_3-(CH_2)_n-C(=O)SCoA$ + $CH_3-C(=O)SCoA$

Figure 6.3 β-oxidation reactions.

The series of reactions 1–4 is repeated, and a saturated fatty acid with an even number of carbons can be converted to acetyl-CoA with no further complications, except that different acyl-CoA dehydrogenases are used as the carbon chain of the acyl group gets shorter. Unsaturated fatty acids can be dealt with by a combination of isomerases and reductases (requiring NADPH), depending on the position of the double bonds and their configuration. The β-oxidation of odd-chain fatty acids leads to the formation of propionyl-CoA, which must be dealt with by a distinct series of reactions:

(1) propionyl-CoA + ATP + CO_2 → (*S*)-methylmalonyl-CoA + ADP + P_i

(2) (*S*)-methylmalonyl-CoA → (*R*)-methylmalonyl-CoA

(3) (*R*)-methylmalonyl-CoA → succinyl-CoA

Reactions 1–3 are catalyzed by a carboxylase, a racemase and a mutase, respectively. The methyl-malonyl-CoA mutase is an unusual enzyme requiring 5-deoxyadenosylcobalamin (coenzyme B_{12}) as a cofactor; this cofactor contains Co(III) in the center of a ring resembling heme and has the only carbon–metal bond known in biology. Vitamin B_{12} deficiency was first shown in 1926 to cause pernicious anemia, and humans must acquire it from their food (meat). Uptake and transport in the bloodstream requires specific proteins, cobalamins. The solution of the crystal structure of cobalamin by D. Hodgkin was a notable achievement in crystallography in 1956, which was later dwarfed only by the determination of the crystal structure of whole proteins. Cobinamide, a precursor in cobalamin biosynthesis, has an extremely high affinity for cyanide, and it has been proposed to be used in cyanide detoxification (5).

While the acetyl-CoA formed during β-oxidation could enter the Krebs cycle as expected, its actual fate in liver may be different. A process referred to as ketogenesis converts acetyl-CoA into compounds called ketone bodies, which include acetone, acetoacetate, and β-hydroxybutyrate. The solubility of ketone bodies makes them readily transportable by the blood to other organs and tissues where they serve as alternate metabolic fuels. For example, when fats are mobilized during starvation, ketone bodies can substitute for glucose in the brain.

Although the biosynthesis of fatty acids takes place in the cytosol, and its details get short shrift in this book on mitochondria, it is appropriate and relevant to discuss one of the first steps required for this cyctosolic sequence to take place: Acetyl-CoA has to get out of the mitochondria. This is achieved by making citrate in the mitochondria, which can be exported by a transport system for tribarboxylic acids. In the cytosol, acetyl-CoA is regenerated by ATP-citrate lyase:

citrate + ATP + CoA ⇒ acetyl-CoA + ADP + P_i + oxaloacetate

6.4 THE UREA CYCLE

The Krebs cycle ketoacids α-ketoglutarate and oxaloacetate can be withdrawn to form amino acids by transamination or added to the cycle by the easily reversible reaction, and it may be worth mentioning that mammalian cells in tissue culture thrive in media with added glutamine even though it could be considered a nonessential amino acid. It enters the Krebs cycle as a-ketoglutarate, thus providing nitrogen in a transamination reaction. This nitrogen is recaptured when oxaloacetate is converted to aspartate (and asparagine). In fibroblasts these two nonessential amino acids are derived almost exclusively from glutamine via the operation of the Krebs cycle (4, 6).

A most significant role of mitochondria in amino acid metabolism is the participation in the urea cycle. The urea cycle was the first metabolic cycle discovered and described by Krebs five years before his more famous citric acid cycle. The cycle is crucial for the breakdown of excess amino acids and the conversion of excess nitrogen into urea, the excretable form of nitrogen in many terrestial animals. The major activity takes place in the liver, from where urea is deliverd via the bloodstream to the kidneys for excretion in urine.

As expected, there is one reaction creating a precursor, carbamoyl-phosphate, which can enter the cycle by condensation with a cycle intermediate, and products including urea are generated in various steps in the cycle. Only some reactions of the cycle take place in mitochondria; it is completed in the cytosol. The complete cycle is shown in Figure 6.4. The two reactions taking place in mitochondria are the formation of carbamoyl phosphate from ammonia and bicarbonate, driven by the hydrolysis of ATP, and the condensation of carbamoyl phosphate with ornithine to form citrulline. Carbamoyl phosphate synthetase I, the enzyme required for step 1, is allosterically regulated by *N*-acetylglutamate, made from glutamate and acetyl-CoA. Increased glutamate concentrations signal excess amino acids (nitrogen), and an increase in *N*-acetylglutamate activates carbamoyl phosphate synthesis.

6.5 BIOSYNTHESIS OF HEME

Studies on respiration, cytochromes, and heme leading to the elucidation of the role of mitochondria have been intricately linked for over a century. In hemoglobin in the blood, the heme group is used to bind and transport oxygen. In the electron transfer chain, heme groups associated with peptides form the cytochromes. Another important cytochrome is the cytochrome P_{450}, an enzyme system used primarily in the liver for metabolizing various toxic compounds such as carcinogens. An intermediate in heme biosynthesis is a tetrapyrrole derivative (see below), a conjugated, planar ring structure without the metal ion, and in plants this intermediate is the precursor of chlorophyll.

The use of isotope tracers made it possible in the late 1940s to investigate the precursors for heme biosynthesis, and the whole pathway was established

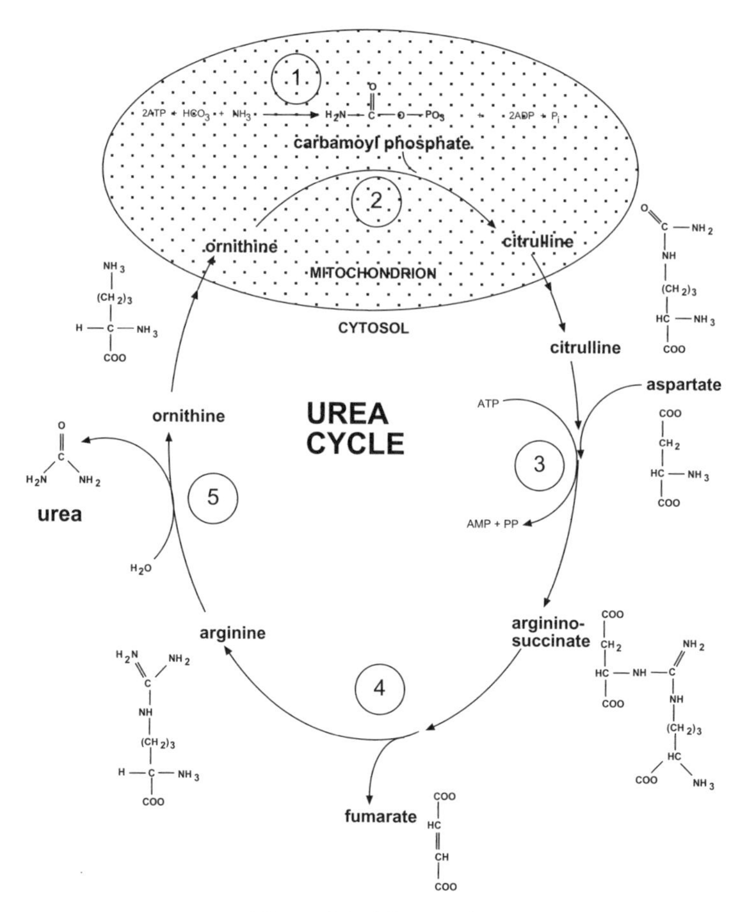

Figure 6.4 Urea cycle.

in the following decade (Figure 6.5). It starts in the mitochondria with a Krebs cycle intermediate, succinyl-CoA, and the formation of δ-aminolevulinic acid (ALA), involving the rate-limiting enzyme ALA synthase. It was described earlier how ALA protein levels can be controlled by regulating the efficiency with which ALA mRNA is translated (see also below). ALA leaves the mitochondria to be converted to coproporphyrinogen III in a complex series of reactions. An oxidative modification of the ring structure is coupled to its transport back into mitochondria, where the synthesis of heme is completed with another oxidation and the final addition of the iron to the center of the porphyrin ring. It it noteworthy that the latter reactions require molecular oxygen and thus are sensitive to oxygen concentrations.

ALA synthase is the key regulatory enzyme in the liver, but may not play this role in immature erythrocytes, another tissue in which heme biosynthesis is prominent. Under these conditions, ALA synthase may be feedback inhibited by its product. Its import into mitochondria may be controlled. It is also

A Glycine + Succinyl-CoA ⟶ δ-Aminolevulinic acid

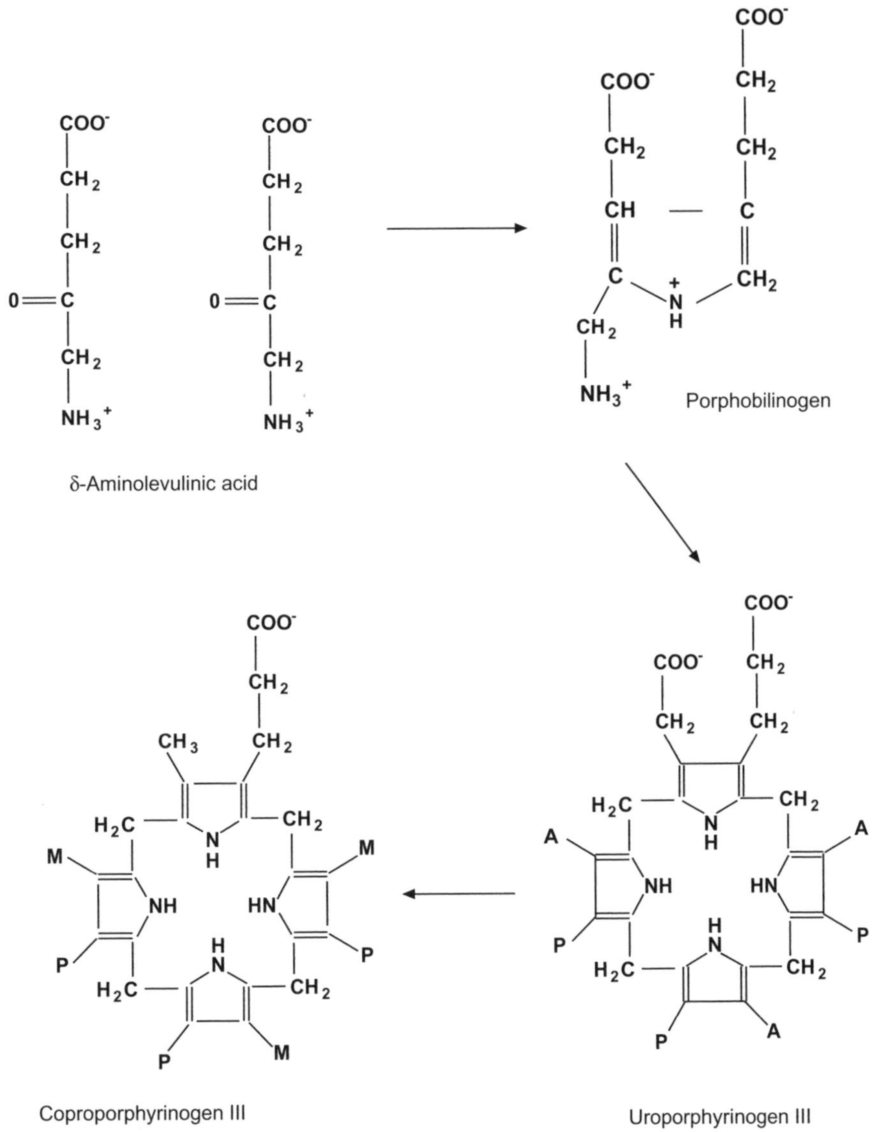

Figure 6.5 (A,B) Biosynthesis of heme. (C) Heme substituents in type a, b, and c cytochromes.

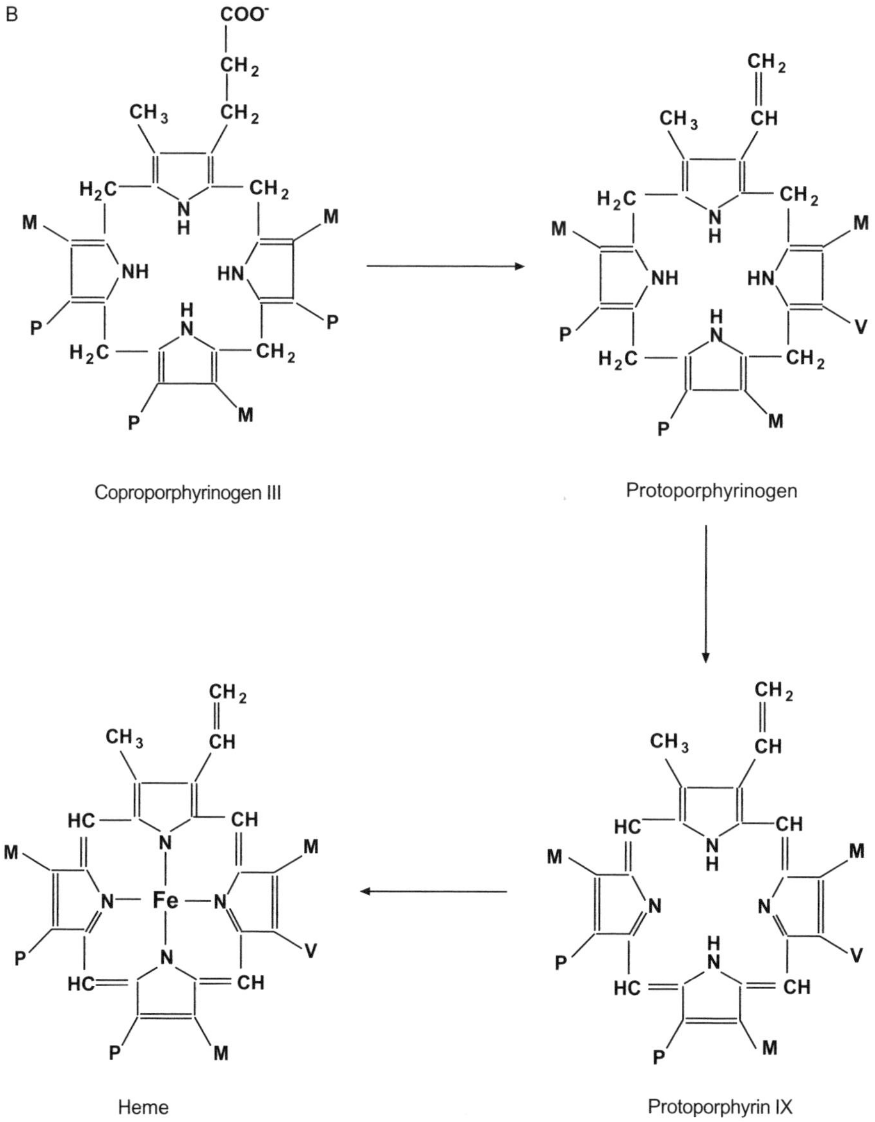

B

Coproporphyrinogen III

Protoporphyrinogen

Heme

Protoporphyrin IX

Figure 6.5 (*Continued*)

among the group of proteins that are associated with iron transport, storage, and metabolism (see page 324; Section 6.8.) and that are encoded by mRNAs containing iron-responsive elements (IRE) in their untranslated regions. When iron is deficient, the cytosolic aconitase is in an open conformation (the iron–sulfur center is dissociated) and capable of blocking the translation of ferritin and ALA mRNAs by binding tightly to the IRE. This mechanism allows the biosynthesis of heme to be adjusted to the availability of iron (Fe(II)) in the cell and mitochondria.

C

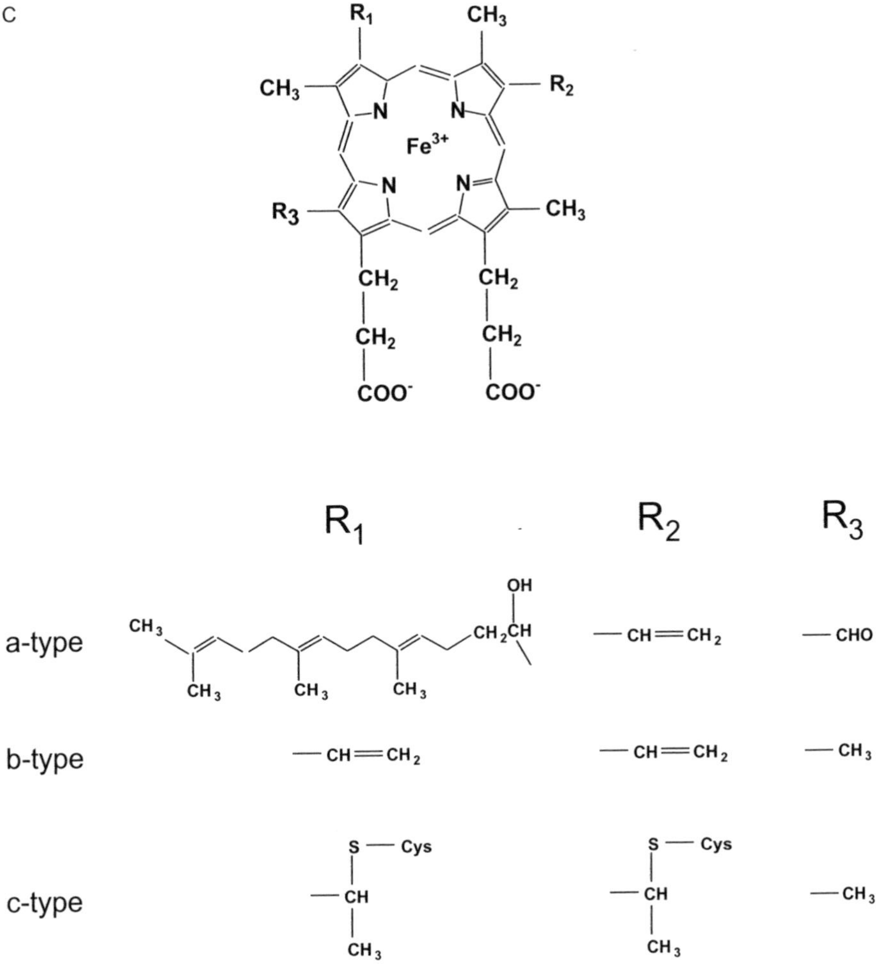

Figure 6.5 (*Continued*)

As described in an earlier chapter, the expression of genes encoding mitochondrial proteins in yeast is sensitive to oxygen. More specifically, the need for molecular oxygen in the final oxidations in the pathway makes heme synthesis sensitive to oxygen. In turn, heme can induce a number of genes required for respiration, control of oxidative damage, and sterol synthesis, and it can repress others (for review see reference 7).

As shown in Figures 6.5A and 6.5B, the porphyrin rings have various groups attached, making the whole structure asymmetrical. In their final form in a-, b-, and c-type cytochromes the Fe (II/III) ions are found in distinct microenvironments controlling their redox potential, not only because the peptides differ, but also because the substituents on the rings vary considerably (Figure

6.5C). In a-type cytochromes the heme is characteristically derivatized with an isoprenyl side chain. In c-type cytochromes the heme is covalently linked to cysteine side chains of the peptide.

6.6 CARDIOLIPIN AND LIPID BIOSYNTHESIS/METABOLISM

A comprehensive review of the lipid composition of mitochondria has been published by Daum (8). It is generally agreed that the inner membrane has one of the highest protein : lipid ratios of all biological membranes studied, but in spite of the dense packing of protein complexes and the various carriers and import machinery, the membrane nevertheless behaves as a fluid mosaic (9, 10). One of the most noteworthy aspects of the inner mitochondrial membrane is that it is the exclusive location for the lipid diphosphatidylglycerol, also named cardiolipin. This lipid has long been believed to be absolutely essential for the functioning of various complexes and integral membrane proteins. Some doubts were raised from the initial studies of a mutant yeast strain about the "absolute" requirement. In the first report of a yeast mutant with the cardiolipin synthase gene, *CLS1*, knocked out, and no cardiolipin found in any of the membranes, it may have been a surprise to find the cells viable in both fermentable and nonfermentable carbon sources. Purified, functional preparations of such complexes have inevitably contained cardiolipin, suggesting that this lipid has a special affinity for such proteins. More discussion on the need of this lipid for proper function of these complexes is found in Chapter 5. A comprehensive and expert review of the biosynthesis and biological significance of cardiolipin has been written by Schlame et al. (11) The proposed functions of this unique lipid have also been broadened to include a participation in mitochondrial protein import, protein folding, and assembly—that is, in the assembly and maintenance of the inner membrane. However, most of these roles of cardiolipin must remain highly speculative. In this discussion the focus will be on the synthesis of this lipid and some other characteristics of its composition.

Cardiolipin is found in the membranes of eubacteria such as *Escherichia coli*, as well as in the mitochondrial membranes of all eukaryotes studied so far, but not in archaebacteria (12–14). It is an unusual phospholipid with four acyl chains. The synthesis in bacteria involves the combination of two molecules of phosphatidyl glycerol with the elimination of one glycerol molecule. In contrast, in eukaryotes (liver, *Saccharomyces cerevisiae, Neurospora crassa*, bean sprouts, and mussels) the two starting substrates have been shown to be phosphatidylglycerol (PG) and CDP-diacylglycerol (see references 11, 13, and 15 for reviews and listing of additional references). The phosphatidyl group from CDP-diacylglycerol is transferred to PG by the enzyme cardiolipin synthase (Figure 6.6). The evolution of the reaction in eukaryotes from the biosynthetic pathway in bacteria has been postulated to be a consequence of the different levels of the precursor present in bacteria and mitochondria (15). In

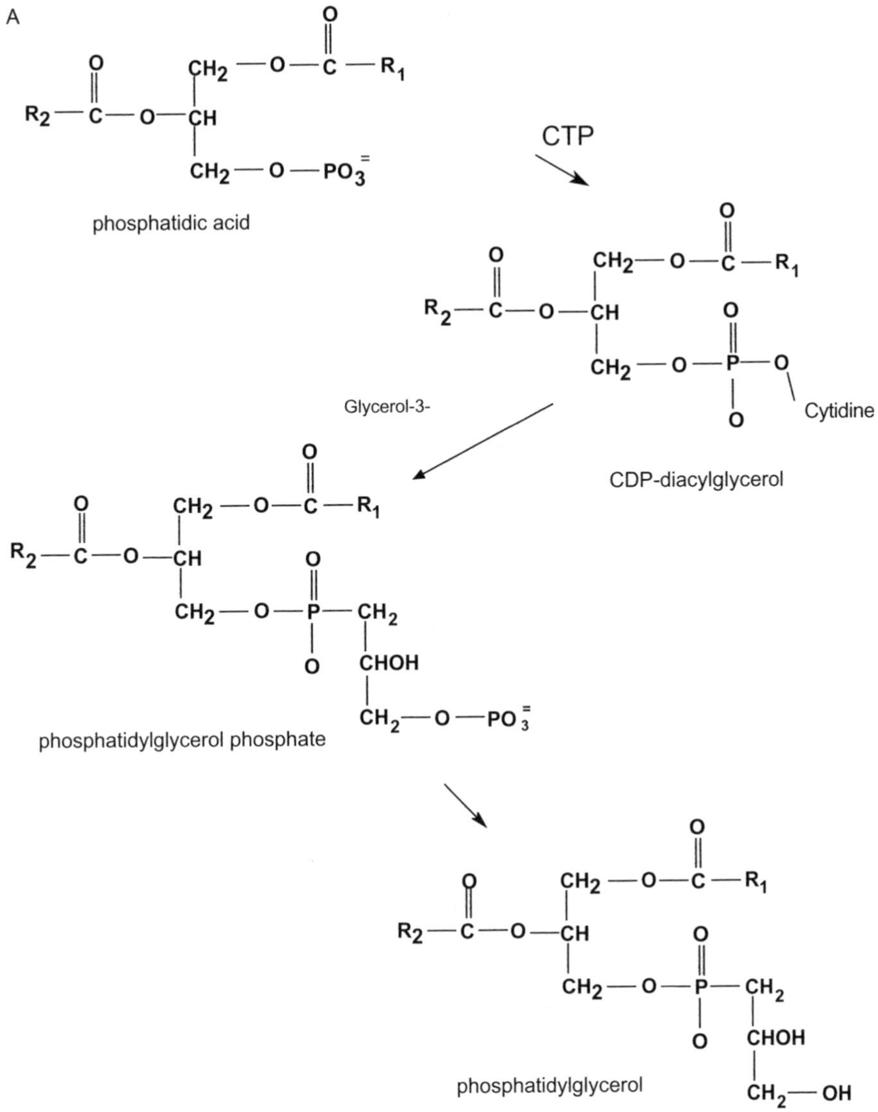

Figure 6.6 (A,B) Cardiolipin biosynthesis.

the former, phosphatidyl glycerol is abundant, and a simple, reversible trans-esterification can be driven toward cardiolipin by substrate availability. In mitochondria the precursor may be limited, and the forward reaction is driven by the irreversible cleavage of a high-energy anhydride bond in phosphatidyl-CMP (CDP-diacylglycerol). Hence, the mitochondrial reaction is highly exergonic. Properties of this enzyme from various mitochondrial preparations have been described (13, 15, 16). Although it had long been suspected to be present

B

Cardiolipin

$$
\begin{array}{l}
R_1-\overset{O}{\overset{\|}{C}}-O-CH_2 \qquad\qquad\qquad\qquad\qquad\qquad\qquad\qquad CH_2-O-\overset{O}{\overset{\|}{C}}-R_1 \\
R_2-\underset{O}{\underset{\|}{C}}-O-CH \qquad\qquad\qquad\qquad\qquad\qquad\qquad\qquad CH-O-\underset{O}{\underset{\|}{C}}-R_2 \\
\qquad\qquad\quad CH_2-O-\overset{O}{\overset{\|}{\underset{O}{\underset{|}{P}}}}-O-CH_2-\underset{HO\ \ H}{C}-CH_2-O-\overset{O}{\overset{\|}{\underset{O}{\underset{|}{P}}}}-O-CH_2
\end{array}
$$

Figure 6.6 (*Continued*)

exclusively in mitochondria, a detailed study has localized it to the matrix side of the inner membrane in rat liver mitochondria (17); that is, it is an integral membrane protein in this membrane with a large hydrophilic domain exposed to the matrix. The enzyme has been solubilized with the help of detergents and purified to a significant degree from yeast and mammalian sources. Several of the properties of this enzyme appear to be similar in all eukaryotes examined, in support of the monophyletic origin of mitochondria (13), but there are also some distinctions between the yeast enzyme and the mammalian enzyme of as yet uncertain significance. Most notably, the molecular mass of the yeast enzyme has been reported to be 20–30 kDa, in contrast to a molecular mass of ~50 kDa for the mammalian enzyme.

Schlame and co-workers (11, 13, 15, 16) also addressed the problem of the molecular acyl species found in cardiolipin of various species. In all organisms a high percentage of symmetrical acyl species was found—that is, species with identical diacylglycerol moieties. Acyl chains either were identical or differed only by two carbons or one double bond, respectively. Minimum energy conformations were calculated by these authors, but they point out that the conformation in a real membrane and in association with protein complexes may be quite different. Not surprisingly, the models suggest all acyl chains to be arranged more or less in parallel, and in a highly cartoon-like fashion one could represent cardiolipin as a single polar head group with four aliphatic chains (Figure 6.6B). Another notable property of cardiolipin is that it is an anionic phospholipid.

The Barth syndrome is an X-linked cardiomyopathy due to a defect in a protein called tafazzin (gene: TAZ). This enzyme is involved in a remodeling reaction of cardiolipin following the *de novo* biosynthesis of this lipid. Remodeling leads to cardiolipin highly enriched in linoleic acid. In the absence of tafazzin, cardiolipin levels are significantly reduced, and the remaining cardiolipin has a more diverse set of acyl chains (18, 19). Model systems for the study of tafazzin have been established in yeast (20) and in *Drosophila* (21), where the tafazzin mutation causes reduced motor activity and morphologically abnormal mitochondria in flight muscles.

Since cardiolipin is synthesized on the inside of the inner membrane and is therefore most likely found in the inner leaflet of the bilayer, one has to ask whether it can be distributed beyond this particular location. A "flippase" would be required to translocate it to the outer leaflet. On the other hand, in the immediate vicinity of the multisubunit integral membrane protein complexes where cardiolipin is found, the distinction between outer and inner leaflets may be meaningless. Movement to the outer membrane is even less likely but still not definitely excluded by experiments. Since mitochondria may fuse with each other, but not (normally) with other membranous organelles in a cell, the exclusive presence of cardiolipin in mitochondria can be explained.

The problem is less settled for the precursor phosphatidylglycerol (PG). It is derived from the precursor phopatidylglycerophosphate (PGP) (catalyzed by a phosphatase). PGP is made from CDP-diacylglycerol, and PGP synthase has also been localized in mitochondria. Such a finding would explain the presence of cardiolipin and PG in mitochondria, but PG appears to be a minor constituent of other intracellular membranes of mammalian cells, and it may be necessary to postulate a second, cytoplasmic isozyme for PGP synthase (22).

The biosynthesis of the other common phospholipids found in mitochondria is not exclusive, and detailed reviews on phospholipid metabolism are beyond the scope of this treatise (8, 12, 14). If the synthesis of these lipids takes place in the smooth ER, a mechanism must be found for distributing these lipids to other cellular locations. Vesicular traffic from the ER via the Golgi to the plasma membrane, as well as vesicular traffic in both directions between these membranes and various intermediate compartments, can be used to rationalize the distribution of lipids within a cell. Nevertheless, mechanisms must be postulated to explain the nonuniform distribution of lipids such as cholesterol in these membranes. Less attention appears to have been devoted to the delivery of lipids to mitochondria: Are specialized shuttle vesicles involved? Are lipids delivered individually by means of specialized carrier proteins? "Confluencies" between the outer mitochondrial membrane and the endoplasmic reticulum have been observed (see reference 23 for a review). It is not clear whether it is a transient state, and whether it serves to import components into mitochondria, or whether it represents an early step in degradation. When mitochondria are vitally stained with the cyanine dye $DiOC_6$, the dye is first accumulated by mitochondria, but subsequently it can be observed in the ER after the mitochondria have lost their elongated shape (23).

A few generalizations and basic principles are expertly reviewed by Daum (8), with special reference to mitochondria, and more recently by Kent (24) and Dowhan (14) from a more general perspective. While the lipid content of mitochondria is somewhat variable as expected, differing between tissues as well as between organisms, it is clear that the major phospholipids of mitochondria in addition to cardiolipin (10–20%) are phosphatidylethanolamine, PE (20–40%), and phosphatidylcholine, PC (35–50%). The other phospholipids are present at a level generally less than 5%, and cholesterol is present only in traces. These generalizations apply to vertebrate tissues, as well as to a variety of microorganisms and plant mitochondria. The latter also contain a variety of free and derivatized sterols (see reference 8 for review and original references). A further distinction can be made between the outer and inner membranes of mitochondria, and a few attempts have been made to characterize the differences. The presence of cardiolipin in the outer membrane is still controversial, since contamination is a possibility. Other differences in the ratio of PE:PC, or the relative amounts of phosphatidylinositol, PI, have been observed, but the functional significance is still obscure.

The origin and site of synthesis of mitochondrial lipids are areas still under investigation and in a state of evolution (25). Conclusions are somewhat subject to the success of fractionation schemes—most notably, fractionation of mitochondria from microsomes and other vesicular structures, identified by marker enzymes. Phosphatidylethanolamine is made in mitochondria from the decarboxylation of phosphatidylserine, PS. A search for the cellular location (in yeast) of phosphatidylserine synthase has identified a microsomal subfraction, but the absence of typical microsomal marker enzymes has led the authors to propose a novel particle population. One is forced to conclude that PS is imported into mitochondria, but some of the PE made must also be exported to other cellular membranes where it is found. Alternatively, a second PS decarboxylase may exist in a cytosolic membrane, and indications for its existence are found from the study of a yeast mutant defective in the mitochondrial PS decarboxylase (see reference 24 for discussion).

While cell fractionation studies have proved their worth over the decades, the fractionation of subcellular membranous structures has been a technical challenge because of microheterogeneity, and results are subject to misinterpretation when contamination cannot be ruled out. In the future, one can expect that model organisms like yeast will allow systematic gene disruptions to explore how the absence of a particular enzyme (and hence the product from this pathway) affects growth and other cellular activities. An example of the unexpected type of result derived from a genetic approach is the finding that unsaturated fatty acids are essential for the distribution of mitochondria to the daughter cells in the yeast *Saccharomyces cerevisiae*, presumably because the necessary shape changes (elongation) are dependent on the fluidity of the mitochondrial membranes (26).

6.7 BIOSYNTHESIS OF UBIQUINOL (COENZYME Q)

The biosynthesis of ubiquinol was elucidated some time ago from studies in *Escherichia coli* and yeast *Saccharomyces cerevisiae*. Genetic studies were critical, since CoQ-deficient strains of each organism could be isolated, and intermediates accumulating at different stages in the biosynthetic pathway gave clues about the sequence of reactions. Although the pathways differ slightly in some intermediate steps in prokaryotes and yeast, 10 genes have been defined in each organism by the isolation of eight complementation groups (*coq1–coq10* in yeast). The CoQ-deficient yeast strains are respiration-deficient and fail to grow on nonfermentable carbon sources, but they do grow on glucose (27). The laboratory of Clarke has made a number of significant contributions to the detailed functional analysis of a number of different coq mutants in yeast (28–32). The deficiency can in many cases be overcome by the addition of exogenous Q_2.

The reactions defined by mutations in CoQ/ubiquinone-deficient bacteria or yeast are the final reactions unique for the biosynthesis of ubiquinone. Key starting materials in this pathway are *p*-hydroxybenzoic acid and polyisoprene diphosphate. The final polyprenyl tail contains n isoprene units, where *n* varies between organisms. In mammalian mitochondria $n = 10$, and hence the compound is also referred to as Q_{10}. The Q_{10} derivative is extremely insoluble in water, and in *in vitro* experiments as well as in therapeutic applications derivatives with shorter tails are often used.

The synthesis of the polyprene tail begins in the well-known pathway from acetyl-CoA via hydroxymethylglutaryl-CoA , mevalonate, geranyl-pyrophosphate, and farnesyl-pyrophosphate, the initial reactions established first by K. Bloch in the biosynthesis of cholesterol. It has now become clear that farnesyl-pyrophosphate is a key intermediate and branch-point in several very important pathways leading to cholesterol synthesis, dolichol synthesis, ubiquinone synthesis and protein isoprenylation (see references 33 and 34 for recent reviews). The reactions up to the synthesis of farnesyl-pyrophosphate are also referred to as the mevalonate pathway, and they are illustrated in Figures 6.7A and 6.7B. Three acetyl-CoA molecules are first combined successively to form hydroxymethylglutaryl-CoA, the precursor for mevalonate. Phosphorylations and a decarboxylation yield isopentenyl-pyrophosphate (IPP), which is reversibly isomerized to dimethylallyl-pyrophosphate (DMA-PP). A condensation of IPP and DMA-PP yields geranyl-pyrophosphate (GPP), and the addition of another isopentene unit leads to farnesyl-pyrophosphate (FPP). The enzymes are soluble cytosolic enzymes, with one exception: The enzyme HMG-CoA reductase is an integral membrane protein in the endoplasmic reticulum. It has received a considerable amount of attention as a key regulatory enzyme in cholesterol biosynthesis. It may therefore also influence the rate of synthesis of the other products at the end of the various branches, but while its regulatory role in cholesterol biosynthesis is undisputed, modulating its activity by dietary conditions or inhibitors has been

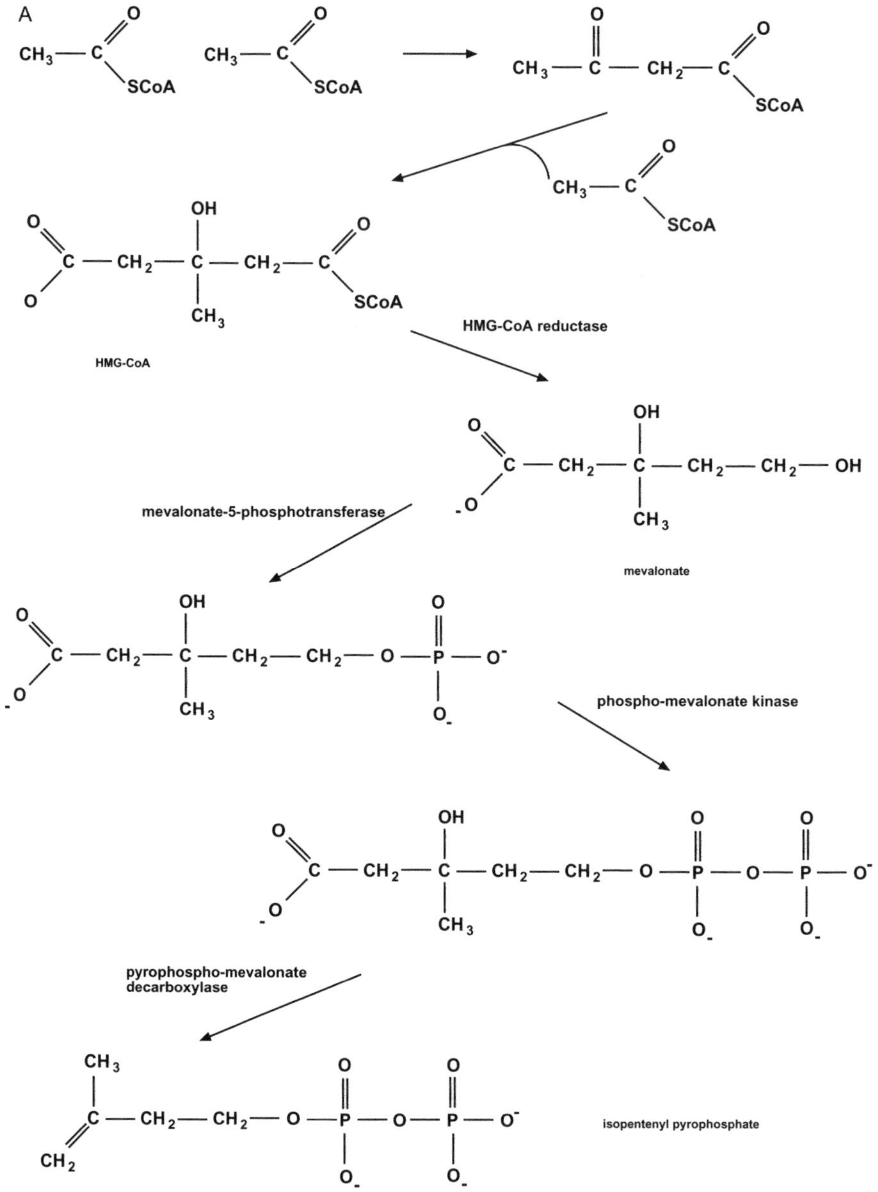

A

HMG-CoA reductase

HMG-CoA

mevalonate-5-phosphotransferase

mevalonate

phospho-mevalonate kinase

pyrophospho-mevalonate decarboxylase

isopentenyl pyrophosphate

Figure 6.7 (A–D) Ubiquinone, cholesterol, and Dolichol synthesis.

found to have no significant influence on the biosynthesis of dolichol, ubiquinone, or the isoprenylation of proteins. This apparent dilemma has been addressed by several authors, and a plausible model referred to as the "flow diversion hypothesis" has been postulated. According to this hypothesis, most

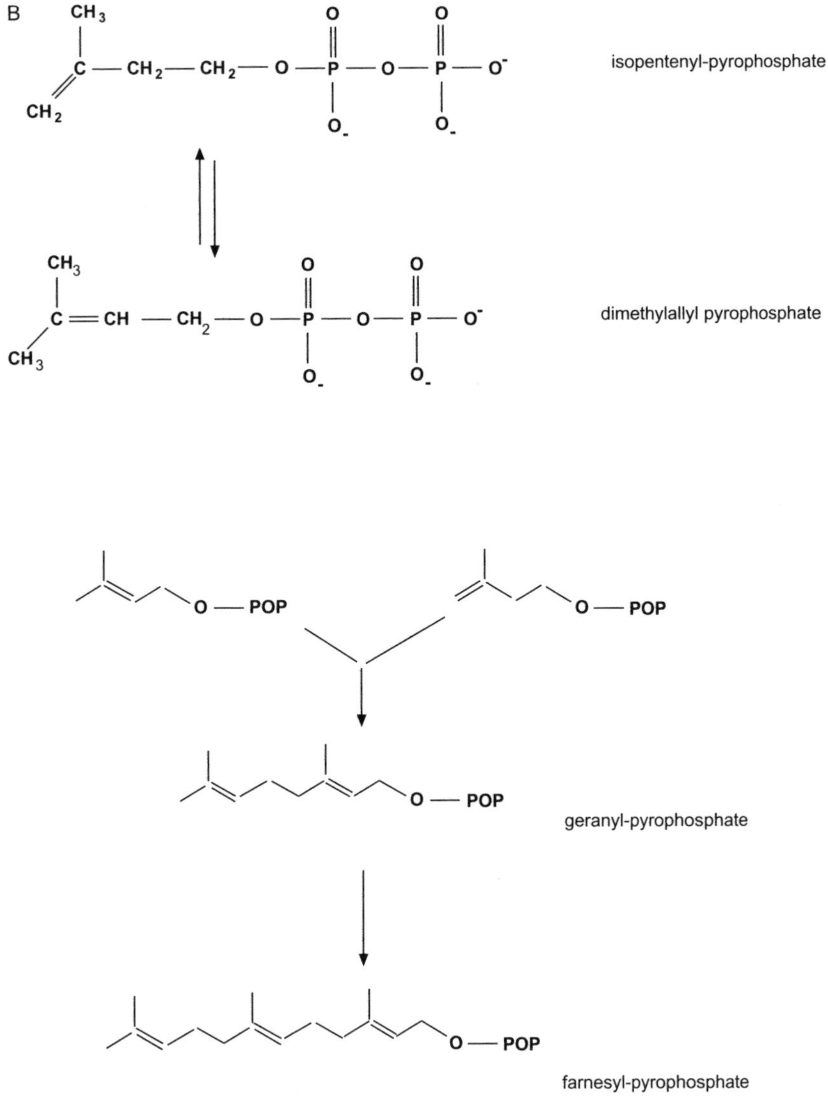

Figure 6.7 (*Continued*)

of the enzymes involved in the branch-point reactions have a high affinity for FPP, and flux into these branches can occur even at low FPP concentrations. The branch-point enzyme squalene synthase, on the other hand, has a low affinity for FPP, and thus cholesterol biosynthesis will occur only when high concentrations of FPP are achieved by an activation of HMG-CoA reductase.

Protein isoprenylation with FPP or GGPP (see Figure 6.7C) as isoprenoid donors is now recognized to be widespread in eukaryotic cells. Proteins such

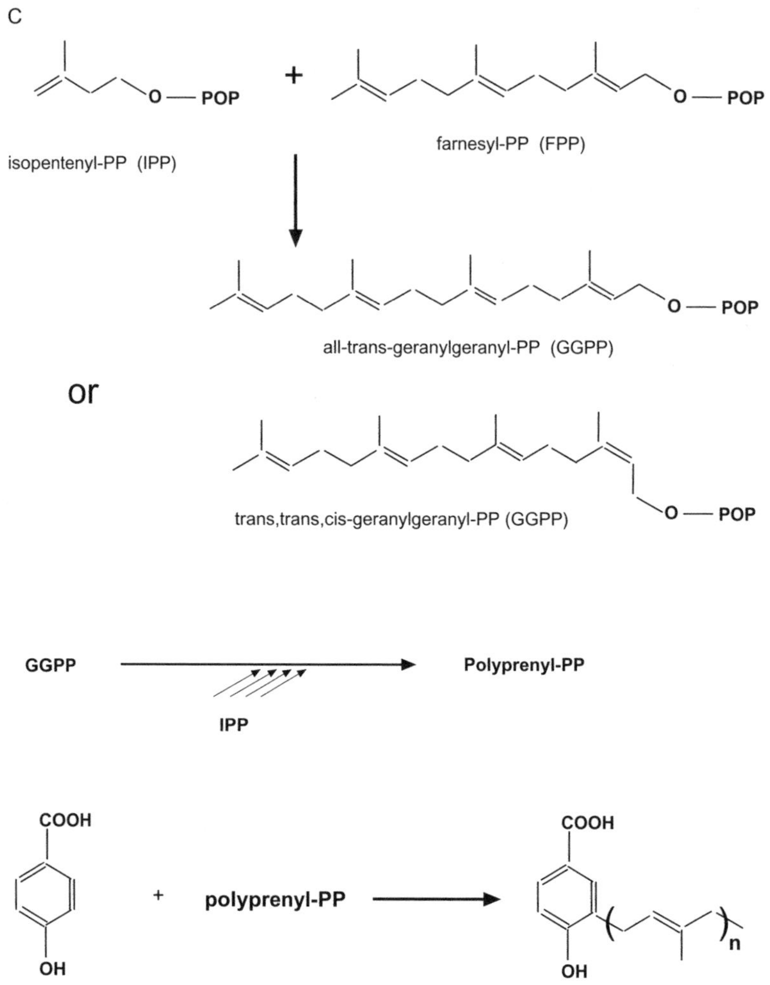

C

O — POP + O — POP

isopentenyl-PP (IPP)

farnesyl-PP (FPP)

O — POP

all-trans-geranylgeranyl-PP (GGPP)

or

trans,trans,cis-geranylgeranyl-PP (GGPP) O — POP

GGPP → Polyprenyl-PP

IPP

COOH

OH

+ polyprenyl-PP →

COOH

OH

n

Figure 6.7 (*Continued*)

as the ras protein can be derivatized with long hydrophobic prenyl side chains that serve to anchor them on membranes at specific subcellular locations. A full discussion of this subject is beyond the scope of this chapter. As shown in Figure 6.7C, geranyl-geranyl-pyrophosphate (GGPP) is made from FPP and IPP by enzymes named GGPP synthases (a subset of prenyltransferases). There are at least two types, and the product can be either the all-*trans*-GGPP, or the *trans, trans, cis*-GGPP. These enzymes differ in their intracellular distribution; they may be cytosolic or membrane-associated; they have been found in mitochondria, peroxisomes, and the ER; either Zn^{2+} or Mg^{2+} may be absolutely required. The cytosolic enzyme appears to produce only all-*trans*-GGPP, while *trans, trans, cis*-GGPP is produced by membrane-bound enzymes.

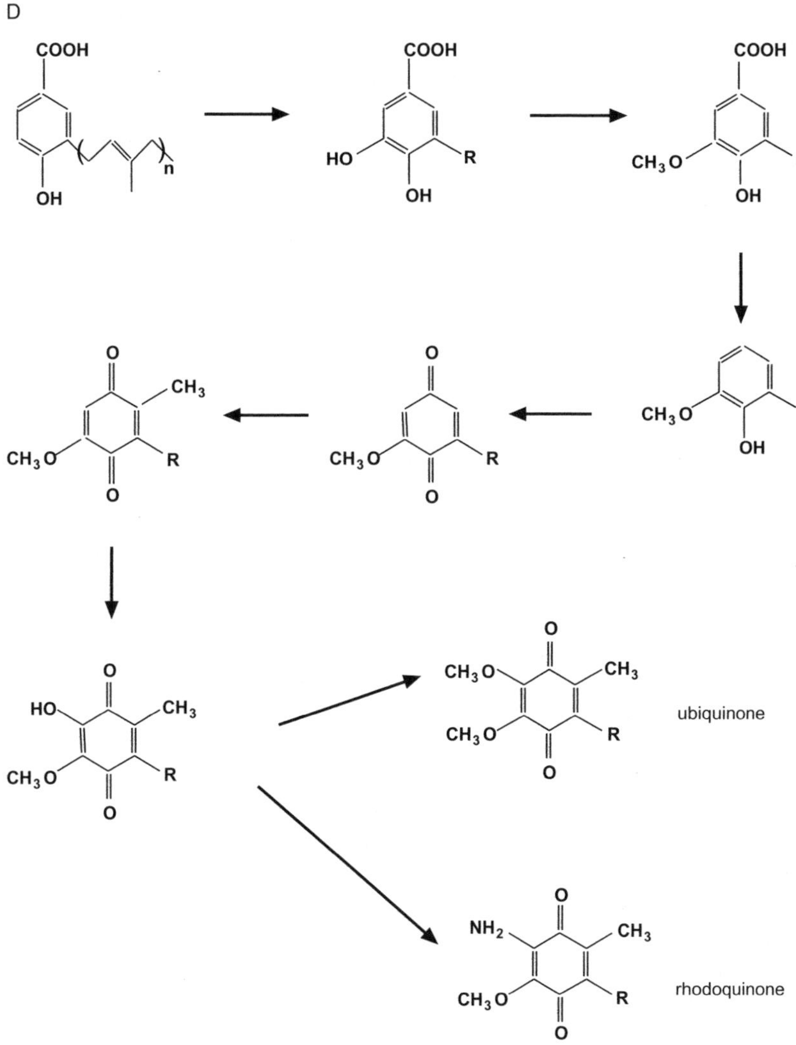

Figure 6.7 (*Continued*)

Additional prenyltransferase reactions with IPP as donor serve to elongate the chain, but the all-*trans*- and the *trans, trans, cis*-GGPP have different fates. The latter is used for the synthesis of dolichol, containing between 16 and 23 isoprenoid units in animal cells. Dolichol is phosphorylated and esterified to oligosaccharides to form an essential intermediate in glycoprotein syntheses on the ER. This is another long story not to be pursued here. The all-*trans*-GGPP has been clearly established as the substrate for protein isoprenylation; and it is also likely to be the substrate for *trans*-prenyltransferase, leading to the synthesis of all-*trans*-polyisoprenoid-PP. Consisting of about 9 or 10 units, the polyisoprenyl group can be transferred to *p*-hydroxybenzoic acid in the first distinct step leading to ubiquinone (Figures 6.7C and 6.7D).

The subcellular localization of the distinct prenyl transferase enzymes could be used to control the destination of the various products, and also serve in establishing separate regulatory mechanisms. Cell fractionation studies in conjunction with genetic analyses are not yet complete to present a final picture. As long as ubiquinone was thought to be exclusively present in the inner mitochondrial membrane, the presence of the biosynthetic enzymes in mitochondria could be rationalized. However, it has since been established that ubiquinone is found in extra-mitochondrial membranes as well. Are there multiple enzymes for ubiquinone synthesis, or can this molecule be exported from mitochondria? The more abundant species is found to be the reduced form (ubiquinol), and there appears to be agreement that ubiquinol has an important function as an antioxidant (34). Its abundance excedes α-tocopherol (vitamin E) by a factor of 10 in animal cells.

An interesting problem presents itself in certain eukaryotic microorganisms that have complex life cycles and may include a stage that exists in an aerobic environment (carrying out normal respiration and oxidative phophorylation), and another stage living in an environment requiring drastic adjustments in metabolism and even anearobic electron transfer (35). One such example of a parasite, the liver fluke, *Fasciola hepatica*, has been studied in some detail. The free-living stage (aerobic) oxidizes succinate and transfers electrons via UQ_{10} to complex III of the electron transfer chain. The parasitic stage oxidizes NADH via complex I and the reduction of fumarate to succinate by fumarate reductase. The mobile carrier between complex I and FRD is another quinone, rhodoquinone (RQ) (36, 37). Experiments with [2-^{14}C]-mevalonate demonstrated that UQ and RQ can be synthesized *de novo* in these organisms. Rhodoquinone has an amino group substituted on the quinone ring, and this modification must be part of the developmental program of this organism. The substitution alters the redox potential of the RQ/RQH_2 redox couple that appears necessary to allow electron flow to occur in the reverse direction in FRD compared to complex II (38).

6.8 BIOSYNTHESIS OF Fe–S CENTERS

Iron–sulfur clusters were discovered in the early 1960s (39), and for several decades it was assumed that they formed spontaneously. Research during the past decade has revealed that the process is surprisingly complex, and mitochondria were found to play an essential and central role in it. The studies were greatly stimulated by research on the biosynthesis of Fe–S clusters in bacteria (40), as well as by the use of yeast as a model organism for genetic and biochemical approaches. In turn, insights from studies in yeast can be extrapolated to higher eukaryotic systems including humans with increasing confidence. Many of the components involved in the machinery have been highly conserved in evolution. This has greatly facilitated the search for, and the identification of, the relevant genes/proteins in higher eukaryotes from knowledge about the corresponding genes in prokaryotes or in other eukaryotes.

Bacteria are capable of Fe–S cluster assembly by three distinct systems (NIF, ISC, SUF), one or two of which may be present in the same cells depending on the strain. The NIF machinery is most specific and exclusively responsible for the biosynthesis of nitrogenase required for nitrogen fixation. The SUF machinery may be induced (?) under iron-limiting or stress conditions, but it may also be the only system present—for example, in cyanobacteria. The majority of cellular Fe–S proteins are generated with the help of the ISC machinery that may therefore be considered a "housekeeping" system. It is the system inherited by eukaryotic cells from their bacterial symbionts, with modifications and additions made in the course of evolution. In the following the ISC system of eukaryotes will be the focus of attention.

Fe–S clusters in proteins can serve a variety of functions (40, 41). In the present context, their role in electron transfer and redox reactions is the most prominent, but they also serve in enzymatic reactions in substrate binding and activation (without electron transfers), or they may have a purely structural role in stabilizing the tertiary structure of a protein. They may serve in iron storage (ferredoxins) or in iron-sensing (cytoplasmic aconitase). As the list is growing, it is important to recognize that Fe–S proteins are found in at least three distinct compartments of a eukaryotic cell: mitochondria, cytosol, and nucleus. Only one example of a nuclear Fe–S protein is known so far (DNA glycosylase). A larger list of cytoplasmic Fe–S proteins includes a variety of metabolic enzymes, and two that deserve special mention. The cytosolic aconitase alias "iron-regulatory protein 1" (IRP-1) has already been introduced as the sensor of intracellular iron (page 312). When intracellular iron becomes deficient, its Fe–S cluster dissociates, producing a protein with high affinity for an iron-responsive element (IRE) localized in the 5′UTR or 3′UTR of a subset of mRNAs. Their stability or translatability is controlled by this mechanism, and thus the level of key enzymes/proteins associated with uptake, storage, and utilization of iron (e.g., heme biosynthesis) is regulated. The ABC protein Rli1is essential for the biogenesis of ribosomes, RNA processing, and translation initiation in yeast (42, 43). It is one of the most conserved proteins in evolution, and under some conditions in yeast it may be the only indispensable Fe–S protein (see below).

An authoritative and comprehensive reviews of Fe–S protein biogenesis in eukaryotes by one of the pioneering laboratories in this field has appeared recently (41). The reader is referred to this review and a complementing review emphasizing the bacterial systems (40), but the key features of the eukaryotic system are also concisely summarized by Johnson et al. A summary and some discussion is presented here with reference to Figure 6.8 (from reference 41).

The process can be decomposed into several distinct stages, beginning with import of Fe^{2+} into mitochondria. The ISC assembly machinery is made up of

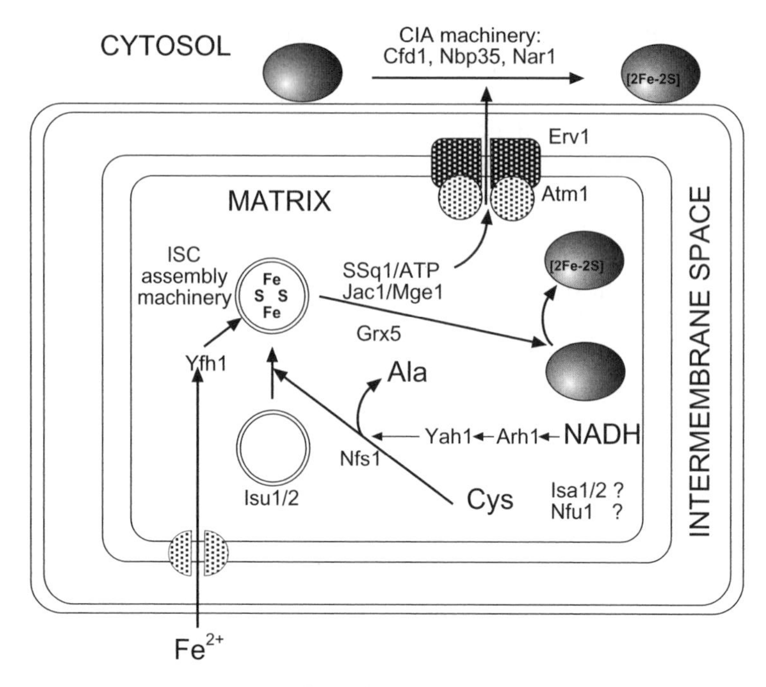

Figure 6.8 Biosynthesis of [Fe–S] centers (adapted from reference 41).

so-called scaffolding proteins (Isu1,2) to which the two constituents of a Fe–S cluster are delivered. Yfh1 is the yeast homologue of human frataxin; its function in yeast is to deliver Fe^{2+} to the ISC assembly machinery. Defects in this protein lead to an impairment of Fe–S cluster biosynthesis, mitochondrial iron accumulation, heme deficiency, and oxidative stress. In humans, frataxin deficiency causes a neurodegenerative disease, Friedreich's ataxia (44). The second branch in the lower part of the figure is concerned with the production and delivery of the sulfide (S^{2-}). The sulfide is produced from cysteine by a pyridoxal phosphate-dependent cystein desulfurase (Nsf1p). A more speculative aspect of this pathway is the role of the Yah1 and Arh1 proteins. Lill and Muhlenhoff propose an electron transport chain for electrons provided by NADH and transferred to the desulfurase and to elemental sulfur (S^0) to form the sulfide (S^{2-}) ion. Intriguingly, in yeast these proteins are also required, together with Cox10p and Cox15p, for the synthesis of heme a in complex IV, while homologous proteins in the mammalian adrenal gland are associated with steroidogenesis from cholesterol.

The Fe–S cluster thus formed and bound to the scaffold (consisting of Isu1p, Isu2p, and possibly other proteins) now has to be transferred to apoproteins in the mitochondria, where they may remain as soluble proteins (e.g., aconitase), or they will have to be incorporated into the various complexes of the electron transport chain. The temporal sequence of these events is not clear. For example, there are three Fe–S clusters in the SDHB subunit of complex

II, and it remains to be established whether these clusters are incorporated before or after association of this subunit with the flavoprotein, SDHA, and before or after the association of SDHB with the two integral membrane proteins SDHC and SDHD. Elucidation of the assembly pathway of the peripheral subcomplex of complex I with 7 or 8 Fe–S clusters in several different subunits can be anticipated to be even more challenging. Genetic studies in yeast have identified additional components believed to act in the transfer of the Fe–S cluster from the initial scaffold to diverse target proteins. SSq1p and Jac1p are chaperone proteins, and Grx5 is glutaredoxin. A precise mechanism for their involvement has not been formulated, and ATP hydrolysis is a likely driving force for various association/dissociation steps. A role for the gene products from the *ISA1/2* and *NFU1* genes in yeast can be inferred from the defects in the activity and synthesis of various mitochondrial Fe–S proteins resulting from a depletion of Isa1p or Isap2. Nfu1-like proteins have been demonstrated to be required for Fe–S protein biosynthesis in plant mitochondria, chloroplasts and cyanobacteria. In yeast a deletion of Nuf1 has no phenotype, but synthetic lethality with other ISC genes (e.g., *ISU1*) suggests an auxiliary function.

As mentioned above, Fe–S proteins are also found in the cytosol and even in the nucleus. It appears now that Fe–S cluster formation is initiated in mitochondria and all mitochondrial ISC-assembly proteins are required for the biosynthesis of the cytosolic and nuclear proteins. A current picture includes an ISC-export machinery in the inner mitochondrial membrane and a CIA machinery in the cytosol (Figure 6.8). Components of each have been identified in yeast by genetic approaches. A mutation in the *ATM1* gene was found to cause leucine auxotrophy due to the absence of an essential cytosolic enzyme (and Fe–S protein) in the leucine maturation pathway. *ATM1* encodes an ATP binding cassette transporter in the inner mitochondrial membrane. Further support for a role of this transporter comes from the discovery that mutations in the corresponding human gene ABCB7 are the cause of X-linked sideroblastic anemia and cerebellar ataxia, an iron storage disease. It is hypothesized that an Fe–S cluster produced by the ISC-assembly machinery is transferred to a stabilizing factor that is then exported by the Atm1 transporter, but the precise nature of the exported substrate remains to be established. Indirect experiments have suggested that the exported Fe-cluster is stabilized by a peptide. The Erv1p in the export machinery is a sulfhydryl oxidase localized in the intermembrane space, and deletion of the *ERV1* gene is lethal. Another essential component of the export machinery is glutathione, but its precise role is still unknown.

The cytosolic machinery for Fe–S protein biosynthesis (CIA) includes several components identified only recently. The highly conserved eukaryotic proteins Cfd1p and Nbp1p belong to a class of soluble P-loop NTPases, and the Nar1p is related to the bacterial iron-only hydrogenases. Many details of the molecular mechanism for cytosolic and nuclear Fe–S protein assembly will be the subject of future experiments.

Lill and colleagues have raised the question, Why is Fe–S protein maturation essential in yeast, even when glucose is abundant for fermentation and rich media provide all (?) essential metabolites? They conclude that the only known essential yeast Fe–S protein is the cytoplasmic Rli1p, required for ribosome biogenesis and function. It is one of the most highly conserved proteins in evolution and likely to be equally essential in other eukaryotes. This leads to the further proposal that the only absolutely indispensable function of mitochondria in yeast is Fe–S protein biogenesis. What about eukaryotes that lack mitochondria? Such organisms (*Giardia intestinalis, Trichomonas vaginalis, cryptosporidium parvum*; see chapter 2) have been shown to harbor subcellular organelles called hydrogenosomes or mitosomes. When complete genomic sequences were available, genes for oxidative phosphorylation and reactions of the citric acid cycle were missing. However, numerous genes encoding proteins of the mitochondrial ISC-assembly complex were found by amino acid sequence comparisons, suggesting that an ISC-like complex is present in these organisms. Furthermore, some of these proteins have now been localized in the hydrogenosomes of *T. vaginalis* and in mitosomes of *G. intestinalis*, and it is a good guess that these organelles are the site of Fe–S cluster assembly. Most, if not all, of the Fe–S clusters will have to be exported for incorporation into cytosolic proteins, since no Fe–S proteins are known in mitosomes. A consensus is emerging that these organelles are derivatives of ancestral mitochondria. They appear to have evolved independently several times during the evolution of organisms adapting to anaerobic environments (45–47). It may be concluded that the biosynthesis of Fe–S clusters is the most important function of mitochondria that must be conserved under all conditions.

A highly succinct summary of the most significant features of Fe–S cluster biogenesis has also been provided by Johnson et al. (40):

- Participation of cysteine desulfurases in the mobilization of sulfur
- A requirement for molecular scaffolds for assembly of [Fe–S] clusters
- The participation of molecular chaperones
- Intact cluster transfer from scaffolds to target proteins
- Specialized [Fe–S] cluster biosynthetic machinery for specific targets
- Specialized [Fe–S] cluster biosynthetic machinery to accommodate stress conditions
- Feedback regulation of [Fe–S] cluster biosynthesis and regulation in response to environmental conditions
- Compartmentalization of [Fe–S] cluster biosynthetic machinery
- Intracellular trafficking of preformed [Fe–S] clusters
- Mechanisms for the protection or repair of [Fe–S] proteins (not discussed in this review)

Many details of each of these mechanisms, along with the genes/proteins involved, can be expected to emerge in the years to follow. It can hoped that studies will succeed in reconstituting major steps *in vitro* with isolated and pure components to gain insight into the precise interaction of various enzymes, scaffolds, chaperones, and transporters.

6.9 TRANSPORT OF SMALL SOLUTES INTO AND OUT OF MITOCHONDRIA

6.9.1 Introduction

Although numerous references to the transport of various ions across the mitochondrial inner membrane have been made in the sections above, it seems worthwhile to focus on this problem once more as a distinct problem. The following ions and small solutes feature prominently in discussions of mitochondrial functions: H^+, P_i, ADP/ATP, Fe^{2+}, Cu^+, Zn^{2+}, Ca^{2+}, NAD^+/NADH, and a large variety of small organic compounds that are metabolites associated with biochemical reactions. The first three have received a maximum of attention because of their role in oxidative phosphorylation and the Mitchell hypothesis. Fe^{2+}, Cu^+, and Zn^{2+} ions have been encountered as constituents in the electron transport chain (cytochromes, iron–sulfur centers, complex IV) and in metallo-proteases. Mitochondria are prominently mentioned in the context of the regulation of intracellular calcium; they may constitute a reservoir or buffer, and/or their activities may themselves be regulated by Ca^{2+} ions (48, 49). NAD^+/NADH are not transported freely across mitochondrial membranes, and most biochemistry textbooks emphasize the importance of shuttle systems in the transport of reducing equivalents into mitochondria. Nevertheless, there is an intra-mitochondrial pool of NAD^+/NADH, and concentrations may or may not be similar to those in the cytosolic pool; a limited transport of this cofactor must be possible. Finally, amino acids, fatty acids, acetate, glycerol-phosphate, keto acids, and many other metabolites must be able to enter or exit mitochondria. It seems appropriate to consider these transport systems and shuttles in the present chapter.

The flux of protons through the inner membrane has been discussed extensively in the previous sections. Protons are pumped out of the matrix by the operation of the electron transport chain, and they return via complex V and through various "leakage" channels, uncouplers, symporters, and antiporters. In *in vitro* systems, protons can be exchanged for potassium ions by the use of the ionophore Nigericin. The elucidation of the precise path of a proton through the various complexes remains a challenging problem. An interesting mitochondrial proton pump that has received less attention (at least in textbooks) is the nicotinamide nucleotide transhydrogenase. The bovine enzyme is an integral membrane protein made up of two identical monomers of 109,065 kDa. An N-terminal domain (430 residues) and a C-terminal domain (200 residues) flank a hydrophobic central domain that can be predicted to have 14 transmembrane helices. The hydrophilic end-domains extend into the matrix, where the N-terminal domain binds NAD(H), and the C-terminal domain binds NADP(H). In the process of transhydrogenation involving a hydride transfer from one cofactor to the other, protons are pumped across the inner membrane.

$$\mathrm{NADH + NADP + nH^{+}_{out} \Leftrightarrow NAD + NADPH + nH^{+}_{in}}$$

As discussed in more detail in a review by Hatefi and Yamaguchi (50), this transhydrogenase is a unique proton pump and a potentially powerful model system to understand how a simple protein can translocate protons across a membrane driven by conformational transitions that in turn are induced by a chemical reaction on one side of the membrane and by the different binding energies for substrates and products.

6.9.2 Porin Alias VDAC

It is generally assumed that pores in the outer membrane formed by an abundant mitochondrial protein named VDAC (or porin) make the outer membrane highly permeable to most small molecules (<4000–5000 kDa). Thus, the inner mitochondrial membrane is the major permeability barrier between the cytosol and the mitochondrial matrix, and the outer membrane is usually ignored when considering small molecules. VDAC stands for "voltage-dependent anion channel," and this term immediately becomes confusing if one has the image of "pores" with a fixed diameter pore size. There is a generic definition of porins as membrane proteins that form a channel through which molecules of limited size can diffuse passively. They are found prominently in bacterial outer membranes forming β-barreled structures. Most of the porin types form trimeric structures, and they exhibit some specificity so that a variety of porins are typically found in a bacterial outer membrane. In many discussions a β-barreled protein (VDAC) in the mitochondrial outer membrane is considered to be very similar, but it should be pointed out that the mitochondrial porin-like protein has no primary sequence homology to bacterial porins. In contrast to the bacterial form, it is believed that a single monomer can form the functional channel, although aggregation in the membrane has been observed (see reference 51). There are still arguments over whether association of monomers (dimers or tetramers) is necessary for forming a channel.

Crystal structures for bacterial porins have strongly influenced the speculations about the VDAC channel and its selectivity. The latest interpretation based on protease studies, peptide-specific antibodies, and site-directed mutagenesis suggests a channel composed of 13 β strands and one α helix (51). The authors call the single α helix "aesthetically unappealing" but rationalize it by the need to have a gating mechanism associated with a large conformational change.

VDAC proteins have been reconstituted into planar lipid bilayers for electrophysiological studies of single-channel conductance, ion selectivity, and voltage gating parameters. For a detailed discussion the reader is referred to the work by Colombini and co-workers (51). How are these properties related to function? It is suggested that the channel can exist in various functional states differing in selectivity between anions and cations. In the open state the selectivity for anions is not very large, while cations are favored by the closed state. It is argued that pore size is one major determinant with regard to non-electrolytes (e.g., PEG 3400), but the electrostatic profile within the channel can have an influence on the passage of charged molecules such as ATP or NAD^+. Thus, in the closed state the channel may prevent ATP/ADP exchange with the cytosol. The question becomes, What controls the open and closed states of the VDAC channel? Several suggestions are summarized in the review by Rostovtseva et al. (51), among them phosphorylation. Another group (52) has elevated VDAC to the "governator of global mitochondrial function both in health and in disease." As has been pointed out, the lack of a specific and potent pharmacological inhibitor of this channel has made physiological studies difficult.

VDAC has also assumed a prominent place in discussions of early events in apoptosis, where it has been implicated in the formation of an even larger pore allowing the escape of cytochrome c from the IMS (51). Either it could become part of the pore, or its behavior could cause outer membrane permeabilization by an unknown mechanism. Alternatively, VDAC has been hypothesized to stack on top of the adenine nucleotide transporter (ANT) to form a contact site and a continuous pore through the outer and inner membrane. Such a structure could be responsible for the permeability transition and breakdown of $\Delta\Psi$ at some stage during apoptosis. This topic will be expanded on in the section on apoptosis.

6.9.3 The ADP/ATP Translocator

The ADP/ATP translocator (adenine nucleotide translocase, ANT) is an interesting integral membrane protein with a rather unique function in mitochondria. Its properties, shared with the uncoupling protein, UCP-1 and with what is believed to be a large family of mitochondrial carriers, have already been emphasized (53). To repeat: The protein is an integral membrane protein with a molecular mass of ~32 kDA (297 residues in the bovine protein). There are three repeats of about 100 amino acids, each of which contains two shorter motifs representing transmembrane helices. The protein therefore is localized in the membrane with six transmembrane helices, and both the N terminal and the C terminal are on the same side, facing the intermembrane space. The loops on the matrix side contain an exceptionally polar combination of about 40 side chains. There are also signature motifs with short sequences conserved in all members of the family. These have been exploited very successfully in blast searches of whole genomes to identify novel family members (54). Biochemical and biophysical experiments had suggested that a dimeric structure makes up the functional carrier, with a twofold axis of rotational symmetry

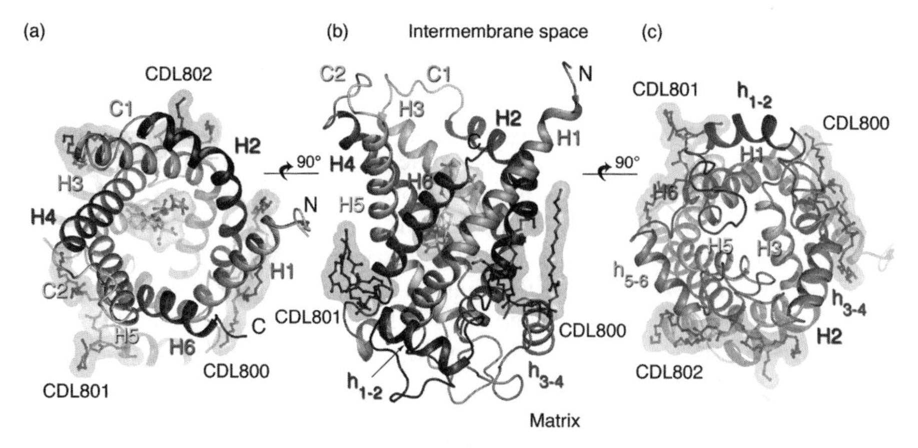

Figure 6.9 X-ray structure of ANT (from reference 57 with permission).

that has also been proposed to form the channel through which ADP/ATP translocation occurs. A structure at a resolution of 2.2 Å by X-ray crystallography in complex with the inhibitor carboxyatractyloside (see below) was published in 2003, representing the first high-resolution structure of a member of this large carrier family (Figure 6.9) (55–57). It has since become a model structure for predicting and understanding other members of this mitochondrial carrier family (see page 334). Six transmembrane α-helices are clearly seen, but they are severely tilted relative to each other and relative to the orthogonal axis through the membrane. In the crystal the unit cell contained only the monomer. Furthermore, the six helices form a barrel-shaped structure with the postulated nucleotide binding site. These results clearly contradict the idea of a dimeric structure with 12 helices required for the formation of the pore through the membrane. It is therefore possible that the proteins are aggregated as dimers (or even higher-order oligomers) while each monomer still functions independently (55, 57). The crystal structure suggests a cavity on one side lined with side chains that are presumably responsible for the specificity of the carrier. It is thought to represent one conformation of the carrier, as stabilized by the inhibitor.

A distinguishing feature of the translocator/carrier and UCP family is the interaction with purine nucleotides. In the UCP-1 the nucleotide is a regulatory ligand controlling unidirectional transport of other small ions (H^+/OH^-, or Cl^-), while in the ADP/ATP translocator the purine nucleotide is in fact the substrate and activator of translocation. UCP-1 binds the regulatory ligand only on the outside (cytoplasmic side) of the inner membrane. The translocator must bind nucleotides on either side, and the net reaction (ADP imported, or ATP exported) is driven by the membrane potential, $\Delta\Psi$, since ATP has one more negative charge than ADP. The function of this protein is therefore intimately linked to the overall process of oxidative phosphorylation, although its control coefficient may be modest (58). Nevertheless, a more severe dysfunc-

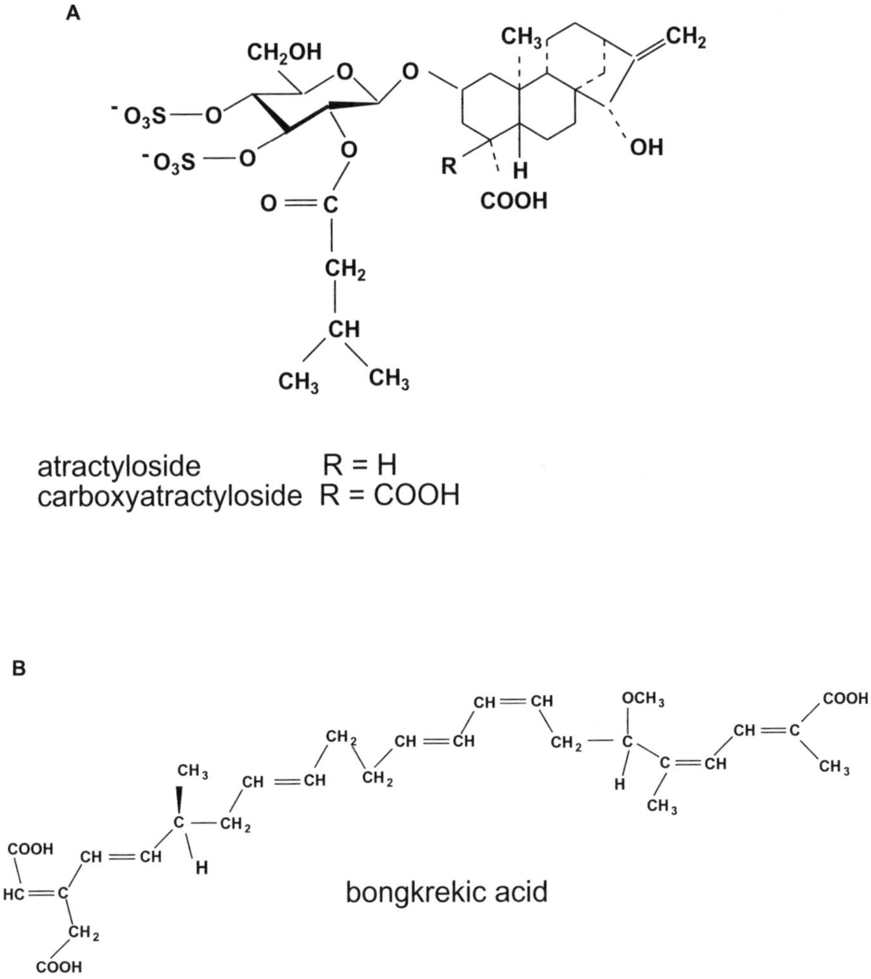

A

atractyloside R = H
carboxyatractyloside R = COOH

B

bongkrekic acid

Figure 6.10 (A,B) Two highly specific inhibitors employed in the purification of the ADP/ATP translocator.

tion may contribute to pathological effects that may be implicated in neuromuscular diseases (59).

Two highly specific inhibitors have been employed in the study and in the purification of the ADP/ATP translocator (Figure 6.10). These inhibitors do not bear much resemblance to each other or to ATP, except for the presence of three or four negative charges. Binding sites for the inhibitor and ATP are similar or overlapping, since binding of one precludes the binding of the other. Carboxyatractyloside is derived from plants, and bongkrekic acid has been isolated from certain *Pseudomonas* strains. Their structures are shown in Figure 6.10). They differ in their ability to penetrate the inner membrane.

Atractyloside is restricted to the outside and hence binds to the cytoplasmic side of the carrier, whereas the bongkrekic acid can enter the mitochondria, and it binds only to the matrix side. The use of radiolabeled or fluorescent inhibitors can help in estimating the number of carrier sites per mitochondrion (it is generally a very abundant protein—up to 15% of total protein). With the atractyloside bound, the carrier can be stabilized during purifications and protected against enzymatic degradation (see reference 53 for review and earlier references). This form has also been successfully used for the crystallization and X-ray diffraction experiments (56, 57) (Figure 6.9).

Most interesting and revealing, however, has been the observation that no ternary complexes have been found; that is, the inhibitors not only prevent the binding of ATP, or ADP, but both inhibitors cannot bind at the same time, even though they bind to opposite sides of the carrier. The interpretation by Klingenberg was that the carrier is alternately open for the substrate on one side or the other. It is postulated to exist in two conformational states; and during the transition from one state to the other, the substrate is translocated across the membrane. In fact, it was further hypothesized that the transition between the two states (c state and m state) is possible only with the substrate (either ATP or ADP) present ("induced transition fit theory of carrier transport catalysis"). According to this model, the inhibitors can bind to the open faces of two opposing "ground states" of the carrier, stabilizing a conformation that is incapable of binding the adenine nucleotide. The crystal structure is consistent with this model, but it would clearly be strengthened considerably, if a crystal structure of the alternate conformation binding a nucleotide or bongkrecic acid on the matrix side could be solved.

6.9.4 The Mitochondrial Carrier Protein Family

The previous sections have emphasized a variety of reactions and pathways localized in the mitochondrial matrix. It is clear that many intermediates have to transported into mitochondria, and others have to be exported. Because metabolism is thus compartmentalized and because the inner membrane must maintain a membrane potential, the transport of metabolites must be tightly controlled. Mechanisms for transporting various metal ions have already been described. Most/all metabolites are transported by a large group of mitochondrial transport proteins that can be grouped into 37 subfamilies with a total of 75 members in humans (54). Several prominent family members have already been introduced and discussed in a different context: The adenine nucleotide transporter (ANT) and its isoforms are responsible for the import of ADP and the export of ATP. A phosphate transporter brings phosphate from the cytosol into the matrix for ATP synthesis. Multiple isoforms of uncoupling proteins (UCP1, UCP2, . . .) were identified after the first member of the group was found to be a key factor in thermogenesis in brown adipose tissue. The role of the others appears to be more subtle (see page 266 or see section 5.6.2). It is now clear that most of these transporters have a structure very similar to that

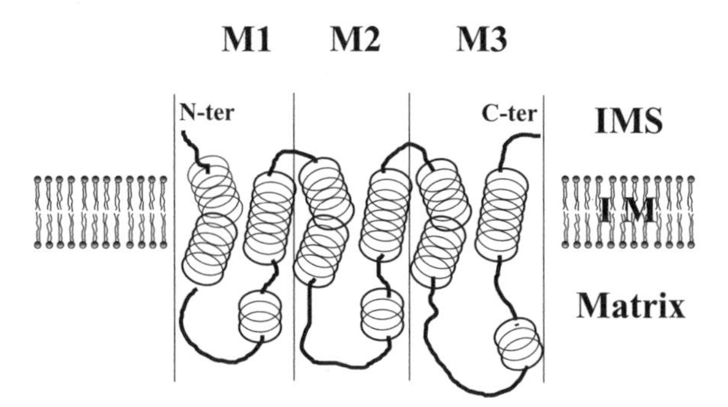

Figure 6.11 Structure of a phosphate transporter. (Adapted from reference 54.)

of the ANT: a molecular mass of 30 kDa, three degenerate repeats of ~100 amino acids, six transmembrane helices, N and C terminals in the intermembrane space, two short loops in the IMS, and three larger loops in the matrix (Figure 6.11). This topology, originally predicted from the sequence and hydropathy profiles, has been confirmed by a high-resolution structure (2.2 Å) determined from X-ray diffraction (see page 332; also see references 55 and 56) A blast of the human genome with 23 residue long sequences from 12 different regions of 20 known human mitochondrial transporters has predicted 75 such transporters (54). Some of these are significantly larger due to N-terminal extensions forming a calcium binding site. Others are much smaller, with partial sequences and truncations at either the N terminal or the C terminal. Their functions, if any, remain to be established, although at least one has been found to be expressed in mitochondria. Highly conserved sequence and structure elements are constrained by the need to form homodimers for their activity.

Only 21 of the human mitochondrial transport proteins have been identified with regard to their specificity and function. There is likely to be a certain degree of redundancy, leading to an estimate of a possible total of 39 functions (60). Known and expected functions include the import/export of amino acids including citrulline and ornithine for the urea cycle, keto acids (pyruvate), di- and tricarboxylic acids (malate, citrate, α-ketoglutarate), carnitine and acyl carnitine (fatty acid oxidation), nucleotides for DNA and RNA metabolism, coenzyme A, NAD^+, FAD, *S*-adenosyl methionine, and more (see Table 6.1). The specificity of the transporter for a metabolite is not predictable from the amino acid sequence, and it has to be established from genetic experiments (gene knockouts) or biochemical experiments. The latter typically involve overexpression of a carrier in bacteria or yeast, followed by purification and reconstitution in liposomes.

6.9.5 Cation Transport

The most significant cations other than protons are Na^+, K^+, Ca^{2+}, Mg^{2+}, and the trace metals Fe^{2+}, Cu^{2+}, Zn^{2+}, and others. Because of its special importance

TABLE 6.1 Carrier Proteins in the Inner Mitochondrial Membrane

Substrate 1	Substrate 2
Antiporters	
Malate	Phosphate
ATP	ADP
ATP	dNTP
S-Adenosyl-methionine	*S*-Adenosyl-homocysteine
Carnitine	Acyl-carnitine
Ornithine	Citrulline
Glutamate	Aspartate
Malate	α-Ketoglutarate
Malate	Citrate
Uniporters (or antiporters with one proton)	
Pyruvate	
Glutamate	
Phosphate	
Glutamine	
NAD+	
FAD	
Coenzyme A	

in specific regulatory mechanisms, the uptake and release of Ca^{2+} will be considered under its own heading below. One primary mechanism of cation uptake/exchange involves antiporters, and several such proteins with distinct specificities have been described (61, 62). Na^+ ions can traverse the inner membrane via an apparently unregulated Na^+/H^+ antiport, or a slightly more complex Na^+/Ca^{2+}antiport (see below). The Na^+ ion concentration is generally low in the cytosol, and it is also kept low in the mitochondrial matrix. The presence of the Na^+/H^+ antiport keeps the electrochemical Na^+ gradient equal to the H^+ gradient. K^+ ions have a major role in maintaining mitochondrial volume, and their uptake is achieved by three mechanisms: an electroneutral K^+/H^+ antiporter that may be regulated by free Mg^{2+} in the matrix, a K^+ channel activated by ATP (63), and a leakage driven by the high electrochemical gradient across the inner membrane (64). The purified mitK_{ATP} channel has been studied in synthetic lipid bilayer to prove that regulatory nucleotides (and also acyl-CoA esters) bind to the cytosolic side of the membrane (65, 66). Thus, a balance must be maintained between (a) an outward flux via the antiporter driven by the proton gradient and (b) an inward flux through the leak and channel driven by the electrochemical gradient. The precise physiological significance of the ATP-sensitive channel remains to be fully explained, but it has been suggested that it serves in regulated energy dissipation at times of low ATP demand by the cell, when the flow and metabolism of reducing equivalents between matrix and cytosol has to be maintained. Mitochondrial volume has been shown to influence the rate of electron transport and respiration (61, 67). The mitK_{ATP} channel has received particular attention in relation to cardiac function (64, 68, 69). Based on co-purification and re-constitution experiments Garlid's group has proposed a possible signaling mechanism from the observed association of the mitK_{ATP} channel with the protein kinase C ε. Various activators of this kinase could also activate K^+ flux in liposomes with these two components (70).

Uniports for Na^+ and K^+ have also been identified, but their characteristics and functional importance are still under investigation. Similarly, the passage of Mg^{2+} ions through the inner mitochondrial membrane is still poorly understood (71). It appears that mitochondria take up Mg^{2+} through membrane leakage channels driven by the membrane potential, rather than by a specific transporter (72). Interesting speculations have centered on the possibility of hormonal control of $[Mg^{2+}]$ in both the cytosol and matrix.

6.9.5.1 Transport of Calcium and Its Physiological Role Intramitochondrial free calcium $[Ca^{2+}]_m$, and the mechanisms associated with its control have been examined extensively, since its recognition as a potential mediator of a variety of metabolic activities. Additional importance is derived from the recognition that certain pathological conditions may be associated with abnormal Ca^{2+} levels in mitochondria (49, 73). The nature of the transporters, the bioenergetics of transport, and the regulation of Ca^{2+} transport all have been investigated since the discovery of mitochondrial Ca transport in the early 1960s (74). The existing membrane potential favors influx, and efflux is energetically uphill. An expert discussion of the energetics of Ca^+ transport is presented in a review by Gunter and Gunter (48). From typical values for $\Delta\Psi$ and internal and external $[Ca^{2+}]$, it is estimated that ~29 kJ/mole Ca^{2+} is required to get Ca^{2+} out of mitochondria. From another point of view, this considerable free energy difference explains why mitochondria have been thought of as intracellular "sinks" or reservoirs for Ca^{2+}, leading to the challenge to understand how they contribute to the control to the overall cytosolic level of this ion. This idea has been challenged in the past decade from the discovery that intracellular Ca^{2+} is largely found in the ER, from where it is mobilized by inositol triphosphate (IP_3). Mitochondria became the target for Ca^{2+} signaling rather than the origin of regulatory Ca^{2+} (75).

A major player in Ca^{2+} transport in vertebrate mitochondria is a uniporter that can also transport Mn^{2+}, Ba^{2+}, Fe^{2+}, Pb^{2+} in decreasing order of selectivity, although, curiously, Mg^{2+} uptake may not involve this route. The conclusion that it is a uniporter rests on a variety of experiments summarized in reference 48. The membrane potential (negative inside) is essential, favoring the net charge transfer of 2; that is, Ca^{2+} diffuses down the Ca^{2+} electrochemical gradient, thereby diminishing the membrane potential. Detailed kinetic studies have been interpreted to show the existence of two sites for Ca^{2+} on the uniporter: one for activation and the second for the actual transport. In other words, Ca^{2+} itself modulates the activity of this uniporter and can inhibit its own uptake; experiments show that it is a calmodulin-mediated process (see reference 75 for a further discussion and references). Another interesting observation is that this uniporter does not appear to allow Ca^{2+} ions to pass down a chemical gradient alone; that is, a membrane potential is required to induce a conformation necessary for transport.

Two distinct Ca^{2+} efflux mechanisms in vertebrate mitochondria are responsible for protection against the buildup of toxic levels of Ca^{2+} in the matrix. It was a challenge to prove that the Na^+-independent mechanism was not an artifact or simply a reversed uniport (reviewed by in reference 48). It now appears established that Ca^{2+} efflux is accompanied by proton influx, and a ratio $1Ca^{2+}/2H^+$ appears to be favored. However, bioenergetic considerations conclude that it cannot be a simple passive exchanger, and some additional input of energy from the electron transport chain must be postulated. The energy coupling mechanism is still obscure. An Na^+-dependent mechanism can be distinguished on the basis of certain ion selectivities (Sr^{2+} but not Mn^{2+}), stimulation by Na^+, spermidine, and spermine, and inhibition by a variety of inhibitors. Current thinking favors an electrogenic $Ca^{2+}/3\ Na^+$ exchanger with the evidence pointing to a $\Delta\Psi$-dependent electrophoretic antiport (76).

Our understanding of the physiological role of mitochondrial Ca^{2+} transport is still expanding (49, 75, 77). As stated above, mitochondria can constitute a buffering compartment with a role in modulating cytosolic calcium, $[Ca^+]_{cytosol}$. A relevant question was whether transport in and out of mitochondria was sufficiently rapid to keep up with rapid, pulsed changes in cytosolic $[Ca^{2+}]$, found, for example, in beating cardiomyocytes or in nerve cells engaged in active signaling (exocytosis triggered by elevated $[Ca^+]_{cytosol}$. A second major insight was that two prominent mitochondrial dehydrogenases (pyruvate dehydrogenase and α-ketoglutarate dehydrogenase) are activated by $[Ca^{2+}]m$. It should be noted that the critical concentrations are those of the free Ca^{2+}. Estimates are that only one part of 10^3 of total mitochondrial Ca^{2+} is free, and this fact must also be considered in the context of the buffering capacity of mitochondria. The stimulation of the dehydrogenases increases the NADH/NAD^+ ratio and hence oxidative phosphorylation. One can consider a plausible scenario in which increased $[Ca^+]$cytosol in stimulated skeletal or cardiac muscle is rapidly coupled to increased $[Ca^{2+}]_m$ and hence increased rates of oxidative phosporylation. Such a mechanism could serve to signal increased energy demand in the cytosol to the OXPHOS machinery.

The use of fluorescent Ca^+ indicators (e.g., fura-2) developed by Tsien and colleagues (77a) has revolutionized the measurement of $[Ca^+]_{cytosol}$, and has even been applied to isolated mitochondria. Subsequently, recombinant aequorin has been targeted to mitochondria to serve as a sensitive indicator of $[Ca^{2+}]_m$ (78). The combination of fura-2 and mitochondrial aequorin in single cells permitted simultaneous measurements of $[Ca^+]_{cytosol}$ and $[Ca^{2+}]_m$, with intriguing results (reviewed in reference 79 and 80). Raising $[Ca^+]_{cytosol}$ by influx from extracellular Ca^{2+} affected $[Ca^{2+}]_m$ less and at a slower rate than an increase in $[Ca^+]_{cytosol}$ by a release (IP$_3$-induced) from intracellular stores

(ER). These observations can now be explained by observations made in the light/fluorescent microscope showing that the ER and mitochondria can form intracellular "synaptic" regions. At these junctions the IP_3-stimulated Ca^{2+} release channel in the ER is physically close to the mitochondrial Ca^{2+} uniporter, and the ion can be taken up directly into the mitochondria without significant spreading into the cytosol (81, 82). An interesting paper by Rapizzi et al. (83) demonstrates that overexpression of VDAC can enhance the transfer of Ca^{2+} from the ER to mitochondria. As a consequence, cytoplasmic Ca^{2+} microdomains are generated within the cellular architecture that fluctuate both in space and in time. Subcellular Ca^{2+} concentrations are dynamically heterogeneous from one location to another. They can thus regulate a diverse group of cellular events stimulated by various extracellular factors (77).

6.9.6 The Mitochondrial Permeability Transition

The mitochondrial permeability transition (MPT) was originally defined as sudden increase in the inner membrane permeability to solutes of molecular mass less than ~1500 Da. It is now believed to be due to the opening of a mega channel, which is referred to in some publications as the mitochondrial permeability transition pore (MTP). It has received particular attention in recent years because pore opening is sensitive to cyclosporin A (see below) and because it has been implicated in the pathway of apoptosis (Chapter 8). The transition is detectable and measured with isolated mitochondria in protein-free buffers by an uptake of radioactive molecules such as sucrose, by a change in optical density due to swelling, and even by patch-clamp techniques (84, 85). The original observations were made with animal mitochondria (beef heart, rat liver), but recent publications suggest that yeast mitochondria are also capable of forming a transition pore. However, the regulation of this pore in yeast may be different (86).

Several broad questions can be raised, and answers to these questions are beginning to accumulate. The discussion will focus on the following issues: (1) What components of the inner membrane constitue the pore? (2) What mechanisms regulate the opening and closing of the pore? (3) What is the physiological significance of the operation of this megachannel? To some extent the answers are interrelated.

The exact composition of the pore is still not completely clear, except for a general agreement that the adenine nucleotide translocase (ANT) is a participant in channel formation, perhaps by forming complexes with outer membrane proteins to create contact sites and hence to stabilize and/or enlarge the channel. The peripheral benzodiazepine receptor and VDAC in the outer membrane have been implicated, and there is speculation about the inclusion of Bcl2, cardiolipin synthase, 3-*b*-hydroxysteroid isomerase, and phospholipid hydroperoxide glutathione peroxidase (87). On the cytosolic side, hexokinase may also be associated, and on the matrix side the participation of cyclophilin D is suggested by the sensitivity to cyclosporin.

The channel can be opened when Ca^{2+} is accumulated in the matrix by exposure of mitochondria to high concentrations of Ca^{2+} (generally above physiological levels), and the sensitivity to Ca^{2+} is heightened under various conditions including oxidative stress, adenine nucleotide depletion, elevated phosphate concetrations, and low membrane potenial, $\Delta\Psi$? The role of nucleotides points to an involvement of ANT, but a more direct indication is the effect of inhibitors of ANT on the MPT. Inhibitors that stabilize the "c" conformation of ANT (e.g., atractyloside) favor the MPT, while inhibitors stabilizing the "m" conformation (e.g., bongkrekic acid) also inhibit pore opening. A high $\Delta\Psi$ or a low pH prevents the transition (88). Of special interest was the observation that submicromolar concentrations of cyclosporin A are very effective in keeping the pore from opening. The target for cyclosporin is the matrix peptidyl-proline cis–trans isomerase, a member of the cyclophilin family. One interpretation was that the (soluble) matrix isomerase may interact with an integral membrane protein, and when triggered by Ca^{2+}, the pore opening is induced. Cyclosporin A would interfere with this required interaction. A further speculation was that the relevant IMP was the ANT (89). The idea was supported by the observed attachment of cyclophilin to the inner membrane under certain conditions (oxidative stress, thiol reagents, increased matrix volume), which favor or enhance pore opening (90, 91). Even though the MPT makes an appearance in a large fraction of publications related to apoptosis, necrosis, oxidative stress, and other pathologies, there is still no clear distinction between the "MPT" and the "MTP." The first describes a phenomenon, while the second implies the existence of a structure that has never been isolated and whose composition is hypothesized on the basis of indirect studies with various inhibitors. The suggestion has even been made that the "pore" is constituted of unfolded integral membrane proteins damaged by ROS (e.g., oxidation of sulfhydryls, etc.) (92).

The combined observations have served to support a variety of potential functions for the MTP, which may not be exclusive (87): (1) a voltage sensor, (2) a matrix pH sensor, (3) a divalent cation sensor (role in calcium homeostasis (e.g., 93)), (4) a sensor of adenine nucleotide concentrations, (5) a sensor of the redox state of the pyridine nucleotide pool, and (6) a thiol sensor (redox status of glutathione). The context in which the MPT is most commonly encountered is the triggering and progression of apoptosis, a subject that will be expanded in a following chapter.

It is clear that a prolonged opening of a number of pores, or the simultaneous opening of many pores, would deenergize and inactivate mitochondria rapidly, and it may be that such an event is associated with the irreversible breakdown of $\Delta\Psi$ observed in the course of apoptosis (see Chapter 8). However, at this time it is still a matter of much speculation as to whether the MTP makes a significant contribution to mitochondrial activities under physiological conditions. It could be imagined as a safety valve under stressfull conditions, but the lack of specificity for any small-molecular-weight solute precludes any fine-tuning for specific ions or solutes.

REFERENCES

1. Klausner, R. D., Rouault, T. A., and Harford, J. B. (1993) *Cell* **72**, 19–28.
2. Philpott, C. C., Klausner, R. D., and Rouault, T. A. (1994) *Proc. Natl. Acad. Sci. USA* **91**, 7321–7325.
3. Hentze, M. W. and Kühn, L. C. (1996) *Proc. Natl. Acad. Sci. USA* **93**, 8175–8182.
4. Scheffler, I. E. (1986) Biochemical genetics of respiration-deficient mutants of animal cells. In Morgan, M. J., editor. *Carbohydrate Metabolism in Cultured Cells*, Plenum Press, London, pp. 77–109.
5. Broderick, K. E., Potluri, P., Zhuang, S., Scheffler, I. E., Sharma, V. S., Pilz, R. B., and Boss, G. R. (2006) *Exp. Biol. Med.* **231**, 641–649.
6. Lanks, K. W. and Li, P.-W. (1988) *J. Cell Physiol.* **135**, 151–155.
7. Zitomer, R. S. and Lowry, C. V. (1992) *Microbiol. Rev.* **56**, 1–11.
8. Daum, G. (1985) *Biochim. Biophys. Acta* **822**, 1–42.
9. Hackenbrock, C. R., Chazotte, B., and Gupte, S. S. (1986) *J. Bioenerg. Biomembr.* **18**, 331–368.
10. Chazotte, B. and Hackenbrock, C. R. (1991) *J. Biol. Chem.* **266**, 5973–5979.
11. Schlame, M., Rua, D., and Greenberg, M. L. (2000) *Prog. Lipid Res.* **39**, 257–288.
12. Hostetler, K. Y. (1982) Polyglycerophospholipids: Phosphatidylglycerol, diphosphatidylglycerol and bis(monoacylglycero)phosphate. In Hawthorne, J. N. and Ansell, G. B., editors. *Phospholipids*, Elsevier Biomedical Press, Amsterdam, pp. 215–261.
13. Schlame, M., Brody, S., and Hostetler, K. Y. (1993) *Eur. J. Biochem.* **212**, 727–735.
14. Dowhan, W. (1997) *Annu. Rev. Biochem.* **66**, 199–232.
15. Schlame, M. and Greenberg, M. L. (1997) *Biochim. Biophys. Acta Lipids Lipid Metab.* **1348**, 201–206.
16. Schlame, M. and Hostetler, K. Y. (1997) *Biochim. Biophys. Acta Lipids Lipid Metab.* **1348**, 207–213.
17. Schlame, M. and Haldar, D. (1993) *J. Biol. Chem.* **268**, 74–79.
18. Hauff, K. D. and Hatch, G. M. (2006) *Prog. Lipid Res.* **45**, 91–101.
19. Xu, Y., Malhotra, A., Ren, M., and Schlame, M. (2006) *J. Biol. Chem.* **281**, 39217–39224.
20. Li, G., Chen, S., Thompson, M. N., and Greenberg, M. L. (2007) *Biochim. Biophys. Acta* **1771**, 432–441.
21. Xu, Y., Condell, M., Plesken, H., Edelman-Novemsky, I., Ma, J., Ren, M., and Schlame, M. (2006) *Proc. Natl. Acad. Sci USA* **103**, 11584–11588.
22. Ohtsuka, T., Nishijima, M., and Akamatsu, Y. (1993) *J. Biol. Chem.* **268**, 22908–22913.
23. Bereiter-Hahn, J. and Voth, M. (1994) *Microsc. Res. Tech.* **27**, 198–219.
24. Kent, C. (1995) *Annu. Rev. Biochem.* **64**, 315–343.
25. Zinser, E., Sperka-Gottlieb, C. D., Fasch, E. V., Kohlwein, S. D., Paltauf, F., and Daum, G. (1991) *J. Bacteriol.* **173**, 2026–2034.
26. Stewart, L. C. and Yaffe, M. P. (1991) *J. Cell. Biol.* **115**, 1249–1257.
27. Tzagoloff, A. and Dieckmann, C. L. (1990) *Microbiol. Rev.* **54**, 211–225.

28. Baba, S. W., Belogrudov, G. I., Lee, J. C., Lee, P. T., Strahan, J., Shepherd, J. N., and Clarke, C. F. (2004) *J. Biol. Chem.* **279**, 10052–10059.
29. Barkovich, R. J., Shtanko, A., Shepherd, J. A., Lee, P. T., Myles, D. C., Tzagoloff, A., and Clarke, C. F. (1997) *J. Biol. Chem.* **272**, 9182–9188.
30. Barros, M. H., Johnson, A., Gin, P., Marbois, B. N., Clarke, C. F., and Tzagoloff, A. A. (2005) *J. Biol. Chem.* **280**, 42627–42635.
31. Do, T. Q., Hsu, A. Y., Jonassen, T., Lee, P. T., and Clarke, C. F. (2001) *J. Biol. Chem.* **276**, 18161–18168.
32. Gin, P., Hsu, A. Y., Rothman, S. C., Jonassen, T., Lee, P. T., Tzagoloff, A., and Clarke, C. F. (2003) *J. Biol. Chem.* **278**, 25308–25316.
33. Olsen, R. E. and Rudney, H. (1983) *Vitamins and Hormones* **40**, 1–43.
34. Grunler, J., Ericsson, J., and Dallner, G. (1994) *Biochem. Biophys. Acta* **1212**, 259–277.
35. Tielens, A. G. M. (1994) *Parasitol. Today* **10**, 346–352.
36. Van Hellemond, J. J., Luijten, M., Flesch, F. M., Gaasenbeek, C. P. H., and Tielens, A. G. M. (1996) *Mol. Biochem. Parasitol.* **82**, 217–226.
37. Van Hellemond, J. J., Klockiewicz, M., Gaasenbeek, C. P. H., Roos, M. H., and Tielens, A. G. M. (1995) *J. Biol. Chem.* **270**, 31065–31070.
38. Tielens, A. G. M. and Van Hellemond, J. J. (1998) *Parasitol. Today* **14**, 265–271.
39. Beinert, H., Holm, R. H., Munck, E., and Münck, E. (1997) *Science* **277**, 653–659.
40. Johnson, D. C., Dean, D. R., Smith, A. D., and Johnson, M. K. (2005) *Annu. Rev. Biochem.* **74**, 247–281.
41. Lill, R. and Muhlenhoff, U. (2005) *Trends Biochem. Sci.* **30**, 133–141.
42. Kispal, G., Sipos, K., Lange, H., Fekete, Z., Bedekovics, T., Janaky, T., Bassler, J., Aguilar Netz, D. J., Balk, J., Rotte, C., and Lill, R. (2005) *EMBO J.* **24**, 589–598.
43. Yarunin, A., Panse, V. G., Petfalski, E., Dez, C., Tollervey, D., and Hurt, E. (2005) *EMBO J.* **24**, 580–588.
44. Alper, G. and Narayanan, V. (2003) *Pediatr. Neurol.* **28**, 335–341.
45. Van Der Giezen, M. and Tovar, J. (2005) *EMBO Rep.* **6**, 525–530.
46. van der Giezen M., Tovar, J., and Clark, C. G. (2005) *Int. Rev. Cytol.* **244**, 175–225.
47. Tielens, A. G. M., Rotte, C., Van Hellemond, J. J., and Martin, W. (2002) *Trends Biochem. Sci.* **27**, 564–572.
48. Gunter, K. K. and Gunter, T. E. (1994) *J. Bioenerg. Biomembr.* **26**, 471–485.
49. Hansford, R. G. (1994) *J. Bioenerg. Biomembr.* **26**, 495–508.
50. Hatefi, Y. and Yamaguchi, M. (1996) *FASEB J.* **10**, 444–452.
51. Rostovtseva, T. K., Tan, W., and Colombini, M. (2005) *J. Bioenerg. Biomembr.* **37**, 129–142.
52. Lemasters, J. J. and Holmuhamedov, E. (2006) *Biochim. Biophys. Acta* **1762**, 181–190.
53. Klingenberg, M., Winkler, E., and Huang, S. (1995) *Methods Enzymol.* **260**, 369–389.
54. Wohlrab, H. (2005) *Biochim. Biophys. Acta* **1709**, 157–168.
55. Pebay-Peyroula, E. and Brandolin, G. (2004) *Curr. Opin. Struct. Biol.* **14**, 420–425.

56. Pebay-Peyroula, E., Dahout-Gonzales, C., Kahn, R., Trezeguet, V., Lauquin, G. J. M., and Brandolin, G. (2003) *Nature* **426**, 39–44.
57. Nury, H., Dahout-Gonzalez, C., Trezeguet, V., Lauquin, G. J. M., Brandolin, G., and Pebay-Peyroula, E. (2006) *Annu. Rev. Biochem.* **75**, 713–741.
58. Ciapaite, J., Bakker, S. J., Diamant, M., van Eikenhorst, G., Heine, R. J., Westerhoff, H. V., and Krab, K. (2006) *FEBS J.* **273**, 5288–52302.
59. Dahout-Gonzalez, C., Nury, H., Trezeguet, V., Lauquin, G. J., Pebay-Peyroula, E., and Brandolin, G. (2006) *Physiology (Bethesda)* **21**, 242–249.
60. Kunji, E. R. (2004) *FEBS Lett.* **564**, 239–244.
61. Garlid, K. D. (1994) *J. Bioenerg. Biomembr.* **26**, 537–542.
62. Brierley, G. P., Baysal, K., and Jung, D. W. (1994) *J. Bioenerg. Biomembr.* **26**, 519–526.
63. Ardehali, H. and O'Rourke, B. (2005) *J. Mol. Cell Cardiol.* **39**, 7–16.
64. O'Rourke, B., Cortassa, S., and Aon, M. A. (2005) *Physiology (Bethesda.)* **20**, 303–315.
65. Yarov-Yarovoy, V., Paucek, P., Jaburek, M., and Garlid, K. D. (1997) *Biochim. Biophys. Acta* **1321**, 128–136.
66. Paucek, P., Yarov-Yarovoy, V., Sun, X., and Garlid, K. D. (1996) *J. Biol. Chem.* **271**, 32084–32088.
67. Garlid, K. D. (1996) *Biochim. Biophys. Acta* **1275**, 123–126.
68. Szewczyk, A., Skalska, J., Glab, M., Kulawiak, B., Malinska, D., Koszela-Piotrowska, I., and Kunz, W. S. (2006) *Biochim. Biophys. Acta* **1757**, 715–720.
69. Costa, A. D., Quinlan, C. L., Andrukhiv, A., West, I. C., Jaburek, M., and Garlid, K. D. (2006) *Am. J. Physiol. Heart Circ. Physiol.* **290**, H406–H415.
70. Jaburek, M., Costa, A. D., Burton, J. R., Costa, C. L., and Garlid, K. D. (2006) *Circ. Res.* **99**, 878–883.
71. Jung, D. W. and Brierley, G. P. (1994) *J. Bioenerg. Biomembr.* **26**, 527–536.
72. Jung, D. W., Panzeter, E., Baysal, K., and Brierley, G. P. (1997) *Biochim. Biophys. Acta* **1320**, 310–320.
73. Leo, S., Bianchi, K., Brini, M., and Rizzuto, R. (2005) *FEBS J.* **272**, 4013–4022.
74. DeLuca, H. F. and Engstrom, G. W. (1961) *Proc. Natl. Acad. Sci. USA* **47**, 1744–1750.
75. Putney, J. W.Jr, . and Thomas, A. P. (2006) *Curr. Biol.* **16**, R812–R815.
76. Jung, D. W., Baysal, K., and Brierley, G. P. (1995) *J. Biol. Chem.* **270**, 672–678.
77. Rizzuto, R. and Pozzan, T. (2006) *Physiol. Rev.* **86**, 369–408.
77a. Tsien, R. Y. and Poenie, M. (1986) *Trends in Biochem. Sci.* **11**, 450–455.
78. Rizzuto, R., Simpson, A. W. M., Brini, M., and Pozzan, T. (1992) *Nature* **358**, 325–327.
79. Sheu, S. S. and Jou, M. J. (1994) *J. Bioenerg. Biomembr.* **26**, 487–493.
80. Bianchi, K., Rimessi, A., Prandini, A., Szabadkai, G., and Rizzuto, R. (2004) *Biochim. Biophys. Acta* **1742**, 119–131.
81. Rizzuto, R., Pinton, P., Carrington, W., Fay, F. S., Fogarty, K. E., Lifshitz, L. M., Tuft, R. A., and Pozzan, T. (1998) *Science* **280**, 1763–1766.
82. Rutter, G. A. and Rizzuto, R. (2000) *Trends Biochem. Sci.* **25**, 215–221.

83. Rapizzi, E., Pinton, P., Szabadkai, G., Wieckowski, M. R., Vandecasteele, G., Baird, G., Tuft, R. A., Fogarty, K. E., and Rizzuto, R. (2002) *J. Cell Biol.* **159**, 613–624.

84. Szabo, I., Bathori, G., Wolff, D., Starc, T., Cola, C., and Zoratti, M. (1995) *Biochim. Biophys. Acta* **1235**, 115–125.

85. Bernardi, P. and Petronilli, V. (1996) *J. Bioenerg. Biomembr.* **28**, 131–138.

86. Jung, D. W., Bradshaw, P. C., and Pfeiffer, D. R. (1997) *J. Biol. Chem.* **272**, 21104–21112.

87. Kroemer, G., Zamzami, N., and Susin, S. A. (1997) *Immunol. Today* **18**, 44–51.

88. Scorrano, L., Petronilli, V., and Bernardi, P. (1997) *J. Biol. Chem.* **272**, 12295–12299.

89. Halestrap, A. P., Woodfield, K. Y., and Connern, C. P. (1997) *J. Biol. Chem.* **272**, 3346–3354.

90. Connern, C. P. and Halestrap, A. P. (1996) *Biochemistry* **35**, 8172–8180.

91. Connern, C. P. and Halestrap, A. P. (1994) *Biochem. J.* **302**, 321–324.

92. Kowaltowski, A. J., Castilho, R. F., and Vercesi, A. E. (2001) *FEBS Lett.* **495**, 12–15.

93. Halestrap, A. P., Griffiths, E. J., and Connern, C. P. (1993) *Biochem. Soc. Trans.* **21**, 353–358.

7

MITOCHONDRIAL MUTATIONS AND DISEASE

Summary

When mitochondrial DNA was discovered, it could be anticipated that there could be mutations, and an explanation was provided for "cytoplasmic mutations" and non-mendelian inheritance first observed in *Saccharomyces cerevisiae*. Since yeast could proliferate under anaerobic conditions, mtDNA mutations affecting respiration and oxidative phosphorylation could be isolated and characterized. In humans the detection of such mutations seemed inconceivable for two reasons: (1) Respiration-deficient humans would be expected to be dead, and (2) with >1000 mtDNAs per cell it was difficult to imagine how a mutation in one of these genomes could ever be expressed as a phenotype. In the late 1980s, human pedigrees with mtDNA mutations and a phenotype were discovered, and the term "mitochondrial disease" was introduced. Soon it was recognized that such diseases were more common than expected; symptoms ranged from mild to severe, and combinations of symptoms made it a challenge to classify these diseases into distinct categories. However, neuropathies, myopathies, or a combination of these are typical. Initially, mitochondrial diseases were characterized by a partial defect in oxidative phosphorylation (an energy deficiency) due to mtDNA mutations, but it is immediately obvious that nuclear mutations could also lead to defective electron transport and ATP synthesis. Such mutations should follow mendelian genetics in a pedigree. Severe mutations are lethal (in early embryogenesis/development), and affected patients must still have partial function of the affected enzyme/complex. With >1000 proteins in mitochondria, one could in principle have thousands of "mitochondrial diseases." Here

Mitochondria, Second Edition, by Immo E. Scheffler

the term will be applied in a limited sense to mutations that affect oxidative phosphorylation and energy metabolism. Included in this category are diseases that are the result of problems in mtDNA replication and maintenance. Numerous other defects in mitochondrial proteins are the cause of "inborn errors of metabolism" or "metabolic diseases" that will not be discussed in this chapter.

There has also been an increasing awareness over the last two decades that mitochondria and mitochondrial defects are implicated in age-dependent neurological degeneration (Alzheimer's disease, Parkinson's disease, Friedreich's ataxia, and others). The current state of knowledge will be summarized. Finally, the phenomenon of aging (senescence) has attracted attention for a long time and for obvious reasons. There are clearly multiple facets of this problem and the underlying causes. Mitochondria are thought to play a prominent role as a major source of reactive oxygen species (ROS); an accumulation of mutations in mtDNA and damage to mitochondrial proteins due to ROS provides a rationale for the diminishing capacity of mitochondria and for the accumulation of cellular debris that can overwhelm the cellular garbage disposal machinery (proteasome).

7.1 GENERAL INTRODUCTION

Human genetic diseases due to defective mitochondrial enzymes can be grouped broadly into two major classes: (1) Nuclear mutations causing defects in enzymes affecting specific metabolic pathways have been studied extensively for more than three decades. Many inborn errors of metabolism are due to nuclear mutations and hence enzyme deficiencies in metabolic reactions taking place in the mitochondrial matrix. As an example, one can mention ornithine transcarbamylase deficiency, one of several defects causing disorders in the urea cycle. A discussion of such metabolic diseases is not intended here. (2) The present chapter will emphasize the mutations affecting the capacity of cells for respiration and the rate of oxidative phosphorylation. Severe defects in energy metabolism would be expected to be lethal, but leaky mutations resulting in partial deficiencies are now recognized as the major cause of a broad spectrum of neurological and other diseases.

Of greatest interest for the present discussion are mutations in mitochondrial DNA. The large number of mtDNAs per cell requires the consideration of conditions such as homoplasmy and heteroplasmy, defined as the presence of at least two populations of mtDNA molecules, one being normal (wild type) and the other being mutated. The fraction of mutated mtDNAs can range from barely detectable to 100%, and hence the expression of the defect can range between undetectable and severe. Consequently, the symptoms and their severity vary widely. The mechanisms giving rise to such mixtures of genomes are intriguing and still relatively obscure. Distinctions also arise from the possibility of mutations in mitochondrial (structural) genes encoding, for example, (a) a subunit of cytochrome oxidase and (b) mutations in a mitochondrial tRNA gene, affecting mitochondrial protein synthesis and hence the biogenesis of multiple complexes.

Previous chapters have made frequent reference to nuclear and mitochondrial mutations that have had global effects on mitochondrial biogenesis. Mutations affecting DNA replication and transcription, RNA processing, and translation of mitochondrial mRNAs, as well as specifically nuclear mutations that cause defects in the protein import machinery of mitochondria, were discussed in the context of the mechanisms that operate in mitochondrial proliferation, fusion and fission, and the assembly of the mitochondrial electron transport chain. The mutations were of interest because they illuminated the role of specific gene products in the biogenesis of mitochondria. Obviously, such mutations also affect the capacity for respiration. In this chapter the focus will initially be on mutations in the mtDNA affecting the capacity for oxidative phosphorylation, and hence one of the major functions of the mitochondria as the powerhouse of the cell. The existence of mtDNA mutations was first recognized in cell culture with lower eukaryotes such as yeast, and to a very limited extent with mammalian cells in culture. During the past decade a new field of "mitochondrial medicine" has been created from the discovery that a complex and variable set of symptoms in humans could be caused by partial deficiencies in oxidative phosphorylation due to mitochondrial mutations. It should be stated at the outset that nuclear mutations can also have a direct effect on oxidative phosphorylation, since both genomes contribute to the complexes of the mitochondrial electron transport chain. While the well-known rules of Mendelian inheritance will apply to such nuclear mutations, mitochondrial mutations passed along the maternal lineage pose a much more significant challenge to the genetic counselor. The severity of the resulting phenotype will depend on the degree of heteroplasmy that can vary widely among siblings, as will be discussed below.

Included in this chapter are discussions of several diseases such as Alzheimer's disease, Parkinson's disease, Huntington's disease, and amyotrophic lateral sclerosis (ALS), where partially defective mitochondria have been implicated, but the question is still unresolved as to whether the mitochondrial defect is a primary cause or a secondary consequence of a more fundamental defect within specific (neuronal) tissues. An intriguing aspect of these diseases is their delayed onset, similar to the delayed onset of mitochondrial diseases resulting from mtDNA mutations. Investigations of programmed cell death, apoptosis, have linked the pathway to mitochondrial components, and they have established a significant role of mitochondria in determining the fate of a cell. Research into mechanisms responsible for normal aging (senescence) in multicellular organisms has led to the formulation of very plausible and stimulating hypotheses, proposing an accumulation of mitochondrial mutations and a gradual deterioration of OXPHOS capacity as a fundamental cause of aging.

Finally, this chapter includes a summary and discussion of the salient features of fungal senescence and male sterility in plants, because mitochondria and mitochondrial genomes (and mitochondrial plasmids) are strong genetic determinants in the expression of these phenomena.

7.2 IN CELL CULTURE

7.2.1 Mitochondrial Mutations in Microorganisms

For very familiar practical and theoretical reasons the budding yeast, *Saccharomyces cerevisiae*, has become one of the most widely used model systems in which to study eukaryotic molecular genetics, cell biology, and even cell physiology. Growth requirements are minimal, and growth will occur under a wide variety of conditions. The yeast cell has all or most of the typical subcellular organelles and compartments of a eukaryotic cell. Its genome is relatively small, and the complete sequences of all of its genome has been available for some time. Large numbers of cells can be easily cultured in suspension or on plates. Mutant selections on a relatively large scale are possible, and clones are readily isolated. A system of mating and sexual reproduction has made the standard genetic methodology applicable for mapping and complementation analysis. Most readers will not need to be reminded of the enormous advantages of yeast as a model system, and it may at times be more interesting to ask, What biological problems cannot be tackled with the help of yeast? Even prions appear to be found in yeast (1). The complex I of the ETC is not present in the common yeasts, *S. cerevisiae* and *S. pombe*, but the yeast *Yarrowia lipolytica* promises to fill this gap (2, 3).

Studies on plant variants early in this century had noted non-Mendelian (maternal) types of inheritance, a phenomenon that became known as cytoplasmic inheritance (4, 5). Mutations in chloroplast DNA were eventually demonstrated to be responsible for these original observations. The field received a major boost when cytoplasmic inheritance was discovered in yeast (6), and a large variety of yeast mutants classified as *pet* mutants, *mit* mutants, ρ^- or ρ^0 mutants has been useful in making pioneering discoveries on mitochondrial inheritance, mitochondrial functions, oxidative phosphorylation, and so on. A detailed description of all of these would constitute a book all by itself (7). Only some general principles and insights will be discussed in the following section and illustrated with some specific examples.

A property of yeast cells is that they can grow under a variety of different conditions. Such conditions are likely to be encountered in the wild as well, and the ability of yeast to adapt to different defined conditions in the laboratory continues to amaze and to provide systems for the study of gene regulation in many laboratories. Of primary significance for the present discussion is the ability to adapt to the carbon source in the medium and to the availability of oxygen. The expression of multiple genes is adjusted to prevailing conditions. When a fermentable carbon source and oxygen are present, many genes involved in transport, metabolism of other sugars, and, most significantly, the biosynthesis of the entire electron transport chain and ATPsynthase are repressed. This phenomenon, referred to as "glucose repression," has been the subject of numerous reviews (8–12). In a previous chapter on nuclear mitochondrial interactions, many aspects of this phenomenon relevant to mitochondrial biogenesis have already been presented.

Here it is simply relevant to emphasize that wild-type yeast cells can grow on media either rich in glucose (a fermentable carbon source) or deficient in glucose (in the presence of nonfermentable carbon sources such as acetate, glycerol, or ethanol). In combination with replica plating, it is therefore straightforward to isolate mutant clones that can grow on glucose but are unable to grow in the presence of glycerol alone. Among the possible mutants, respiration-deficient mutants (res-) form a large group; and because they tend to form small (petite) colonies under defined conditions of glucose deprivation, they are routinely referred to as *pet* mutants. Collectively, they have been expertly described and reviewed by Dujon (7) and Tzagoloff and Dieckmann (13). Mutations giving rise to this phenotype can be subdivided into nuclear and mitochondrial mutations. Nuclear mutations are local and generally affect single, specific genes whose function has been exemplified in numerous instances in earlier chapters of this book. The mitochondrial mutations can again be subdivided into point mutations affecting single functions (referred to as *mit*$^-$) and large deletions of mtDNA (termed cytoplasmic petites or rho$^-$ or ρ^-) (14). Mutants completely lacking mtDNA are termed rho^0 (ρ^0) strains, and they can be obtained by prolonged culture of yeast cells in glucose and sublethal concentrations of ethidium bromide. Since the deletions in rho$^-$ mutants include tRNA genes most of the time, they are incapable of mitochondrial protein synthesis and thus lack the entire electron transport chain. When *mit*$^-$ mutants are crossed with rho$^-$ mutants, marker rescue by recombination can occur, if the *mit* mutation is in a region of the genome still present as a wild-type sequence in the rho$^-$ mitochondria. Respiration-competent diploid cells arise at high frequency by this mechanism, which involves recombination rather than complementation. A recombination event is necessary to generate wild-type diploid cells, because it is found that "merodiploid" or heteroplasmic mitochondria are not maintained under routine culture conditions. Instead, rapid mitotic segregation of mitochondrial genomes takes place resulting in a loss of potentially complementing sequences (14).

A particularly useful new method in yeast mitochondrial genetics has been the development of the transformation of mitochondria by high-velocity bombardment of yeast cells with tungsten microprojectiles to which cloned DNA sequences are adsorbed. By this method, DNA constructs made *in vitro* can be introduced into mitochondria of rho^0 strains to form "synthetic rho$^-$" strains. After mating with rho$^+$ strains, the DNA constructs can recombine with the mtDNA and deliberate mutations can be introduced into mtDNA. Details of the procedure can be found in the original publications and authoritative reviews (14, 15). Using this approach, Fox and colleagues have been able to perform site-directed mutagenesis to study sequences at the 5′ end of mitochondrial mRNAs encoding subunits of complex IV that are required for efficient translation and insertion of the proteins into the inner membrane (16).

7.2.2 Mitochondrial Mutations in Mammalian Cells in Culture

Thirty years ago it would probably have been considered impossible to obtain mammalian cell mutants in tissue culture with severe defects in respiration. Attempts to grow mammalian cells under anaerobic conditions had consistently failed, and since normally such cells would not have experienced severe anaerobiosis, it was believed that oxidative phosphorylation was an essential activity in the energy metabolism of such cells. There had been a great deal of speculation about an altered energy metabolism in cancer versus normal cells since the days when O. Warburg had proposed that tumor cells had partially reverted to a more "primitive" type of energy metabolism in which glycolysis played a larger role (17), but no specific enzyme or activity had been shown to be affected in cancer cells, which would have explained the elevated secretion of lactate by cells grown under normal aerobic conditions. An informative, entertaining and highly personal review of the literature on this subject up to about 1975 has been written by Racker (18). The Warburg effect in cancer cells has recently had a notable revival (see, for example, references 19–21).

Serendipity played a large role in the isolation and characterization of the first respiration-deficient Chinese hamster fibroblast mutant cells in tissue culture (22–24). The original mutant cell line was eventually found to have a defect in complex I of the mitochondrial electron transport chain, making it unable to oxidize NADH in the mitochondria. The elevated levels of NADH, in turn, inhibited the Krebs cycle enzyme α-ketoglutarate dehydrogenase by a simple feedback mechanism (24). The mutation was clearly established to be a nuclear mutation, and a complementing gene was mapped on the X-chromosome of mammals (25).

A complete understanding of the phenotype of this first mutant allowed the Scheffler laboratory to devise a selection scheme for the isolation of additional mutants of Chinese hamster fibroblasts in tissue culture with severe defects in respiration (26). Complementation tests were carried out by pairwise fusions, and the mutants were sorted into seven complementation groups (27). Similar mutants were subsequently isolated in two other laboratories (28, 29) in the course of their search for Chinese hamster cell mutants in tissue culture which would reproduce the phenotype of human cells from galactosemic patients. It was expected that defects in galactokinase or other enzymes in the Leloir pathway would make cells unable to grow in galactose. Clearly, respiration-deficient mutants of the type described above would resemble galactokinase mutants. Reactions of the Leloir pathway are too slow to sustain the high rate of glycolysis required by such mutants. Reviews summarizing the earlier efforts in the various laboratories have been written by Scheffler (30) and Whitfield (28).

A potentially very interesting mutant is a mutant with a severe defect in mitochondrial protein synthesis (31–33). DNA replication and maintenance are unaffected, in contrast to similar mutants of yeast. RNA transcription and processing are also normal, but the failure to translate the mitochondrial mRNAs makes some of these quite unstable, and their steady-state levels are far below normal. cDNAs for mitochondrial elongation factors (mtEF-Tu and EF-Ts) cannot complement the mutation, and a search is in progress to clone the complementing cDNA from a mammalian library and to identify a perhaps novel factor required for translational initiation in mammalian mitochondria.

The biochemical characterization of the specific defect in many of these mutants has been slow. Cloning mammalian genes by complementation with genomic or cDNA libraries is unpredictable and time-consuming. The mutant cells lacking succinate dehydrogenase activity were shown to have a point mutation in the nuclear gene for the anchor protein C_{II-3}, yielding a truncated peptide and resulting in the failure to assemble a functional complex II (34, 35). Five complementation groups of mutants defective in complex I were first characterized biochemically, and the genes for three of these groups were shown to be X-linked in 1982 (25). More recently, two of these mutated X-linked genes were identified: (a) the NDUFA1 gene encoding the MWFE subunit and (b) the NDUFB11 gene encoding the ESSS subunit (36–38). The finding of the third group suggests the existence of a gene encoding either an as yet unidentified subunit of complex I, or an essential assembly factor (see Chapter 5).

The demonstration that mammalian fibroblasts can proliferate in tissue culture in the absence of respiration created a basis for the efforts to obtain mutant cells in tissue culture equivalent to the ρ^0 mutants of yeast. Several groups have been successful in such efforts. The first report described ρ^0 chicken cells (39), and subsequently ρ^0 human tumor cells (40, 41), ρ^0 mouse cells, and ρ^0 primary human fibroblasts have been successfully isolated. The methodology is based on first defining a medium that makes respiration obsolete and then using specific inhibitors of mitochondrial DNA replication to deplete cells of mtDNA over a prolonged period of time. Ethidium bromide at sublethal concentrations was used effectively with human tumor cells, but it was less effective with mouse cell lines. Other inhibitors investigated are dideoxy cytidine (ddCTP inhibits γ-DNA polymerase in mitochondria) or bis-intercalating antitumor drugs such as ditercalinium (42).

The ρ^0 human cell lines have found very interesting applications in the study of mitochondrial mutations isolated from human patients, since they can be repopulated with mitochondria from the same species by fusion with enucleated cytoplasts. Thus, mutant mitochondria can be introduced into ρ^0 cells with a different nuclear genome, and effects of mitochondrial mutations can be clearly separated from potential effects of a specific nuclear background. It should be noted that experiments of this type have also given the clearest indication that mitochondria from one species are not compatible with the nuclear genome from another species (43). A more detailed description of such experiments and their usefulness will be presented in a later section.

While yeast rho$^-$ mutants are generated spontaneously at a high frequency, mammalian cells in culture appear to have a stable mitochondrial genome over many generations, and even a mutant cell line with no mitochondrial protein synthesis (31–33) has maintained an intact mtDNA. If there are mutations or deletions, they are present at a very low frequency. In the early 1970s, when mammalian cells were beginning to be regarded as microorganisms from which useful and informative mutants could be isolated, a challenging question was whether mitochondrial mutations could also be isolated. Few people thought about mtDNA, but those who did were aware that there were thousands of copies per cell. *A priori*, one would therefore predict that a single mutation on one mtDNA molecule in the entire population would not lead to a phenotype, and a phenotype could arise only if there was a mechanism for segregation. It was quite a surprise when the Eisenstadt laboratory announced in 1974 that chloramphenicol resistance in a mouse cell line A9 was cytoplasmically inherited (44). The targets for chloramphenicol were mitochondrial ribosomes, and there was an expectation from experience with bacteria and yeast that one or more mutations in the large rRNA could lead to antibiotic resistance. A year earlier the same laboratory had described chloramphenicol-resistant HeLA cells, but the proof of cytoplasmic inheritance required enucleated cells (by a technique that had just been introduced) and the fusion of such cytoplasts containing the mutated mitochondria with nucleated cells containing wild-type mitochondria. The resulting cells have been referred to as "cybrids". Curiously, the first paper reporting cytoplasmic inheritance concluded that "CAP resistance . . . may be encoded in mtDNA, or possibly in one of the other types of cytoplasmic DNAs reported in mammalian cells . . . (spcDNA, microsome-associated DNA, informational or I-DNA and membrane associated cmDNA)." Of course the latter types of DNA have now been lost in history, since they most likely were simply contaminants derived from nuclear DNA.

The method of obtaining such mutants is worth recounting. Cells were treated for about a day with ethidium bromide, followed by incubation with 50 μg/ml of chloramphenicol. Two and a half months later, colonies appeared which could be cloned and propagated in the presence of the drug. Clearly, at the later stages the cells had a mixed population of mtDNAs with the molecules containing the mutation in the majority. The cells were heteroplasmic. It is necessary to postulate that during the prolonged exposure to the antibiotic the cells initially did not proliferate, and a majority of the cells in fact died. By a mechanism that is still quite obscure, in some cells a buildup of mutated mtDNA relative to wild-type mtDNA must have occurred. As the proportion increased, mitochondrial protein synthesis could proceed, and eventually the cells gained the level of energy metabolism necessary to sustain cell divisions. With hindsight, one might also wonder why such cells did not become rho^0 cells, but one must presume that the media chosen would not have supported such cells (glucose level, no uridine or pyruvate present).

The original discoveries were followed by a series of interesting and informative studies on intra- and interspecific hybrids and cybrids (45). It was shown that the mitochondrial and nuclear genome must be from the same organism to be compatible and to yield viable cells. For example, when CAP-resistant moue mitochondria were transferred into Chinese hamster cells, the cybrids had a very limited life span. When interspecies hybrids were made and selected directly in chloramphenicol, the chromosomes of the parent with the resistant mitochondria were retained. In some examples this represents a reversal of the usual loss of human chromosomes from a rodent–human hybrid cell line. By making intraspecific hybrids, it was possible to make cells that were deliberately heteroplasmic and had a roughly known proportion of wild-type and mutant mtDNAs. Continued culture under selective or nonselective conditions allowed an assessment of the rates of segregation of these genomes from each other under selective pressure or in its absence. So-called xeno-mitochondrial cybrids have been studied in which the nuclear genes are human and the mitochondrial genome is derived from other primates (46). A functional electron transport chain can be produced when the mtDNA is from the chimpanzee or gorilla, but not when the mtDNA is from the orangutan.

The first success in producing xenomitochondrial mice harboring trans-species mitochondria on a *Mus musculus domesticus* (MD) nuclear background has been reported. First, xenomitochondrial ES cell cybrids were produced by fusing *Mus spretus* (MS), *Mus caroli* (MC), *Mus dunni* (Mdu), or *Mus pahari* (MP) mitochondrial donor cytoplasts and rhodamine 6-G treated CC9.3.1 or PC4 ES cells. Homoplasmic (MS, MC, Mdu, and MP) and heteroplasmic (MC) cell lines were injected into MD embryos, and liveborn chimeric mice were obtained (MS/MD 18 of 87, MC/MD 6 of 46, Mdu/MD 31 of 140, and MP/MD 1 of 9 founder chimeras, respectively). Chimeric founder females were mated with wild-type MD males producing homoplasmic offspring with low efficiency (5%) (47). Heteroplasmic mice carrying wild-type and mutated mtDNA have also been reported (48, 49).

Other antibiotics were used successfully in other laboratories to obtain resistant cell lines in which drug resistance was cytoplasmically inherited (50–53). Oligomycin (or rutamycin) resistance could arise from a mitochondrial mutation in the "oligomycin-sensitivity-conferring protein" of complex V (ATPsynthase), and antimycin resistance resulted from a mutation in the cytochrome b gene encoded by mtDNA. In all cases the cytoplasmic inheritance had to be carefully documented, since other, nuclear alterations (affecting uptake of the drug, for example) could also give rise to such phenotypes. Later, the mutation responsible for oligomycin resistance in Chinese hamster ovary cell mutants was shown to be in the ATPase 6 gene on the mitochondrial DNA (54), and the formation of an altered form of this subunit was also demonstrated by immunological methods (55).

In conclusion, mutations on mammalian mtDNA were first definitively demonstrated in tissue culture cells. The astonishment and continued puzzlement arises not from the existence of such mutations, but about the mechanisms that operate on a population of approximately 1000 mitochondrial genomes/cell to segregate and accumulate a sufficiently large fraction of mutant mtDNAs for a phenotype to be expressed. The extreme selective pressure exerted during the culture with the drugs is responsible for the ultimate selection, but the shift in the mtDNA populations takes weeks or months, while cell division is arrested or extremely slow. This issue will become relevant again in the discussion of human mitochondrial diseases and similar or related mechanisms occurring *in vivo*.

7.3 MOLECULAR GENETICS OF HUMAN MITOCHONDRIAL DISEASES

7.3.1 Introduction

Until about 1988, human patients with myopathies, neuropathies, or encephalomyopathies had attracted attention mainly of neurologists, pathologists, and histologists. Abnormal staining of muscle biopsies viewed in the light microscope had been detected ("ragged red fibers"), and later electron microscopic investigations had indicated that such abnormalities were associated with morphologically abnormal mitochondria. The term mitochondrial encephalomyopathies (MEM) described a combination of symptoms and morphological characteristics with significant variability and severity, which made a genetic analysis complicated and difficult to interpret. Was it a complex genetic trait? Nevertheless, as examples accumulated and reliable pedigree studies became possible with larger samples, there were indications that MEM may follow a pattern of maternal inheritance. The maternal inheritance of human mtDNA had been firmly established in 1980 (56), after related studies had reached the same conclusion in horse–donkey hybrids (56a), in the rat (56b, 56c), in the white-footed mouse *Peromyscus polionotus* (56d), and in frogs, *Xenopus laevis* (56e), and *Drosophila melanogaster* (56f). The ideas converged dramatically in 1988 with two publications describing mutations in mtDNA being linked to "mitochondrial diseases." In one report, Wallace et al. (57) described a point mutation in the ND4 gene (complex I) in several patients with Leber's hereditary optical neuropathy (LHON). The other, by Holt et al. (58), described deletions in mtDNA in patients with mitochondrial myopathies. It has been difficult to keep up with the field ever since. A flood of publications from these and other laboratories rapidly established mtDNA mutations as the genetic basis for a variety of clinical syndromes, in the process defining concepts and generalizations, but also raising new questions of fundamental importance for understanding mitochondrial inheritance and the phenotypic expression of mitochondrial mutations (59–62). To date (October 2006), there have been reports on 126 pathogenic mtDNA point mutations affecting the function of rRNAs and tRNAs (63, 64), and 103 point mutations in the thirteen mt genes encoding polypeptide (D. Wallace, personal communication). It is not the intent in this chapter to discuss these mutations exhaustively. Select examples will be used to illustrate some fundamental and general principles.

The idea of uniparental transmission of mtDNA has been challenged in several publications, for species in which intact sperm is found in the egg cytoplasm after fertilization. Back crosses between *Mus musculus* and *Mus spretus* (an interspecific cross) had yielded offspring in which a very small proportion of paternal mtDNA (0.01–0.1%) could be detected by sensitive PCR techniques (65). This prompted a reexamination for the much more common intraspecific crosses between mammals, from which the authors concluded that in intraspecific crosses (*Mus musculus*) the paternal mtDNA is eliminated at the late pronucleus stage (66). Such a mechanism requires a species-specific recognition system by which egg cytoplasm identifies and destroys sperm mitochondria and mtDNA (67–71). Paternal transmission in rare circumstances may be irrelevant for the present discussion, but it will become a potential issue again in the discussion of mitochondria and human evolution (72).

7.3.2 Maternal Versus Sporadic Inheritance

As one of the pioneers in these investigations pointed out (73), it is perhaps "ironic that the first mtDNA defects identified, large deletions, are not usually inherited"; that is, the observation of maternal inheritance of MEM, though a clue for other disorders, was not relevant in the first studies reported by Holt et al. (58). Patients with MEM had muscle biopsies, and a substantial fraction of such patients (9/25) exhibited heteroplasmy in muscle: Up to 7 kb of mtDNA were deleted in 20–70% of the population of mtDNAs. A detailed description of the size of the deletions and the region of the mitochondrial genome affected will be given in a later section. At this point it is to be stressed that there was heteroplasmy, but this heteroplasmy was found originally only in muscle, not in cells obtained from blood. The defects were not maternally inherited. The generation of deletions and the accumulation of deleted mtDNAs in specific tissues may have been a postzygotic event.

As more patients with mtDNA deletions and rearrangements were identified, a fairly complex genetic picture emerged (73, 74). Only one family was found in which the same deletion was transmitted over three generations. In another family a mother and daughter with similar symptoms had different deletions. The deletions found in several patients could not be detected in their mothers. The child of one patient did not have any deletions detectable by Southern blotting. It is important to recognize that Southern blots may be quite insensitive in the detection of a small fraction of deleted mtDNA molecules. PCR-based methods are very suitable, if primers are chosen which flank a relatively large deletion. Efficient amplification occurs only when the deletion is present, since in the normal mtDNA the primers are too far apart for standard PCR conditions. In general, this group of mtDNA deletions and the associated symptoms are classified as spontaneous, because there is no apparent Mendelian pattern of inheritance and there is also no predictable maternal transmission over several generations.

The most likely explanation for the observed distributions of mtDNAs in a pedigree is that the deletions occurred either during oogenesis, leading to the formation of the occasional oocyte with a subpopulation of mutated mtDNAs, or very early during embryogenesis. If the deletion occurred before fertilization, one would expect a widespread distribution of deleted mtDNAs in different tissues. On the other hand, such a distribution could be strongly influenced by unequal segregation of wild-type and mutant mtDNAs during embryogenesis, or by a selection against cells with a large proportion of mutated mtDNAs and hence partially defective mitochondria. It is conceivable that mtDNAs with substantial deletions are replicated more rapidly and thus have a selective advantage over normal mtDNAs. Such an explanation may account for several observations: (1) Depending on the syndrome and its severity, high proportions of deleted mtDNA were found frequently in the muscle and brain (after autopsy, where possible), while other tissues (e.g., liver or blood) had a lower proportion, sometimes only detectable by PCR. (2) The fraction of deleted mtDNAs increases with age in muscle and possibly the brain. In a situation where the total number of muscle fibers does not change appreciably after birth, mitochondrial DNA replication must be dissociated from cell division, selection against defective cells ceases, and shorter mtDNA molecules gain a replicative advantage. Direct tests of such a hypothesis are becoming possible through the use of ρ^0 human cell lines that can be repopulated with mitochondria from enucleated cells with a known proportion of deleted mtDNAs. In one such experiment the fraction of deleted mtDNAs increased from 40–50% to about 80% in a period of 10 weeks. (3) There is a wide range of age of onset of neurological symptoms from between birth to the age of 60. The factor determining the age of onset could be the percentage of mutated mtDNAs present in certain tissues at birth, or the rate at which mutated mtDNAs accumulate relative to normal mtDNAs during the period after birth.

A most impressive example of a mtDNA mutation accumulating with time in muscle has been reported by Weber et al. (75). A patient, healthy and athletically active as a teenager, experienced the first symptoms at the age of 34 (joint pains and muscle weakness), and she was subsequently observed and examined over a period of 12 years, during which the symptoms progressively worsened to the point where she became dependent on a wheelchair, and suffered from dysphagia to solids and liquids. Ragged red fibers and COX-negative fibers were observed to increase from 2% and 4.5%, respectively, to 18% and 68%, respectively, over the 12-year period. Finally, mtDNA sequencing established a single-point mutation in the $tRNA^{Leu(CUN)}$ gene at position 12320, affecting the TΨC loop at a conserved site. It is difficult to argue that mtDNA with a point mutation has a replicative advantage. The mutation is found in heteroplasmic cells and is restricted to skeletal muscle, the tissue that is clinically and biochemically affected. Within muscle, it was highest in the COX-deficient fibers. In the muscle biopsy specimens that had been collected between 1983 and 1995, the fraction of mutant mitochondrial genomes increased from 0.7 to 0.9. It was absent in fibroblasts and white blood cells. Speculation is that the mutation occurred postzygotically and may already have been restricted to specific cell lineages, including myoblasts. However, the significant expansion of the mutation in skeletal muscle is dramatically demonstrated here for a postmitotic tissue.

The patient has daughters over the age of 20, and examining them might be reassuring to confirm that it is indeed a somatic mutation in the mother and hence absent in the offspring, but they have opted not to be tested at this time. This is not an atypical situation encountered in genetic counseling.

The same female patient provided biopsy material to support another very intriguing observation made in a number of laboratories (76). In patients with myopathies caused by mt mutations, even when the mutated mtDNA was present in large excess, progenitor satellite cells contain no, or only very low levels of, mutated mt genomes. Such satellite cells can form regenerated muscle following induction of necrosis. The authors were successful in using a local anesthetic, bupivicaine hydrochloride, to induce a localized necrosis, and three weeks later the same region of muscle showed clear muscle regeneration, but now a very significant portion of fibers was COX-positive, and in such fibers the levels of mutated mtDNA was very low or undetectable (76). The stimulation of muscle regeneration therefore offers the prospect of reversing severe biochemical and physiological defects in such patients, as an alternative to gene therapy. For obvious reasons, the obstacles to gene therapy with mitochondrial DNA are even greater than those with nuclear genes.

In a pioneering experiment, mitochondrial DNA in normal human oocytes has been investigated by Chen et al. (77). Using sensitive PCR-based techniques, these investigators were able to detect rearranged mtDNAs in single oocytes. Among ~100,000 normal mtDNAs less than 0.1% had deletions/duplications or other structural changes responsible for abnormal PCR products. For most practical reasons, one would consider this homoplasmy. But can such a minority population of mtDNAs become the templates from which the embryo is equipped with mutated mitochondrial genomes, possibly differentially in different tissues? A further discussion of this subject will be resumed below. The consideration of mtDNA in oocytes has led to an interesting additional speculation. MtDNA deletions have been found to accumulate in an age-related fashion in tissues with slowly or nondividing cells. Oocytes in humans are also arrested (at meiosis I) for a prolonged period of time (years), and it is conceivable that mtDNA deletions might arise during this time. The sample size may be too small at this stage, but one could ask whether the incidence or risk of mitochondrial diseases due to mtDNA deletions could increase with advancing maternal age (78).

7.3.3 Mapping mtDNA Deletions/Rearrangements

A significant effort has been expended to date to map the mtDNA deletions precisely and to define the deletion junctions at high resolution by sequencing. The mapping of the deletions has been pursued with several goals in mind. First, it is desirable to establish precisely which genes or gene functions are affected and to relate such deletions to measurements of residual activities of the mitochondrial electron transport chain. Second, one can ask whether the deletions are entirely random, or whether there are common regions that are most frequently deleted. And finally, if there is a "common deletion," do the deletion junctions provide any evidence for a likely mechanism by which such deletions could be generated?

The results are provocative, but not uncomplicated to interpret. Not surprisingly, few deletions have been found so far which delete either one of the origins of replication on either strand, and transcriptional promoters have not been found to be missing. These are the important cis-acting elements required for replication, and their loss would also lead quickly to the loss of the mutated mtDNA. Exceptions can be found in the case of a maternally inherited mtDNA deletion (10.4 kb) (79), as well as in four patients in which the heavy-strand promoter and the adjacent binding site for the transcription factor mtTFA have been deleted (73, 80). Initiation of DNA replication requires transcription from the light-strand promoter, and its continued presence makes the propagation of the deleted mtDNA possible. The two promoters are close together, and deletions starting within the interval are rare. As predicted, and in confirmation of earlier conclusions (see Chapter 4), transcripts dependent on the light-strand promoter were abundant in these muscle fibers, but transcripts from the other strand were missing, regardless of whether the genes were deleted or not (80).

The junctions and flanking regions of mtDNA deletions have been investigated in more than 150 patients (as of 1993 (73)) after amplification by PCR and sequencing. Significantly, in about half of the patients there is a "common deletion" of 4.9-kb mapping in the interval between bp 8000 and bp 13,600. More precisely, the sequence up to 8482 is generally retained, and the sequence following 13,460 is also present. Further inspection of the human mtDNA sequence reveals a 13-bp sequence (8470–8482) at one breakpoint region that is directly repeated at the other breakpoint region (13447–13459). Loop formation and recombination between the direct repeats is one mechanism suggested by these observations. One plausible mechanism for the formation of such molecules starts with mtDNA dimers that are formed during replication, followed by a recombination event generating the deletion. This model is strengthened by additional data which show that direct repeats of 3 to 13 bp are found flanking the deleted sequences in 80–90% of all the patients examined (73). Although the model is attractive, it should be considered in light of the knowledge that recombination is generally thought to be absent or very infrequent in mammalian mitochondria, as evidenced also by the lack of active repair systems of the type familiar to us from the nucleus (81). Therefore, recombination events giving rise to deleted mtDNAs are rare, and, fortunately, patients suffering the consequences are also relatively rare. Another interesting question is whether the postulated recombination activity is restricted to or particularly active at specific stages in the life cycle of an individual—for example, during rapid mitotic divisions preceding oogenesis or during the rapid early zygotic divisions in the embryo. Such a restricted period of activity might help explain the origin and distribution of these deletions in families and in different tissues of an affected individual. An alternative to the recombination mechanism for generating deletions is slippage and mispairing during replication of the mtDNA. The issue remains unsettled at this time.

With the mitochondrial genome so packed with genes, every major deletion clearly includes one or more genes, and it is nowadays straightforward to predict the consequences at the biochemical level. It should be emphasized that patients with the diagnostic symptoms have been found in which mtDNA deletions were not detectable. Obviously, very small deletions or even single nucleotide changes can cause serious problems. Such single nucleotide changes will be discussed below for the cases where maternal inheritance is observed, but clearly, small changes can also occur spontaneously by replication errors, and so far in the discussion the focus has been on spontaneous errors observed in individual patients and, rarely, in multiple offspring from the same mother.

A description of the pathologies and symptoms and their relationship to mitochondrial functions will be presented later. It need not be stressed that individuals would not be alive if at least some of their mitochondrial genomes were not intact and capable of sustaining the biogenesis of the complexes for oxidative phosphorylation. A generalization that can be made, therefore, is that heteroplasmy is the rule in patients with mtDNA deletions .

7.3.4 mtDNA Point Mutations and Maternal Inheritance

It has been mentioned before that numerous families with incidents of mitochondrial encephalomyopathies show maternal inheritance, a strong hint that an inheritable mitochondrial mutation is involved. The first report with strong experimental support for this genetic basis of the transmission of a disease appeared in 1988 by Wallace et al. (57). The group at Emory University sequenced most of the mtDNA of a patient with Leber's hereditary optic neuropathy (LHON) and compared it with the published nucleotide sequence for human mtDNA. It is important to point out here that a total of 25 "mutations" were found—that is, nucleotide differences between the Cambridge sequence and the sequence from an African American LHON patient. Which is the deleterious mutation and which represents a harmless single nucleotide polymorphism (SNP or "snip")? An answer to this question continues to be crucial for all correlations of mitochondrial point mutations with a particular disease (symptom), since it is not yet possible to recreate the same mutation by recombinant DNA technology to verify that it alone is responsible for the defect. In 1988 the rationale for identifying the true mutations causing the disease was as follows: Two of the nucleotide changes were in the rRNA genes and were thought to be of little consequence. Fifteen of the changes were base changes that would not have caused any change in the amino acid sequence of the encoded peptides. No changes were found in any of the tRNA genes in this patient (however, see below). This narrowed the choice to eight potential sites. The extreme variability in metazoan mtDNA sequences, between closely related species, and even within a species has been noted (see Chapters 4 and 8). Phylogenies can be constructed providing very valuable information on evolution, and because of the speed of the mitochondrial clock, a time scale of tens of thousands of years can be resolved (see Section 8.2, "Human Evolution"). Investigators had only begun to become aware of the variability among human populations, but sufficient sequence data were available in 1988 to suggest that five of the changes were likely to be of no consequence since they were also found in other humans not suffering the neuropathy (LHON). Two additional nucleotide changes were found in various ethnic groups and/or in other members of the Georgia pedigree unaffected by LHON. This left the change at nucleotide 11,778 as the likely candidate mutation. A codon change replaces an arginine by a histidine in subunit 4 (ND4 gene) of complex 1. The homologous protein has an invariant arginine at that site in species ranging from fungi to fruit flies to humans. Luckily, the nucleotide change also eliminated an SFaNI restriction site found in normal mtDNA, and it became possible to check this site quickly in several other LHON patients of diverse ethnic origin. All LHON patients in the limited sample investigated in 1988 had this mutation.

A later analysis employed a parsimony computer program PAUP to construct a number of plausible phylogenetic trees for European, African, African-American, and Asian populations, based on nucleotide changes in mtDNA (82). Details will be discussed in a more general context of human evolution in one of the following chapters, but here it is relevant that the authors demonstrated that the SfaNI mutation at position 11778 appears to be quite recent, has occurred twice, and in each case was associated with LHON. It was unlikely, therefore, that the mutation at 11778 was a silent or neutral mutation.

In the intervening years a number of other syndromes with maternal inheritance have been investigated at the molecular level and found to result from single nucleotide changes in mtDNA. Point mutations were found in structural mitochondrial genes as well as in rRNA and tRNA genes (Table 7.1).

Before describing LHON and other defined clinical syndromes with maternal inheritance in more detail from a biochemical and physiological standpoint, some important distinctions between single-point mutations and larger deletions must be emphasized. Individuals with point mutations in their mtDNA are generally heteroplasmic, but frequently they are homoplasmic, that is, all mtDNAs examined contain the mutation. This astonishing observa-

TABLE 7.1 Mitochondrial Diseases Resulting from Mutations in mtDNA

Clinical Description	Inheritance	Ragged Red Fibers	Defect in mtDNA
Kearns–Sayre syndrome (KSS)	Sporadic	+	Single deletion or insertion
Incomplete KSS	Sporadic	+	Single deletion
Pearson syndrome	Sporadic	+	Single deletion or insertion
Leber's hereditary optic neuropathy (LHON)	Maternal	–	Point mutation in ND4, ND1, CYTB, COI
Neuropathy, ataxia, retinitis pigmentosa (NARP)	Maternal	–	Point mutation in ATPase 6
MELAS	Maternal	+	Point mutation $tRNA^{Leu(UUR)}$
MERRF	Maternal	+	Point mutation $tRNA^{Lys}$
MIMyCa	Maternal	+	Point mutation $tRNA^{Leu(UUR)}$
Sensorineural hearing loss	Maternal		12 S rRNA, tRNA

tion immediately raises two questions, one less challenging, the other difficult to answer at this time. How can such individuals survive? And how does homoplasmy for the mutation come about?

The first question can be answered theoretically and experimentally simply by showing that the point mutation leads to a "leaky" phenotype, a protein or a tRNA that is still partially functional. Another possibility is that the mutation in a tRNA affects the rate or efficiency of maturation of individual mRNAs from the polycistronic transcript with no effect on tRNA function. While this answer may satisfy temporarily, it immediately generates more questions, especially in the case of homoplasmy. First, one of the outstanding and most challenging questions in the whole field is to explain why only certain tissues are affected; in the most extreme case, why, for example, do LHON patients suffer the central vision loss (although other symptoms have also been observed)? Second, the optic atrophy (similar to other syndromes discussed here) is delayed in onset that is variable among affected individuals in a pedigree. Third, the severity of the vision loss is also variable. Finally, pedigrees harboring the more severe mutations in a homoplasmic state can include family members that are not affected at all.

The distinction between severe and mild mutations is in part based on an assessment of the amino acid substitution and the effect it might have on the residual activity of the peptide. A Leu-to-Pro change observed in one pedigree with homoplasmy is expressed in 80% of the individuals of the family, presumably because of a drastic alteration in the conformation of the peptide. Where amino acids are conserved in evolution, functional or spatial restraints are indicated, and changes at such positions are expected to be more serious. Changes found in nonconserved regions of the peptide are difficult to evaluate in the absence of a complete structure of the complex.

In more formal genetic terminology, one can speak of the **expressivity** of the mutation in individuals within affected families and the **penetrance** of the mutation in the general population. Penetrance is defined as the probability of detecting a particular combination of genes (mutations) when they are present (in the population or in a pedigree). This may depend on the criteria and means of observation (measurement). In other words, the degree of penetrance of a gene is a measure of the fraction of individuals possessing the mutation and showing manifestations of the mutation. Expressivity refers to the degree to which a particular phenotype due to a particular mutation manifests its presence. "Expressivity refers to the type of manifestation of a gene that is penetrant" (82a, page 290). Both parameters are most likely dependent on other genes, and in the present discussion the expression of mitochondrial mutations will depend on a combination of nuclear genes, often referred to as modifier genes.

While the possibility of the existence of homoplasmy in viable individuals can be explained on the basis of a leaky mutation, the development of homoplasmy for a mutation in a rare pedigree in the human population is more difficult to explain. In almost all normal individuals, ~99.9% of mtDNAs are identical with no harmful mutations. In its simplest form the problem can be stated as follows: A rare point mutation occurs in a mtDNA in germ-line cells of a female. At the time of origin the mutated mtDNA is one among hundreds, if not thousands, of normal mtDNAs. How many generations does it take to accumulate a significant fraction of mutated mtDNAs (heteroplasmy)? How quickly is homoplasmy achieved?

There are indications that a significant change in the mitochondrial genotype can occur within a few generations, or even a single generation, in mammals. Hauswirth, Laipis, and colleagues (83, 84) have taken advantage of a common polymorphism in a herd of Holstein cows to demonstrate that a switch between alleles differing by point mutations was frequently almost complete from one generation to the next. In other words, homoplasmy with respect to a specific base defining the alleles was achieved in the change and transmission from mother to offspring.

7.3.5 Mitochondria and Oogenesis

To escape from this conundrum, investigators have proposed the idea of a "bottleneck": a very small number of mtDNAs, or even a single mtDNA, serve as templates for replication to populate the oocytes from which the mtDNA is transmitted to the offspring (85–87). A critical review has challenged the use of the term "bottleneck" because it can be confusing in its meaning and has been used inconsistently by different authors (88). Instead, the term "sampling and amplification" has been suggested by these authors. They also point out that while either random drift or a complex sampling process can readily explain a rapid segregation of mitochondrial genomes, the real puzzle is the stable heteroplasmy observed in some pedigrees. A female with a relatively small fraction of mutated mtDNAs can have offspring with a significantly higher proportion of mutated mtDNAs, if the mutated DNA is preferentially amplified. But unless there is bias in the sampling, in the majority of her offspring the heteroplasmy should be reduced.

At what stage in gametogenesis or embryogenesis is the sampling and amplification mechanism operational? *A priori*, several stages could be considered (Figure 7.1). One should be aware that the human egg has an estimated 200,000 mtDNAs (77) and that as much as one-third of the total DNA of an oocyte is mtDNA. There is certainly no "bottleneck" at this level, in the sense that only a small number of mitochondria are transmitted by the oocyte. After fertilization and prior to implantation, a number of rapid zygotic (mitotic) divisions occur (step 1 in Figure 8.1), apparently without any further mtDNA replication, in which the pool of oocyte mitochondria is distributed between the daughter cells and the number of mtDNAs per cell is reduced to the normal range of ~1000 or less (89). At a relatively early stage in development a primordial germ cell lineage is set aside (step 2) to migrate to the ovary, and within the ovary a subpopulation of diploid cells (oogonia) is formed by repeated mitotic divisions (step 3). The oogonia increase significantly in size, and a multiplication in the number of mitochondria could accompany this growth (step 4). Meiosis and maturation leads to one ovum (and polar bodies) from one oogonium. (It should be noted that in mammals, meiosis I is arrested, possibly for years, and completed only at periodic intervals when eggs mature. Furthermore, meiosis II is subsequently arrested and completed only if fertilization by a mature sperm takes place.) It has also been suggested that a series of zygotic divisions with nuclear but no mtDNA replication could reduce the number of mtDNAs to a much smaller number ("the bottleneck"), accompanied by large stochastic variations in mitochondrial genotype between different cells. In this hypothesis, segregation is a postzygotic event. Thus, after 12–15 zygotic divisions without mtDNA replication, oogonia could have mtDNAs numbering in the range of 50–56. If the subsequent amplification of mtDNA molecules during the maturation of the primary oocyte from an oogonial precursor cell started from such a small population, or it involved only a small subpopulation of a larger number of mtDNA templates ("sampling and amplification"), shifts in the distribution of "alleles" could be achieved quite rapidly.

To restate the two plausible hypotheses, sampling and amplification could occur in the process of formation of an oocyte from an oogonial cell, or a reduction in mtDNA copy number could occur by simple dilution during the formation of the germ cell lineage in the early embryo. A third hypothesis has proposed that the sampling may occur during a series of mitotic divisions leading to a large population of oogonia (step 3). Finally, it is conceivable that the partitioning of the embryonic cells between the inner cell mass and the

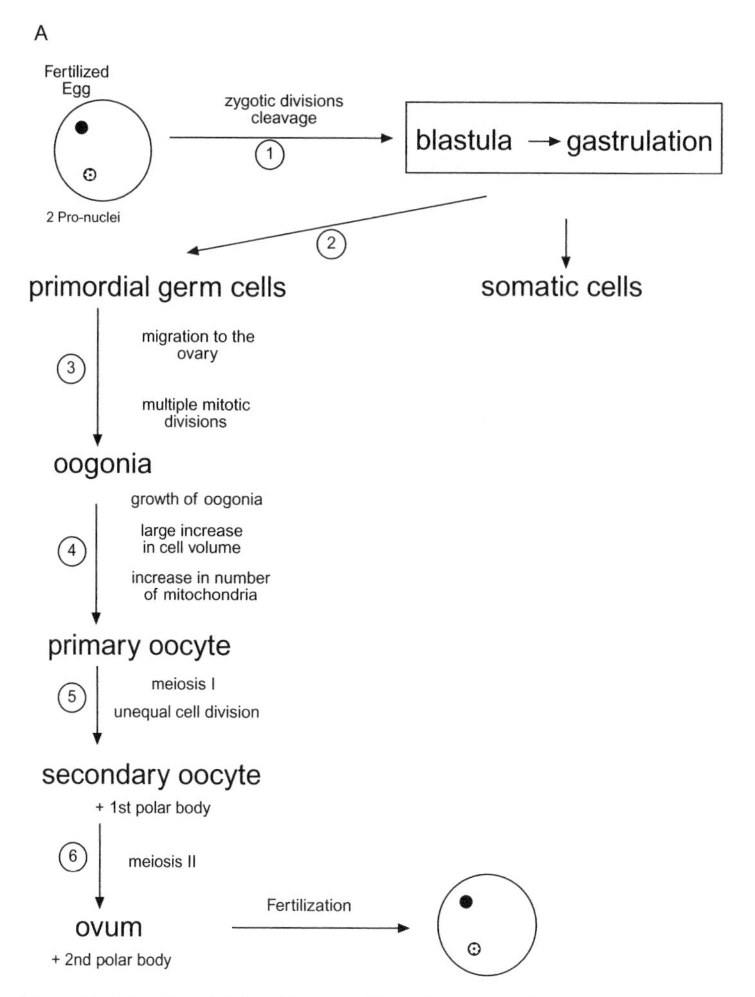

Figure 7.1 (A) Mitochondrial DNA amplification and relationship to oogenesis. The fertilized egg contains an estimated 100,000 mtDNAs. These may be distributed during the early zygotic divisions (step 1) to yield cells with considerably smaller populations of mtDNA. Early in development, germ cells are differentiated from somatic cells (step 2). A very limited number of mtDNAs could be replicated during step 3 (expansion of the precursor pool of oogonia by mitotic divisions), or in step 4 when oogonia increase in size and prepare for meiosis. (B) A model for the origin of the bottle neck based on the assumption that the first n zygotic divisions occur without any mtDNA replication, diluting the population of mtDNA molecules to a small number per cell at the time when germ line cells are set aside. This small number then expands again during meiosis and formation of the mature oocyte.

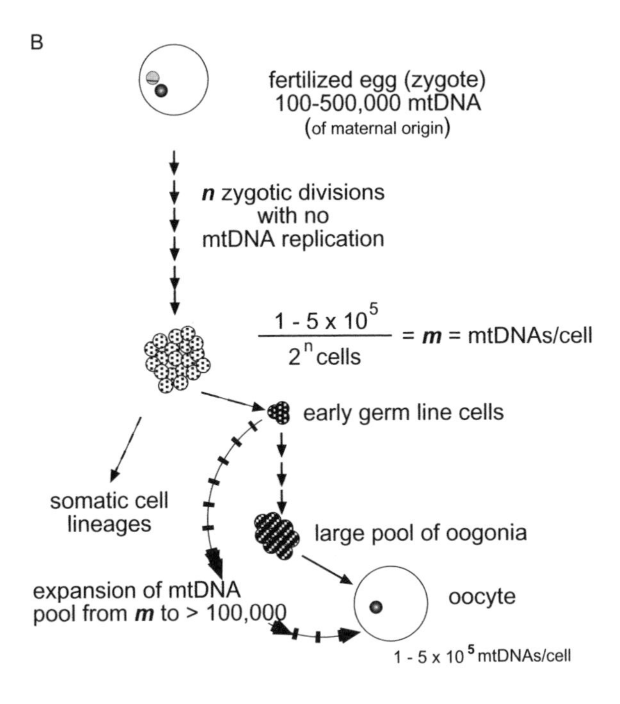

Figure 7.1 (*Continued*)

extraembryonic tissues is accompanied by an unequal partitioning of mtDNAs which could be responsible for the segregation and reduction of mtDNA genotypes in the ICM. A review by Chinnery et al. (90) presents these hypotheses with more references to the original literature, and it also illuminates some interesting differences in oogenesis between mice and humans. They illustrate that precise quantitative predictions for genetic counseling based on mouse models may be difficult.

Several experimental systems have been developed in recent years to achieve "cracks in the bottleneck" (85) Heteroplasmic mouse strains have now been created to be used to understand the mechanisms for segregation of genetically marked mtDNAs (48) (49, 91, 92). Advances in the manipulation of mouse embryos have been instrumental in creating the kind of experimental conditions allowing a direct observation of the segregation of mtDNAs in heteroplasmic mice. Jenuth et al. (93) produced mouse embryos by the fusion of cytoplasts containing mitochondria of one haplotype with one-cell embryos of a strain with another mitochondrial haplotype. For simplicity the mtDNA genotypes will be referred to as NZB and BALB, respectively. The cytoplasts were obtained from one cell embryos by using a glass pipette to draw up and pinch off cytoplasm surrounded by plasma membrane. These were injected under the zona pellucida of recipient one-cell embryos and electrofused. Transplantation into pseudopregnant females produced female and male animals whose mtDNA could be analyzed from tail biopsies. Five founder females were produced with 3.1–7.1% of the donor mtDNA genotype. When a female with 5% donor (NZB) mtDNA was crossed to a BALB male, 28 offspring were analyzed and found to contain, on average, 6.7% NZB mtDNA. However, in individual offspring the proportion of donor (NZB) mtDNA varied from 0% to 29.6%. A rigorous statistical analysis of 191 F_1 offspring from all five crosses, along with 176 offspring from several more back-crosses with F_1 females, indicated a large variance of haplotype frequencies. It was concluded that the results supported the idea that germ-line segregation is a stochastic process capable of producing significant fluctuations from an original distribution of genotypes.

In mice approximately 50 primordial germ cells can be identified in 7.5-day mouse embryos. When several (3–12) such primordial germ cells from a given embryo were analyzed, the coefficient of variation was very small, suggesting that a given proportion of donor mtDNAs received in the zygote was maintained during the earliest stages up to the establishment of the germ cell lineage. These primordial germ cells, when properly located on the genital ridge, divide and give rise to oogonial cells, from which approximately 25,000 primary oocytes are derived in a series of synchronous divisions. Based on these experimental results, Jenuth et al. (93) propose that segregation of mtDNA variants must occur during the mitotic cell divisions that precede meiosis in the ovary.

At this point, one can only speculate about details of the process (86, 93). Several aspects of mtDNA replication must be considered (see Chapter 4). Unlike the case of the nuclear genome, mtDNA replication is relaxed; that is, in a given population of mtDNAs, some may replicate multiple times during the cell cycle, while others do not replicate at all. This still leaves the question, How many tDNA molecules are selected for replication, and is there any kind of selection for a subpopulation? There is evidence that localization within a cell can influence mtDNA replication, with proximity to the nucleus being a determining parameter (94). Employing mathematical modeling using population genetic theory, along with assuming that segregation is the result of genetic drift, one can calculate the effective number of segregating units, N, from the observed variance in the nth generation, if the initial distribution of genomes is known. With an estimated 15 cell divisions required to produce the total population of primary oocytes in the mouse ovary, Jenuth et al. (93) calculated that N is in the range of 200.

7.3.6 Clinical Aspects of Mitochondrial DNA Mutations

In humans, the distinction between variants with point mutations and variants with large deletions is an important one. The fact that changes are occurring post-fertilization is clearly indicated by the differences in heteroplasmy observed between muscle tissue and blood cells in patients with mtDNA deletions. Homoplasmic cells with deleted mtDNA would not be expected to be viable. Whatever the starting position in the fertilized egg (zygote), an unresolved issue is whether there is genetic selection in somatic cells in specific tissues, or simply random genetic drift due to relaxed mtDNA replication.

7.3.6.1 MtDNA Deletions: Kearns–Sayre Syndrome and Pearson Syndrome It need not be stressed that homoplasmic null mutations would be lethal. Two distinct mechanisms can lead to partial deficiency. In heteroplasmic cells the most severe mutation, a complete deletion of one or more genes, can be partially compensated by the presence of normal mitochondrial genomes. In homoplasmic cells a mutation must be less severe—for example, by causing a missense substitution in a protein which reduces its activity. In either case, one expects the tissue to suffer from an inadequate production of ATP by oxidative phosphorylation; and one may now ask, What is it that distinguishes the various "mitochondrial diseases" from each other. *A priori*, one would expect the severity of the disease to depend on the degree to which a particular function is compromised. In the same vein, one would expect the deficiency to be most deleterious in tissues such as the brain and muscle which are energetically the most demanding. In practice, the situation is much more complicated in ways that continue to challenge our understanding of the basic phenomena.

Large-scale rearrangements or deletions of mtDNA are associated with two clinical conditions that share overlapping symptoms. The first, recognized by Kearns and Sayre in 1958 (95) includes among the symptoms retinitis pigmentosa, external ophthalmoplegia, a complete heart block. Morphological alterations in the brain were also noted at autopsy, and these have since been confirmed as the rule rather than an exception or artifact. The disease is characterized by a childhood onset, progressive ophthalmoparesis, pigmentary retinopathy, increased propensity of complete heart block (which could lead to sudden death), and neurological symptoms including incoordination, mental retardation, and episodic coma. Growth retardation and hearing loss are common, and ragged red fibers are found in almost all affected patients. An unbalanced and insufficient energy metabolism is indicated by elevated lactate and pyruvate levels in plasma. Named after its original discoverers, the syndrome is now referred to as the Kearns–Sayre syndrome (KSS). As explained earlier, the occurrence of KSS is sporadic not familial, tissues are heteroplasmic with respect to wild type and rearranged mtDNA, and both sexes are affected at about equal frequencies.

A second syndrome named after its discoverer is Pearson syndrome (96). The affected infants had pancytopenia and altered pancreatic exocrine function, and it was fatal in some cases, but others survived into adulthood and developed clinical features of KSS. When molecular genetic analyses became possible, lesions in mtDNA were found very similar to those found in KSS. How is this syndrome survivable, only to be converted into the second severe condition? Speculation centers around cell population dynamics in the bone marrow, where cells with the greater proportion of mutated mtDNAs have a disadvantage during rapid cell divisions and may be eliminated; that is, in this tissue there may be a selection for cells with the highest proportion of normal mtDNAs. At the same time, in tissues with nondividing cells like muscle or the brain there may be a selection at the level of mtDNA replication which favors the deleted mtDNAs, and ultimately an oxidative phosphorylation deficiency is developed concomitant with encephalomyopathy. Patients diagnosed in infancy with Pearson syndrome do not have ragged red fibers, whereas muscle biopsies showing an abundance of ragged red fibers also show deleted mtDNAs in excess.

Patients with Pearson syndrome and those KSS patients with the most severe symptoms have detectable heteroplasmy not only in muscle, but also in fibroblasts and leukocytes, tissues easily available from biopsies. At autopsy, brain and liver tissue (and other tissues) become available, with the brain showing pronounced heteroplasmy or an abundance of mutated mtDNAs, while liver may have a smaller proportion. On the one hand, the ubiquitous presence of mutated mtDNA is an indication that the mutation was present in the oocyte before fertilization, but the variability in the ratio of normal to mutated DNA in different tissues is indicative of perhaps two selection mechanisms based on (1) the fitness of individual cells in rapidly dividing or renewing tissue (blood, liver), and the replicative advantage of mutated mtDNA molecules in nondividing tissue. As a result, these distributions may vary between patients and, most significantly, may shift with age. Delayed onset and variable expressivity may be explained quite satisfactorily in general terms, but more definite predictions at the time of the first diagnosis may not be possible. Even a determination of the proportion of mutated DNA in a muscle biopsy (which muscle?) may not have any predictive value, since there is no correlation between this fraction and the clinical severity.

It is commonly assumed that there are several mtDNAs per mitochondrion, and that intra-mitochondrial complementation occurs. Continuous fission and fusion of mitochondria would be expected to contribute to a mixing of the two genomes. However, such a picture, derived from observations in tissue culture, may be misleading in tissues such as muscle, where the highly structured cytoplasm and the cytoskeleton (sarcomeres) may prevent extensive diffusion and mixing of mitochondria.

The severity and the precise combination of symptoms may also be dependent on the nature of the rearrangement and the specific segment of mtDNA missing in heteroplasmic cells. Nevertheless, for reasons explained above, patients with identical deletions may exhibit different clinical abnormalities. Almost any deletion of some length will cause the loss of several genes, including structural genes encoding essential subunits for complexes I, II, and IV, alone or in combination, and almost always the loss of at least one tRNA gene. Thus, very specific predictable biochemical activities will be compromised, but the loss of one or more tRNA genes will also diminish the capacity for mitochondrial protein synthesis, and hence the level of all complexes of the electron transport chain may be affected. A more detailed discussion of several specific cases can be found in the review of Harding (73). Superimposed on the effects of the specific deletions are (a) mechanisms of post-transcriptional regulation such as mRNA stability and the control of steady-state levels of individual mRNAs and (b) translational control mechanisms (initiation?) that may influence the half-life of an mRNA. If such mechanisms are tissue-specific, a simple consideration of heteroplasmy and gene dose is definitely not sufficient to predict the full scope of clinical symptoms and their severity.

7.3.6.2 Familial Mitochondrial DNA Depletion A rare disorder with an inevitably fatal outcome in early childhood has been studied in several families. Healthy parents have produced children suffering a severe depletion of mtDNA in specific tissues, most frequently liver or muscle. First recognized by Moraes et al. (97) as a novel genetic abnormality, the defect appears to reside in a nuclear gene, since no mitochondrial mutations could be detected.

7.3.6.3 Point Mutations Table 7.1 lists five distinct clinical conditions resulting from point mutations in mtDNA which are maternally inherited. They are: LHON, NARP, MELAS, MERRF, MIMyCa, and sensorineural hearing loss. A genetically interesting and challenging feature of these diseases is the frequent finding of homoplasmy (see discussion in Section 7.3.5); and a perhaps oversimplified aspect of each mutation is that it is "leaky," which in biochemical terms implies that some function is retained. A detailed consideration of each condition and patients having the mutation leave many other questions still unanswered.

7.3.6.2.1 Leber's Hereditary Optical Neuropathy (LHON) Leber's hereditary optic neuropathy (LHON) has as the defining feature the sudden onset of bilateral central vision loss. The condition was first described by Leber in 1871. Onset is usually delayed until young adulthood (20–24 years of age), but may occur as early as age 5. Although bilateral, vision may be affected first in one eye, followed by involvement of the other eye weeks or months later. Individuals are generally healthy, although the mutation is found in all tissues. Detailed histological and anatomical studies have established the specific degeneration of the ganglion cell layer and the optic nerve. Although vision loss is permanent in most cases, a number of LHON patients have shown improvement, sometimes relatively soon after the diagnosis, sometimes after several years. Disedema and subtle alterations in retinal vessels are part of the diagnosis, but curiously, the clinical condition is not followed rapidly by the appearance of other neurologic abnormalities, although patients (and family members) have exhibited some or all of the following: hyperreflexia, Babinski signs, incoordination, peripheral neuropathy, and cardiac conduction abnormalities. How this highly restricted pattern of degeneration of the optic nerve is initiated is still a matter of debate, which has recently been summarized by Howell (98). The pathogenesis of LHON is not yet well understood. A detailed discussion requires the description of the microanatomy of the optic nerve head, the region around the lamina cribosa, and the possibility of a "chokepoint" in the optic nerve where axoplasmic transport of mitochondria may be impaired. The resulting swelling may accentuate ischemia and ultimately neurodegeneration. One plausible model is that increased or altered mitochondrial ROS production renders the retinal ganglion cells vulnerable to apoptotic cell death (99). Patients do not have ragged red fibers in muscle, and lactate levels in blood and cerebrospinal fluid are normal.

Members of some families, predominantly females, have been reported to be at risk for developing symptoms resembling multiple sclerosis (MS). The risk of LHON is inherited from the mother, typical of mitochondrial diseases, and the mother may or may not have been affected. The primary mutation (see below) in most LHON patients is homoplasmic, but patients with heteroplasmy have also been found. There is incomplete penetrance even in the homoplasmic population, and the current view is that the primary mutation predisposes toward LHON, but secondary genetic and environmental etiological factors contribute to the development of the disease (98, 100). Eighty-five percent or more of the patients are male (100). Historically, this fact was thought to indicate X-linkage, with the appearance in a larger-than-expected fraction of females requiring an explanation. Today the maternal inheritance is firmly established, challenging us to explain what genetic factor(s) contribute to the biased expression of this phenotype. An X-linked susceptibility locus has been proposed to play a role. However, while the issue has not been definitively resolved, recent studies have failed to obtain evidence for such a locus on the X chromosome. On the other hand, three X-linked genes required for complex I biogenesis have been identified (36).

In their review, Brown and Wallace (101) have described 16 point mutations associated with LHON, in subunits ND1, ND2, ND4, and ND6 of complex I, in CYT b, and in the CO1 subunit of complex IV. These authors express the view is that mtDNA LHON mutations "represent a continuum of severity of OXPHOS defects with each imparting a proportional increased risk for LHON." The issue is controversial, and Howell (100) emphasizes that the vast majority of LHON families have one of three primary pathogenic mutations in mtDNA at nucleotides 3460 (ND1 gene), 11778 (ND4 gene), and 14484 (ND6 gene). The most abundant mutation in the ND4 gene at nucleotide 11778 converts a highly conserved arginine to a histidine. Other, less common mitochondrial mutations may have a secondary etiological role, or they may appear in rare "nonclassical" LHON families suffering from optic neuropathy and additional neurological abnormalities. Another perspective is that mild "LHON mutations" may act synergistically. This is deduced from the finding that at least four "LHON mutations" by themselves do not cause loss of vision, but in combination with other "LHON mutations" the phenotype is expressed in almost 100% of individuals carrying the two mutations. The probability of such double mutants would at first seem extremely small, until it is realized that some of these mutations are present in a substantial fraction of the general population (0.1% to 10%, probably variable in different ethnic groups) (102). Additional data on population genetics and history will be required to establish whether other rare mutations are mildly pathogenic, or secondary, or of no consequence at all. It will be important to resolve whether LHON is caused exclusively by mutations affecting complex I.

While there is agreement that the primary mutations in the ND1, ND4, and ND6 genes represent the major risk factor in LHON, there is more controversy about the actual effect of these mutations on the rate of electron transport and oxidative phosphorylation. It would seem straightforward to look at the specific activity of complex I (the NADH-CoQ oxidoreductase) in mitochondria from LHON patients and normal controls. Nevertheless, results in different laboratories have been contradictory, depending on what was measured (summarized by reference 100). One type of measurement involves oxygen consumption with NADH-linked substrates, and respiration with succinate is used as a control and for normalizations. Care must be taken to measure complex I activity by demonstrating the sensitivity to rotenone. Experiments can be performed with white blood cells and platelets from patients, or after construction of cybrid lines containing mitochondria from LHON patients and nuclei from ρ^0 lines. Another approach is to measure the pyruvate : lactate ratio as an indicator of the NADH/NAD balance in fibroblasts. The method has pitfalls discussed by Howell (100).

Why does a reduced activity of complex I in homoplasmic patients cause LHON, while a reduction in electron transport and OXPHOS capacity in a patient with heteroplasmy and mtDNA deletions cause KSS? In homoplasmic LHON patients there is no mechanism for a further shift toward a worse condition in individual cells or tissues. In contrast, in KSS patients, sampling of available tissue may indicate 50% heteroplasmy, for example, but this is an average, and individual cells in the brain or myotubes in muscle may be more severely compromised, causing clinical symptoms. It is clear that much remains to be learned about the distinction between localized defects in individual complexes of the electron transport chain and an overall reduction in several or all complexes due to a mtDNA deletion. The metabolic consequences are almost certainly more complex that a simple reduction in ATP supply.

7.3.6.2.2 Neuropathy, Ataxia, Retinitis Pigmentosa (NARP) If there is a "bottleneck" in oxidative phosphorylation, it can easily be imagined in complex V, as already indicated by the well-known phenomenon of coupling in mitochondria and the dependence of electron transport on the ADP concentration. A more rigorous statement is that ATP turnover (synthesis and consumption) has the highest control coefficient of all the reactions of OXPHOS, and changes in this rate would be expected to have the largest effect on the overall rate of oxidative phosphorylation (see Chapter 5). One may, therefore, be able to rationalize that mutations in ATP synthase (complex V) subunits are rare, and in the one reported case the symptoms are relatively broad. A mutation in subunit 6 (nucleotide 8993; Leu $\rightarrow$ Arg) leads to neuropathy, ataxia, and retinitis pigmentosa (NARP). Since the structure of complex V is beginning to become resolved at high resolution (see Chapter 5) the amino acid substitution can be related to the observed inhibition in ATP synthesis: An additional positive charge is positioned in the proton channel. In fact, the inhibition appears to be severe, and NARP patients are heteroplasmic, with the severity of the symptoms related to the fraction of mutated genomes. At one end of the spectrum are mild neurological signs such as migraine headaches and retinitis pigmentosa, and at the other end one observes necrotizing encephalopathy or Leigh syndrome (102–104). A pedigree has been described in which the mother was mildly affected (87% mutant mtDNA), but all four offspring had >90% mutant DNA and were severely affected; two daughters died of Leigh syndrome at an early age.

7.3.6.2.3 MELAS, MERRF, and MIMyCa Point mutations responsible for mitochondrial encephalomyopathy with lactic acidosis and stroke-like episodes (MELAS), myoclonus epilepsy and ragged red fibers (MERRF), and maternally inherited disorder with adult-onset myopathy and cardiomyopathy (MIMyCa) are found in specific tRNA genes. The MERRF mutation ($tRNA^{Lys}$) is heteroplasmic. Although the symptoms vary from mild (hearing loss, mitochondrial myopathy, electrophysiological changes) to severe (uncontrolled myoclonic jerking, dementia, cardiomyopathy, nephropathy, neurosensory hearing loss, altered VER and EEG), there is no strict correlation with the percentage of mutant mtDNAs. In the view of Wallace et al. (105), age strongly influences the expression of the mutation. This group has become the strongest advocate of the hypothesis that an accumulation of mitochondrial mutations in somatic cells with age causes a general decline in OXPHOS capacity with age, possibly at different rates in different tissues. For each tissue there is a critical threshold below which the energy needs of the cells can no longer be met. In an individual starting out with a lower capacity due to a $tRNA^{Lys}$ mutation, the drop in capacity with age will reach the threshold sooner, and symptoms start to appear. The nervous system and muscle with the highest minimal requirements are predictably the first tissue to be affected (102). Based on the observations in pedigrees, an individual with >90% of mutant tRNALys could nevertheless be normal at a young age, but would deteriorate as he/she aged. This hypothesis has to some extent prompted research and is now supported by data showing an increase in mtDNA mutations with normal aging in various tissues (see Section 7.4.2 below).

7.3.6.2.4 Sensorineural Hearing Loss Deafness can arise from many causes, including overexposure to rock concerts and iPods. An interesting genetic case are specific mutations in the 12S rRNA gene that lead to aminoglycoside-induced and nonsyndromic deafness (106–108). The investigation of several large pedigrees from different ethnic origins has clearly established maternal inheritance and the 12S rRNA alteration as the primary factor, which is, however, not sufficient by itself to produce the phenotype. Guan and his colleagues have postulated that modifier genes in the nucleus must contribute to the expression of the phenotype. A very exhaustive and thorough genotypic analysis has focused on a gene (TRMU) encoding a mitochondrial protein required for tRNA modification (post-transcriptional) (106). The role of this gene product had first been elucidated in bacteria. The same group has also identified a deafness-associated mitochondrial tRNA(Ser(UCN)) T7511C mutation, which becomes highly penetrant in families in conjunction with homoplasmic ND1 T3308C and tRNA(Ala) T5655C mutations (109).

7.3.7 Nuclear Mutations and Mitochondrial Disease

7.3.7.1 Defective Electron Transport Chain The term "mitochondrial disease" was at first associated with pathologies arising from mutations in mtDNA, and therefore with partial defects in energy metabolism. However, it is clear that a substantial number of nuclear genes encode subunits of the complexes of the oxidative phosphorylation system, or proteins that are essential for the biogenesis of mitochondria and the assembly of such complexes. Mutations in these genes could cause similar phenotypes/pathologies, and they are therefore included in the discussion. On the other hand, many "inborn errors of metabolism" result from the compartmentalization of metabolism and from defective mitochondrial enzymes. Such "metabolic disorders" resulting from nuclear mutations will not be further discussed here. Adding up all the nuclear-encoded subunits of complexes I–V would suggest ~70 nuclear genes contributing to the biogenesis of the ETC and ATP synthase. The number of assembly factors required is still unknown. One can add to this number (1) the genes/gene products required for mtDNA replication, maintenance, repair; (2) gene products required for transcription and processing of the polycistronic transcripts; and (3) gene products for the mitochondrial translation machinery. Thus, at least 200 genes contribute directly to the biogenesis of the OXPHOS machinery, and defects in such genes could lead directly to an "energy crisis." Many more genes are needed for mitochondrial protein import, metabolite exchange between the cytosol and the matrix, and the biosynthesis of essential cofactors and prosthetic groups (hemes, Fe–S centers, ubiquinone). A deficiency would probably affect energy metabolism, but may cause a more complex pathology. The term "mitochondrial disease" is at risk of becoming less specific since so many functions are associated with mitochondria.

Some of these genes are known to be X-linked. A severe deficiency would most likely be embryonic lethal in males. In X-linked genes, only missense mutations encoding partially functional proteins would be expected in affected males. It is interesting to speculate whether X-inactivation in a heterozygous female might be biased during development to lead to a functional OXPHOS system in surviving/selected cells. Random X-inactivation would lead to a deficiency at the cellular level, and mosaicism may not be compatible with a viable tissue. Severe autosomal mutations appear to be rare. In a homozygous condition, one would most likely find embryonic lethality. Thus, patients, even newborns with a short life expectancy, must have some residual OXPHOS activity. How much is detectable can depend on the tissue in which it is measured (and which tissue is available for autopsy from a live patient). Dominant autosomal mutations in any of the above-mentioned genes have not been reported. Large pedigrees for patients with severe mitochondrial diseases due to nuclear mutations are rare.

A significant number of patients have now been characterized with an OXPHOS deficiency due to nuclear mutations affecting the activity of either an isolated complex or the entire electron transport chain (see references 59, 60, and 110–112 for reviews). Mutations in genes encoding structural proteins (subunits) generally cause a deficiency in a specific complex, but there are exceptions. Patients with an isolated complex I deficiency represent a majority of such patients (113–122). In most or all of such patients a reduced complex I activity results from a reduced level of the assembled complex I, rather than from a reduced specific activity of the complex. It is noteworthy that in a large fraction of such patients the defective gene has not been identified, despite the determined efforts by several laboratories where all known nuclear and mitochondrial genes for subunits of complex I were sequenced. A majority of such patients are male, and the possibility of an as yet unidentified X-linked gene required for complex I assembly or activity must be considered. The existence of such an X-linked gene is suggested by experiments in the Scheffler laboratory, where Chinese hamster mutant cells lacking complex I could be complemented with a wild type mammalian X chromosome (36). These mutants constitute a third complementation group distinct from the mutants with defective *NDUFA1* (MWFE) or *NDUFB11* (ESSS) genes.

Patients with a partial complex II activity have also been described by several laboratories (123–125). While some of these have the expected symptoms from an OXPHOS deficiency, mutations in the *SDHB, SDHC*, and *SDHD* genes can cause the formation of an unusual class of tumors called paragangliomas, a tumor of the carotid body (the main arterial chemoreceptor that senses oxygen levels in the blood) (126–130). Isolated complex III deficiencies are very rare. There is one report of a deletion in the nuclear gene *UQCRB* encoding the human ubiquinone-binding protein of complex III (QP-C subunit or subunit VII) in a consanguineous family (131). Among the patients with an isolated complex IV deficiency, several defective nuclear genes have been characterized that do not encode subunits of the mature complex. The observed low activity of cytochrome oxidase was proved to be due to missing assembly factors. As described above, many genes/proteins have been shown to be required for complex IV assembly. In human patients with Leigh syndrome, severe mutations have been described in the *SURF1* gene (132–135), and other patients have been identified with mutated *SCO1* or *SCO2* genes (59, 136). The products of these genes have similar but not redundant functions in the delivery of copper to the complex. In contrast to patients with a complex I deficiency, the relevant complex IV-related genes are found to have homologues in yeast, and thus any suspected mutation or deficiency can be tested in this model system to verify its effect on complex IV assembly/activity. Nuclear mutations responsible for complex V deficiencies have not been detected in humans so far.

7.3.7.2 MtDNA Maintenance and Replication Although mtDNA replication is relaxed (not tightly coupled to nuclear DNA replication), there must be controls that maintain the copy number in a range appropriate for a particular tissue or organ. A number of patients have now been characterized with mutations affecting this pool of mtDNAs. Symptoms include progressive external ophthalmoplegia (PEO), exercise intolerance, ataxia, cardiomyopathy, gastrointestinal symptoms, and more (see references 59 and 137–139). The symptoms appear between 18 and 40 years of life and are associated with extensive and increasingly large proportions of mtDNA deletions. Such deletions, detectable by Southern blotting, are not found in all tissues, and it appears that they arise in nondividing tissues such as muscle, in contrast to leukocytes. The deletions are the result of defective DNA replication, which can be caused by mutated enzymes of the replication machinery itself (DNA polymerase γ, or the helicase/primase TWINKLE) or by perturbations in the mitochondrial nucleotide pool(s) involving enzymes such as the thymidine phosphorylase (140) or transporters such as the adenine nucleotide translocator (141).

There are also syndromes resulting from mtDNA depletion. Children healthy at birth may develop severe hepatopathy, myopathy, nephropathy, or encephalopathy from a tissue-specific loss of mtDNA down to levels of ~10% of normal. Mitochondria from such a patient could be introduced into ρ^0 cells and repopulate such cells with mtDNA (142). Thus, the unaffected parents were believed to be heterozygous for a recessive mutant allele, a conclusion strengthened by the finding of two affected sons and an affected daughter in a consanguinous marriage (143). The restriction of the loss of mtDNA to specific tissues is still much of a puzzle, especially since variable tissue involvement has been reported in a single family (137).

Because affected families are rare and usually small, linkage studies are difficult. One approach has been to select candidate genes and carry out haplotype analysis with closely linked satellite markers. Among the candidates, the genes for the mitochondrial transcripion factor (mtTFA), the nuclear respiratory factor 1 (NRF-1), the mitochondrial single-stranded DNA binding protein (mtSSBP), and endonuclease G were examined, but mtTFA remained the only candidate, if inheritance is indeed recessive (143). A limited or biochemical analysis has again implicated perturbed mitochondrial dNTP pools. In a small number of families, defects were identified either in the deoxyguanosine kinase (144) or in the mitochondrial thymidine kinase (TK2) (138, 145). The puzzling aspect of these findings is that certain tissues (skin, blood) are completely spared. It may be noted in this context that HIV patients being treated with drugs such as azidovine (nucleoside analogues) may also suffer from mtDNA depletion in various tissues after prolonged treatment. The mtDNA populations recover after withdrawal of the drug.

7.3.7.3 Friedreich's Ataxia The discussion of mitochondrial diseases cannot be exhaustive in this volume, but two more syndromes caused by very specific mitochondrial protein defects should be mentioned here. Friedreich's ataxia is the most common hereditary ataxia, and it is considered a nuclear-encoded mitochondrial disease because it is clearly caused by a deficit in the mitochondrial protein frataxin (146–149). The deficit results from a reduced level of frataxin mRNA in the cytosol. The frataxin gene on chromosome 9q13 has an unstable GAA trinucleotide repeat in the first intron; an expanded repeat leads to inefficient splicing. The severity of the phenotype is correlated with the size of the expanded repeat. Although the corresponding yeast gene had been identified and used to reproduce the phenotype in this model organism, it has nevertheless been a challenge to characterize the precise role of frataxin in mitochondria. It is generally agreed that frataxin plays a role in mitochondrial iron homeostasis and may be the donor of iron in the synthesis of heme and of iron–sulfur clusters. A deficiency reduces the supply of these important constituents of the electron transport complexes and hence reduces ATP production by oxidative phosphorylation. However, further complications arise. Frataxin may bind surplus iron (it binds six Fe^{3+} per monomer); and, when depleted, iron accumulates in mitochondria. This surplus/free iron can facilitate the interconversion of ROS into the highly reactive hydroxyl radical by the Fenton reaction and thus accentuate oxidative stress. In other words, frataxin contributes to the antioxidant defense, in addition to its iron chaperone function in heme and ISC synthesis. There is also a contrary point of view suggesting that increased superoxide production could not explain by itself the FRDA cardiac pathophysiology (150).

Oxidative stress caused by frataxin deficiency leads to the selective loss of dorsal root ganglia neurons, cardiomyocytes, and pancreatic beta cells. Why these nondividing cells are particularly sensitive to frataxin loss is still a puzzle, since frataxin is widely expressed in all cells and tissues. From this selective cell loss, one can rationalize the cardiomyopathy and diabetes that form part of the clinical manifestations of Friedreich's ataxia. The disease is relatively common because the trinucleotide GAA expansion is an event occurring at a higher frequency compared to other mutations in nuclear genes.

7.3.7.4 Deafness and Dystonia Syndrome (Mohr–Tranebjaerg Syndrome) DDS is an X-linked recessive disorder (151). Sensorineural hearing loss, dystonia, mental deterioration, paranoid psychotic features, and optic atrophy are indicative of progressive neurodegeneration. The corresponding human gene was identified and cloned by molecular–genetic techniques after careful mapping, and when the amino acid sequence of the human DDP1 protein was "blasted" against other proteins in the whole protein database, it was found to be homologous to the Tim8p of yeast (152–154). DPP1 (alias TIMM8a) assembles with TIMM13 in a 70-kDa complex in the intermembrane space and assists in the integration of the hTim23 protein into the inner membrane (see Chapter 4). The elucidation of the mechanism was greatly helped by molecular genetic manipulations of the homologous genes in yeast (*TIM8, TIM13*). The DPP1 protein was not present in fibroblasts from a patient with this syndrome, presumably because a missense mutation, Cys66Trp, abolished a conserved CX(3)C' motif, preventing the protein from assembly with TIMM13 (as predicted from similar studies in yeast). Thus, a satisfactory molecular explanation can be provided for the observed symptoms. On the other hand, one might have predicted that the loss of hTim8 and hence the effect on the import of Tim23 would have an effect on a much wider range of tissues.

7.3.8 Conclusion

A migraine headache or muscle fatigue should not cause an immediate panic and provoke thoughts about mutations affecting mitochondrial functions, but a chronic condition does deserve attention. A little over two decades ago, physicians may have had a distant memory of mitochondria from the basic biochemistry course in their first year in Medical School. Today, histochemical, biochemical, and molecular tests of mitochondrial dysfunction have become almost routine in the analysis of a broad spectrum of neuropathies and myopathies. Both nuclear and mitochondrial mutations can contribute to defective mitochondrial energy metabolism. Severe nuclear mutations in homozygous individuals would almost certainly be lethal. If heterozygotes have no increased fitness or selective advantage (as may be the case in sickle cell anemia and possibly in the case of cystic fibrosis) and homozygotes are not viable, the frequency of specific alleles in the population may remain small. However, the large number of genes required for an active oxidative phosphorylation system represents a large target for mutational damage, and hence the overall frequency of individuals with mitochondrial diseases is potentially very large. The number may also depend on where one draws the line between a headache, a fatigue, and exercise intolerance, and a disease. Ask your doctor, as the TV ads advise.

7.4 MITOCHONDRIAL DNA AND AGING

> The most important inventions in evolution are sex and death.
>
> —F. Jacob

7.4.1 Introduction

The achievements of the last decade have revitalized research on mitochondria, not only because of new insights gained into the genetic basis of mitochondrial function and dysfunction, but also because of the realization of their vital role in maintaining a healthy individual. It is, with hindsight, perhaps not too surprising that mitochondria as the "powerhouse" of the cell would play a critical role in the health (function and survival) of cells. Turnover of all components is a constant activity, consuming as much as 50% of the total energy supply, and in the broad sense turnover is clearly related to "maintenance and repair." When the energy supply becomes limiting, the infrastructure slowly deteriorates, and more severe consequences may be triggered by catastrophic failures at localized sites. This view, particularly as related to the process of aging, is undoubtely very simplistic. Nevertheless, the concept that aging and mitochondrial dysfunction may be related has had a powerful appeal to a number of investigators in the past two decades (62, 155–167). The citations listed here include some influential "historical" papers and a sampling of the most recent reviews. One of the most prominent journals in the biological sciences has recently acknowledged that "aging research come of age" and devoted a whole issue to this subject (168). The present chapter will attempt to review some of the hard facts and follow up on some of the more promising speculations. A discussion of the many aspects of senescence is clearly beyond the scope of this treatise. Familiar characteristics associated with aging are a decline of muscle power and often, but not always, diminished mental capacities. From the well-established cause-and-effect relationship between clinically significant encephalomyopathies and mitochondrial DNA mutations, it is tempting to look for a similar relationship between normal aging and a reduced capacity for respiration and oxidative phosphorylation.

Some clear distinctions should be made at the outset. Attempting to define a single cause for normal aging may be presumptious. However, it is probably not too far-fetched to claim that a diminution of mitochondrial functions may either be central to the process or be a strong contributor influencing the timing and the severity of the overall deterioration. The challenge is to set up a plausible chain of events and to establish a cause-and-effect relationship. A credible scenario gaining acceptance is based on the postulate of a slow and stochastic buildup of damage caused by reactive oxygen species (ROS) that are a normal byproduct of respiration. Superoxide, hydrogen peroxide, and hydroxyl radicals are constantly being formed. Estimates are that up to 4% of the oxygen consumed in mitochondria is converted to reactive oxygen species. This number has recently been challenged and reduced by an order of magnitude, but it is still recognized to be a significant risk of damage to macromolecules (163). Their potential harmfulness can be judged from the co-evolution of a set of enzymatic defence mechanisms such as superoxide dismutase, catalase, and peroxidases, enzymes which destroy these reactive species before damage can be done. A number of experiments with transgenic flies and mice have demonstrated quite convincingly that knocking down these enzymes can reduce the life span, while increased doses of such enzymes can prolong the life expectancy (169, 170). A word of caution: The life extension of fruit flies by this means was successful with short-lived strains of *Drosophila melanogaster*, but failed with naturally long-lived strains. Additional substances (e.g., vitamin E) can act as free radical scavengers, and they have been proposed for megavitamin therapies. Although the largest proportion of oxygen utilized by a cell is by the mitochondrial electron transport chain, cytochrome P-450, cytosolic oxidases, and the oxidations in the peroxisome generate reactive oxygen species outside of the mitochondria, and the protective functions are found in multiple compartments.

The reactive oxygen species can damage DNA, proteins, or lipids. Thus, it is at least conceivable that the primary cumulative damage is to mitochondrial lipids (e.g., cardiolipin), altering membrane fluidity and ultimately causing defects in electron transport and respiration; as a result, the generation of reactive oxygen species may be accelerated. Eventually, defense mechanisms and repair systems are overwhelmed and damage to mtDNA becomes permanent. Alternatively, one could consider mtDNA to be the primary target of superoxide or hydroxyl radicals; and as more and more mtDNAs sustain mutations in critical coding regions, complexes in the electron transport chain become less efficient or inactive, leading to a decline in mitochondrial function. H. T. Jacobs reviewed the experimental support for this hypothesis (171), concluding that a rigorous test of the hypothesis remains to be undertaken, but would require a direct manipulation of the rate of mtDNA mutagenesis, to test whether this could alter the kinetics of aging. In evaluating these alternatives (among still others), attention must be paid to the relative rates of turnover of the major mitochondrial constituents. Generalizations applying to all differentiated cells and tissues are likely to be inappropriate.

Very elegant, informative, and provocative studies were published by Trifunovich and colleages in the past few years (172–174). Mouse strains were constructed with homozygous mutations in the mitochondrial DNA polymerase γ gene. The mutations had no effect on the polymerase activity, but eliminated the exonuclease (proofreading) activity of the enzyme. The result was a three- to fivefold increase in point mutations in mtDNA, in addition to increased deletions which was correlated with a significantly reduced life span (median life span of 48 weeks compared to 2 years of controls). Death was preceded by many of the typical, age-related phenomena such as weight loss, hair loss, osteoporosis, curvature of the spine, anemia, heart disease, and reduced fertility. On the one hand, this study appeared to establish quite conclusively a strong connection between the accumulation of mutations in mtDNA and aging. On the other hand, a follow-up study by the same laboratory claimed to refute the favorite hypothesis that mtDNA mutations induced by ROS were the cause of aging. The mtDNA mutator mice developed severe respiratory chain dysfunction, but there was no increased production of ROS and no evidence for increased oxidative stress (measured by protein carbonylation levels), aconitase activity (sensitivity of Fe–S to ROS), or levels (induction) of scavenger enzymes for ROS. To quote these authors: "The premature aging phenotypes in mtDNA mutator mice are thus not generated by a vicious cycle of massively increased oxidative stress accompanied by exponential accumulation of mtDNA mutations. We propose . . . that respiratory chain dysfunction per se is the primary inducer of premature aging in mtDNA mutator mice." The emphasis here might be placed on the idea of a vicious cycle, also referred to as the "error catastrophy," originally proposed by L. E. Orgel to explain cellular senescence in tissue culture and aging (the Hayflick phenomenon) with a focus on nuclear genes. This hypothesis poses that initial mutations in polymerases cause an increase in mutations in other genes on mtDNA, leading to additional faulty proteins which continue to aggravate the damage to the point of the final catastrophy.

In the above example the mutator polymerase probably swamps any mutational damage by ROS occurring over a normal life span. Nevertheless, it should be noted that mutations diminishing OXPHOS activity do not necessarily and inevitably lead to increased ROS production, as is frequently assumed in many publications. Very similar conclusions were reached by Kujoth et al. (175) with a very similar but independently constructed mutator mouse model.

7.4.2 Accumulation of mtDNA Damage and Normal Aging

The idea that the accumulation of mtDNA damage might be part of, or even responsible for, normal senescence is a relatively novel concept. If it was expressed before, it would have been pure speculation, since the technical means for testing it were not available until the late 1980s. The first prerequisite was the determination of the human mtDNA sequence. The second challenge was to find, characterize, and quantitate abnormal mtDNA molecules present at low abundance in a population of 1000–3000 molecules per cell. Restriction mapping could detect only large rearrangements or deletions in heteroplasmic populations when the abnormal molecules were a homogeneous subpopulation and present as a significant fraction (10%?). A heterogeneous mutant population with many different point mutations, each present at the <1% level, is not so easily detectable.

As in so many other areas, the polymerase chain reaction has had an enormous impact on the analysis of mitochondrial genomes. Taking a cue from the observations in patients with myopathies, it became possible to focus the initial attention on relatively large deletions, and specifically the ~5-kb "common deletion" between the repeats at 8470 and 13447 frequently found in Kearns–Sayre patients (see Section 7.3.6.1). By the appropriate choice of oligonucleotide primers, such deletions can be detected when present in just a few mtDNA molecules (0.1% to 1%). Following the initial observations in the laboratory of Linnane (176) (and see references 157 and 158 for reviews), several laboratories have investigated the incidence of age-related deletions in a variety of human tissues including brain, heart, kidney, liver, assorted muscle and blood (157, 158, 177–182).

Several generalizations can be made. (1) As suspected, there is an age-related increase in the frequency of mtDNA deletions. (2) Deletions are detected readily in nondividing tissue such as brain and skeletal muscle, while in cells with a relatively short half-life (e.g., blood cells), few, if any, mtDNA deletions have been detected. There must be a selection against an accumulation of damage in rapidly dividing stem cell populations, or, alternatively, metabolically deficient cells are eliminated even more rapidly than by normal turnover. The underlying assumption, that mtDNA turns over even in nondividing cells, is based on an observation made as long ago as 1969 (183), but there is a dearth of information on how fast this turnover occurs in such cells. (3) Different deletions may accumulate in different tissues of the same individual. Such an observation lends strong support to the hypothesis that the observed mutations are the result of stochastic processes in somatic tissue. (4) The range and nature of the deletions vary from individual to individual; however, as discussed in a different context, the presence of short repeats in the mt genome may be responsible for the mechanism generating the deletions by recombination or "slip-replication" (184), and therefore the deletions are not entirely random. (5) The tissues of an individual accumulate deletions at different rates; that is, they do not "age" uniformly. (6) Even within a particular tissue such as skeletal muscle or cardiac muscle, a tissue mosaic pattern has been observed by histochemical staining reactions. For example, staining for cytochrome oxidase activity in cardiomyocytes in a large collection of specimens has shown rather uniform staining in samples from young subjects, but in samples from older individuals the incidence of weakly stained or even unstained myocytes was roughly related to the age of the individual (185). While provocative and consistent with present thinking, it should be stressed that the variable levels of cytochrome oxidase activity have not been related directly to mtDNA mutations.

Sequencing of PCR-amplified segments (clones) of mtDNA from a given tissue is a tedious and time-consuming process, and hence the accumulation of point mutations in aging tissues is less well documented and the rate of generation of such mutations is still more difficult to determine (186).

Apart from errors resulting from illegitimate recombination or replication, mtDNA damage is very likely also caused by chemical reactions, and it has been suggested by numerous authors that reactive oxygen radicals generated as a byproduct of respiration may make a very substantial contribution to mtDNA damage. Free radicals are known to be potent mutagens that can induce base changes and modifications. It has been suggested that three factors make mtDNA particularly vulnerable: (1) The mtDNA is localized close to the inner membrane, where free radicals are produced; (2) mtDNA is not extensively condensed and protected by histones; and (3) the DNA repair activity is limited (see Chapter 4). The resulting point mutations (after replication) may be difficult to find (see above), but the reaction with free radicals (superoxide and peroxide) may also leave modified bases in the DNA which can be detected by sensitive biochemical means. Several reviews on the subject have been written by Ames and colleagues (187–189). A representative example is 8-hydroxy-guanosine, which has been shown to accumulate in mtDNA of the human diaphragm in individuals of advanced age (190). Approximately 0.5% of the guanine residues had been converted to 8-hydroxy guanine in an 85-year-old individual. In a related study, the fraction of derivatized guanines was 0.87% in the brain of a 90-year-old. Covalent DNA alterations termed I-compounds have also been described by Gupta et al. (191) in the liver of aged rats.

Covalent modifications of DNA in the nucleus are readily detected by DNA repair systems, but there is agreement that some types of DNA repair cannot be found in mammalian mitochondria. A continuous exposure to free radicals is therefore expected to be damaging over the long run, and the extent of such damage would be expected to be more severe in actively respiring tissue compared to less active tissue. This idea has been extended in an investigation of the effects of physical activity and age on mitochondrial function (192), and specifically in the muscle from elderly athletes (193). These authors make the important comment that the observed decline in respiratory chain function in human skeletal muscle from older individual is significantly greater than the observed level of actual mutations. In a carefully selected and matched group of individuals, there was no correlation between oxidative metabolism and chronological age. Instead, the results suggested that physical activity contributes greatly to the maintenance of OXPHOS capacity; in other words, exercise may counteract and mask the effect of accumulated mtDNA damage.

In order to assess the contribution of mutated mtDNA molecules in the population present in "old cells" (i.e., cells derived from older individuals), several informative experiments have been performed with the help of ρ^0 mutant mammalian cells (194, 195). Such cells do not harbor any mtDNA of their own, and they can be repopulated with mitochondria under investigation by fusion with an enucleated donor cell. Such an experiment has the additional advantage that it removes the nucleus from the old donor cell, and hence eliminates the influence of nuclear genes. The results of an elaborate series of fusions and analyses have recently been published by Laderman et al. (195). Donor cytoplasts were from individuals ranging from 20 weeks of age (fetal) to more than 100 years. Almost 200 clones were investigated with respect to growth rate and respiration rate. Every clone was also analyzed to determine the mtDNA levels per cell, since parallel experiments had shown that the "transformation" of ρ^0 cells with mtDNA is quite variable and slow; and oxygen consumption measurements had to be compared with, if not normalized to, the mtDNA content of the cells. Thus, several clones derived from the same mtDNA donor were also compared, and considerable variation in respiratory capacity as well as mtDNA levels were observed. The final evaluation of the data required careful statistical analysis, because at least two distinct age-related phenomena were observed. First, the number of mtDNA molecules in the cybrids was lower, when the enucleated donor cells were from older individuals. Second, the same subgroup of cybrids also exhibited a lower respiration rate. Notably, however, there was no correlation between oxygen consumption and mtDNA content of individual clones. The authors speculated about an increase in age-dependent mtDNA mutations that could affect the functional properties of the complexes, the translational machinery of the mitochondria (rRNAs or tRNAs), or even promoter sequences controlling transcriptional efficiency. Such damage would not necessarily affect DNA replication. On the other hand, mutations in a region crucial for DNA replication could have affected the efficiency of transforming the ρ^0 cells with mtDNA.

These results clearly support the hypothesis of a functional deterioration of mtDNA with age, since presumably all mitochondrial proteins encoded by nuclear genes and lipids from the old enucleated cell will have been replaced at the time of the analysis. One may also wonder why not all transformants derived from old cells were partially respiration-deficient, and the answer is most likely that aging is accompanied by stochastic mtDNA mutations with some cells being much less affected than others. Finally, it should be noted that any such *in vitro* experiments require the establishment of clonal cell populations of sufficient size to be able to perform the analyses, and hence the cells must grow, albeit at more or less reduced rates. Selections and segregation were considered by the authors by following the long-term behavior of some of the clones, which had initially all been analyzed 8–18 weeks after fusion. Additional changes were observed in some cases.

Parenthetically, it should be mentioned that the ρ^0 cells were derived from 143B.TK$^-$ cells, an immortal human osteoblastoma cell line. Do such tumor cell lines accumulate mtDNA mutations after extensive subculture *in vitro*, or are such mutations continuously subject to counterselection? A comparison of the mtDNA sequence of a 143B.TK$^-$ cell with the mtDNA sequence in the original tumor and in normal somatic tissue would perhaps be interesting. Alternatively, mtDNA sequences from HeLa cell strains kept for decades in different laboratories would also be worthy of investigation to learn why mtDNA mutations do not seem to accumulate significantly in rapidly proliferating cells.

7.4.3 Neurodegenerative Diseases

A large literature exists describing mitochondrial dysfunction and oxidative stress as a significant factor in neurodegenerative diseases, many of which develop with age or have a pronounced delay of onset (196, 197). The most prominent among them include Alzheimer's disease. Parkinson's disease, Huntington's disease, and amyotrophic lateral disease (ALS). Since many mitochondrial diseases, especially those due to mutations in mtDNA, also have a delayed onset, it was tempting to look for an excess of mtDNA damage in those patients, or even for preexisting mtDNA mutations in pedigrees with a familial incidence of these diseases (198–201). The subject has remained controversial and several large surveys have failed to find mtDNA mutations as a *primary* cause of these diseases (202). There is a strong emerging consensus that mitochondrial energy metabolism, oxidative stress, ROS production, cell death (apoptosis), and neurodegeneration are all part of a larger picture, where the major uncertainties may be the following: What are the primary defects? What is a definitive cause-and-effect relationship leading to the particular pathology? Why is the observed neurodegeneration frequently restricted to a limited domain of the nervous system? Are there specific mtDNA haplotypes that predispose toward or increase the risk of initiating a neurodegenerative disease in combination with nuclear mutations?

7.4.3.1 Parkinson's Disease Much remains to be learned about this relatively common disorder, affecting approximately 1 in 40 individuals over a lifetime. Clinical symptoms include bradykinesia, rigidity, and tremor. The most specific and characteristic pathological condition is the loss of dopaminergic cells in the substantia nigra of the brain. It can be stated at the outset that there is no evidence for maternal inheritance in PD, and the analysis of mtDNA in brain has not revealed any major abnormalities such as deletions or rearrangements, or consistent point mutations that can be associated with PD (see below).

One of the most intriguing observations hinting at a mitochondrial involvement in PD was made serendipitously. An illegally produced "designer drug," MTPT (1-methyl-4-phenyl-1,2,3,6-tetrahydropyridine), an injectable pethidine analogue, was found to be neurotoxic and to induce parkinsonism in users (203, 204). Further experimental support for this analysis came from the induction of very similar symptoms in primates. It was discovered that the drug specifically destroyed dopaminergic cells in the substantia nigra. MPTP is converted to the active metabolite 1-methyl-4-phenylpyridinium (MPP^+) by monoamine oxidase in glial cells, and MPP^+ is transported into dopaminergic neurons by the dopamine re-uptake pathway. It can be concentrated in mitochondria at millimolar concentrations by a process dependent on a membrane potential, and there it acts as a specific inhibitor of complex I. A more detailed discussion of MPP^+ and its inhibition of complex I can be found in Chapter 5. For the present discussion it suffices to state that inhibition of complex I activity has been postulated to be responsible for dopaminergic cell death.

A second promising approach has employed the inhibitor rotenone at sublethal concentrations to create a rat model with many of the typical symptoms of Parkinsonism (205–207).

What makes drug-induced parkinsonism so relevant is the substantial body of evidence showing that in the substantia nigra of PD patients, complex I activity is reduced 30–40% relative to controls, but this reduction is not found in other brain regions (reviewed in reference 208). Mitochondrial abnormalities, and specifically reductions in complex I activity, have also been investigated in muscle and platelets of PD patients. The results and their interpretation are still incomplete in establishing the primary cause (209, 210).

Since 7 of 45 polypeptides of complex I are encoded by mtDNA, looking for mitochondrial mutations in PD patients became imperative. A few reports on more detailed mtDNA sequence analyses have appeared, including at least one with an examination of the entire mtDNA sequence of several patients (211). One was a PD patient, but three others had overlapping symptoms for PD and Alzheimer's disease (AD). The conclusion was that there was no single mtDNA mutation shared by all the patients. In particular, earlier studies at

Emory University and by Hutchin and Cortopassi (212) had found an apparent prevalence (7- to 200fold increase) of a mutation at position 4336 ($tRNA^{Gln}$) in AD or AD + PD patients, compared to race- and age-matched controls, and phylogenetic analysis. Some patients were found to be variants in ND1 or ND5. As the sample size increased, the risk factor for AD associated with the 4336 mutation actually decreased (212). Howell et al. (202) have critically reviewed the evidence for mtDNA mutations in PD (and AD), with the conclusion that there is no convincing evidence for a (primary) role of mtDNA mutations in these neurodegenerative disorders. Other large-scale studies have provided evidence that certain mtDNA haplogroups may be responsible for a reduction in risk of idiopathic PD (213).

Some very promising leads have originated from the genetic analysis of pedigrees (often consanguineous) with familial Parkinsonism (see references 214 and 215 for reviews). A series of nuclear genes (PARK1, PARK2, . . . PARK8) are being recognized to be associated with the familial form of the disease, and evidence is accumulating that an interaction of environmental toxins with the products of these genes may create oxidative damage and mitochondrial dysfunction leading to cell death (apoptosis). The function of the proteins encoded by these genes is very suggestive. A hallmark of the disease is the accumulation of Lewy bodies. The main component of these Lewy bodies is a protein, α-synuclein (PARK1/4), normally involved in synaptic vesicle formation. Certain dominant point mutations yield a protein that is postulated to be toxic in the substantia nigra when it forms protofibrils. There are claims that a mutated α-synuclein, when overexpressed, can interact with and damage mitochondria directly (216). Furthermore, α-synuclein null mice are resistant to the toxic effect of MPTP (217). The protein parkin (PARK2) is an E3 ligase, implicating a defect in ubiquitin-mediated protein turnover in the proteasome as a step in the development of the disease. For example, parkin-dependent ubiquitination may be required for the removal of oxidatively damaged proteins. It has been localized to the outer mitochondria membrane, but is also placed in the matrix in order to account for its interaction with the PINK1 kinase (see below). DJ1 is another gene/protein identified in two consanguineous families with early onset familial PD. Its function is suggested by its membership in a large superfamily, but it remains to be more precisely defined as either a redox sensor or an antioxidant protein, a chaperone, or a protease. It has also been tentatively localized in the mitochondrial matrix. The focus was also clearly shifted to mitochondria when several pedigrees revealed a role of the phosphatase and tensin homologue (PTEN)-induced kinase (PINK1) (218, 219). This serine–threonine kinase has been localized in mitochondria (220).

Several of the above genes have been studied using mice and *Drosophila* as model systems. One can observe symptoms such as reduced life span, flight muscle degeneration, apoptosis in various tissues, and male sterility. Upregulation of genes related to oxidative stress is observed, and in flies a parkin deficiency causes a reduced expression of mitochondrial complexes such as complex I and IV. An *in vivo Drosophila* model with the *Drosophila* equivalent of the PINK1 gene knocked out showed phenotypes similar to those observed with a parkin deficiency including defects in mitochondrial morphology (221). In summary, one can identify two pathways to parkinsonism: protein misfolding and a defective (or overwhelmed) ubiquitin-proteasome apparatus, and mitochondrial dysfunction. How do they interact or converge to cause the specific dopaminergic degeneration? Some more detailed speculations can be found in the reviews by Abou-Sleiman et al. (214), Gandhi and Wood (222), and Kwong et al. (197). One plausible idea is that oxidative stress and excess ROS production leads to an abundance of abnormal proteins that overwhelm the cellular garbage disposal machinery (the proteasome); alternatively, the proteasome could be starved for ATP. In either case, abnormal proteins then form the aggregates characteristic of such degenerating neurons.

7.4.3.2 Alzheimer's Disease Alzheimer's disease (AD) is an increasingly common form of dementia in our population in which the life expectancy has been raised significantly over the past decades. It is a progressive disease that is linked to the premature death of cholinergic neurons in the brain. The primary event responsible for this destruction has remained elusive, although it is likely that multiple genetic and environmental factors can either trigger it or influence the age of onset and the rate of progression. The neuropathology of AD includes loss of neurons and synapses as well as the extracellular deposition of plaques (made up mostly of amyloid β peptide, Aβ) and accumulation of intracellular neurofibrillary tangles (consisting of hyperphosphoryted microtubule-associated protein tau). In the pedigrees where there is early onset and a familial pattern of inheritance, the products of several nuclear genes have been implicated. Autosomal dominant mutations have been mapped on chromosomes 21 (APP), 14 (PSEN-1), and 19 (ApoE), and the β-amyloid precursor protein (APP) as well as the presenilin 1 (PSEN-1) and 2 (PSEN-2) proteins are currently subject to intense research efforts. APP is cleaved sequentially by β and γ secretases to produce Aβ. The presenilins are constituents of the secretase complex. The identification of these genes from familial AD in humans has directed efforts to the study of the corresponding genes (and mutations) in transgenic mice (see reference 223) for a critical review of mouse models for AD). Such mouse models do not recapitulate the complete AD phenotype. The most interesting observation is that amyloid β peptide accumulation alone does not cause apoptosis (loss) of neuronal cells, but may initiate a secondary processes with pathological consequences. One earlier review (224) had concluded that overproduction and increased aggregation of the amyloid β peptides may be an obligatory early step in the development of the disease, followed by deposition of diffuse plaques which, in turn, activate glia and astrocytes to release cytokines and acute-phase proteins. Direct neurotoxicity of the amyloid peptides or the induced "inflammatory" changes lead to numerous metabolic and cytoskeletal alterations, including changes in calcium homeostasis, and injury to mitochondrial functions. A fundamental question is therefore whether mitochondrial

(OXPHOS) dysfunction is a secondary effect or one of the primary causes. A number of authors have reported a direct effect of amyloid β on mitochondrial functions ((225, 226), but one has to reconcile such claims with the fact that amyloid β plaques are extracellular. In a recent review of AD, Lin and Beal (196) summarize a bewildering array of observations that suggest either that oxidative stress causes Aβ accumulation in cells or that it alters APP processing. Alternatively, APP has been reported to have a dual ER/mitochondrial targeting sequence, and overexpression can clog the mitochondrial protein import machinery. Finally, Aβ is claimed to be able to inhibit a variety of enzymes in the mitochondrial matrix (α-ketoglutarate dehydrogenase and cytochrome oxidase), thus causing free radical production. How Aβ gets into the mitochondrial matrix is left unanswered.

In the case of the more common, late-onset AD a genetic component has been indicated by some tantalizing observations, but the uncertainty is indicated by the statement that "the lack of a family history is a negative risk factor for AD." One possible risk factor identified to date is the prevalence of the apolipoprotein E type 4 allele among affected individuals (see reference 224 for a review). The delayed onset and age-related incidence of the disease has made it very tempting to establish a connection to mitochondrial mutations. Indeed, the existence of a maternal relative with the disease is associated with an increased risk of AD, indicative of a link to maternal inheritance and mitochondria. Numerous studies have attempted to solidify the basis for a role of the mitochondrial genotype in AD. The results are intriguing, but the evidence is still inconclusive or negative. As mentioned above, a recent critical review of the relevant literature by Howell et al. (202) found little evidence in support of a role of mtDNA mutations in the development of PD and AD. A clear distinction should be made between (a) mtDNA mutations inherited from the mother and leading to a predisposition toward the development of AD (or an early onset) and (b) mtDNA mutations acquired by somatic cells over a lifetime. Arguments in favor of a role of the acquired mutations have been made emphatically by Wallace and colleagues (201, 227), by Beal (164), and by others (228, 229). Much attention was once attracted by a study correlating mutations in two mitochondrial cytochrome oxidase genes with late-onset AD (230). Several results were striking and even puzzling. A follow-up study in two independent laboratories concluded that the presumed COX mutations were found in a pseudogene of mtDNA transferred to the nucleus (231, 232). In a related but less extensive study the mitochondria from 33 AD patients and 30 age-and sex-matched controls were examined. The activity of four complexes (I–IV) were assayed in mitochondria obtained directly from lymphocytes. None of the patients had respiratory chain enzyme activities below the normal range. Thus, the relationship of cytochrome oxidase activity to AD remains controversial.

7.4.3.3 Huntington's Disease Huntington's disease is well-characterized at the genetic level as an autosomal dominant disease with 100% penetrance (233). The responsible gene (IT-15) has been mapped on chromosome 4, and the mutation belongs to the class of trinucleotide repeat polymorphisms

(CAG), first identified in fragile X individuals (and also found in Friedreich's ataxia; see above). The resulting abnormal protein, huntingtin (HTT), with an expanded polyglutamine tract near the N terminal, must represent a gain in function to explain the dominant phenotype. Its normal function is still not completely clear, and the mechanism of the gain in function (or enhancement of a normal function) is also obscure at this time. It appears to interact with calmodulin in the presence of calcium. Inappropriate apoptosis triggered by abnormal huntingtin has been proposed (234). Its expression in a wide variety of tissues including brain, testes, liver, heart and skeletal muscle, and lung, among others, suggests that it has a very general function outside of the nervous system, but selective neuronal cell death remains to be explained. It is interesting and revealing that transgenic mice expressing only exon I with the polyglutamine tract display most of the neurological symptoms. Thus, it has been suggested that huntingtin is proteolytically processed to produce the pathogenic peptide when the glutamine tract exceeds a certain length.

A possible relationship to mitochondrial dysfunction is deduced very indirectly from the late onset of the disease, and possibly from the neurodegeneration observed in affected individuals. Measurements of biochemical activities in mitochondria of HD patients have also been suggestive, but here, even more than in PD and AD, one has to approach the problem from the point of view that the mitochondrial deficiencies are clearly secondary. Studies measuring OXPHOS enzymatic activities in selected brain tissue have to be interpreted with care, since the problem of normalization is not a trivial one. If a subpopulation of cells degenerate and disappear, the remaining cells may have normal activities. A more reliable determination is the measurement of an intrinsic property, namely the fraction of mtDNAs in a given tissue with point mutations or deletions. Seeking to ascertain whether the progression of HD is correlated with an accelerated accumulation of somatic mtDNA mutations, and specifically the mtDNA4977 deletion (the "common deletion"), Horton et al. (235) indeed found that HD patients have an-order-of-magnitude-higher levels of this deletion in the frontal and temporal lobes of the cerebral cortex compared to age-matched controls. There was no statistically meaningful difference in the corresponding comparison for the occipital lobe and the putamen. The latter tissue is atrophied in the most advanced stages of the disease. The apparent paradox is hypothesized to be explained by the death of the neurons containing the highest levels of mutations. With such a scenario, any analysis must be made at a time when the integrity of mtDNA or the activity of mitochondrial enzymes is already in decline, but before the damage is sufficient to kill the cells.

A biochemical link between mutated huntingtin and energy metabolism is indicated by the following intriguing observations. The polyglutamine tract of huntingtin has been shown to bind to glyceraldehyde-3-phosphate dehydrogenase, a key enzyme in glycolysis. It is not yet clear whether this interaction is directly responsible for the hypometabolism of glucose in HD brains (236). In conjunction with decreased activities of complexes II and III in mitochondria of HD brain (caudate), this excess metabolism of glucose causes lactic acidosis and elevated lactate/pyruvate ratios in HD cerebrospinal fluid. Animal models resembling HD have been induced by the complex II inhibitors malonate or 3-nitropropionic acid (237). Most recently, a study describes how the pathogenic polyglutamate tract induced preferential loss of complex II by a post-transcriptional mechanism (238). Others have claimed that the pathogenic polyQ peptide increases ROS production by mitochondria (oxidative stress) (239) or an impairment of ATP production (240) without a reduction in OXPHOS complexes. Closely related studies with an additional emphasis on NMDA signaling and the role of Ca^{2+} in mitochondrial energy metabolism in striatal cells have been published by Seong et al. (241). Finally, abnormal huntingtin and p53 have been demonstrated to form a "deadly alliance" (242, 243) and perhaps link activities in mitochondria with activities in the nucleus leading to the HD pathology. Speculations have centered around mutant HTT binding p53, up-regulation of BAX, mitochondrial membrane depolarization. and apoptosis (196).

7.4.3.4 Amyotrophic Lateral Sclerosis (ALS) ALS (also known as Lou Gehrig's disease) results from the degeneration of upper and lower motor neurons in the cortex, brainstem, and spinal cord. The clinical conseuences are progressive weakness, atrophy, and spasticity of muscle tissue. It is a fatal neurodegenerative disease with a characteristic delayed onset. The vast majority of cases (~90%) are sporadic, and very little is understood about the mechanism triggering the onset of the disease. About 10% of the cases are familial, and 20% of these (i.e., 2% of the total) are caused by specific mutations in the SOD1 gene encoding Cu/Zn-superoxide dismutase. This scavenger enzyme is generally found in the cytosol, while the Mn-superoxide dismutase (SOD2) is located in the mitochondrial matrix. Recent observations from several laboratories have suggested that (specific) mutant forms of SOD1 can be found associated with the outer mitochondrial membrane and/or with the intermembrane space, or even with the mitochondrial matrix, somehow leading to aberrant ROS production, and so on. Cleveland and colleagues (244) attribute its effect on a clogging of the import machinery, but in a recent review they hypothesize that a combination of several mechanisms (protein misfoldiong, oxidative damage, defective axonal transport, excitotoxicity, insufficient growth factor signaling, and inflammation) contribute to the motoneuron damage that can spread to neighboring non-neuronal cells (245). Binding to Bcl-2 and aggregation in spinal cord mitochondria could also explain the death of these neurons (246).

7.5 MITOCHONDRIA AND APOPTOSIS

Although cell death can occur as the result of a variety of causes, including accidental failure of an incubator, it is "programmed cell death" that has attracted enormous attention in recent years. The program referred to in this context is a genetically controlled pathway and mechanism that can eliminate unwanted cells under three major circumstances: (1) development and homeostasis, (2) as a defense mechanism against genetically damaged and hence potentially tumorigenic cells, and (3) natural senescence and aging. The phenomenon is primarily found in multicellular, eukaryotic organisms, although it has been argued that cell death may have evolved in single-celled organisms as a defense strategy (247). By committing suicide, a cell may eliminate itself as a host for a parasitic invader and hence protect related, neighboring cells from the spread of the pathogen. In higher organisms, programmed cell death or apoptosis is now recognized as an essential aspect of development, and many examples could be cited. Many neurons die within a developing brain, self-reactive T-lymphocytes are eliminated during the development of the immune system (248), the morphogenesis of limbs requires cell death as part of the shaping of the final limb such as a hand, and a most detailed description of the cells destined to die has been worked out for all the lineages in a developing worm, *C. elegans*. The immune system can take advantage of and activate an existing program to kill infected cells. In the adult, apoptosis assures that there is turnover and repair in many tissues without continued growth, and certain other tissues may expand or contract reversibly in response to hormones and cytokines. Another aspect of apoptosis is that this program is inhibited or inactivated in tumor cells. Hence some genes playing a role in this mechanism were initially identified as oncogenes in tumor cells. Overexpression of bcl-2 in hematopoietic tumor cells, for example, inhibits cell death normally induced in growth factor-deprived B-cells and others. Inducing apoptosis specifically in tumor cells has become a major new strategy in the fight against cancer, as opposed to inhibition of tumor cell proliferation (249).

A number of isolated systems were initially studied without an obvious connection, but now the emphasis is on (a) the common mechanisms found in many cells across a broad spectrum of species and (b) homologous genes encoding functions associated with apoptosis. For example, treatment of lymphocytes with glucocorticoids has long been known to initiate a suicide program, but the generality of the mechanism was not recognized until later. A major boost to the field also came from the observation that very specific cells in the developing nematode *C. elegans* would die during development and that specific mutations could identify genes (*ced* = cell death abnormal) required for this process. This observation stimulated the search for homologous genes in other organisms and made the nematode a powerful model system for studies of programmed cell death. A detailed description of the physiological processes in different organisms is beyond the scope of this treatise. It is to be reemphasized that the programmed cell death is orchestrated by genes within the affected cell, and that common genetic pathways may have evolved over a wide phylogenetic range. Nevertheless, since most of this research has been performed with mammalian cells, the emphasis in the following discussion will be on mammalian systems. A most relevant aspect of the phenomenon for the present discussion has been the well-documented involvement of mitochondria or mitochondrial functions.

It has become useful to subdivide the process of apoptosis into several phases or stages. The final phase encompasses a number of characteristic changes such as cell shrinkage, chromatin degradation and nuclear fragmentation, and loss of plasma membrane integrity. It appears that these changes are essential for rapid recognition of the dying cell by its neighbors, which phagocytose the remnants and efficiently eliminate any debris that would otherwise cause an inflammatory response by the immune system. As a consequence, "dead" cells from apoptosis are not readily seen *in vivo* in intact tissue, and the description of this terminal phase rests largely on observations in tissue culture.

The first phase includes the description of the various stimuli or signals that induce apoptosis in different cells. The stimulus may be an external one, received by a cell surface receptor, or an internal one, as the result of the action of a drug, toxin, or radiation damage. Of course, drugs, toxins, and radiation are also ultimately external influences, but the difference between these two mechanisms becomes apparent during the following phase: the signal transduction, which leads to changes in metabolic state, induction of various enzymes (proteases, nucleases), and the morphological changes preceding cell death. Needless to say, these two signaling pathways may converge at some level, but at the same time the above description should not give the impression of a simple, universal and invariable pathway operating in all cells under all conditions. Deprivation of growth factors (e.g., interleukin-3 for hematopoietic progenitor cells) induces apoptosis; that is, unoccupied receptors can stimulate apoptosis. Failure to stimulate germinal center B cells by antigens causes them to die. Transforming growth factor β and tumor necrosis factor can trigger apoptosis in various cell types. Many more examples could be given (247, 250). It should be clear that these signal transduction pathways include most, if not all, of the following known reactions, intermediates, and interactions: tyrosine kinases, phosphatases, steroid receptors, inositol phosphates, transcriptional activators and factors, and so on.

There is usually a significant time interval (several hours) between the delivery of the first stimulus and the observation of morphological changes in the targeted cells. It is also during this interval that thresholds or checkpoints are reached beyond which the process becomes irreversible. Many details of events occurring during this effector phase remain elusive and are the subject of a very active research.

As mentioned above, the nematode, *C. elegans*, continues to be a useful model system, and it was in this model system that the first genes specifically associated with apoptosis were isolated (ced1, ced-2, ced-3, . . . , ced-10, etc.). Many of these genes were subsequently shown to be required for the disposal of the dead cell(s), and hence their detailed description will be omitted here. The Ced-3 gene, however, was identified to encode a cysteine protease, and many different homologues have since been found in mammalian cells. The first was interleukin 1β converting enzyme (ICE), and hence this group of aspartate-specific cysteine proteases was initially referred to as ICE*x*, where *x* is a distinguishing subscript. More recently, this family of proteases has been referred to as caspases. In a simplified view, apoptosis is accompanied by extensive proteolytic cleavage, and there is in fact a cascade of such cleavages, since these ICE enzymes exist as inactive precursors (pro-caspase) that must be cleaved after a specific aspartate residue before they can be assembled into heterotetramers and active proteases. The other targets of these proteases are still largely unknown, although it was intriguing that one of these proteases (CPP32) can cleave the poly-(ADP ribose) polymerase (PARP), an enzyme involved in DNA repair. However, PARP knockout mice develop normally. The ICE proteases are constitutively expressed in many cells, and their activation is a post-translational event, although some additional transcriptional activation following the death stimulus may also occur. It is apparent, therefore, that there must be an upstream signal for post-translational activation which remains to be more clearly defined in mammalian cells. Certain broad-range protease inhibitors can prevent apoptosis in response to common stimuli, but other highly specific and restricted inhibitors may interfere with the signaling pathway initiated from only a limited number or a unique inducer. Apoptosis induction may have "common" as well as "private" pathways.

The ced-9 mutation in nematodes led to the characterization of the corresponding nematode gene and the potential functional identification of the human homologue Bcl-2. The Bcl-2 protein was localized in the outer mitochondrial membrane (although see below), and this finding was one of the first links to mitochondria. The mammalian Bcl-2 gene was initially implicated in tumorigenesis because it was found to be the location of a breakpoint (18q21) in a reciprocal translocation found persistently in human B-cell tumors (251–254). Bcl-2 protein became overexpressed under the influence of an immunoglobin promoter. A whole family of bcl-2-like proteins (Bcl-2, Bcl-x_L, Mcl-1, Bfl-1, A1, . . .) were subsequently defined and characterized as inhibitors of apoptosis by transfection studies and overexpression in various cells (254). Other family members have also been localized in the outer mitochondrial membrane (255). In addition, factors promoting apoptosis have also been found: Bax, Bcl-x_S, Bad, Bak, Bik, . . . and so on (see below).

Before discussing Bcl-2 and mitochondria in more detail, some other approaches related to the study of apoptosis and mitochondria must be introduced. Active, healthy mitochondria exhibit a transmembrane potential ($\Delta\Psi_m$) across the inner mitochondrial membrane which is of fundamental importance in the discussion of the mechanism of oxidative phosphorylation (see Chapter 5). A pH and electrochemical gradient across the inner membrane is set up as the result of electron transport through the electron transport chain, with excess positive charge on the outside—that is, in the intermembrane space. The discovery was made some years ago that certain lipophilic dyes such as rhodamine 123, 3,3′-dihexyloxacarbocyanine iodide (DiOC6(3)), 5,5′,6,6′-tetrachloro-1,1′,3,3′-tetraethylbenzimidazolcarbocyanine iodide (JC-1), and chloromethyl-X-rosamine (CMXRos) (see Chapter 5 for a more detailed description) can accumulate in the matrix. Under controlled conditions the fluorescence intensity can be correlated with the magnitude of the membrane potential ($\Delta\Psi_m$), and these dyes have therefore been used as convenient and powerful indicators of mitochondrial $\Delta\Psi_m$. When such dyes were applied to cells undergoing apoptosis, a fall in $\Delta\Psi_m$ was observed as one of the relatively early events in many different cell types, prior to DNA fragmentation. In fact, the collapse of the transmembrane potential was correlated in time with the point of no return—that is, the point at which apoptosis could no longer be reversed by withdrawal of the stimulus (256). The breakdown of the $\Delta\Psi_m$ can be shown to involve the so-called mitochondrial "permeability transition" (PT) (257). Briefly, it is believed to be the result of the opening of a PT pore, which can be measured electrophysiologically by the opening of a "megachannel," large enough to allow solutes of up to 1500 Da to pass through along a concentration gradient. This PT pore is hypothesized to be regulatory with diverse functions such as voltage sensing, thiol sensing, redox sensing, pH sensing, sensing of adenine nucleotide levels, and regulation of calcium homeostasis. Normally, only a brief opening or transient assembly of a small number of such pores may be required, but the massive opening of many such pores will collapse the $\Delta\Psi_m$, stop oxidative phosphorylation, arrest the import of proteins into mitochondria, and induce leakage of cytochrome c out into the cytosol (see below). Thus, a long-term, massive opening of the PT pores can be expected to do permanent damage to mitochondria, and the PT was postulated to be a central, rate-limiting event in the apoptotic cascade (255–257). In simple terms, it made sense to think that turning off the power supply was a logical first step in programmed cell death. A study of respiration deficient mammalian cell mutants has shown that such cells die within a day when they are switched from a high glucose medium to medium containing galactose. In the absence of oxidative phosphorylation, cells are sustained by a high rate of glycolysis (30) which is not possible with galactose (see Section 1.2).

On the other hand, several experiments in cell free systems suggested a more complex picture. Isolated nuclei can be induced to undergo apoptosis when exposed to cytoplasmic extracts from cells in the process of apoptosis. Controversy existed about whether such extracts can be free of mitochondria or whether enrichment with mitochondria accelerates the process. Subsequently, the presence of "damaged" mitochondria was shown to be required, and it was hypothesized that such mitochondria release an apoptogenic protein (apoptosis inducing factor, AIF). This protein has been characterized as an ~50-kDa intermembrane protein, with protease or protease activating activity but no nuclease activity, which can induce apoptosis in isolated nuclei more efficiently than ceramide, another mitochondrial constituent apparently released by such organelles (255).

A major breakthrough from *in vitro* studies of characteristic events associated with apoptosis (DNA fragmentation in isolated nuclei, activation of caspases) was the discovery of the crucial role of cytochrome c, initially based on the fractionation of the cytoplasm and further demonstrated by immunodepletion and reconstitution experiments (258, 259). These and subsequent studies have suggested that the cytochrome c is involved in triggering the cleavage and activation of DEVD-specific caspases. In short, cytochrome c is now known to interact with Apaf-1, leading to the formation of the multimeric "apoptosome" in the presence of ATP or dATP (260, 261). The apoptosome activates the initiator caspase, caspase 9, which subsequently cleaves and activates the effector caspases 3 and 7. Many more studies have since defined cytochrome c-dependent activation of apoptosis and cytochrome c-independent pathways, the latter requiring proteins such as AIF and endonuclease G. These proteins and others (Smac, Omi, adenylate kinase-2) are found in the intermembrane space of mitochondria.

A central question that was raised immediately following these observations was, How do these proteins get released into the cytosol? Much potential confusion was generated by attempts to link the mechanism of cytochrome c release to the breakdown of the $\Delta\Psi$ and the permeability transition (PT). The release of cytochrome c requires the permeabilization of the outer membrane, while the PT occurs in the inner membrane. A very clear formulation of the problem and a partial resolution of this problem can be found in the authoritative recent review by Green and Kroemer (262). The authors clearly distinguish between mitochondrial outer-membrane permeabilization (MOMP) and the PT causing a breakdown of $\Delta\Psi$. Which one comes first? It depends.

In one scenario the PT is associated with an equilibration/flux of ions and water, causing the mitochondria to swell and leading eventually to the rupture of the outer membrane. The PT-associated MOMP could be prevented by specific inhibitors such as bongkrekic acid and cyclosporin A, which target the adenine nucleotide transporter (ANT) and the protein cyclophilin D (CyP-D, a peptidyl-prolyl-cis-trans isomerase), respectively. Thus, ANT and cyclophilin D became postulated constituents of the PT pore, although this pore has has not been isolated and well-hdefined even today. In current thinking the PT pore is opened by Ca^{+2} overloads, and prolonged opening causes cell death by necrosis (263). CyP-D-knockout mice develop normally and show no protection against a range of apoptotic stimuli, suggesting that the PT pore does not play a role in most forms of apoptosis (263). There is also recent evidence for a PT pore forming in the absence of ANT, and a participation of the porin (VDAC) channel in the outer membrane has been considered (262).

The more dominant mechanism for MOMP nowadays excludes the PT and events at the inner membrane, and a breakdown of $\Delta\Psi$ probably is a later event following cytochrome c release and caspase activation (Figure 7.2). In this mechanism the anti-apoptotic Bcl-2 proteins (Bcl-2, Bcl-x_L, Bcl-W. . . .) and the pro-apoptotic Bax family members (Bax, Bak, Bad, Bid . . .) play a very prominent role. Two problems must be addressed: (1) What constitutes the channel in the outer membrane through which proteins of considerable size (>50 kDa) can escape from the IMS to the cytosol? (2) What mechanisms induce the formation of such channels that are obviously not present in normal

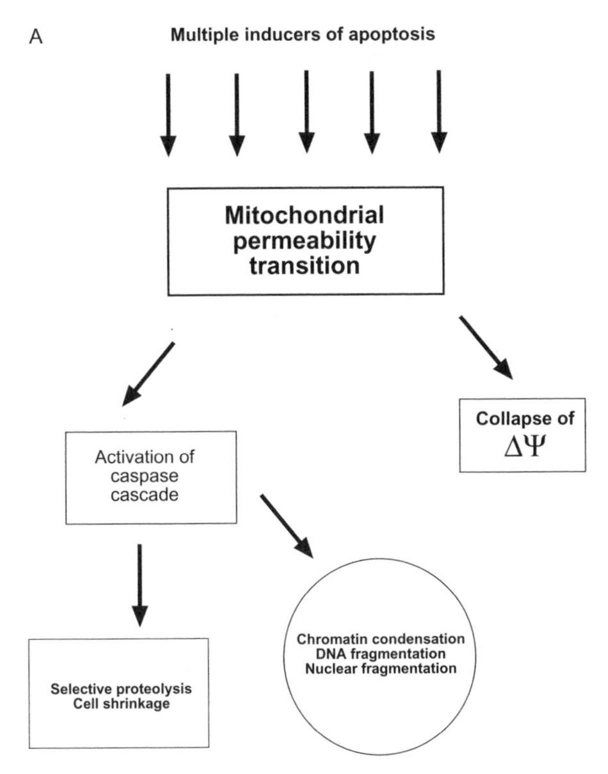

Figure 7.2 (A) Schematic view of signaling in apoptosis. In this view the mitochondrial permeability transition takes a central position on which multiple early pathways converge. (B) (Left) In this view the outer membrane permeabilization occurs before the collapse of the membrane potential (see reference 262 for a recent review). (Right) A less-defined mechanism leading to OMP and a permeability transition involving the inner membrane.

mitochondria? The literature on these questions is enormous, and it is a challenge to review for the uninitiated. Controversies and/or uncertainties still abound. The interested reader should start with some recent reviews (260, 261) and plunge into the literature. Based on *in vitro* studies with isolated outer membranes or vesicles made from mitochondrial lipids, it is now believed that oligomeric assemblies of Bax or Bak, in combination with active Bid (tBid), form the channel (264, 265). The possibility of a participation of porin (VDAC) has also been considered. However, before the monomeric, cytosolic Bax or Bad can insert into the outer membrane and form multimers, it must be activated by interaction with activated Bid (tBid, formed by caspase-8-mediated cleavage of Bid) and by other mechanisms such as phosphorylation and protein modifications by reactions with ROS (266). The anti-apoptotic factors interact with Bax or Bad to prevent activation, membrane insertion, and aggregation

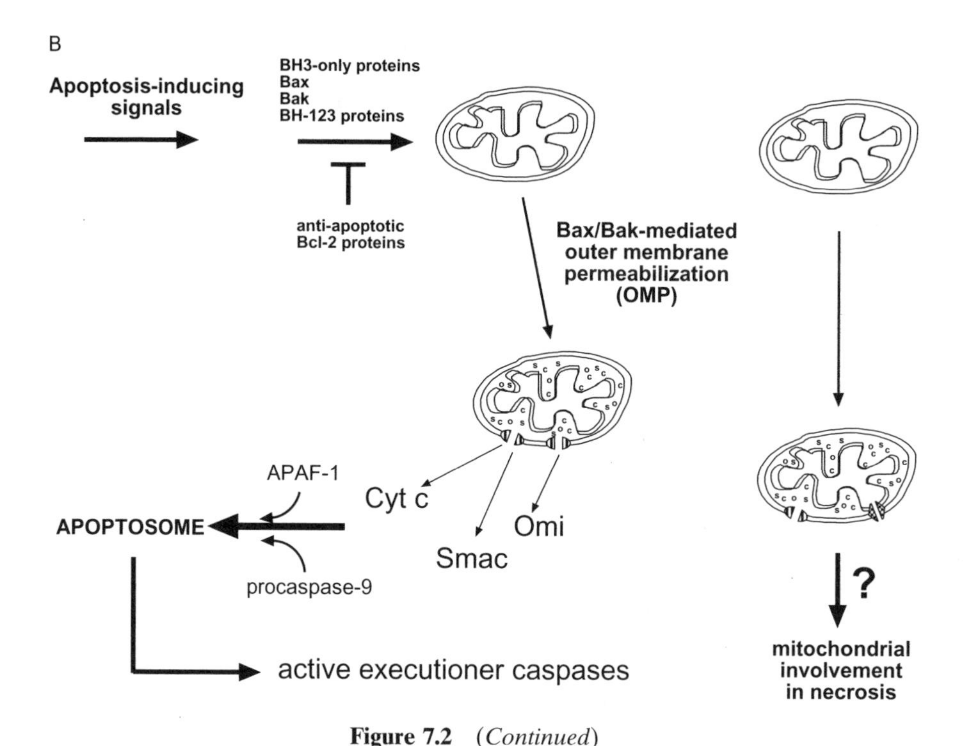

Figure 7.2 (*Continued*)

into channels. Bax has a C-terminal membrane-spanning region that is folded into a hydrophobic groove in the monomer. Upon activation, this C-terminual region becomes exposed for insertion into the outer-membrane. A truncation of the Bcl-2 protein deleting its C-terminal transmembrane (anchor) sequence makes the protein incapable of suppressing apoptosis induced by several agents. Furthermore, this anchor sequence can be replaced by the corresponding anchor sequence from the yeast outer-membrane protein Mas70 (now Tom . . .), and Bcl-2 targeting and function are restored. An important experiment has shown that Bcl-2 can block apoptosis even in cells that lack mitochondrial DNA and are unable to carry out oxidative phosphorylation (267). This is not to say, however, that they lack a membrane potential, since such cells and mutant cells completely defective in mitochondrial protein synthesis (31, 33) are perfectly capable of mitochondrial protein import.

High-resolution structures for several of these proteins are available (see reference 261 for a summary). A similarity to the pore-forming domains of diphtheria toxin and the bacterial colicins became apparent, supporting the conclusions from the studies performed with liposomes. The anti-apoptotic members have four conserved BH domains, named after the defining region in Bcl-2—that is, Bcl-2 homology domains (BH1, BH2, BH3, BH4); Bax/Bad family members lack the BH4 domain, and Bid is a BH3-only protein. The relative ratios and the interaction of these proteins determine the susceptibility to MOMP. Many studies with transfected cells and transgenic mice support such a view. Bid (and other BH3-only members) serve as "upstream sentinels" ready to receive and interpret death and survival signals (260).

The examination of mice in which either the Bcl-2, the Bcl-x, or the Bax genes were knocked out has yielded a complex picture. Bcl-2 –/– mice are born as viable pups but die at a few weeks of age. Bcl-x –/– mice die at ~day 13 of embryonic development, due to massive cell death in the central nervous system among other tissues. Bax –/– mice appear healthy, with lymphoid hyperplasia and sterility of males due to lack of sperm cell production. Overall, such studies support a lineage-specific role for these genes, but details remain to be clarified, and for the moment no significant relationship to mitochondrial function or activity can be seen from such studies in intact mice. It may have come as a surprise that live, normal-looking mice were born.

A new development that remains to be fully explored promises to shed new light on the mechanism of IMS protein release in the process of apoptosis. Up till now, Bax and Bak proteins were thought to be involved exclusively in apoptosis, by relocating from the cytosol and from a position on the ER to induce MOMP. A poineering new study by Karbowski et al. (268) has implicated these proteins in the normal and highly dynamic process of mitochondrial fusions. These results and some related investigations by Delivani et al. (269) indicate that Bax and Bad coordinate both fusion and fission of mitochondria in healthy cells. Karbowski and colleagues found that Bax/Bak double knockout mice have fragmented but still functional mitochondria. In contrast, Delivani et al. showed that overexpression of the anti-apoptotic protein Bcl-xL (or the nematode homologue CED-9) in mammalian cells promoted mitochondrial fusion. A working model that will undoubtedly become more complex in the future has Bax/Bak act on the factor Mnf2 which is required for fusion. The absence of Bax/Bak in DKO mice can be compensated for by overexpression of Mnf2. At the same time, Bax/Bak are postulated to inhibit a factor that is involved in fission, a process that proceeds unchecked in the absence of these pro-apoptotic factors.

It may be too early to dismiss the idea that Bax monomers form aggregates in the outer membrane that constitute the channel for the escape of cytochrome c and other IMS proteins. The channels demonstrated in liposomes may be *in vitro* artifacts, and so far they have been shown to allow small ions and fluorescent dyes to pass through, but the passage of a 50-kDa protein may be a greater challenge. Since we do not understand the process of mitochondrial fusion and fission in much detail (see Chapter 3), we cannot readily explain the release of the IMS proteins during apoptosis by invoking mechanisms active in fusion. However, it should be noted that fusion, fission, and MOMP all involve major perturbations of the outer membrane (and the inner membrane). Under conditions where normal fusion and fission occur, the process may be sufficiently orderly/controlled to prevent the loss of any proteins from the IMS, but excess recruitment of pro-apoptotic Bax may lead to large enough perturbations to allow leakage of such proteins. The process could be autocatalytic as downstream caspases become activated.

7.6 FUNGAL SENESCENCE

Many fungi appear to be able to divide mitotically for an almost unlimited period of time, growing into large clones whose size may be limited only by physical parameters. In the wild, *Armillaria bulbosa* colonies in excess of 0.5-km diameter have been found. On the other hand, some strains in the laboratory exhibit the phenomenon of senescence, that is, they cannot be serially transferred indefinitely in culture. In a few prominent cases the phenomenon has been investigated genetically. Since mitochondrial genomes were found to play a large role in the process, it has been tempting to draw parallels with aging in humans; but as our understanding has advanced, the resemblance is only at the physiological level: the inability of cells to undergo further mitotic divisions and a generalized cellular decay at the end. The best-studied examples have been expertly reviewed by Griffiths (270, 271), and some salient features are appropriate for the present chapter. Although nuclear genes can play a role, a common theme is that senescence in fungi is associated with cytoplasmic inheritance. A triggering event appears to be an alteration in the mitochondrial genome which eventually leads to a severe mitochondrial degeneration and dysfunction and hence senescence.

There has been a renewed interest in using yeast as a model system for life-span research, in connection with calorie restriction and the role of the gene SIR2 (encoding a NAD-dependent deacetylase (272). A full discussion is beyond the scope of this book. It should be noted that there are two life spans to be considered for yeast: (a) a replicative life span limiting the number of times a given mother cell can produce buds and (b) a chronological life span, which is a measure of the capacity of stationary cultures to remain viable over time. Piper (273) discusses the relevance of studies with yeast to our understanding of aging in more complex organisms.

The existence of mitochondrial plasmids in many organisms, and fungi in particular, has already been discussed in Section 4.1.6. Even at relatively high copy numbers, they may act as relatively innocuous molecular parasites, expressing functions primarily devoted to their own autonomous replication, but they can also contribute to the phenomena of senescence and hypovirulence. In *Neurospora crassa* and *Neurospora intermedia*, plasmids isolated from wild strains have been shown to be responsible for senescence. The kalilo plasmid (~9 kb) isolated originally in Hawaii causes *N. intermedia* to stop growing, and in the terminal stages, abnormalities in cytochrome spectra and content, deficiencies in large ribosomal subunits, and loss of cyanide-sensitive respiration are observed. It is now clear that the plasmid can insert at one of several locations in the mitochondrial DNA. There may be more than one insertion in the process of senescence, and an mtDNA population may evolve until at death a predominant insert is observed. A similar picture is seen with the maranhar plasmid isolated in an *N. crassa* strain in India, as well as with other plasmids found in a large sample of natural isolates (274). While there appears to be no doubt that the kalilo and maranhar plasmids are the responsible agents, other plasmids in *Neurospora* are neutral. The challenge therefore is to understand why some plasmids can and do insert themselves in the mtDNA. According to Griffiths, the "initial insertion event is probably best viewed as a mistake." The mistake, however, is irreversible and ultimately lethal. Does it have any biological significance, especially outside of the laboratory?

Historically, senescence in the fungus *Podospora anserina* was one of the first to attract attention, since all natural isolates died, but they each exhibited distinct and reproducible life spans. Cytoplasmic inheritance was also observed, but heterologous, extragenomic mitochondrial plasmids cannot be detected. Instead, as senescence progresses, circular DNA molecules derived from the mtDNA are observed. The triggering event is unclear, but the net effect is that one of several different segments of mtDNA (labeled α, β, γ, etc.) becomes excised, associated as head-to-tail multimers, and circularized to form a series of senDNAs. At the same time, intact mtDNA molecules disappear. The most common sequence is found in a-senDNA, which consists of monomers identical in sequence to the first intron of the CO1 gene (cytochrome oxidase). It may be significant that this group II intron encodes a protein with homology to reverse transcriptase, but generalizations of the mechanisms for senescence in *Podospora anserina* must include an explanation for the β-, γ-, δ-, etc. senDNAs, which are not restricted to introns and do not encode obvious functions (270). Curiously, a senescing culture can be rejuvenated by treatment with ethidium bromide, as if the intercalator were capable of selecting for wild-type mtDNA molecules, since the sen plasmid present at the time of addition is eliminated. When ethidium bromide is removed again, senescence recurs in association with the appearance of a different senDNA type.

Numerous studies have sought to identify strains with different life spans, and to identify nuclear or mitochondrial mutations responsible for this variation. Details can be found in the reviews by Griffiths (270, 271) and by Bertrand (275). For example, a short-lived strain dies only on solid medium, but not in liquid culture. After repeated transfers in liquid cultures, immortal cultures could be obtained. These were female sterile and incapable of transmitting the immortality as males. Mitochondrial mutations were likely responsible for the altered phenotype. A publication on *Podospora anserina* (276) can serve for an interested reader as an entry point into the recent literature.

In conclusion, senescence in *Podospora anserina* can be linked to mitochondrial DNA instability and mitochondrial energy metabolism; but here also the triggering mechanism, the transition from the nonsenescent (juvenile) state to the senescent state, remains to be elucidated. As in the case of the *Neurospora* plasmids, the true biological significance of the phenomenon of this induced aging remains another puzzle.

7.7 CYTOPLASMIC MALE STERILITY IN PLANTS

Cytoplasmic male sterility (CMS) has been presented briefly in a different context in Chapter 4. The integration of specific plasmids into mtDNA can cause CMS in maize (277). Another genetic mechanism, independent of the existence of extragenomic plasmids, will be illustrated here with a select number of examples. It should be noted that the phenomenon has been observed in more than 150 plant species. It is widely exploited commercially as an expedient method to produce pollen-sterile plants (278). Some authors have referred to it as a "molecular arms race between two conflicting genomes with different modes of inheritance (279, 280). The involvement of the mitochondrial genome in all cases points to a crucial role of mitochondrial function(s) in the development of the male reproductive system and pollen formation in particular. The following discussion will give the reader a flavor of the kinds of molecular mechanisms being studied. It cannot be exhaustive, but the reader will find much up-to-date information in an expert review by Chase (281).

The exceptionally large mitochondrial genomes in plants have already been discussed (Chapter 4) in terms of their high recombination rate, thought to include a population of large circular mtDNA molecules, each representing a portion of the genome. Specific "recombination repeat" sequences typically greater than 2kb are responsible for homologous recombinations capable of permuting the genome without mutating the coding regions. Occasionally, illegitimate (?) recombination events involving much shorter repeat sequences can occur, and they may be responsible for the evolutionary changes in the plant mt genomes. Even more rarely, such recombination events can create new chimeric genes responsible for CMS. The search for such new sequence arrangements in the large plant mtDNA is a formidable task.

L'Homme et al. (282) have thrown considerable light on the situation in the oilseed rape (canola) species *Brassica napus*, and the results may exemplify the general nature of the genetic changes associated with CMS (Figure 7.3). The fertile *B. napus* plant has a cytoplasm (mitochondrial genome) referred to as *cam*; two infertile strains (male sterility) have cytoplasms (mitochondrial genomes), *nap* and *pol*. The *cam* cytoplasm may have originated from the diploid *B. campestris* progenitor of the amphidiploid *B. napus*. The *nap* cytoplasm makes plants male-sterile only when certain nuclear *Rf* s (restorer of fertility) are missing, but the *pol* cytoplasm makes most *B. napus* cultivars male sterile. The whole mtDNA sequence is not yet available, but many segments containing known genes have been identified and physical maps of restriction fragments are available. The task was to find differences between the *cam, nap*, and *pol* mt genomes.

Several lines of evidence have pointed at the vicinity of the *atp6* gene as the region differing in the *cam, pol*, and *nap* mtDNA. Detailed restriction mapping, sequencing, and the analysis of chimeric transcripts found in the various strains has yielded a picture of the arrangement of sequences in the

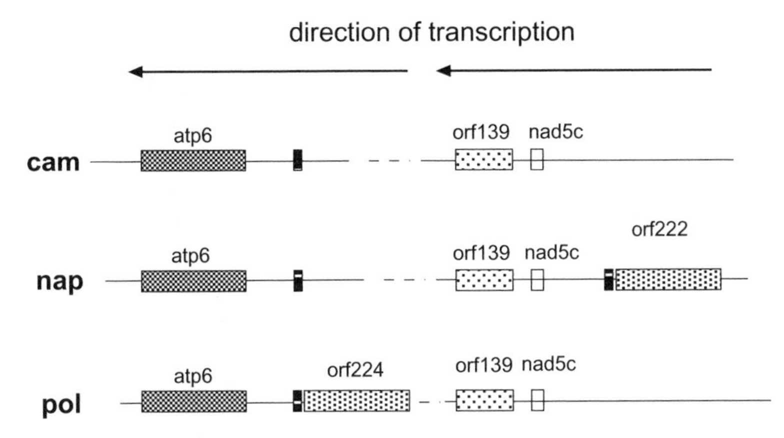

Figure 7.3 Schematic illustration of rearrangements in the mt genome of *Brassica napus* associated with cytoplasmic male sterility (282). In the male-sterile *nap* and pol mt genomes, insertions have occurred upstream of the atp6 locus, with reference to the same locus in the male-fertile *cam* genome. The shaded boxes represent open reading frames; *orf224* and share a 10-bp repeat at the 3′ end (small black box), and they otherwise have 85% nucleotide sequence similarity. The trans-spliced exon nad5c is indicated by the small open box. The dotted lines represent unique sequences.

three mtDNAs, which confirms the postulated differences at this location and is consistent with the observed CMS phenotypes.

In *pol* cytoplasms the *atp6* gene is co-transcribed with a chimeric gene *orf224*. The *orf224* sequence is contained within a 4.5-kb insert upstream of the *atp6* gene which is absent in the *cam* mtDNA. A very similar copy of this 4.5k-kb insert was also found in the *nap* mtDNA near the *atp6* locus, but it was further upstream of *atp6,* and only monocistronic *atp6* transcripts were found in this cytoplasm. Detailed sequencing of the 4.5-kb insert in the *nap* mtDNA revealed an ORF with the same initiation and termination codons as in *orf224*, but with only 85% sequence similarity and two fewer codons. It was designated orf222. Transcripts detected with the *orf222* probe were larger than expected, and further analysis showed co-transcription with two additional ORFs, one of which could be identified as the trans-spliced exon c of the *nad5* gene, while the other (orf139) remains unknown.

Details of this painstaking analysis can be found in the original publication (282), but the reader should come away with a sense of the types of rearrangements encountered in plant mitochondrial genomes which can lead to a CMS phenotype. Needless to say, the authors had to prove that expression of the *orf222/nad5/orf139* is indeed responsible for the male sterility. This was accomplished by probing Northern blots of RNA from male sterile, fertile, and restored cytoplasm with a variety of probes, including *orf222, nad5c*, and sequences from sites where a previous analysis by Palmer and Herbon (283) had shown differences between *nap* and *cam* mtDNAs. Over 40% of the *Brassica* mitochondrial genome were surveyed, with the result that the orf222/nad5c/orf139 region is the only one affected by nuclear restoration. The result is consistent with the general conclusion that all CMS phenotypes can be correlated with differences in the organization, rearrangements and possibly small deletions of plant mitochondrial genomes (284). The transposition of the 4.5-kb fragment may have involved a 10-bp repeat found upstream of atp6 and at the downstream end of orf 222 and orf224.

In recent years at least 14 genes causing CMS have been characterized as ORFs arising from a reassortment of mitochondrial promoter sequences with various mtDNA sequences (281). They exist in the presence of a full complement of respiratory genes and are generally considered to be gain of function mutations.

The fact that illegitimate recombination between plant mt genomes can create chimeric genes and novel transcripts may no longer surprise. However, the study by L'Homme and colleagues also indicated that a simple transposition of the 4.5-kb fragment is not sufficient, and further sequence changes had to accumulate. Presumably there must then also exist a mechanism to convert the plant to a homoplasmic condition, or at least to a heteroplasmic state where the mutation is expressed. One study in which this question was specifically addressed found a very low level of the "wild-type" (fertile) mtDNA in the sterile cytoplasm of sunflower. However, the analysis of heteroplasmy is potentially more complicated for plant mtDNAs because of the frequent recombinations generating aberrant recombinant molecules (sublimons) at substoichiometric, low levels (see reference 285 for discussion and references). Finally, the biologically most interesting question is, which genes are affected, and how does the disruption of a mitochondrial gene specifically cause male sterility—that is, abnormal or aborted anther or pollen development? Analyses in this and other plants lead to the hypothesis that the CMS phenotype is the direct result of expression of the chimeric genes, rather than due to an effect on transcription, splicing, or translation of the normal genes. Nevertheless, no significant sequence homology has been detected in CMS-associated genes from different plant species, which would point at a particular protein/function being required. Although *orf224* and *orf222* have extensive homology, morphological, developmental, and genetic differences distinguish the two *Brassica* CMS systems. For example, the nuclear *Rf* genes restoring male fertility are different for *pol* and *nap*. The diversity of mechanisms giving rise to CMS is also illustrated by the existence of multiple types of CMS in maize and other plants (281, 286).

Results that are closer to answering questions raised above come from a study of male sterility in the sunflower. In this case as well, an unusual ORF (*orf522*) had been created and associated with CMS. It is long enough to encode a 15-kDa peptide, and this peptide had been identified in sterile lines. In a recent study, Moneger and colleagues showed that a dominant nuclear restorer gene can suppress this 15-kDa peptide significantly in a tissue-specific manner; that is, the reduction is observed only in male florets. Furthermore, run-on transcription experiments with isolated mitochondria from this specific tissue and comparisons with steady-state levels of the transcript in sterile and restored hybrid lines strongly suggest that the nuclear *Rf* product regulated *orf522* transcript levels post-transcriptionally (285, 287, 288).

The 15-kDa ORF522 peptide also shows no homology with a 13-kDa mitochondrial peptide (URF13) from maize CMS-T or the 25-kDA mitochondrial peptide from the *pcf* locus in petunia, both of which have been implicated in the mechanism leading to CMS in these plants. (One should be reminded again that protein coding genes in plant mitochondria are edited, and the nature of the specific base change(s) must be known before sequence comparisons can be made.) However, these proteins all appear to be relatively hydrophobic, and there is speculation that their expression and integration into the mitochondrial inner membrane may interfere with OXPHOS.

Several scattered observations are relevant for explaining male sterility. First, a 20- to 40-fold increase in the relative number of mitochondria has been observed in maize anthers (289, 290). Second, the levels of transcripts of a select number of mitochondrial genes (*cob, atp6, atp9*) have been determined to be elevated in microspores compared to leaves in maize (291). Finally, isolated mitochondria from different tissues of maize and sugar beet synthesize organ-specific polypeptides (292, 293). Thus, mitochondrial activity and function clearly appear to be significant factors in pollen formation. In other tissues a diminished capacity for OXPHOS due to mtDNA rearrangements may be less influential, or OXPHOS may not even be affected if the CMS-associated ORF is not expressed.

Another advance in the quest for understanding the molecular basis for CMS was added by an elegant study from Mackenzie's laboratory (294). In CMS, common bean rearrangements in the mtDNA have created two new ORFs, orf239 and orf98, and a 27-kDA peptide encoded by *orf239* was found exclusively in the reproductive tissues. This and other evidence made the *orf239* peptide a prime suspect for a primary role in CMS; to prove it, the authors expressed the *orf239* sequence after transformation into tobacco plants. The gene was engineered to express the 27-kDa protein with or without a mitochondrial targeting sequence; specific promoters were attached to restrict the expression of the chimeric genes to developing microspores. The perhaps unexpected results were that male sterility and defective pollen development were observed regardless of whether the mitochondrial targeting sequence was present. The presence of a cryptic targeting sequence in the *orf239* peptide was excluded on theoretical grounds, and the conclusion was that pollen development could be interrupted by a protein expressed in the cytoplasm. Immunoelectron microscopy was used to localize the protein exclusively in developing microspores, and there it appeared to be in the cell wall of developmentally arrested microspores. The transgene was dominant with incomplete penetrance, possibly due to the less-than-perfect timing of expression from the heterologous promoter.

Two major results emerge from this study. First, a CMS-related gene product may act without compromising mitochondrial function. Second, in the original CMS common bean the 27-kDa protein is made inside the mitochondria, and hence it has to be exported to interact with the extramitochondrial target. The next stage will address whether export is from intact mitochondria by a novel pathway or whether the protein can be dumped into the cytosol by mitochondrial lysis.

CMS in plants is an excellent model system to study nuclear–mitochondrial interactions. A number of nuclear genes have been identified (restorer of fertility genes) that can suppress or restore the ability to form functional pollen. Some of these have not been shown to endcode proteins targeted to mitochondria where they are implicated in RNA metabolism (see reference 281 for a review). The process of retrograde signaling has been defined as the regulation of the expression of nuclear genes by signals from a cellular organelle, and accumulating evidence suggests that flower development is subject to such mechanisms (281). It remains to be established whether the signal comes from an energy insufficiency, or other signals, and even reactive oxygen species (ROS) have again been introduced into the discussion.

REFERENCES

1. Wickner, R. B. (1996) *Annu. Rev. Genet.* **30**, 109–139.
2. Brandt, U. (2006) *Annu. Rev. Biochem.* **75**, 69–92.
3. Abdrakhmanova, A., Zickermann, V., Bostina, M., Radermacher, M., Schagger, H., Kerscher, S., and Brandt, U. (2004) *Biochim. Biophys. Acta* **1658**, 148–156.
4. Correns, C. (1909) *Z. Verebungsl.* **1**, 291.
5. Baur, E. (1909) *Z. Vererbungsl.* **1**, 330.
6. Slonimski, P. P. and Ephrussi, B. (1949) *Ann. Inst. Pasteur* **77**, 47.
7. Dujon, B. (1981) Mitochondrial Genetics and Function. In Strathern, J. N., Jones, E. W., and Broach, J. R., editors. *The Molecular Biology of the Yeast Saccharomyces. Life Cycle and Inheritance*, Cold Spring Harbor Press, Cold Spring Harbor, NY, pp. 505–635.
8. Johnston, M. and Carlson, M. (1993) Regulation of carbon and phosphate utilization. In Broach, J., Jones, E. W., and Pringle, J., editors. Cold Spring Harbor Press, Cold Spring Harbor, NY, pp. 193–281.
9. Trumbly, R. J. (1992) *Mol. Microbiol.* **6**, 15–21.
10. Brown, T. A., Evangelista, C., and Trumpower, B. L. (1995) *J. Bacteriol.* **177**, 6836–6843.
11. Ronne, H. (1995) *TIG* **11**, 12–17.
12. De Winde, J. H. and Grivell, L. A. (1993) *Prog. Nucleic Acids Res.* **46**, 51–91.
13. Tzagoloff, A. and Dieckmann, C. L. (1990) *Microbiol. Rev.* **54**, 211–225.
14. Fox, T. D., Folley, L. S., Mulero, J. J., McMullin, T. W., Thorsness, P. E., Hedin, L. O., and Costanzo, M. C. (1990) *Methods Enzymol.* **194**, 149–168.

15. Mittelmaier, T. M. and Dieckmann, C. L. (1995) *Mol. Cell. Biol.* **15**, 780–789.
16. Williams, E. H. and Fox, T. D. (2003) *RNA* **9**, 419–431.
17. Warburg, O. (1956) *Science* **123**, 309–314.
18. Racker, E. (1976) *A New Look at Mechanisms in Bioenergetics*, Academic Press, New York.
19. Schulz, T. J., Thierbach, R., Voigt, A., Drewes, G., Mietzner, B., Steinberg, P., Pfeiffer, A. F., and Ristow, M. (2006) *J. Biol. Chem.* **281**, 977–981.
20. Kroemer, G. (2006) *Oncogene* **25**, 4630–4632.
21. Mathupala, S. P., Ko, Y. H., and Pedersen, P. L. (2006) *Oncogene* **25**, 4777–4786.
22. Scheffler, I. E. (1974) *J. Cell. Physiol.* **83**, 219–230.
23. DeFrancesco, L., Werntz, D., and Scheffler, I. E. (1975) *J. Cell. Physiol.* **85**, 293–306.
24. DeFrancesco, L., Scheffler, I. E., and Bissell, M. J. (1976) *J. Biol. Chem.* **251**, 4588–4595.
25. Day, C. and Scheffler, I. E. (1982) *Somat. Cell Genet.* **8**, 691–707.
26. Ditta, G. S., Soderberg, K., and Scheffler, I. E. (1976) *Somat. Cell Genet.* **2**, 331–344.
27. Soderberg, K., Mascarello, J. T., Breen, G. A. M., and Scheffler, I. E. (1979) *Somat. Cell Genet.* **5**, 225–240.
28. Whitfield, C. D. (1985) Mitochondrial mutants. In Gottesman, M. M., editor. *Molecular Cell Genetics*, John Wiley and Sons, New York, pp. 545–588.
29. Maiti, I. B., Comlan de Souza, A., and Thirion, J. P. (1981) *Somat. Cell Genet.* **7**, 567–582.
30. Scheffler, I. E. (1986) Biochemical genetics of respiration-deficient mutants of animal cells. In Morgan, M. J., editor. *Carbohydrate Metabolism in Cultured Cells*, Plenum Publishing Co., London, pp. 77–109.
31. Burnett, K. G. and Scheffler, I. E. (1981) *J. Cell Biol.* **90**, 108–115.
32. Ditta, G. S., Soderberg, K., and Scheffler, I. E. (1977) *Nature* **268**, 64–67.
33. Au, H. C. and Scheffler, I. E. (1997) *Somat. Cell Mol. Genet.* **23**, 27–35.
34. Soderberg, K., Ditta, G. S., and Scheffler, I. E. (1977) *Cell* **10**, 697–702.
35. Oostveen, F. G., Au, H. C., Meijer, P.-J., and Scheffler, I. E. (1995) *J. Biol. Chem.* **270**, 26104–26108.
36. Scheffler, I. E., Yadava, N., and Potluri, P. (2004) *Biochim. Biophys. Acta* **1659**, 160–171.
37. Potluri, P., Yadava, N., and Scheffler, I. E. (2004) *Eur. J. Biochem.* **271**, 3265–3273.
38. Au, H. C., Seo, B. B., Matsuno-Yagi, A., Yagi, T., and Scheffler, I. E. (1999) *Proc. Natl. Acad. Sci. USA* **96**, 4354–4359.
39. Desjardins, P., de Muys, J.-M., and Morais, R. (1986) *Somat. Cell Mol. Genet.* **12**, 133–139.
40. King, M. P. and Attardi, G. (1989) *Science* **246**, 500–503.
41. Hayashi, J.-I., Ohta, S., Kikuchi, A., Takemitsu, M., Goto, Y., and Nonaka, I. (1991) *Proc. Natl. Acad. Sci. USA* **88**, 10614–10618.
42. Inoue, K., Ito, S., Takai, D., Soejima, A., Shisa, H., LePecq, J. B., Segal-Bendirdjian, E., Kagawa, Y., and Hayashi, J.-I. (1997) *J. Biol. Chem.*

43. Hayashi, J.-I., Tagashira, Y., Yoshida, M. C., Ajiro, K., and Sekiguchi, T. (1983) *Exp. Cell Res.* **147**, 51–61.
44. Bunn, C. L. D., Wallace, D. C., and Eisenstadt, J. M. (1974) *Proc. Natl. Acad. Sci. USA* **71**, 1681–1685.
45. Wallace, D. C. and Eisenstadt, J. M. (1979) *Somat. Cell Genet.* **5**, 373–396.
46. Kenyon, L. and Moraes, C. T. (1997) *Proc. Natl. Acad. Sci. USA* **94**, 9131–9135.
47. Trounce, I. A., McKenzie, M., Cassar, C. A., Ingraham, C. A., Lerner, C. A., Dunn, D. A., Donegan, C. L., Takeda, K., Pogozelski, W. K., Howell, R. L., and Pinkert, C. A. (2004) *J. Bioenerg. Biomembr.* **36**, 421–427.
48. Inoue, K., Nakada, K., Ogura, A., Isobe, K., Goto, Y., Nonaka, I., and Hayashi, J.-I. (2000) *Nature Genet.* **26**, 176–181.
49. Sligh, J. E., Levy, S. E., Waymire, K. G., Allard, P., Dillehay, D. L., Nusinowitz, S., Heckenlively, J. R., MacGregor, G. R., and Wallace, D. C. (2000) *Proc. Natl. Acad. Sci. USA* **97**, 14461–14466.
50. Breen, G. A. M. and Scheffler, I. E. (1980) *J. Cell Biol.* **86**, 723–729.
51. Harris, M. (1978) *Proc. Natl. Acad. Sci. USA* **75**, 5604–5608.
52. Howell, N. and Sager, R. (1979) *Somat. Cell Genet.* **5**, 833–845.
53. Lichtor, T. and Getz, G. S. (1978) *Proc. Natl. Acad. Sci. USA* **75**, 324–328.
54. Breen, G. A., Miller, D. L., Holmans, P. L., and Welch, G. (1986) *J. Biol. Chem.* **261**, 11680–11685.
55. Holmans, P. L. and Breen, G. A. (1987) *Somat. Cell Mol. Genet.* **13**, 347–353.
56. Giles, R. E., Blanc, H., Cann, R. M., and Wallace, D. C. (1980) *Proc. Natl. Acad. Sci. USA* **77**, 6715–6719.
56a. Hutchison, C. A., Newbold, J. E., Potter, S. S., and Edgell, M. H. (1974) *Nature* **251**, 536–538.
56b. Kroon, A. M., de Vos, W. M., and Bakker, H. (1978) *Biochem. Biophys. Acta* **519**, 269–273.
56c. Hayashi, J. I., Yonekawa, H., Gotoh, O., Watanabe, J., and Tagashira, Y. (1978) *Biochem. Biophys. Res. Commun.* **83**, 1032–1038.
56d. Avise, J. C., Giblin-Davidson, C., Laerm, J., Patton, J. C., and Lansman, R. A. (1979) *Proc. Natl. Acad. Sci. USA* **76**, 6694–6698.
56e. Dawid, I. B. and Blackler, A. W. (1972) *Dev. Biol.* **29**, 152–161.
56f. Reily, J. G. and Thomas, C. A. Jr. (1980) *Plasmid* **3**, 109–115.
57. Wallace, D. C., Singh, G., Lott, M. T., Hodge, J. A., Schurr, T. G., Lezza, A. M. S., Elsas, L. J., II, and Nikoskelainen, E. K. (1988) *Science* **242**, 1427–1431.
58. Holt, I. J., Harding, A. E., and Morgan Hughes, J. A. (1988) *Nature* **331**, 717–719.
59. Scheffler, I. E. (2004) The Human OXPHOS System: Structure, Function, Physiology. In Smeitink, J. A. M., Sengers, R. C. A., and Trijbels, J. M. F. *Oxidative Phosphorylation in Health and Disease*, Landes Bioscience, Georgetown, TX, pp. 1–27.
60. Scheffler, I. E. (2004) Mitochondrial Dysfunction in Genetic Diseases. In Fuchs, J., Podda, M., and Packer, L., editors. *Redox–Genome interactions In Health and Disease*, Marcel Dekker, New York, pp. 235–262.
61. Wallace, D. C. (2005) *Annu. Rev. Genet.* **39**, 359–407.

62. Schapira, A. H. (2006) *Lancet* **368**, 70–82.

63. Levinger, L., Morl, M., and Florentz, C. (2004) *Nucleic Acids Res.* **32**, 5430–5441.

64. Florentz, C. (2002) *Biosci. Rep.* **22**, 81–98.

65. Gyllensten, U., Wharton, D., Josefsson, A., and Wilson, A. C. (1991) *Nature* **352**, 255–257.

66. Kaneda, H., Hayashi, J.-I., Takahama, S., Taya, C., Fischer Lindahl, K., and Yonekawa, H. (1995) *Proc. Natl. Acad. Sci. USA* **92**, 4542–4546.

67. Nishimura, Y., Yoshinari, T., Naruse, K., Yamada, T., Sumi, K., Mitani, H., Higashiyama, T., and Kuroiwa, T. (2006) *Proc. Natl. Acad. Sci.* **103**, 1382–1387.

68. Sutovsky, P. (2003) *Microsc. Res. Tech.* **61**, 88–102.

69. Sutovsky, P., Van Leyen, K., McCauley, T., Day, B. N., and Sutovsky, M. (2004) *Reprod. Biomed. Online.* **8**, 24–33.

70. Thompson, W. E., Ramalho-Santos, J., and Sutovsky, P. (2003) *Biol. Reprod.* **69**, 254–260.

71. Hayashida, K., Omagari, K., Masuda, J. I., Hazama, H., Kadokawa, Y., Ohba, K., and Kohno, S. (2005) *Cell Biol. Int.* **29**, 472–481.

72. Ankel-Simons, F., Cummings, J. M., and Cummins, J. M. (1996) *Proc. Natl. Acad. Sci. USA* **93**, 13859–13863.

73. Harding, A. E. (1993) Spontaneous errors of mitochondrial DNA in human disease. In DiMauro, S. and Wallace, D. C., editors. *Mitochondrial DNA in Human Pathology*, Raven Press, New York, pp. 53–62.

74. Schon, E. A., Bonilla, E., and DiMauro, S. (1997) *J. Bioenerg. Biomembr.* **29**, 131–149.

75. Weber, K., Wilson, J. N., Taylor, L., Brierley, E., Johnson, M. A., Turnbull, D. M., and Bindoff, L. A. (1997) *Am. J. Hum. Genet.* **60**, 373–380.

76. Clark, K. M., Bindoff, L. A., Lightowlers, R. N., andrew, R. M., Griffiths, P. G., Johnson, M. A., Brierley, E. J., and Turnbull, D. M. (1997) *Nature Genet.* **16**, 222–224.

77. Chen, X., Prosser, R., Simonetti, S., Sadlock, J., Jagiello, G., and Schon, E. A. (1995) *Am. J. Hum. Genet.* **57**, 239–247.

78. Zenke, F. T., Zachariae, W., Lunkes, A., and Breunig, K. D. (1993) *Mol. Cell. Biol.* **13**, 7566–7576.

79. Ballinger, S. W., Shoffner, J. M., Hedaya, E. V., Trounce, I., Polak, M. A., Koontz, D. A., and Wallace, D. C. (1992) *Nat. Genet.* **1**, 11–15.

80. Moraes, C. T., Andreetta, F., Bonilla, E., Shanske, S., DiMauro, S., and Schon, E. A. (1991) *Mol. Cell. Biol.* **11**, 1631–1637.

81. Shadel, G. S. and Clayton, D. A. (1997) *Annu. Rev. Biochem.* **66**, 409–435.

82. Singh, G., Lott, M. T., and Wallace, D. C. (1989) *N. Engl. J. Med.* **320**, 1300–1305.

82a. Sutton, H. E. (1988) An Introduction to Human Genetics, 4th edition; Harcourt Brace Jovanovich, Publishers, San Diego, p. 290.

83. Laipis, P. J., Van de Walle, M. J., and Hauswirth, W. W. (1988) *Proc. Natl. Acad. Sci. USA* **85**, 8107–8110.

84. Ashley, M. V., Laipis, P. J., and Hauswirth, W. W. (1989) *Nucleic Acids Res.* **17**, 7325–7331.

85. Poulton, J. (1995) *Am. J. Hum. Genet.* **57**, 224–226.
86. Clayton, D. A. (1996) *Nature Genet.* **14**, 123–125.
87. Chinnery, P. F. (2002) *Mitochondrion* **2**, 149–155.
88. Lightowlers, R. N., Chinnery, P. F., Turnbull, D. M., and Howell, N. (1997) *Trends Genet.* **13**, 450–455.
89. Piko, L. and Taylor, K. D. (1987) *Dev. Biol.* **123**, 364–374.
90. Chinnery, P. F., Thorburn, D. R., Samuels, D. C., White, S. L., Dahl, H. H., Turnbull, D. M., Lightowlers, R. N., and Howell, N. (2000) *Trends Genet.* **16**, 500–505.
91. Battersby, B. J. and Shoubridge, E. A. (2001) *Hum. Mol. Genet.* **10**, 2469–2479.
92. Battersby, B. J., Redpath, M. E., and Shoubridge, E. A. (2005) *Hum. Mol. Genet.* **14**, 2587–2594.
93. Jenuth, J. P., Peterson, A. C., Fu, K., and Shoubridge, E. A. (1996) *Nature Genet.* **14**, 146–151.
94. Davis, A. F. and Clayton, D. A. (1996) *J. Cell Biol.* **135**, 883–893.
95. Kearns, T. P. and Sayre, G. P. (1958) *Ophthalmology* **60**, 280–289.
96. Pearson, H. A., Lobel, J. S., Kocoshis, S. A., et al. (1979) *J. Pediatr.* **95**, 976–984.
97. Moraes, C. T., Shanske, S., Tritschler, H.-J., Aprille, J. R., andreetta, F., Bonilla, E., Schon, E. A., and DiMauro, S. (1991) *Am. J. Hum. Genet.* **48**, 492–501.
98. Howell, N. (2003) *Dev. Ophthalmol.* **37**, 94–108.
99. Carelli, V., Rugolo, M., Sgarbi, G., Ghelli, A., Zanna, C., Baracca, A., Lenaz, G., Napoli, E., Martinuzzi, A., and Solaini, G. (2004) *Biochim. Biophys. Acta* **1658**, 172–179.
100. Howell, N. (1997) *Vision Res.* **37**, 3495–3507.
101. Brown, M. D. and Wallace, D. C. (1994) *Clin. Neurosci.* **2**, 138–145.
102. Wallace, D. C. and Lott, M. T. (1993) Maternally inherited diseases. In DiMauro, S. and Wallace, D. C., editors. *Mitochondrial DNA in Human Pathology*, Raven Press, New York, pp. 63–83.
103. Tatuch, Y., Christodoulou, J., Feigenbaum, A., et al. (1992) *Am. J. Hum. Genet.* **50**, 852–858.
104. Shoffner, J. M., Fernhoff, M. D., Krawiecki, N. S., et al. (1992) *Neurology* **42**, 2168–2174.
105. Wallace, D. C., Zheng, X., Lott, M. T., Shoffner, J. M., Hodge, J. A., Kelley, R. I., Epstein, C. M., and Hopkins, L. C. (1988) *Cell* **55**, 601–610.
106. Guan, M. X., Yan, Q., Li, X., Bykhovskaya, Y., Gallo-Teran, J., Hajek, P., Umeda, N., Zhao, H., Garrido, G., Mengesha, E., Suzuki, T., Castillo, I. D., Peters, J. L., Li, R., Qian, Y., Wang, X., Ballana, E., Shohat, M., Lu, J., Estivill, X., Watanabe, K., and Fischel-Ghodsian, N. (2006) *Am. J. Hum. Genet* **79**, 291–302.
107. Guan, M. X. (2004) *Ann. N. Y. Acad. Sci.* **1011**, 259–271.
108. Zhao, H., Young, W. Y., Yan, Q., Li, R., Cao, J., Wang, Q., Li, X., Peters, J. L., Han, D., and Guan, M. X. (2005) *Nucleic Acids Res.* **33**, 1132–1139.
109. Li, X., Fischel-Ghodsian, N., Schwartz, F., Yan, Q., Friedman, R. A., and Guan, M. X. (2004) *Nucleic Acids Res.* **32**, 867–877.
110. Smeitink, J. A., Zeviani, M., Turnbull, D. M., and Jacobs, H. T. (2006) *Cell Metab.* **3**, 9–13.

111. Janssen, A. J., Trijbels, F. J., Sengers, R. C., Wintjes, L. T., Ruitenbeek, W., Smeitink, J. A., Morava, E., van Engelen, B. G., Van den Heuvel, L. P., and Rodenburg, R. J. (2006) *Clin. Chem.* **52**, 860–871.

112. Van den Heuvel, L. and Smeitink, J. (2001) *Bioessays* **23**, 518–525.

113. Dinopoulos, A., Smeitink, J., and Ter Laak, H. (2005) *Acta Neuropathol. (Berl.)* **110**, 199–202.

114. Ugalde, C., Janssen, R., Van den Heuvel, L. P., Smeitink, J. A., and Nijtmans, L. G. (2004) *Hum. Mol. Genet.* **13**, 659–667.

115. Smeitink, J. A., van den Heuvel, L. W., Koopman, W. J., Nijtmans, L. G., Ugalde, C., and Willems, P. H. (2004) *Curr. Neurovasc. Res.* **1**, 29–40.

116. Van Der Westhuizen, F. H., Van den Heuvel, L. P., Smeets, R., Veltman, J. A., Pfundt, R., Van Kessel, A. G., Ursing, B. M., and Smeitink, J. A. (2003) *Neuropediatrics* **34**, 14–22.

117. Triepels, R. H., Van den Heuvel, L. P., Trijbels, J. M., and Smeitink, J. A. (2001) *Am. J. Med. Genet.* **106**, 37–45.

118. Loeffen, J. L., Smeitink, J. A., Trijbels, J. M., Triepels, R. H., Sengers, R. C., and Van den Heuvel, L. P. (2000) *Hum. Mutat.* **15**, 123–134.

119. Benit, P., Beugnot, R., Chretien, D., Giurgea, I., Lonlay-Debeney, P., Issartel, J. P., Corral-Debrinski, M., Kerscher, S., Rustin, P., Rotig, A., and Munnich, A. (2003) *Hum. Mutat.* **21**, 582–586.

120. Chretien, D., Benit, P., Chol, M., Lebon, S., Rotig, A., Munnich, A., and Rustin, P. (2003) *Biochem. Biophys. Res. Commun.* **301**, 222–224.

121. Ogilvie, I., Kennaway, N. G., and Shoubridge, E. A. (2005) *J. Clin. Invest.* **115**, 2784–2792.

122. Procaccio, V. and Wallace, D. C. (2004) *Neurology* **62**, 1899–1901.

123. Briere, J. J., Favier, J., Ghouzzi, V. E., Djouadi, F., Benit, P., Gimenez, A. P., and Rustin, P. (2005) *Cell Mol. Life Sci.* **62**, 2317–2324.

124. Rustin, P. and Rotig, A. (2002) *Biochim. Biophys. Acta* **1553**, 117–122.

125. Ackrell, B. A. (2002) *Mol. Aspects Med.* **23**, 369–384.

126. Chew, S. L. (2001) *Clin. Endocrinol. (Oxf)* **54**, 573–574.

127. Senior, K. (2000) *Mol. Med. Today* **6**, 183.

128. Douwes, D. P., Hogendoorn, P., Kuipers-Dijkshoorn, N., Prins, F., Van Duinen, S., Taschner, P., Van Der, M. A., and Cornelisse, C. (2003) *J. Pathol.* **201**, 480–486.

129. Baysal, B. E., Willett-Brozick, J. E., Lawrence, E. C., Drovdlic, C. M., Savul, S. A., McLeod, D. R., Yee, H. A., Brackmann, D. E., Slattery, W. H. 3rd, Myers, E. N., Ferrell, R. E., and Rubinstein, W. S. (2002) *J. Med. Genet.* **39**, 178–183.

130. Astuti, D., Latif, F., Dallol, A., Dahia, P. L., Douglas, F., George, E., Skoldberg, F., Husebye, E. S., Eng, C., and Maher, E. R. (2001) *Am. J. Hum. Genet.* **69**, 49–54.

131. Haut, S., Brivet, M., Touati, G., Rustin, P., Lebon, S., Garcia-Cazorla, A., Saudubray, J. M., Boutron, A., Legrand, A., and Slama, A. (2003) *Hum. Genet.*

132. Tiranti, V., Hoertnagel, K., Carrozzo, R., Galimberti, C., Munaro, M., Granatiero, M., Zelante.L., Gasparini, P., Marzella, R., Rocchi, M., Bayona-Bafaluy, M. P., Enriquez, J. A., Uziel, G., Bertini, E., Dionisi-Vici, C., Franco, B., Meitinger, T., and Zeviani, M. (2001) *Am. J. Hum. Genet.* **63**, 1621.

133. Tiranti, V., Galimberti, C., Nijtmans, L., Bovolenta, S., Perini, M. P., and Zeviani, M. (1999) *Hum. Mol. Genet.* **8**, 2533–2540.

134. Poyau, A., Buchet, K., Bouzidi, M. F., Zabot, M. T., Echenne, B., Yao, J., Shoubridge, E. A., and Godinot, C. (2000) *Hum. Genet.* **106**, 194–205.

135. Zhu, Z., Yao, J., Johns, T., Fu, K., De Bie, I., Macmillan, C., Cuthbert, A. P., Newbold, R. F., Wang, J., Chevrette, M., Brown, G. K., Brown, R. M., and Shoubridge, E. A. (1998) *Nat. Genet.* **20**, 337–343.

136. Papadopoulou, L. C., Sue, C. M., Davidson, M. M., Tanji, K., Nishino, I., Sadlock, J. E., Krishna, S., Walker, W., Selby, J., Glerum, D. M., Van Coster, R., Lyon, G., Scalais, E., Lebel, R., Kaplan, P., Shanske, S., De Vivo, D. C., Bonilla, E., Hirano, M., DiMauro, S., and Schon, E. A. (1999) *Nat. Genet.* **23**, 333–337.

137. Suomalainen, A. and Kaukonen, J. (2001) *Am. J. Med. Genet.* **106**, 53–61.

138. Saada-Reisch, A. (2004) *Nucleosides Nucleotides Nucleic Acids* **23**, 1205–1215.

139. Ponamarev, M. V., Longley, M. J., Nguyen, D., Kunkel, T. A., and Copeland, W. C. (2002) *J. Biol. Chem.* **277**, 15225–15228.

140. Nishino, I., Spinazzola, A., and Hirano, M. (2001) *Neuromuscul. Disord.* **11**, 7–10.

141. Kaukonen, J., Juselius, J. K., Tiranti, V., Kyttälä, A., Zeviani, M., Comi, G. P., Keränen, S., Peltonen, L., and Suomalainen, A. (2000) *Science* **289**, 782–785.

142. Bodnar, A. G., Cooper, J. M., Holt, I. J., Leonard, J. V., and Schapira, A. H. V. (1993) *Am. J. Hum. Genet.* **53**, 663–669.

143. Spelbrink, J. N., Van Galen, M. J. M., Zwart, R., Bakker, H. D., Rovio, A., Jacobs, H. T., and Van den Bogert, C. (1998) *Hum. Genet.* **102**, 327–331.

144. Mandel, H., Szargel, R., Labay, V., Elpeleg, O., Saada, A., Anbinder, Y., Berkowitz, D., Hartmen, C., Barak, M., Eriksson, S., and Cohen, N. (2001) *Nature Genet.* **29**, 337–341.

145. Saada, A., Shaag, A., Mandel, H., Nevo, Y., Eriksson, S., and Elpeleg, O. (2001) *Nature Genet.* **29**, 342–344.

146. Lodi, R., Tonon, C., Calabrese, V., and Schapira, A. H. (2006) *Antioxid. Redox. Signal.* **8**, 438–443.

147. Gakh, O., Park, S., Liu, G., Macomber, L., Imlay, J. A., Ferreira, G. C., and Isaya, G. (2006) *Hum. Mol. Genet.* **15**, 467–479.

148. Napoli, E., Taroni, F., and Cortopassi, G. A. (2006) *Antioxid. Redox Signal.* **8**, 506–516.

149. Alper, G. and Narayanan, V. (2003) *Pediatr. Neurol* **28**, 335–341.

150. Seznec, H., Simon, D., Bouton, C., Reutenauer, L., Hertzog, A., Golik, P., Procaccio, V., Patel, M., Drapier, J. C., Koenig, M., and Puccio, H. (2005) *Hum. Mol. Genet.* **14**, 463–474.

151. Tranebjaerg, L., Schwartz, C., Erikson, H., andreason, S., Ponjavic, V., Dahl, A., Stevenson, R. E., May, M. A. F., and Barker, D. (1995) *J. Med. Genet.* **32**, 257–263.

152. Koehler, C. M., Leuenberger, D., Merchant, S., Renold, A., Junne, T., and Schatz, G. (1999) *Proc. Natl. Acad. Sci. USA* **96**, 2141–2146.

153. Roesch, K., Curran, S. P., Tranebjaerg, L., and Koehler, C. M. (2002) *Hum. Mol. Genet.* **11**, 477–486.

154. Rothbauer, U., Hofmann, S., Muhlenbein, N., Paschen, S. A., Gerbitz, K. D., Neupert, W., Brunner, M., and Bauer, M. F. (2001) *J. Biol. Chem.* **276**, 37327–37334.

155. Linnane, A. W., Marzuki, S., Ozawa, T., and Tanaka, M. (1989) *Lancet* **1**, 642–645.

156. Wallace, D. C. (1995) *Am. J. Hum. Genet.* **57**, 201–223.

157. Cortopassi, G. and Arnheim, N. (1993) Accumulation of mitochondrial DNA mutation in normal aging brain and muscle. In DiMauro, S. and Wallace, D. C., editors. *Mitochondrial DNA in Human Pathology*, Raven Press, New York, pp. 125–136.

158. Nagley, P., Zhang, C., Martinus, R. D., Vaillant, F., and Linnane, A. W. (1993) Mitochondrial DNA mutation and human aging: Molecular biology, bioenergetics and redox therapy. In DiMauro, S. and Wallace, D. C., editors. *Mitochondrial DNA in Human Pathology*, Raven Press, New York, pp. 137–157.

159. Hagen, T. M., Yowe, D. L., Bartholomew, J. C., Wehr, C. M., Do, K. L., Park, J. Y., and Ames, B. N. (1997) *Proc. Natl. Acad. Sci.* USA **94**, 3064–3069.

160. Wallace, D. C. (2001) *Novartis Found. Symp.* **235**, 247–263.

161. Partridge, L. G. D. (2002) *Nature* **418**, 921.

162. Martin, G. M., LaMarco, K., Strauss, E., and Kelner, K. L. (2003) *Science* **299**, 1339–1340.

163. Fridovich, I. (2004) *Aging Cell* **3**, 13–16.

164. Beal, M. F. (2005) *Ann. Neurol.* **58**, 495–505.

165. Terman, A. and Brunk, U. T. (2006) *Antioxid. Redox Signal.* **8**, 197–204.

166. Passos, J. F., Zglinicki, T., and Saretzki, G. (2006) *Rejuvenation Res.* **9**, 64–68.

167. Lenaz, G., Baracca, A., Fato, R., Genova, M. L., and Solaini, G. (2006) *Antioxid. Redox. Signal.* **8**, 417–437.

168. Various authors (2005) *Cell* **120**, 435–567.

169. Schriner, S. E., Linford, N. J., Martin, G. M., Treuting, P., Ogburn, C. E., Emond, M., Coskun, P. E., Ladiges, W., Wolf, N., Van Remmen, H., Wallace, D. C., and Rabinovitch, P. S. (2005) *Science* **308**, 1909–1911.

170. Sun, J., Folk, D., Bradley, T. J., and Tower, J. (2002) *Genetics* **161**, 661–672.

171. Jacobs, H. T. (2003) *Aging Cell* **2**, 11–17.

172. Trifunovic, A. (2006) *Biochim. Biophys. Acta.* **1757**, 611–617.

173. Trifunovic, A., Hansson, A., Wredenberg, A., Rovio, A. T., Dufour, E., Khvorostov, I., Spelbrink, J. N., Wibom, R., Jacobs, H. T., and Larsson, N. G. (2005) *Proc. Natl. Acad. Sci. USA* **102**, 17993–17998.

174. Trifunovic, A., Wredenberg, A., Falkenberg, M., Spelbrink, J. N., Rovio, A. T., Bruder, C. E., Bohlooly, Y., Gidlof, S., Oldfors, A., Wibom, R., Tornell, J., Jacobs, H. T., and Larsson, N. G. (2004) *Nature* **429**, 417–423.

175. Kujoth, G. C., Hiona, A., Pugh, T. D., Someya, S., Panzer, K., Wohlgemuth, S. E., Hofer, T., Seo, A. Y., Sullivan, R., Jobling, W. A., Morrow, J. D., Van Remmen, H., Sedivy, J. M., Yamasoba, T., Tanokura, M., Weindruch, R., Leeuwenburgh, C., and Prolla, T. A. (2005) *Science* **309**, 481–484.

176. Linnane, A. W., Baumer, A., Maxwell, R. J., Preston, H., Zhang, C., and Marzuki, S. (1990) *Biochem. Int.* **22**, 1067–1076.

177. Melov, S., Shoffner, J. M., Kaufman, A., and Wallace, D. C. (1995) *Nucleic Acids Res.* **23**, 4122–4126.

178. Corral-Debrinski, M., Horton, T., Lott, M. T., Shoffner, J. M., Beal, M. F., and Wallace, D. C. (1992) *Nature Genet.* **2**, 324–329.

179. Simonetti, S., Chen, X., DiMauro, S., and Schon, E. A. (1992) *Biochim. Biophys. Acta Mol. Basis Dis.* **1180**, 113–122.

180. Kadenbach, B., Munscher, C., Frank, V., Müller-Höcker, J., and Napiwotzki, J. (1995) *Mutat. Res. DNAging Genet. Instability Aging* **338**, 161–172.

181. Hsieh, R.-H., Hou, J.-H., Hsu, H.-S., and Wei, Y.-H. (1994) *Biochem. Mol. Biol. Int.* **32**, 1009–1022.

182. Chomyn, A. and Attardi, G. (2003) *Biochem. Biophys. Res. Commun.* **304**, 519–529.

183. Gross, N. J., Getz, G. S., and Rabinowitz, M. (1969) *J. Biol. Chem.* **244**, 1552–1562.

184. Shoffner, J. M., Lott, M. T., Voljavec, A. S., Soueidan, S. A., Costigan, D. A., and Wallace, D. C. (1989) *Proc. Natl. Acad. Sci.* USA **86**, 7952–7956.

185. Muller-Hocker, J. (1989) *Am. J. Pathol.* **134**, 1167–1173.

186. Michikawa, Y., Mazzucchelli, F., Bresolin, N., Scarlato, G., and Attardi, G. (1999) *Science* **286**, 774–779.

187. Shigenaga, M. K., Hagen, T. M., and Ames, B. N. (1994) *Proc. Natl. Acad. Sci. USA* **91**, 10771–10778.

188. Ames, B. N., Shigenaga, M. K., and Hagen, T. M. (1995) *Biochim. Biophys. Acta Mol. Basis Dis.* **1271**, 165–170.

189. Chen, Q., Fischer, A., Reagan, J. D., Yan, L.-J., and Ames, B. N. (1995) *Proc. Natl. Acad. Sci. USA* **92**, 4337–4341.

190. Hayakawa, M., Torii, K., Sugiyama, S., Tanaka, M., and Ozawa, T. (1991) *Biochem. Biophys. Res. Commun.* **179**, 1023–1029.

191. Gupta, K. P., van Golen, K. L., Randerath, E., and Randerath, K. (1990) *Mut. Res.* **237**, 17–27.

192. Brierley, E. J., Johnson, M. A., James, O. F. W., and Turnbull, D. M. (1996) *Q. J. Med.* **89**, 251–258.

193. Brierley, E. J., Johnson, M. A., Bowman, A., Ford, G. A., Subhan, F., Reed, J. W., James, O. F. W., and Turnbull, D. M. (1997) *Ann. Neurol.* **41**, 114–116.

194. Hayashi, J.-I., Ohta, S., Kagawa, Y., Kondo, H., Kaneda, H., Yonekawa, H., Takai, D., and Miyabayashi, S. (1994) *J. Biol. Chem.* **269**, 6878–6883.

195. Laderman, K. A., Penny, J. R., Mazzucchelli, F., Bresolin, N., Scarlato, G., and Attardi, G. (1996) *J. Biol. Chem.* **271**, 15891–15897.

196. Lin, M. T. and Beal, M. F. (2006) *Nature* **443**, 787–795.

197. Kwong, J. Q., Beal, M. F., and Manfredi, G. (2006) *J. Neurochem.* **97**, 1659–1675.

198. Wallace, D. C. (1992) *Science* **256**, 628–632.

199. Wallace, D. C., Ruiz-Pesini, E., and Mishmar, D. (2003) *Cold Spring Harb. Symp. Quant. Biol* **68**, 479–486.

200. Wallace, D. C. (2005) *Annu. Rev. Genet.* **39**, 359–407.

201. Loeb, L. A., Wallace, D. C., and Martin, G. M. (2005) *Proc. Natl. Acad. Sci. USA* **102**, 18769–18770.

202. Howell, N., Elson, J. L., Chinnery, P. F., and Turnbull, D. M. (2005) *Trends Genet* **21**, 583–586.

203. Davis, G. C., Williams, A. C., Markey, S. P., Ebert, M. H., Caine, E. D., Reichert, C. M., and Kopin, I. J. (1979) *Psychiatry Res.* **1**, 649–654.

204. Langston, J. W., Ballard, P., Tetrud, J. W., and Irwin, I. (1983) *Science* **219**, 979–980.

205. Panov, A., Dikalov, S., Shalbuyeva, N., Taylor, G., Sherer, T., and Greenamyre, J. T. (2005) *J. Biol. Chem.* **280**, 42026–42035.

206. Sherer, T. B., Betarbet, R., Testa, C. M., Seo, B. B., Richardson, J. R., Kim, J. H., Miller, G. W., Yagi, T., Matsuno-Yagi, A., and Greenamyre, J. T. (2003) *J. Neurosci.* **23**, 10756–10764.

207. Greenamyre, J. T., Betarbet, R., and Sherer, T. B. (2003) *Parkinsonism Relat. Disord.* **9**(Suppl 2), 59–64.

208. Schapira, A. H. V. (1993) Mitochondrial abnormalities in neurodegeneration and normal aging. In DiMauro, S. and Wallace, D. C., editors. *Mitochondrial DNA and Human Pathology*, Raven Press, New York, pp. 159–172.

209. Greenamyre, J. T. and Hastings, T. G. (2004) *Science* **304**, 1120–1122.

210. Greenamyre, J. T., Sherer, T. B., Betarbet, R., and Panov, A. V. (2001) *IUBMB Life* **52**, 135–141.

211. Brown, M. D., Shoffner, J. M., Kim, Y. L., Jun, A. S., Graham, B. H., Cabell, M. F., Gurley, D. S., and Wallace, D. C. (1996) *Am. J. Med. Genet.* **61**, 283–289.

212. Hutchin, T. and Cortopassi, G. (1995) *Proc. Natl. Acad. Sci. USA* **92**, 6892–6895.

213. Pyle, A., Foltynie, T., Tiangyou, W., Lambert, C., Keers, S. M., Allcock, L. M., Davison, J., Lewis, S. J., Perry, R. H., Barker, R., Burn, D. J., and Chinnery, P. F. (2005) *Ann. Neurol.* **57**, 564–567.

214. Abou-Sleiman, P. M., Muqit, M. M., and Wood, N. W. (2006) *Nat. Rev. Neurosci.* **7**, 207–219.

215. Cookson, M. R. (2005) *Annu. Rev. Biochem.* **74**, 29–52.

216. Martin, L. J., Pan, Y., Price, A. C., Sterling, W., Copeland, N. G., Jenkins, N. A., Price, D. L., and Lee, M. K. (2006) *J. Neurosci.* **26**, 41–50.

217. Klivenyi, P., Siwek, D., Gardian, G., Yang, L., Starkov, A., Cleren, C., Ferrante, R. J., Kowall, N. W., Abeliovich, A., and Beal, M. F. (2006) *Neurobiol. Dis.* **21**, 541–548.

218. Valente, E. M., Abou-Sleiman, P. M., Caputo, V., Muqit, M. M. K., Harvey, K., Gispert, S., Ali, Z., Del Turco, D., Bentivoglio, A. R., Healy, D. G., Albanese, A., Nussbaum, R., Gonzalez-Maldonado, R., Deller, T., Salvi, S., Cortelli, P., Gilks, W. P., Latchman, D. S., Harvey, R. J., Dallapiccola, B., Auburger, G., and Wood, N. W. (2004) *Science* **304**, 1158–1160.

219. Petit, A., Kawarai, T., Paitel, E., Sanjo, N., Maj, M., Scheid, M., Chen, F., Gu, Y., Hasegawa, H., Salehi-Rad, S., Wang, L., Rogaeva, E., Fraser, P., Robinson, B., George-Hyslop, P., and Tandon, A. (2005) *J. Biol. Chem.* **280**, 34025–34032.

220. Silvestri, L., Caputo, V., Bellacchio, E., Atorino, L., Dallapiccola, B., Valente, E. M., and Casari, G. (2005) *Hum. Mol. Genet.* **14**, 3477–3492.

221. Clark, I. E., Dodson, M. W., Jiang, C., Cao, J. H., Huh, J. R., Seol, J. H., Yoo, S. J., Hay, B. A., and Guo, M. (2006) *Nature* **441**, 1162–1166.

222. Gandhi, S. and Wood, N. W. (2005) *Hum. Mol. Genet.* **14**, 2749–2755.

223. McGowan, E., Eriksen, J., and Hutton, M. (2006) *TIG* **22**, 281–289.

224. Selkoe, D. J. (1997) *Science* **275**, 630–631.

225. Hauptmann, S., Keil, U., Scherping, I., Bonert, A., Eckert, A., and Muller, W. E. (2006) *Exp. Gerontol.* **41**, 668–673.

226. Keil, U., Bonert, A., Marques, C. A., Scherping, I., Weyermann, J., Strosznajder, J. B., Muller-Spahn, F., Haass, C., Czech, C., Pradier, L., Muller, W. E., and Eckert, A. (2004) *J. Biol. Chem.* **279**, 50310–50320.

227. Wallace, D. C. (2005) *Annu. Rev. Genet.* **39**, 359–407.

228. Zhu, X., Lee, H. G., Casadesus, G., Avila, J., Drew, K., Perry, G., and Smith, M. A. (2005) *Mol. Neurobiol.* **31**, 205–218.

229. Manczak, M., Park, B. S., Jung, Y., and Reddy, P. H. (2004) *Neuromolecular Med.* **5**, 147–162.

230. Davis, R. E., Miller, S., Herrnstadt, C., Ghosh, S. S., Fahy, E., Shinobu, L. A., Galasko, D., Beal, M. F., Howell, N., Parker, W. D., Thal, L. J., and Parker, W. D., Jr. (1997) *Proc. Natl. Acad. Sci. USA* **94**, 4526–4531.

231. Wallace, D. C., Stugard, C., Murdock, D., Schurr, T., and Brown, M. D. (1997) *Proc. Natl. Acad. Sci. USA* **94**, 14900–14905.

232. Hirano, M., Shtilbans, A., Mayeux, R., Davidson, M. M., DiMauro, S., Knowles, J. A., and Schon, E. A. (1997) *Proc. Natl. Acad. Sci. USA* **94**, 14894–14899.

233. Gusella, J. F. and MacDonald, M. E. (2006) *Trends Biochem. Sci.* **31**, 533–540.

234. Nasir, J., Goldberg, Y. P., and Hayden, M. R. (1996) *Hum. Mol. Genet.* **5**, 1431–1435.

235. Horton, T. M., Graham, B. H., Corral-Debrinski, M., Shoffner, J. M., Kaufman, A. E., Beal, M. F., and Wallace, D. C. (1995) *Neurology* **45**, 1879–1883.

236. Schapira, A. H. V. (1997) *Ann. Neurol.* **41**, 141–142.

237. Calkins, M. J., Jakel, R. J., Johnson, D. A., Chan, K., Kan, Y. W., and Johnson, J. A. (2005) *Proc. Natl. Acad. Sci. USA* **102**, 244–249.

238. Benchoua, A., Trioulier, Y., Zala, D., Gaillard, M. C., Lefort, N., Dufour, N., Saudou, F., Elalouf, J. M., Hirsch, E., Hantraye, P., Deglon, N., and Brouillet, E. (2006) *Mol. Biol. Cell* **17**, 1652–1663.

239. Puranam, K. L., Wu, G., Strittmatter, W. J., and Burke, J. R. (2006) *Biochem. Biophys. Res. Commun.* **341**, 607–613.

240. Milakovic, T. and Johnson, G. V. (2005) *J. Biol. Chem.* **280**, 30773–30782.

241. Seong, I. S., Ivanova, E., Lee, J. M., Choo, Y. S., Fossale, E., anderson, M., Gusella, J. F., Laramie, J. M., Myers, R. H., Lesort, M., and MacDonald, M. E. (2005) *Hum. Mol. Genet.* **14**, 2871–2880.

242. La Spada, A. R. and Morrison, R. S. (2005) *Neuron* **47**, 1–3.

243. Bae, B. I., Xu, H., Igarashi, S., Fujimuro, M., Agrawal, N., Taya, Y., Hayward, S. D., Moran, T. H., Montell, C., Ross, C. A., Snyder, S. H., and Sawa, A. (2005) *Neuron* **47**, 29–41.

244. Liu, J., Lillo, C., Jonsson, P. A., Vande, V. C., Ward, C. M., Miller, T. M., Subramaniam, J. R., Rothstein, J. D., Marklund, S., Andersen, P. M., Brannstrom, T., Gredal, O., Wong, P. C., Williams, D. S., and Cleveland, D. W. (2004) *Neuron* **43**, 5–17.

245. Boillee, S., Vande, V. C., and Cleveland, D. W. (2006) *Neuron* **52**, 39–59.

246. Pasinelli, P., Belford, M. E., Lennon, N., Bacskai, B. J., Hyman, B. T., Trotti, D., and Brown, R. H., Jr. (2004) *Neuron* **43**, 19–30.

247. Vaux, D. L. and Strasser, A. (1996) *Proc. Natl. Acad. Sci. USA* **93**, 2239–2244.

248. Williams, G. T. (1994) *Biochem. Cell Biol.* **72**, 447–450.
249. Wyllie, A. H. (1997) *Eur. J. Cell Biol.* **73**, 189–197.
250. Collins, M. K. L. and Rivas, A. L. (1993) *Trends Biochem. Sci.* **18**, 307–309.
251. Haluska, F. G., Tsujimoto, Y., Russo, G., Isobe, M., and Croce, C. M. (1989) *Progr. Nucl. Acid Res. Mol. Biol.* **36**, 269–280.
252. Nowell, P. C. and Croce, C. M. (1990) *Am. J. Clin. Pathol.* **94**, 229–237.
253. Liu, Y., Cortopassi, G., Steingrimsdottir, H., Waugh, A. P., Beare, D. M., Green, M. H., Robinson, D. R., and Cole, J. (1997) *Environ. Mol. Mutagen.* **29**, 36–45.
254. Yang, E. and Korsmeyer, S. J. (1996) *Blood* **88**, 386–401.
255. Kroemer, G., Zamzami, N., and Susin, S. A. (1997) *Immunol. Today* **18**, 44–51.
256. Kroemer, G., Dallaporta, B., and Resche-Rigon, M. (1998) *Annu. Rev. Physiol.* **60**, 619–642.
257. Bernardi, P. (1996) *Biochim. Biophys. Acta* **1275**, 5–9.
258. Liu, X., Kim, C. N., Yang, J., Jemmerson, R., and Wang, X. (1996) *Cell* **86**, 147–157.
259. Kluck, R. M., Bossy-Wetzel, E., Green, D. R., and Newmeyer, D. D. (1997) *Science* **275**, 1132–1136.
260. Danial, N. N. and Korsmeyer, S. J. (2004) *Cell* **116**, 205–219.
261. Yan, N. and Shi, Y. (2005) *Annu. Rev. Cell Biol.* **21**, 35–56.
262. Green, D. R. and Kroemer, G. (2004) *Science* **305**, 626–629.
263. Halestrap, A. P. (2006) *Biochem. Soc. Trans.* **34**, 232–237.
264. Schendel, S. L., Xie, Z., Montal, M. O., Matsuyama, S., Montal, M., and Reed, J. C. (1997) *Proc. Natl. Acad. Sci. USA* **94**, 5113–5118.
265. Antonsson, B., Conti, F., Ciavatta, A., Montessuit, S., Lewis, S., Martinou, I., Bernasconi, L., Bernard, A., Mermod, J. J., Mazzei, G., Maundrell, K., Gambala, F., Sadoul, R., and Martinou, J. C. (1997) *Science* **277**, 370–372.
266. Neuzil, J., Wang, X. F., Dong, L. F., Low, P., and Ralph, S. J. (2006) *FEBS Lett.* **580**, 5125–5129.
267. Jacobson, M. D., Burne, J. F., King, M. P., Miyashita, T., Reed, J. C., and Raff, M. C. (1993) *Nature* **361**, 365.
268. Karbowski, M., Norris, K. L., Cleleand, M. M., Jeong, S. Y., and Youle, R. J. (2006) *Nature* **443**, 658–662.
269. Delivani, P., Adrain, C., Taylor, R. C., Duriez, P. J., and Martin, S. J. (2006) *Mol. Cell* **21**, 761–773.
270. Griffiths, A. J. F. (1992) *Annu. Rev. Genet.* **26**, 351–372.
271. Griffiths, A. J. F. (1995) *Microbiol. Rev.* **59**, 673–685.
272. Guarente, L. and Picard, F. (2005) *Cell* **120**, 473–482.
273. Piper, P. W. (2006) *Yeast* **23**, 215–226.
274. Yang, X. and Griffiths, A. J. F. (1993) *Mol. Gen. Genet.* **237**, 177–186.
275. Bertrand, H. (2000) *Annu. Rev. Phytopathol.* **38**, 397–422.
276. Lorin, S., Dufour, E., and Sainsard-Chanet, A. (2006) *Biochim. Biophys. Acta* **1757**, 604–610.
277. Schardl, C. L., Pring, D. R., and Lonsdale, D. M. (1985) *Cell* **43**, 361–368.
278. Chase, C. D. (2006) *Trends Plant Sci.* **11**, 7–9.

279. Touzet, P. and Budar, F. (2004) *Trends Plant Sci.* **9**, 568–570.
280. Budar, F., Touzet, P., and De Paepe, R. (2003) *Genetica* **117**, 3–16.
281. Chase, C. D. (2007) *Trends Genet.* **23**, 81–90.
282. L'Homme, Y., Stahl, R. J., Li, X.-Q., Hameed, A., and Brown, G. G. (1997) *Curr. Genet.* **31**, 325–335.
283. Palmer, J. D. and Herbon, L. A. (1988) *J. Mol. Evol.* **28**, 87–97.
284. Hanson, M. R. (1991) *Annu. Rev. Genet.* **25**, 461–486.
285. Moneger, F., Smart, C. J., and Leaver, C. J. (1994) *EMBO J.* **13**, 8–17.
286. Levings, C. S., III, and Vasil, I. K. (1995) *The Molecular Biology of Plant Mitochondria*, Kluwer Academic Publishers, Dordrecht.
287. Spassova, M., Moneger, F., Leaver, C. J., Petrov, P., Atanassov, A., Nijkamp, H. J., and Hille, J. (1994) *Plant Mol. Biol.* **26**, 1819–1831.
288. Smart, C. J., Moneger, F., and Leaver, C. J. (1994) *Plant Cell* **6**, 811–825.
289. Warmke, H. E. and Lee, S. L. J. (1978) *Science* **200**, 561–562.
290. Lee, S. L. J. and Warmke, H. E. (1979) *Am. J. Bot.* **66**, 141–148.
291. Moneger, F., Mandaron, P., Niogret, M. F., Freyssinet, G., and Mache, R. (1992) *Plant Physiol.* **99**, 396–400.
292. Newton, K. J. and Walbot, V. (1985) *Proc. Natl. Acad. Sci. USA* **82**, 6879–6883.
293. Lind, C., Hallden, C., and Moller, I. M. (1991) *Plant Physiol.* **83**, 7–16.
294. He, S. C., Abad, A. R., Gelvin, S. B., and Mackenzie, S. A. (1996) *Proc. Natl. Acad. Sci. USA* **93**, 11763–11768.

8

MITOCHONDRIAL DNA SEQUENCING AND ANTHROPOLOGY

Summary

A large number of base changes in mtDNA are neutral, either because they do not cause amino acid substitutions in proteins, or because amino acid substitutions are conservative, or because they occur in noncoding regions of the mitochondrial genome such as the so-called control region (with the exception of some specific promoter or origin sequences). They can be detected as restriction fragment length polymorphisms (RFLPs) or by explicit sequencing.

A large database of such polymorphisms has accumulated over the past decade. These polymorphisms have served in a variety of interesting studies and investigations. All human ethnic groups have highly distinguishing haplotypes. The evolution of these haplotypes and their dispersion over the globe have have given invaluable support to hypotheses about human evolution and the migratory patterns of human populations starting with a subpopulation in Africa which is increasingly acknowledged to constitute the origin of modern humans.

Even within a narrowly defined ethnic group, there are polymorphisms that connect pedigrees and can also serve in forensic investigations. Its small size, complete sequence, and the presence of perhaps thousands of copies per cell make mtDNA a unique target and frequently the only nucleic acid preserved in sufficient quantities to permit PCR amplifications for analysis.

Mitochondria, Second Edition, by Immo E. Scheffler

8.1 INTRODUCTION

In 1987 Cann, Stoneking and Wilson (1) published a paper on the analysis by restriction mapping of mtDNAs from 147 people, drawn from five widely spaced geographic populations (34 Asians, 21 Australian aboriginals, 26 aboriginal New Guineans, 46 Caucasians, and 20 Africans). The scientific impact of this paper was enormous. The application of methods from molecular genetics to the study of primate and human evolution, with a focus on mitochondrial DNA, had been suggested by Brown (2), and several publications by Brown, Wilson, and colleagues had preceded the landmark 1987 paper (3–5). However, the large number and diversity of people represented in this paper provided convincing evidence for the potential strength of this approach in addressing questions about the relatively recent human evolution, about the origins of modern humans, and about the migratory patterns of our ancestors. The short abstract of the paper contained the following eye-catching and provocative statement: "All these mitochondrial DNAs stem from one woman who is postulated to have lived about 200,000 years ago, probably in Africa." Unfortunately, the popular press took this statement literally, and the misconception arose that she was the single ancestor of us all. The catchy headlines derived from this study certainly promoted interest in this field and popularized human evolution among the molecular biologists who can relate more easily to a DNA sequence than to a jaw bone.

In the past two decades, not only have the analyses been vastly refined by the use of many more restriction enzymes, the application of the powerful polymerase chain reaction, and actual sequencing of defined regions, but also human mtDNAs from the remotest areas of the globe have been analyzed, and one can only imagine what it would be like to explain to a Chukchi from north easternmost Siberia or a Yanomama from South America why their mtDNA is of such interest, and how it proves that ultimately they all had descended from a mother in Africa.

The biological and molecular basis for the value and fruitfulness of mitochondrial DNA sequence variations is derived from three facts. There is a strictly maternal mode of inheritance, hence no scrambling of information by recombination during sexual reproduction. The rate of evolution of mtDNA is estimated to be an order of magnitude greater than that of nuclear DNA, thus providing resolution on a time scale that is appropriate for human evolution: A few thousand years are a sufficiently long time to generate informative variants. Finally, mtDNA represents a distinct molecule that has been completely sequenced. It is abundant in cells (high copy number) and is therefore experimentally easily accessible.

The maternal mode of inheritance is an important consideration. There is universal agreement that >>99% of mtDNA in an individual is derived from the mother's egg, and no complications are expected in the analysis of a pedigree of three or more generations. However, phylogeny on an evolutionary time scale could well be influenced by the rare transmission of mtDNA from the paternal side, and a challenge of the current dogma was issued from a broad consideration of fertilization and the fate of the sperm mitochondria (6), and also from some very well-controlled studies with interspecies hybrid mice (7). Since there was evidence for biparental inheritance in interspecies hybrids, it became urgent to prove that this was the exception rather than the rule, and some such proof has been provided recently (8). There is no argument over the observation that sperm mitochondria are found in the zygote shortly after fertilization. The most recent studies suggest, however, that paternal mtDNA is specifically eliminated in intraspecies crosses by a ubiquitination-dependent mechanism (9)–(11). In interspecies crosses, in contrast, the mechanism apparently fails to recognize the parental mitochondria, and therefore they escape destruction and elimination. A rare case of paternal inheritance of mtDNA in humans has been reported (12). The question of biparental inheritance and the possibility of recombination in humans (mammals) continues to be raised periodically, and a recent expert evaluation of the available evidence can be found in the review by Elson and Lightowlers (13). The authors correctly emphasize the importance of the paradigm of mtDNA clonality, and they find none of the published evidence in favor of mtDNA recombination in human lineages to be compelling.

The accumulation of base changes in mtDNA occurs on a time scale sufficiently short to generate polymorphisms within populations of the same broad ethnic category (e.g., Caucasians); there are polymorphisms/haplotypes distinguishing the various human population groups; and there are polymorphisms within a single individual, generated by somatic mutations, or when inherited as heteroplasmy. Heteroplasmy in relation to mitochondrial diseases has been discussed in Chapter 8. In an apparent paradox, most individuals are homoplasmic. Homoplasmy is either (a) the result of a severe "bottleneck" in oocyte development (see Chapter 8) or (b) more apparent than real, depending on the sensitivity with which the analysis or search for heteroplasmy is carried out (14).

8.2 HUMAN EVOLUTION

While sequencing the mtDNAs of large numbers of diverse humans provides hard and undisputed data, the interpretation of the data has created a great deal of controversy. At first it looks deceptively simple. In their original eye-opening paper, Wilson and his students simply compared the number of mutations in various peoples (1). A population that has been in existence and in place for a long time should, upon sampling, reveal a large diversity of mitochondrial genomes, assuming that the molecular clock (nucleotide changes per unit time or per generation) was ticking away steadily. A population established only recently by a small number of founders would be expected to be relatively homogeneous with respect to mtDNA sequences. This was indeed found: Africans had by far the greatest diversity, a finding that was interpreted in terms of a mitochondrial Eve out-of-Africa. A second aspect of the same study was to use estimated rates of the molecular clock to deduce that the total number of mutations observed in the African population would require somewhere between 140,000 and 280,000 years. Criticisms soon appeared, related to several methodological details of the study. Restriction analysis is a crude way of sampling a sequence; and with a small sample size, valuable information could be missed or misinterpreted. African Americans had been used to represent Africans, with a possibility of bias. Finally, the formal methodology to build the phylogenetic tree was questioned.

With time, many of the criticisms were refuted. The sample size and the number of representatives from each geographic region could be increased significantly, including contributions from other groups of researchers. The sequence analysis could be improved by the inclusion of more restriction enzymes and even complete sequences. The mtDNA segment favored in these studies was the control region, a noncoding region containing promoter elements and origins of DNA replication (see Chapter 4) which evolves particularly quickly. The massive amounts of data were analyzed with a newly developed computer program called PAUP, or Phylogenetic Analysis Using Parsimony. Such a program is designed to generate the most parsimonious phylogenetic trees, and this is where the controversy starts. When Wilson and his colleagues followed their original 1987 paper with another much more detailed analysis in 1991 (15), they concluded that our common mtDNA ancestor indeed lived in Africa about 200,000 years ago. Lost in the detail, but discovered by a subsequent critic, was the untenable splitting of the 25!Kung bushmen of Africa on the deepest branches of the tree, in the face of their seemingly obvious relatedness. Recalculations showed that the computer program had not succeeded in finding the most parsimonious tree and that there is no absolutely unique and most parsimonious solution (at least with a limited computation time). Even with the best solutions, the tree has no obvious roots, or roots pointing unambiguously to Africa.

A recent reexamination of the mtDNA data of Cann et al. (1) and of Vigilant et al. (15) was attempted by Wills (16). This highly technical paper introduces a method referred to as "topiary pruning," a term that describes the attempt to reveal the true shape of an ornamental topiary tree by removing the obscuring "new growth." It is argued that a phylogenetic tree may have similar new growth that can obscure significant branches. The "new growth" on a phylogeentic tree may be the result of recent, repeated mutations in different lineages at hypermutable sites. The pruning produces data sets with a reduced number of taxa, and this simplification in turn makes the data sets more amenable to analysis by a variety of statistical methods. A combination of such methods (pruning, weighting, bootstrapping) is expected to yield a tree that is closer to the "true" tree. While Wills is careful to state that the African origin of the human mitochondrial tree is not proved by this approach, the interpretation of the pruned tree adds strength to the argument in favor of an African origin.

There are additional disputes about the calibration of the clock. Is the rate of change the same at all sites where changes have been observed? Do non-coding regions (D-loop) change more rapidly than coding regions? Are the changes in the third positions of the codons occurring at a faster rate? What about the relative rates of transitions versus transversions in a DNA sequence? The challenge was taken up by Hasegawa et al. (17) to predict a more accurate time scale for human mtDNA, making the additional assumption that the divergence of humans from our closest primate ancestor, the chimpanzee, was about 4 million years ago. The emergence of modern humans out of Africa was moved to an even more recent date: 89,000 ± 69,000 years ago. Such a result also put the authors squarely at odds with the multiregional hypothesis of modern human origin, which has gained adherents in recent years. Uncertainties and controversies related to the mutation rate in mitochondria have by no means been resolved to everyone's satisfaction, and as a result the estimates for the age of the mitochondrial Eve range from <100,000 years to almost a million years (see below).

After the African mitochondrial Eve was declared "wounded but not dead yet" (18) in 1992, based on the ambiguity surrounding the construction of phylogenetic trees with mtDNA sequences, supporters of the hypothesis have assembled independent support. The controversy has at times been heated. An attempt by a respected expert to debunk the "myth of Eve" (19) was followed a few issues later by a rebuttal by another expert accusing the first of perpetuating another myth (20). As Ayala states succinctly, the coalescence of the mtDNA sequences from over 100 ethnically diverse individuals to one ancestral sequence, that of the "mitochondrial Eve," is a property of all gene genealogies, but does not mean that all humans descended from a single mother. While Wilson and his colleagues probably also did not mean to be taken literally, the statement had an attention-catching appeal, and the popular press made the most of it. If not a single mother, there is still the question about the size of the pool of humans from which the present-day human groups are descended. The current controversy therefore centers around two or three problems. Where is the mtDNA genealogical tree rooted? Evidence points strongly toward Africa, but it is not yet conclusive. The second question is concerned with the existence of a bottleneck, as well as the size of the bottleneck of those individuals who are the ancestors of all modern humans. Finally, there may have been multiple bottlenecks with subsequent mixing of the surviving populations. The arguments about multiregional models require extensive migrations and interbreeding between populations to account for a single gene pool (19).

Ingenious independent analyses have strengthened the out-of-Africa hypothesis. To establish the true root of the phylogenetic tree, the true human mtDNA ancestor, one must choose an outgroup sequence from a different species more ancient than humans, but still closely related. The chimpanzee mtDNA has served such a purpose; but in the control region used routinely, the rate of change may be too fast and too extreme. The creation of multiple and parallel sequence changes can complicate and obscure the analysis. A novel finding was exploited elegantly by Zischler et al. (21) to identify a superior outgroup for rooting the human mtDNA tree. Following the original discovery of Fukuda et al. (22), several investigators have confirmed that the human nuclear genome contains several hundred copies of sequence fragments closely resembling mtDNA. Such pseudogenes have also been found in other primates and rodents, and their distribution and sequence alterations make them useful molecular fossils in the nucleus (23). Zischler et al. (21) specifically searched for and found a nuclear copy of the human mitochondrial D-loop which had been transferred to the nucleus after the divergence of humans from the chimpanzee, but before the "growth" of the human mitochondrial phylogenetic tree. The nuclear D-loop sequence became a superior out group for rooting the tree in Africa (24).

The mtDNA data assembled by many groups—prominently among them the group at Emory University, as well as Stoneking and colleagues at Pennsylvania State University—have generally supported the following scenario for the expansion of *Homo sapiens* around the globe (25): The migration started out of Africa about 100,000 to 200,000 years ago. It split somewhere in the middle east into a group headed for central and northern Europe (~50,000 years ago), and another group headed toward southern and southeast Asia. Migrations from present day China via the Indonesian Archipelago toward New Guinea and Australia also occurred a long time ago. Somewhat later the populations expanded north along the eastern rim of Asia, and eventually groups reached the Bering Strait to cross into North America. There may have been several crossings of the Bering Strait (see below). Expansion(s) from Alaska to the south populated in order North America, Central America, and South America. Quite recent in human history, another wave of expansion occurred from the islands surrounding New Guinea and North Australia into the northern Pacific (Polynesia and Hawaii).

Of course, many of these deductions were not completely new. Data from paleoanthropology and archeology had already been used to make broadly similar deductions. Linguistic comparisons have also contributed greatly. Nevertheless, mtDNA sequence data can often provide detailed and convincing support. For example, a 9-bp deletion in the small intergenic region between COXII and $tRNA^{Lys}$ is believed to have arisen in Asia about 50,000 years ago. Redd et al. (26) have traced a specific set of control region mutations ("the Polynesian motif") associated with the 9-bp deletion to a Taiwan origin (about 6000–8000 years ago), and from there across the Philippines, Indonesia, along the coast of New Guinea, and finally into Polynesia (26). These molecular data are in excellent agreement with the postulate of a southeast Asian origin of Polynesians based on archeological and linguistic evidence.

The contention about rooting the human phylogenetic tree may have approached a consensus, or at least the African root has acquired the support of the majority. It is yet another issue to date the period of the existence of the mitochondrial Eve. Determining the age of Eve is important, because an accurate estimate of the time span to the present serves in arguments about other theories, specifically the multiple origins theory. Since the fossil record reveals the existence of *Homo erectus* as long as at least a million years ago, and such evidence is found in Europe, the Middle East, and Asia, one is forced to conclude from the genetic evidence that the modern humans (our ancestors) displaced all of these other human populations completely without any detectable genetic exchange. If the lowest estimate for the age of the mitochondrial Eve is accepted (<100,000 years), this scenario would be quite astonishing. On the other hand, if the mitochondrial Eve lived more than 500,000 years ago, a new set of questions and theories can be entertained about multiple origins, shifts in the composition of the human gene pool, and selective pressures that might have contributed to the unusual evolutionary pattern. A thoughtful reexamination of existing data sets by Wills (27) comes to the conclusion that Eve may be between 436,000 and 806,000 years old, an estimate that is also very dependent on the estimate of the time of the split between the human and chimpanzee lineages.

In the past decade there has been an avalanche of sequence data for mtDNA, the Y chromosome, and autosomes from a wide range of human donors. An up-to-date analysis of the conclusions has appeared from Garrigan and Hammer (28). These authors clearly distinguish two epochs in the evolution of *H. sapiens*. The first, starting about 2 million years ago, leads to the emergence of the "anatomically modern human." The second epoch covers the geographic distribution of the modern humans after their emergence from Africa. Sequence analyses are used to construct gene trees, and these in turn can be interpreted to estimate the "time to the most recent ancestor" (TMRCA), with certain assumptions about the molecular clock for accumulating nucleotide changes. The mtDNA tree from the complete sequences of a global sample of humans yields a TMRCA of ~170,000 years. From the tree based on the nonrecombining regions of the Y chromosome the estimate for the TMRCA is roughly half this value, and various explanations have been offered (see reference 28). As more and more sequence data from loci on the X chromosome or autosomes become available, the TMRCA can be pushed back to 500,000–800,000 years and thus include *H. neanderthalensis, H. heidelbergensis*, and other now extinct lineages of *H. sapiens*. Some of these results may contradict the "single origin model" derived from mtDNA and Y chromosome trees, and the multiregional evolution model may gain favor in the future. The interested reader is directed to this highly technical paper for a consideration of alternate models. Some support for the claim that modern humans from Africa actually had "close physical contact" (does that mean sex?) with archaic humans (*H. erectus*) comes from an unexpected direction. A genetic analysis of lice revealed that there are two subspecies, one thought to have lived on *H. erectus* and the other on *H. sapiens*. After a relatively long period of isolation of both the lice and the humans, physical contact between these two groups of humans before *H. erectus* became extinct must have taken place to explain the present-day distribution of these louse subspecies in the world (29, 30).

Mitochondrial DNA sequence analysis has been perfected to the level where analyses have become possible with samples from a 4000-year-old mummy of an Egyptian priest and from a 7000-year-old human brain from a Florida peat bog; and if the bones are ever released for analysis before they are re-buried, there are paleobiologists waiting to obtain samples from a skeleton found on the northwest Washington coast thought to be 9000 years old. One of the most impressive achievements in the study of human evolution by the study of human DNA sequences was reported by Krings et al. in 1997 (31). The group in Munich, with independent confirmation by a group in Pennsylvania, reported mtDNA sequences from a Neanderthal person. The Neanderthals were a race of humans roaming through Europe and Western Asia from about 300,000 to 30,000 years ago. Considerable discussion had centered around the question whether *Homo neanderthalensis* was a distinct and separate species from *Homo sapiens*. Their range overlapped in Europe during the more recent millenia, and the possibility of a mixing of the gene pools by interbreeding had not been convincingly excluded. In the landmark paper published by the Paabo and Stoneking groups the definite conclusion was reached that the Neanderthals went extinct, leaving no trace of their mtDNA in modern humans. Concentrating on the D-loop region, sequence comparisons and phylogenetic analyses were made, including a large number of representatives of modern humans as well as the chimpanzee. The divergence of the Neanderthal from modern humans was estimated from the mtDNA analyses to have occurred about 500,000 years ago, in satisfactory agreement with other estimates based on the archeological record.

Krings et al. (31) is a landmark for its conclusions, but it probably also set the standard for future studies of ancient DNA. The DNA was retrieved from a fossilized bone of an individual living between 100,000 and 30,000 years ago. As discussed in more detail in this paper and elsewhere, measurements of amino acid racemization can give reliable estimates about preservation and the feasibility of obtaining sufficient DNA from a fossil specimen, and the Neanderthal bone may be near the limit of what constitutes a useful source. The quantitative considerations also indicate that only mtDNA sequences can be expected to be recovered for PCR-based amplification, because it is several orders more abundant than nuclear DNA. In a commentary on this paper in the same issue of *Cell*, T. Lindahl discusses the "fiasco of DNA from insects in amber," with reference to studies suggesting that amber is a very poor material for the preservation of DNA (but not of the insect exoskeleton) (32). Thus, claims of DNA amplification from million-year-old insects in amber may continue to stimulate Hollywood, but they cannot stand up to rigorous scrutiny. Because such skepticism is clearly indicated, the Neanderthal mtDNA paper went to extraordinary lengths to include controls and to obtain repeatable results, even when conducted in a distant, second laboratory. Since this paper appeared, more partial sequences from Neanderthal mtDNA have appeared, confirming the original conclusions in favor of the "rapid replacement" model for the origin of modern humans and the absence of any interbreeding between Neanderthals and modern humans (33, 34).

Another, more limited application of mtDNA sequence analysis to human evolution and history is also instructive, but ultimately baffling to the outsider because of still unresolved controversies. Anthropologists have been curious for some time about the following questions: When were the Americas populated by humans, and where did these people originate? Before the advent of molecular genetic analyses, several time-honored approaches had suggested answers, which cannot be discussed here in detail. The more conventional archeological studies use the various artifacts (pottery, tools, weapons) to trace lineages based on designs, patterns, materials used, and so on. A more recent development is the application of the tools of linguistics to the comparative analysis of the languages spoken by the various native Americans (North and South America). Such an analysis had suggested three distinct linguistic divisions: Amerind (spoken by American Indians in North, Central, and South America), Eskimo-Aleut (spoken by the Inuit (Eskimo) and Aleutian Islanders), and Na-Dene (spoken by the by people of the Northeast Coast of Canada and the United States as well as the Navajo and Apache, believed to have migrated south to their present location around 1000 BC). An interpretation was that there were three distinct waves of immigration across the Bering Straits, in agreement with other archeological data, such as distinctive features of molar shapes. Mitochondrial DNA data assembled by Wallace, Torrioni, and colleagues (reviewed in reference 25) not only confirmed the Asian origin of the native Americans, but appeared to lend strong support to the three waves of migration hypothesized by the linguists. Four haplogroups were found based on extensive restriction mapping, and the relative abundance of these groups could be interpreted in terms of three distinct migrations separated in time by thousands of years.

Somewhat ironically, an ex-student of Wallace has collected additional data that challenge the interpretation of his mentor (35). The new findings of all four haplotypes among the three linguistic groups (mtDNA samples from 1300 American Indians) appear to fit better to a scenario in which a single wave of immigration brought women with all four haplotypes across the Bering Strait. Subsequent migrations, along with the possibility of severe temporary reductions in one or more local populations due to environmental influences, may have caused the loss of particular haplotypes (another bottleneck), resulting in the current distributions.

The discussion continues. It is not the intent here to present the definitive conclusion, if there is one, but to make the reader appreciate the wealth and diversity of ideas derived from this research.

There are now well in excess of 2100 mitochondrial DNA sequences in the Human Mitochondrial Genome Database (mtDB) (http://www.genpat.uu.se/mtDB), including more than 1544 complete sequences (36). The database also lists more than 3000 sequence polymorphisms from human population groups all over the globe. This is a treasure trove for human population geneticists and medical researchers interested in correlations between mtDNA haplotypes and phenotypes. The "harvesting of the fruit from the human mtDNA tree" is in full swing (37), with new disciplines, "archeogenetics," "phylogeography," and so on, being created. Another exhaustive database is discussed by Attimonelli et al. (http://www.hmdb.uniba.it) (38).

Regional variation in mtDNA haplotypes has generally been attributed to genetic drift and migration. The question has been raised as to whether some of these sequence variations in coding regions might have functional significance (39) and whether a selective advantage could arise for a population in a specific environment (40). The authors conclude that some mtDNA lineages found in European, Asian, Siberian, and Native North Americans might reflect adaptations to a cold climate. Such an advantage could be achieved by a slight decrease in the coupling of electron transport and ATP synthesis in favor of thermogenesis. However, at this time such a conclusion must remain quite speculative.

8.3 PRIMATE EVOLUTION

Phylogenetic analysis of mitochondrial DNA sequences can of course be pushed back in time to include other species. No significant surprises about evolutionary history may be expected when comparing different vertebrates, or even different mammals. However, for calibrating molecular clocks and for establishing more precisely the time of divergence of the nonhuman apes, a comparison of the mtDNA sequences of humans, chimpanzee, pigmy chimpanzee, gorilla, and orangutan has been very informative. Horai et al. (41) found these genomes to be highly homologous, with a synonymous substitution rate of 3.89×10^{-8}/site per year, if one assumes that the orangutan and the African apes diverged 13 million years ago. In the D-loop region this rate was almost twice as high. The analysis also included three human mtDNA sequences from Africans, Europeans, and Japanese to shed light on the origin on modern humans. From the estimated substitution rates the age of the last common ancestor of modern humans was calculated to be 143,000 ± 18,000 years, and the hypothesis of the African origin of *Homo sapiens sapiens* received additional support.

The most detailed examination of the noncoding control region (human nucleotide positions 16053–16465) of the nine taxa of African or African-derived hominoids (humans, chipanzees, bonobos and gorillas) has been undertaken by Gagneux et al. (42), who examined 1158 unique haplotypes including 811 humans from around the world. A phylogenetic tree was constructed which also included orangutans and the newly established sequence from *Homo neanderthalensis* (Figure 8.1). For details on the computer analysis and related analytical methods, the reader is referred to the original paper. The final tree shown was also subjected to various degrees of topiary pruning

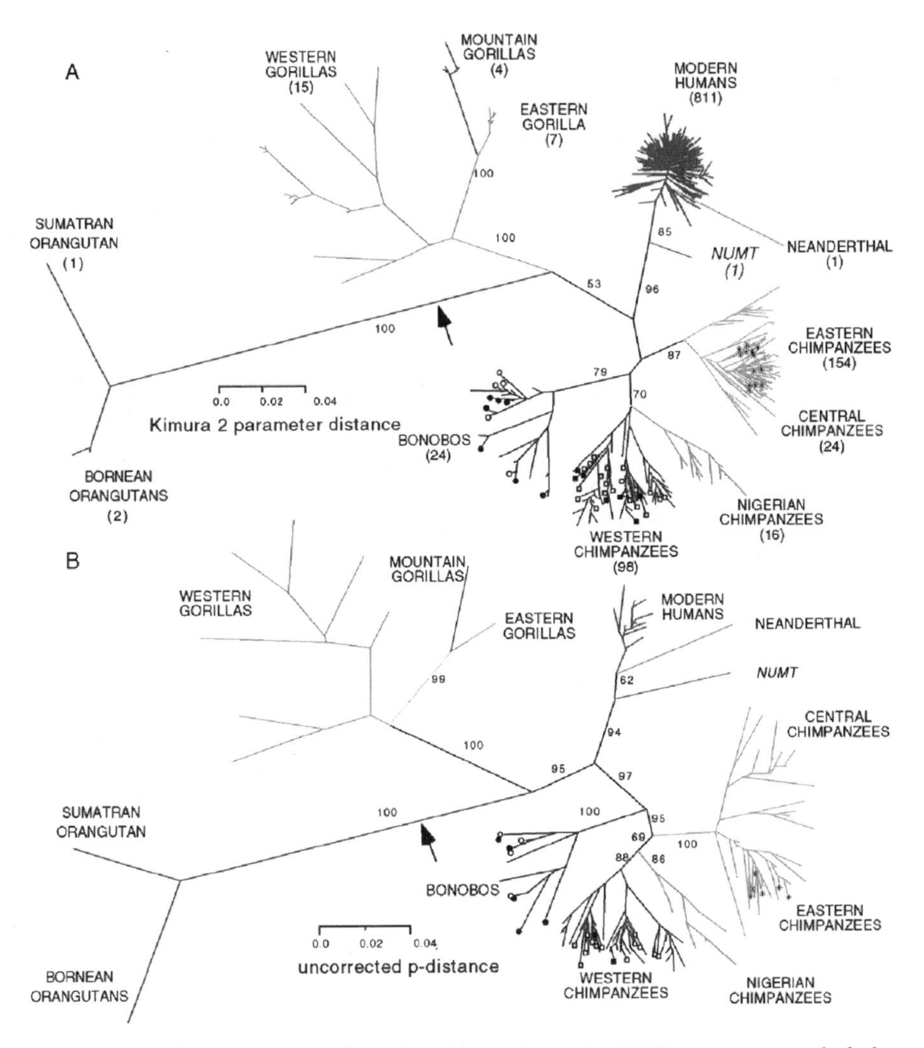

Figure 8.1 Primate evolution based on the analysis of mtDNA. An unrooted phylogram of the neighbor-joining tree of 1158 different mitochondrial control region sequences was constructed (A) and then subjected to topiary pruning (B). See Gagneux et al. (42) for details. [The figure was generously provided by Drs. Gagneux and Wills prior to publication (with permission).]

(16), but only one intermediate level is shown here. The tree illustrates in dramatic fashion some previous conclusions, and it adds some novel insights. It is clearly supportive of the closer relationship between humans and chimpanzees relative to the other great apes. Most remarkably, it reveals deep branches between chimpanzees in various African locations, corresponding to the existence of several subspecies, and a previously unrecognized subclade of chimpanzees in Nigeria and northern Cameroon. There is significantly more variation between the major chimpanzee clades than between the entire human population. The appearance of the tree and specifically the relatively shallow human branches are consistent with the idea of a demographic bottleneck in human evolution—in other words, consistent with the idea that the present human lineage arose from a small effective population of relatively recent origin. A mitochondrial DNA fragment that entered the nuclear genome was also included: It originated after the divergence of humans from the great apes, and even after the divergence of present humans from the Neanderthals.

An interesting corollary to these studies was published by Kenyon and Moraes (43). These authors used human ρ^0 cells (mtDNA-less) in fusions with enucleated cells from the hominoid apes chimpanzee, pigmy chimpanzee, gorilla, and orangutan to produce "xenomitochondrial cybrids." Cybrids with chimpanzee or gorilla mtDNA and a human nucleus could respire and carry out oxidative phosphorylation. In contrast, the orangutan mtDNA and the human nuclear genome were no longer compatible. The results support the notion that the nuclear and mitochondrial genomes have to co-evolve at a significant number of loci in order to maintain the critical interactions between OXPHOS proteins encoded by the nucleus and proteins encoded by the mitochondria. Thus, the more closely related ape mt genomes could be functionally expressed to interact with human nuclear gene products, but the orangutan mt genome had evolved far enough since the distant separation of the lineages to yield mitochondrial proteins which presumably could no longer interact with human proteins in the inner mitochondrial membrane. It will now be a challenge to distinguish simple polymorphisms with no structural–biochemical consequences from those nucleotide changes (and the resulting amino acid substitutions) which contribute to the defective interactions between nuclear and mitochondrial gene products. Unfortunately, the use of recombinant (human–primate) mtDNA molecules is technically still impossible. On the other hand, the available crystal structures for complexes III and IV may guide the search for protein–protein interaction involving nuclear-coded and mitochondrial peptides and may perhaps reveal critical contact sites.

8.4 HUMAN Y CHROMOSOME VARIATION

A crucial, inherent advantage of the mitochondrial genome in the analyses described above is (a) the absence of recombination, facilitating the construction of phylogenies, and (b) the estimation of the time intervals elapsed between the different branch points, assuming that the molecular clock is running at a constant rate. Another chromosome that has long been recognized not to participate in recombination is the Y chromosome. The statement has to be qualified, since a portion at the tip of the short arm, the pseudoautosomal region, does participate in chiasma formation and recombination. The male-specific portion of the Y chromosome, however, offers an alternative and complementary system for the study of human evolution and migratory pathways, and data on polymorphisms on this chromosome have been collected and interpreted extensively. It is self-evident that the Y chromosome is inherited along the paternal lineage, since it is the male-specific sex chromosome containing the critical gene (SRY) encoding the testis determining factor (44). By comparison with mtDNA, the Y chromosome is huge, comprising approximately 60,000 kb (60 Mb) of DNA. The great bulk of Y chromosome DNA, especially the long arm (Yq), is made up of constitutive heterochromatin consisting of various families of repetitive DNA with no obvious genetic information. It has taken some time to define unique loci and to establish polymorphisms at these loci in human populations (45–50).

A detailed discussion of human Y chromosome polymorphisms would go beyond the scope of this monograph. On the other hand, it is worthwhile to consider how far the deductions from the analysis of mtDNA have been confirmed by similar studies with Y chromosome sequences, and what kinds of unresolved discrepancies exist. Independent confirmations of the reconstruction of human evolution are of course a source of intellectual satisfaction, and they strengthen the conclusions drawn. Discrepancies, on the other hand, raise issues about methodology, interpretation, and limitations of computer analyses, but in the long run they can lead to new insights and progress.

A comprehensive and recent description of the geographic distribution of human Y chromosome variation can be found in the publication by Hammer et al. (50), which includes data from 1500 males from 60 population groups, subdivided into 15 major groups based on geographic, linguistic, and historical information. In agreement with other genetic analyses, and specifically the mtDNA data, the Y-chromosome haplotype distributions show that sub-Saharan African populations are distinct from the rest of the world, and they exhibit the highest within-group diversity. Globally, the variants found outside of Africa represent subsets of the haplotypes found within Africa. Subsequent results from a more extensive sequencing of unique regions of the human Y chromosome also supported the idea that "Adam was an African" (51–53). Hammer et al. were careful to point out that while such a within-group and among-group pattern also supports the hypothesis of the origin of modern humans from an older African population—with migrations of subgroups leading to the population of distal sites, along with more limited genetic diversity elsewhere resulting from bottlenecks—it may nevertheless be an overly simplistic view. In the absence of additional information, the genetic data are equally consistent with multiple short-range and long-range migrations, genetic drift due to founder effects, and small population sizes (near extinction). Africa may well have been the origin of more than one migration originating from different parts of Africa. Notably, the Y chromosome data also are consistent with the relatively recent (~100,000 years) origin of our common ancestors, but may be a consequence of a small effective population size throughout the Late Middle Pleistocene, rather than a true definition of the period of origin of *Homo sapiens*. For additional theoretical as well as methodological considerations the reader is referred to several publications from Hammer and colleagues (28, 50), as well as the numerous additional publications cited in these papers.

8.5 FORENSIC APPLICATIONS

> Under other monarchies the male line takes precedence of the female in tracing genealogies, but here the opposite is the case—the female line takes the precedence. Their reason for this is exceedingly simple, and I recommend it to the aristocracy of Europe: They say it is easy to know who a man's mother was, but, etc., etc.
>
> —Mark Twain, "Letters from the Sandwich Islands," describing the Hawaiian Monarchy

Mitochondrial DNA sequences also mutate rapidly enough to create distinctions (polymorphisms) between members of the same ethnic group—that is, close "relatives" on an evolutionary scale, though not relatives in the legal sense. Applications in forensic investigations have been made, and two such applications will be used as examples to illustrate the methodology and the potential limitations.

Attention is focused on the mtDNA control region (D-loop). In 1983 in a relatively small sample of Caucasians and Africans it was first found that there was ~1.7% nucleotide diversity in ~900 basepairs, with most of the differences clustered in two hypervariable regions (54). The 347-basepair hypervariable region (within nucleotides 16023–16388) was later analyzed by Orrego and in 10 additional unrelated Caucasian individuals of European ancestry. In the combined data set from Greenberg et al. (54), Anderson et al. (55), and Orrego and King (55a), 32 sites out of 347 sites compared were found to be variable, and pairwise comparisons showed differences ranging from 1 to 13 nucleotides (average 5.9). A formal statistical analysis revealed that the observed and expected number of sequence differences between pairs of individuals closely approximated a Poisson distribution, which allows one to calculate that the probability of two unrelated Caucasian individuals having an identical sequence of 347 basepairs is 1 in 370 (see reference 56) for a discussion of these results and related issues).

With oligonucleotide primers corresponding to the conserved flanking sequences ($tRNA^{Pro}$ and $tRNA^{Phe}$), it is a routine matter to amplify the D-loop region from very small tissue samples. As described by King (56), a second amplification with nested but unequal amounts of primers can yield a single-strand product ready for DNA sequencing. Optimization of conditions to reduce artifacts from polymerase errors is part of the standard methodology, and mosaics created with unintended target sequences can be detected and eliminated because the normal target is so well known. A successful test analysis was performed with mtDNA from a baby tooth extracted 20 years earlier, and the sequence was matched to mtDNA from the now adult, his sister, and his mother. Baby teeth may be one of the best ways to preserve DNA samples for posterity and future cloning, should that become socially acceptable.

The technology was subsequently applied to a problem illustrating in a marvelous and touching way the application of science to seek justice and uphold human rights. An Interamerican Commission on Human Rights, upon a request by the National Commission on the Disappearance of Persons formed by the newly elected democratic government of Argentina in 1983, assisted the Grandmothers of the Plaza de Mayo and other surviving relatives to be reunited with their kidnapped children. These children had been born to couples and women who subsequently "disappeared"—murder victims of the military junta that ruled Argentina from 1976 to 1983. The children were often given away or adopted by the military or police, but Argentinian courts were willing to promote a reunification of the families, if objective scientific proof could establish the identity of these children. Since mtDNA is maternally inherited, haplotyping of mtDNAs from the children and relevant surviving relatives and specifically maternal grandmothers became a legally accepted method to test kinship in the absence of surviving parents.

While the science was objective and frequently decisive in relating children to grandmothers, a host of legal, ethical, sociological, and psychological problems arose, because by the time the tests became possible, the children were no longer small, they had developed relationships with their foster parents, and it was not always very clear who had the right to know, nor was it clear what to do with the new knowledge. Should a child be informed that the foster parents were among the killers of their parents? Should they be coerced back to their original families? Even the political climate has changed in Argentina and the judicial system is no longer so supportive of the Grandmothers. As King pointed out, once the "children" reach the age of 21 years, they will obtain the legal right to find out their identities if they wish.

Another headline-producing story involving mtDNA sequencing was the recent authentication of the remains of the Tsar Nicholas II of Russia, his wife and children, and some of his closest servants (57, 58). The Russian imperial family had been executed during the revolution and buried in a mass grave in the vicinity of Yekaterinburg. Nine sets of skeletal remains were recovered in 1991, but it remained to prove that these indeed included the remains of the Tsar, his wife, and three of their children. Tissue from the bones yielded sufficient mtDNA to allow amplification and sequencing of D-loops, to be compared with those of living and dead relatives (Figure 8.2). The Tsar's brother was exhumed from his grave in St. Peter and Paul Cathedral in St. Petersburg, where he had rested since 1899. Prince Philip, a known maternal relative, had mtDNA that matched that of the presumed Tsarina. Sequences of the Tsar matched those of two living maternal relatives, but in an additional twist to the story it was found that the Tsar was heteroplasmic at position 16169 (72%C/28%T). There were suspicions of contamination after the first

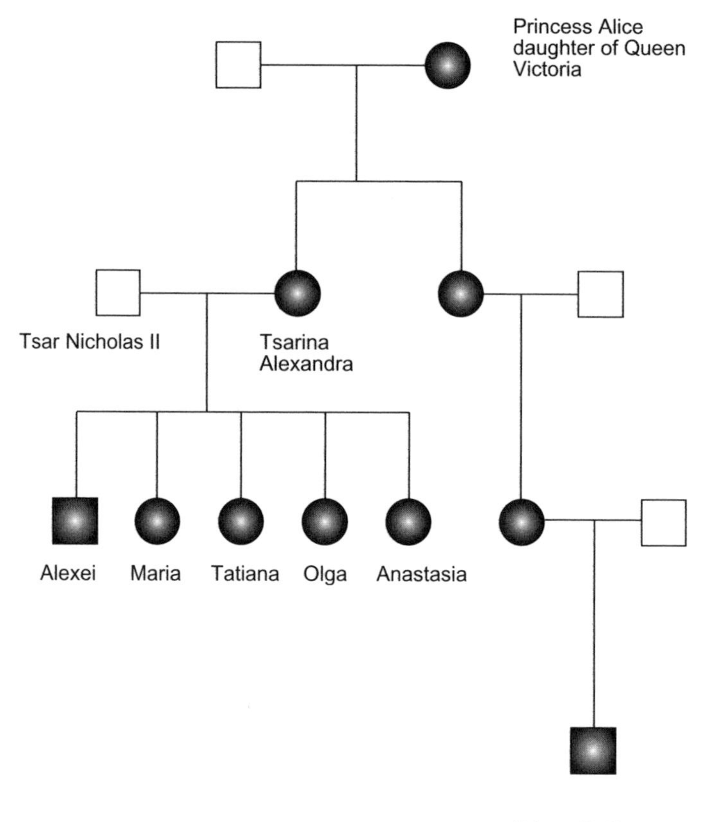

Figure 8.2 Pedigrees of European aristocracy showing maternal inheritance of mtDNA polymorphisms.

publication which could be removed by independent repeats of the sequencing as well as by the finding that the Tsar's brother was also heteroplasmic at the same position (38%C/62%T). Countess Xenia Cheremeteff-Sfiri and the Duke of Fife were homoplasmic, but had the same sequence at the other six positions where substitutions relative to the standard reference sequence had been found (Figure 8.3). The existence of heteroplasmy in two siblings (but with different ratio), along with the progression to homoplasmy within four generational events, is another interesting example illustrating rapid mtDNA segregation. It adds evidence in support of the bottleneck mechanism (see Chapter 7).

The use of mtDNA polymorphisms in forensics has attracted the attention of the FBI and the U.S. Militiary (the Armed Forces DNA Identification Laboratory in Rockville, Maryland). Thus, a considerable number mtDNA control region sequences from hundreds of individuals representing more than 120

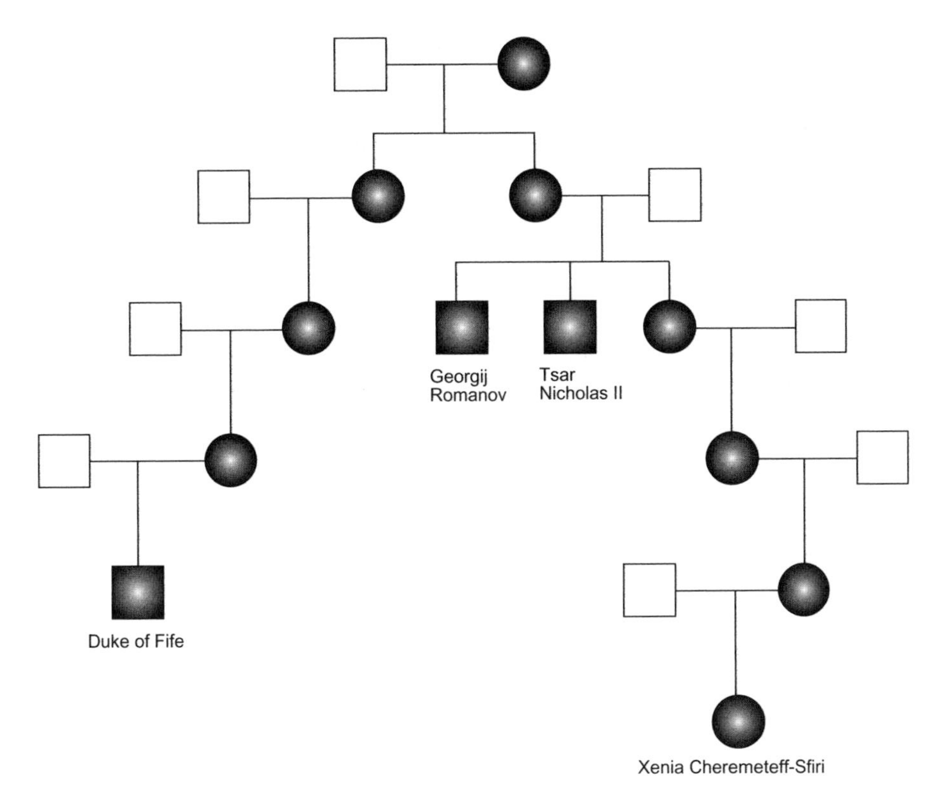

Figure 8.3 Pedigrees of European aristocracy showing maternal inheritance of mtDNA polymorphisms.

families have been analyzed recently, and the results have caused surprise and consternation. Contrary to expectations and previous estimates, the mutation rate was much larger than expected. Instead of a rate of one mutation every 12,000 years (~600 generations), deduced from the differences between the great apes and humans and their divergence ~5 million years ago, the observed differences among contemporary humans yielded an estimate of one mutation every 800 years. The controversy received fuel from additional supporting studies, as well as from studies failing to confirm the high mutation rate (see reference 59 for further discussion). It is clear that the acceptance of a much higher mutation rate would also upset various estimates of landmarks in human evolution (emergence from Africa, expansion of modern Europeans, immigration into North America).

More data will resolve whether the discrepancy is due to "statistical fluctuations" (low sample size). There is also debate about the existence and meaning of "hot spots." If hot spots exist, mutations will be observed over the short term, but over the long term such mutations can revert back to the original sequence, in effect giving an average long-term mutation rate that is significantly slower. Another issue to be resolved is whether the DNA in the control region is not subject to selection pressure. Since it is noncoding, it has been assumed that mutations in this region are neutral. On the other hand, such mutations may in subtle ways affect the initiation of transcription and replication of mtDNA, even if none of the identified sequence elements for protein binding are directly involved.

8.6 FUTURE CHALLENGES

It is possible that the harvesting of information from the human mtDNA tree is almost complete, except for the outermost branches that will yield information about very regional migrations such as those on the Indian subcontinent, in the Canadian Arctic (60), or in the Pacific Islands (61). The challenge will be to account for discrepancies in time estimates derived from the mtDNA tree and from Y chromosome sequences, and ultimately from sequences on autosomes and the X chromosome.

Another challenge will be to provide solid evidence in support of speculations that certain mitochondrial haplotypes may confer a selective advantage under certain environmental conditions and to correlate such haplotypes with specific nuclear polymorphisms. Furthermore, correlations may not be sufficient or convincing unless biochemical and physiological properties of the mitochondria in question can be shown to be responsible for the selection (in the same way in which the sickle cell allele confers resistance to malaria in heterozygotes, or the CFTR Phe508 mutation may allow heterozygotes to survive cholera epidemics).

Clearly, the molecular clock will require continuous re-calibration—for example, for different primate species and perhaps even for different segments of the mt genome (41).

Forensic applications have been illustrated with some spectacular and widely publicized examples. The challenge in the future will be to find "cases" of sufficiently broad interest to warrant publication of the results in journals like *Nature*, since the technology appears to be on a firm footing. Heteroplasmy and rapid fluctuations in the proportion of genomes may add some complications in specific situations.

REFERENCES

1. Cann, R. L., Stoneking, M., and Wilson, A. C. (1987) *Nature* **325**, 31–36.
2. Brown, W. M. (1981) *Ann. NY Acad. Sci.* **361:119–34**, 119–134.
3. Brown, W. M., Prager, E. M., Wang, A., and Wilson, A. C. (1982) *J. Mol. Evol.* **18**, 225–239.
4. Cann, R. L., Brown, W. M., and Wilson, A. C. (1982) *Prog. Clin. Biol. Res.* **103** (Pt A), 157–165.
5. Cann, R. L., Brown, W. M., and Wilson, A. C. (1984) *Genetics* **106**, 479–499.

6. Ankel-Simons, F., Cummings, J. M., and Cummins, J. M. (1996) *Proc. Natl. Acad. Sci. USA* **93**, 13859–13863.
7. Gyllensten, U., Wharton, D., Josefsson, A., and Wilson, A. C. (1991) *Nature* **352**, 255–257.
8. Kaneda, H., Hayashi, J.-I., Takahama, S., Taya, C., Fischer Lindahl, K., and Yonekawa, H. (1995) *Proc. Natl. Acad. Sci. USA* **92**, 4542–4546.
9. Sutovsky, P., Van Leyen, K., McCauley, T., Day, B. N., and Sutovsky, M. (2004) *Reprod. Biomed. Online.* **8**, 24–33.
10. Sutovsky, P. (2003) *Microsc. Res. Tech.* **61**, 88–102.
11. Hayashida, K., Omagari, K., Masuda, J. I., Hazama, H., Kadokawa, Y., Ohba, K., and Kohno, S. (2005) *Cell Biol. Int.* **29**, 472–481.
12. Schwartz, M. and Vissing, J. (2002) *N. Engl. J. Med.* **347**, 576–580.
13. Elson, J. L. and Lightowlers, R. N. (2006) *TIG* **22**, 603–607.
14. Bendall, K. E. and Sykes, B. C. (1995) *Am. J. Hum. Genet.* **57**, 248–256.
15. Vigilant, L., Stoneking, M., Harpending, H., Hawkes, K., and Wilson, A. C. (1991) *Science* **253**, 1503–1507.
16. Wills, C. (1996) *Evolution* **50**, 977–989.
17. Hasegawa, M., DiRienzo, A., Kocher, T. D., and Wilson, A. C. (1993) *J. Mol. Evol.* **37**, 347–354.
18. Gibbons, A. (1992) *Science* **257**, 873–875.
19. Ayala, F. J. (1995) *Science* **270**, 1930–1936.
20. Templeton, A. R. (1996) *Science* **272**, 1363–1364.
21. Zischler, H., Geisert, H., Von Haeseler, A., and Pääbo, S. (1995) *Nature* **378**, 489–492.
22. Fukuda, M., Wakasugi, S., Nomiyama, H., Shimada, K., and Miyata, T. (1985) *J. Mol. Biol.* **186**, 257–266.
23. Perna, N. T. and Kocher, T. D. (1996) *Curr. Biol.* **6**, 128–129.
24. Stoneking, M. and Soodyall, H. (1996) *Curr. Opin. Genet. Dev.* **6**, 731–736.
25. Wallace, D. C. (1995) *Am. J. Hum. Genet.* **57**, 201–223.
26. Redd, A. G., Takezaki, N., Sherry, S. T., McGarvey, S. T., Sofro, A. S. M., and Stoneking, M. (1995) *Mol. Biol. Evol.* **12**, 604–615.
27. Wills, C. (1995) *Evolution* **49**, 593–607.
28. Garrigan, D. and Hammer, M. F. (2006) *Nature Rev. Genet.* **7**, 669–680.
29. Pennisi, E. (2004) *Science* **306**, 210.
30. Reed, D. L., Smith, V. S., Hammond, S. L., Rogers, A. R., and Clayton, D. H. (2004) *PLoS Biol.* **2**, e378.
31. Krings, M., Stone, A., Schmitz, R. W., Krainitzki, H., Stoneking, M., and Paabo, S. (1997) *Cell* **90**, 19–30.
32. Lindahl, T. (1997) *Cell* **90**, 1–3.
33. Caramelli, D., Lalueza-Fox, C., Vernesi, C., Lari, M., Casoli, A., Mallegni, F., Chiarelli, B., Dupanloup, I., Bertranpetit, J., Barbujani, G., and Bertorelle, G. (2003) *Proc. Natc. Acad. Sci.* **100**, 6593–6597.
34. Orlando, L., Darlu, P., Toussaint, M., Bonjean, D., Otte, M., and Hanni, C. (2006) *Curr. Biol.* **16**, R400–R402.
35. Merriwether, D. A. and Ferrell, R. E. (1996) *Mol. Phylogenet. Evol.* **5**, 241–246.
36. Ingman, M. and Gyllensten, U. (2006) *Nucleic Acids Res.* **34**, D749–D751.

37. Torroni, A., Achilli, A., Macaulay, V., Richards, M., and Bandelt, H.-J. (2006) *Trends Genet.* **22**, 339–345.
38. Attimonelli, M., Accetturo, M., Santamaria, M., Lascaro, D., Scioscia, G., Pappada, G., Russo, L., Zanchetta, L., and Tommaseo-Ponzetta, M. (2005) *BMC Bioinformatics* **6**(Suppl 4), S4.
39. Wallace, D. C. (2005) *Annu. Rev. Genet.* **39**, 359–407.
40. Mishmar, D., Ruiz-Pesini, E., Golik, P., Macaulay, V., Clark, A. G., Hosseini, S., Brandon, M., Easley, K., Chen, E., Brown, M. D., Sukernik, R. I., Olckers, A., and Wallace, D. C. (2003) *Proc. Natl. Acad. Sci. USA* **100**, 171–176.
41. Horai, S., Hayasaka, K., Kondo, R., Tsugane, K., and Takahata, N. (1995) *Proc. Natl. Acad. Sci. USA* **92**, 532–536.
42. Gagneux, P., Wills, C., German, J., Tautz, D., Morin, P. A., Boesch, C., Fruth, B., Hohmann, G., Ryder, O. A., and Woodruff, D. S. (1998) *Proc. Natl. Acad. Sci. USA* **96**, 5077–5082.
43. Kenyon, L. and Moraes, C. T. (1997) *Proc. Natl. Acad. Sci. USA* **94**, 9131–9135.
44. Koopman, P., Gubbay, J., Vivian, N., Goodfellow, P., and Lovell-Badge, R. (1991) *Nature* **351**, 117–121.
45. Brookfield, J. F. Y. (1995) *Curr. Biol.* **5**, 1114–1115.
46. Hammer, M. F. (1995) *Nature* **378**, 376–378.
47. Jobling, M. A. and Tyler-Smith, C. (1995) *Trends Genet.* **11**, 449–456.
48. Pääbo, S. (1995) *Science* **268**, 1141–1142.
49. Underhill, P. A., Jin, L., Zemans, R., Oefner, P. J., and Cavalli-Sforza, L. L. (1996) *Proc. Natl. Acad. Sci. USA* **93**, 196–200.
50. Hammer, M. F., Spurdle, A. B., Karafet, T., Bonner, M. R., Wood, E. T., Novelletto, A., Malaspina, P., Mitchell, R. J., Horai, S., Jenkins, T., and Zegura, S. L. (1997) *Genetics* **145**, 787–805.
51. Gibbons, A. (1997) *Science* **278**, 804–805.
52. Ke, Y., Su, B., Song, X., Lu, D., Chen, L., Li, H., et al. (2001) African origin of modern humans in East Asia: A tale of 12,000 Y chromosomes. *Science* **292**, 1151–1153.
53. Templeton, A. R. (2002) *Nature* **416**, 45–51.
54. Greenberg, B. D., Newbold, J. E., and Sugino, A. (1983) *Gene* **21**, 33–49.
55. Anderson, S., Bankier, A. T., Barrell, B. G., de Bruijn, M. H., Coulson, A. R., Drouin, J., Eperon, I. C., Nierlich, D. P., Roe, B. A., Sanger, F., Schreier, P. H., Smith, A. J., Staden, R., and Young, I. G. (1981) *Nature* **290**, 457–465.
55a. Orrrego, C. and King, M.-C. (1990) Determination of familial relationships. In Innis, M., Gelfand, D., Sninsky, J., and White, T., (editors). *PCR Protocols: A Guide to Methods and Applications*, Academic Press, San Diego, pp. 416–426.
56. King, M.-C. (1991) *Mol. Genet. Med.* **1**, 117–131.
57. Gill, P., Ivanov, P. L., Kimpton, C., Piercy, R., Benson, N., Tully, G., Evett, I., Hagelberg, E., and Sullivan, K. (1994) *Nature Genet.* **6**, 130–135.
58. Ivanov, P. L., Wadhams, M. J., Roby, R. K., Holland, M. M., Weedn, V. W., and Parsons, T. J. (1996) *Nature Genet.* **12**, 417–420.
59. Gibbons, A. (1998) *Science* **279**, 28–29.
60. Helgason, A., Palsson, G., Pedesen, H. S., Angulalik, E., Gunnarsdottir, E. D., Yngvadottir, B., and Stefansson, K. (2006) *Am. J. Phys. Anthropol.* **130**, 123–134.
61. Whyte, A. L., Marshall, S. J., and Chambers, G. K. (2005) *Hum. Biol.* **77**, 157–177.

9

MITOCHONDRIA AND PHARMACOLOGY

9.1 INTRODUCTION

Studies on mitochondria have for a long time benefited from the judicious use of specific drugs, antibiotics, inhibitors, and/or analogues that have helped in defining pathways, or in focusing on a particular complex, or even in creating phenocopies of specific mutations. A number of these have been introduced in previous chapters in the context of the reactions or complexes being discussed. The following is intended to be more of a summary and compilation for easy reference and identification of the targets of both the common and less common drugs being employed in this field. Drugs that inhibit specific enzymes in biochemical pathways and metabolic cycles (for example, fluorocitrate) will not be covered exhaustively, unless there is a special relevance to mitochondrial function.

One can subdivide the collection of drugs/inhibitors into several categories. A summary of information with some relevant, recent references is presented in following Table 9.1.

TABLE 9.1 Common Inhibitors Used in Mitochondrial Studies

Name	Target	Reference	Discussion in This Book (page)
1. INHIBITORS OF ELECTRON TRANSPORT			
Rotenone	Complex I		
Amytal	Complex I		
Piericidin A	Complex I		
N-Methylquinolinium derivatives	Complex I	1	
N-Methyl-pyridinium derivatives (MPP^+)	Complex I		
Others	Complex I	2	
Malonate	Complex II		
Antimycin A	Complex III	3, 4	
Quinoline *N*-oxides	Complex III	3, 4	
Myxothiazol	Complex III	3, 4	
Azide, carbon monoxide, cyanide	Complex IV		
2. UNCOUPLERS/IONOPHORES			
Dinitrophenol			
Valinomycin			
Carbonylcyanide-*p*-trifluoromethoxyphenylhydrazone (FCCP)			
3. INHIBITORS OF ATP SYNTHASE			
Oligomycin B	Complex V	5, 6	
Dicyclohexylcarbodiimide (DCCD)	Complex V		
Venturicidin	Complex V		
4. INHIBITORS OF TRANSPORT SYSTEMS			
Atractyloside	ANT	7, 8	
Bongkrekic acid	ANT	7, 8	
5. SENSORS OF MEMBRANE POTENTIAL			
Rhodamine123		9	
3,3′-Dihexylooxacarbocyanine iodide $DIOC_6(3)$		9	
5,5′,6,6′-Tetrachloro-1,1′,3,3′-tetraethylbenzimidazolcarbocyanine (JC-1)		9	
Chloromethyl-X-rosamine (CMXRos)		9	

REFERENCES

1. Miyoshi, H., Inoue, M., Okamoto, S., Ohshima, M., Sakamoto, K., and Iwamura, H. (1997) *J. Biol. Chem.* **272**, 16176–16183.
2. Friedrich, T., Van Heek, P., Leif, H., Ohnishi, T., Forche, E., Kunze, B., Jansen, R., Trowitzsch-Kienast, W., Hofle, G., Reichenbach, H., and Weiss, H. (1994) *Eur. J. Biochem.* **219**, 691–698.

3. Schagger, H., Brandt, U., Gencic, S., and Von Jagow, G. (1995) *Methods Enzymol.* **260**, 82–96.
4. Brandt, U. and Trumpower, B. (1994) *Crit. Rev. Biochem. Mol. Biol.* **29**, 165–197.
5. Walker, J. E., Collinson, I. R., Van Raaij, M. J., and Runswick, M. J. (1995) *Methods Enzymol.* **260**, 163–190.
6. Tzagoloff, A. (1970) *J. Biol. Chem.* **245**, 1545–1551.
7. Klingenberg, M., Winkler, E., and Huang, S. (1995) *Methods Enzymol.* **260:369–89**, 369–389.
8. Klingenberg, M. (1976) *The Enzymes of Biological Membranes: Membrane Transport*, Plenum Publishing Corp., New York.
9. Reers, M., Smiley, S. T., Mottola-Hartshorn, C., Chen, A., Lin, M., and Chen, L. B. (1995) *Methods Enzymol.* **260**, 406–417.

INDEX